D0921542

MTP International Review of Science

Inorganic Chemistry
Series Two

Consultant Editor
H. J. Emeléus, F.R.S.

Publisher's Note

The MTP International Review of Science is an important venture in scientific publishing, which is presented by Butterworths in association with MTP Medical and Technical Publishing Co. Ltd. and University Park Press, Baltimore. The basic concept of the Review is to provide regular authoritative reviews of entire disciplines. Chemistry was taken first as the problems of literature survey are probably more acute in this subject than in any other. Physiology and Biochemistry followed naturally. As a matter of policy, the authorship of the MTP Review of Science is international and distinguished, the subject coverage is extensive, systematic and critical.
The Review has been conceived within a carefully organised editorial framework. The overall plan was drawn up, and the volume editors appointed by seven consultant editors. In turn, each volume editor planned the coverage of his field and appointed authors to write on subjects which were within the area of their own research experience. No geographical restriction was imposed. Hence the 500 or so contributions to the MTP Review of Science come from many countries of the world and provide an authoritative account of progress.
The publication of Inorganic Chemistry Series One was completed in 1972 with ten text volumes and one index volume, and in accordance with the stated policy of issuing regular reviews to keep the series up to date, volumes of Series Two will be published between the autumn of 1974 and the spring of 1975. They will be followed by Series Two of Physical and Organic Chemistry in the period 1975–1976. Volume titles will generally be the same as in Series One but the articles themselves may either cover recent advances in the same subject or deal with a different aspect of the main theme of the volume. In Series Two an index will be incorporated in each volume and there will be no separate index volume.

Butterworth & Co. (Publishers) Ltd.

BIOCHEMISTRY SERIES ONE

Consultant Editors
H. L. Kornberg, F.R.S., *Department of Biochemistry, University of Leicester* and
D. C. Phillips, F.R.S., *Department of Zoology, University of Oxford*

Volume titles and Editors

1 **CHEMISTRY OF MACRO-MOLECULES**
Professor H. Gutfreund, *University of Bristol*

2 **BIOCHEMISTRY OF CELL WALLS AND MEMBRANES**
Dr. C. F. Fox, *University of California*

3 **ENERGY TRANSDUCING MECHANISMS**
Professor E. Racker, *Cornell University, New York*

4 **BIOCHEMISTRY OF LIPIDS**
Professor T. W. Goodwin, F.R.S., *University of Liverpool*

5 **BIOCHEMISTRY OF CARBO-HYDRATES**
Professor W. J. Whelan, *University of Miami*

6 **BIOCHEMISTRY OF NUCLEIC ACIDS**
Professor K. Burton, F.R.S., *University of Newcastle upon Tyne*

7 **SYNTHESIS OF AMINO ACIDS AND PROTEINS**
Professor H. R. V. Arnstein, *King's College, University of London*

8 **BIOCHEMISTRY OF HORMONES**
Professor H. V. Rickenberg, *National Jewish Hospital & Research Center, Colorado*

9 **BIOCHEMISTRY OF CELL DIFFER-ENTIATION**
Dr. J. Paul, *The Beatson Institute for Cancer Research, Glasgow*

10 **DEFENCE AND RECOGNITION**
Professor R. R. Porter, F.R.S., *University of Oxford*

11 **PLANT BIOCHEMISTRY**
Professor D. H. Northcote, F.R.S., *University of Cambridge*

12 **PHYSIOLOGICAL AND PHARMA-COLOGICAL BIOCHEMISTRY**
Dr. H. F. K. Blaschko, F.R.S., *University of Oxford*

PHYSIOLOGY SERIES ONE

Consultant Editors
A. C. Guyton, *Department of Physiology and Biophysics, University of Mississipp Medical Center* and
D. F. Horrobin, *Department of Physiology, University of Newcastle upon Tyne*

Volume titles and Editors

1 **CARDIOVASCULAR PHYSIOLOGY**
Professor A. C. Guyton and Dr. C. E. Jones, *University of Mississippi Medical Center*

2 **RESPIRATORY PHYSIOLOGY**
Professor J. G. Widdicombe, *St. George's Hospital, London*

3 **NEUROPHYSIOLOGY**
Professor C. C. Hunt, *Washington University School of Medicine, St. Louis*

4 **GASTROINTESTINAL PHYSIOLOGY**
Professor E. D. Jacobson and Dr. L. L. Shanbour, *University of Texas Medical School*

5 **ENDOCRINE PHYSIOLOGY**
Professor S. M. McCann, *University of Texas*

6 **KIDNEY AND URINARY TRACT PHYSIOLOGY**
Professor K. Thurau, *University of Munich*

7 **ENVIRONMENTAL PHYSIOLOGY**
Professor D. Robertshaw, *University of Nairobi*

8 **REPRODUCTIVE PHYSIOLOGY**
Professor R. O. Greep, *Harvard Medical School*

INORGANIC CHEMISTRY SERIES TWO

Consultant Editor
H. J. Emeléus, F.R.S.,
Department of Chemistry, University of Cambridge

Volume titles and Editors

1 **MAIN GROUP ELEMENTS—HYDROGEN AND GROUPS I–III**
Professor M. F. Lappert, *University of Sussex*

2 **MAIN GROUP ELEMENTS—GROUPS IV AND V**
Dr. D. B. Sowerby, *University of Nottingham*

3 **MAIN GROUP ELEMENTS—GROUPS VI AND VII**
Professor Viktor Gutmann, *Technical University of Vienna*

4 **ORGANOMETALLIC DERIVATIVES OF THE MAIN GROUP ELEMENTS**
Professor B. J. Aylett, *Westfield College, University of London*

5 **TRANSITION METALS—PART 1**
Professor D. W. A. Sharp, *University of Glasgow*

6 **TRANSITION METALS—PART 2**
Dr. M. J. Mays, *University of Cambridge*

7 **LANTHANIDES AND ACTINIDES**
Professor K. W. Bagnall, *University of Manchester*

8 **RADIOCHEMISTRY**
Dr. A. G. Maddock, *University of Cambridge*

9 **REACTION MECHANISMS IN INORGANIC CHEMISTRY**
Professor M. L. Tobe, *University College, University of London*

10 **SOLID STATE CHEMISTRY**
Dr. L. E. J. Roberts, *Atomic Energy Research Establishment, Harwell*

PHYSICAL CHEMISTRY SERIES TWO

Consultant Editor
A. D. Buckingham,
Department of Chemistry, University of Cambridge

Volume titles and Editors

1 **THEORETICAL CHEMISTRY**
Professor A. D. Buckingham, *University of Cambridge*

2 **MOLECULAR STRUCTURE AND PROPERTIES**
Professor A. D. Buckingham, *University of Cambridge*

3 **SPECTROSCOPY**
Dr. D. A. Ramsay, F.R.S.C., *National Research Council of Canada*

4 **MAGNETIC RESONANCE**
Professor C. A. McDowell, F.R.S.C., *University of British Columbia*

5 **MASS SPECTROMETRY**
Professor A. Maccoll, *University College, University of London*

6 **ELECTROCHEMISTRY**
Professor J. O'M Bockris, *The Flinders University of S. Australia*

7 **SURFACE CHEMISTRY AND COLLOIDS**
Professor M. Kerker, *Clarkson College of Technology, New York*

8 **MACROMOLECULAR SCIENCE**
Professor C. E. H. Bawn, C.B.E., F.R.S., *University of Liverpool*

9 **CHEMICAL KINETICS**
Professor D. R. Herschbach, *Harvard University*

10 **THERMOCHEMISTRY AND THERMODYNAMICS**
Dr. H. A. Skinner, *University of Manchester*

11 **CHEMICAL CRYSTALLOGRAPHY**
Professor J. Monteath Robertson, C.B.E., F.R.S., *University of Glasgow*

12 **ANALYTICAL CHEMISTRY—PART 1**
Professor T. S. West, *Imperial College, University of London*

13 **ANALYTICAL CHEMISTRY—PART 2**
Professor T. S. West, *Imperial College, University of London*

ORGANIC CHEMISTRY SERIES TWO

Consultant Editor
D. H. Hey, F.R.S.,
formerly of Department of Chemistry, King's College, University of London

Volume titles and Editors

1 **STRUCTURE DETERMINATION IN ORGANIC CHEMISTRY**
Professor L. M. Jackman, *Pennsylvania State University*

2 **ALIPHATIC COMPOUNDS**
Professor N. B. Chapman, *Hull University*

3 **AROMATIC COMPOUNDS**
Professor H. Zollinger, *Eidgenossische Technische Hochschule*

4 **HETEROCYCLIC COMPOUNDS**
Dr. K. Schofield, *University of Exeter*

5 **ALICYCLIC COMPOUNDS**
Professor D. Ginsburg, *Technion-Israel Institute of Technology, Haifa*

6 **AMINO ACIDS, PEPTIDES AND RELATED COMPOUNDS**
Professor N. H. Rydon, *University of Exeter*

7 **CARBOHYDRATES**
Professor G. O. Aspinall, *York University, Ontario*

8 **STEROIDS**
Dr. W. F. Johns, *G. D. Searle & Co., Chicago*

9 **ALKALOIDS**
Professor K. Wiesner, F.R.S., *University of New Brunswick*

10 **FREE RADICAL REACTIONS**
Professor W. A. Waters, F.R.S., *University of Oxford*

MTP International Review of Science

Inorganic Chemistry Series Two

Volume 5
Transition Metals—Part 1

Edited by **D. W. A. Sharp**
University of Glasgow

Butterworths · London
University Park Press · Baltimore

THE BUTTERWORTH GROUP

ENGLAND
Butterworth & Co (Publishers) Ltd
London: 88 Kingsway, WC2B 6AB

AUSTRALIA
Butterworths Pty Ltd
Sydney: 586 Pacific Highway 2067
Melbourne: 343 Little Collins Street, 3000
Brisbane: 240 Queen Street, 4000

NEW ZEALAND
Butterworths of New Zealand Ltd
Wellington: 26–28 Waring Taylor Street, 1

SOUTH AFRICA
Butterworth & Co (South Africa) (Pty) Ltd
Durban: 152–154 Gale Street

ISBN 0 408 70594 9

UNIVERSITY PARK PRESS

U.S.A. and CANADA
University Park Press
Chamber of Commerce Building
Baltimore, Maryland, 21202

Library of Congress Cataloging in Publication Data
Sharp, D. W. A.
Transition metals.
(Inorganic chemistry, series two, v. 5) (MTP international review of science)
Includes bibliographies.
1. Transition metal compounds. 2. Complex compounds. I Title. II. Series. III. Series:
MTP international review of science.
QD147.I56 vol. 5, etc. [QD172.T6] 546s [546'.6]
ISBN 0-8391-0204-6 (v. 1) 74-13992

Printed and bound by Butler and Tanner Ltd.,
Frome, Somerset.

Consultant Editor's Note

Following the successful production of the first series of ten volumes on inorganic chemistry in the MTP International Review of Science, a second series has been prepared with the object of keeping inorganic chemists abreast of the many important advances that have taken place in the interim. The original plan has been adhered to in the main, though there are minor variations in the allocation of material to the volumes and in the space devoted to specific topics. The aim remains to provide a comprehensive critical survey of work published in the past few years in each of the main areas of inorganic chemistry. Many experts have collaborated in the production either as authors or as volume editors. I am deeply indebted to them all for the excellent work that they have done and for the way in which they have kept to production schedules. The high standard established in Series One has been fully maintained and I am confident that this publication will be of the utmost value to university teachers, research workers and advanced undergraduates.

Cambridge

H. J. Emeléus

Preface

Volume 5 of Inorganic Chemistry (Series Two) of the MTP International Review of Science is on those aspects of transition element chemistry that do not involve metal–carbon bonds (but including carbides). The contributions are critical reviews of the subjects under consideration, generally covering the two years to 1973 but including older references where these are necessary to place the work in context. In some cases it has been necessary for the authors to limit their reviews to selected areas within their main topics. Where this has been done it is clearly indicated in the introduction to the review.

Readers are reminded that the lanthanides and actinides (Volume 7) and solid state chemistry (Volume 10) are covered in detail elsewhere in this series although there are some references to these topics in this volume. The present volume aims to cover the important groups of binary and ternary compounds and to give some cover of complexes and derivatives of complex anions.

Glasgow D. W. A. Sharp

Contents

1
Halides

J. M. WINFIELD
University of Glasgow

1.1 INTRODUCTION

This review deals with some aspects of recent work on the preparations, structures, thermodynamic properties and reactions of transition metal halides, oxohalides and their complexes with halide ions, other halides, or water. To keep the review to a manageable size, complexes with organic ligands are arbitrarily excluded. However, some aspects of this area are dealt with in other chapters of this volume. Although many reports dealing with the vibrational and electronic spectra and the magnetic properties of halides have appeared during the past 2 years, space does not permit an extensive treatment of these topics. However, some aspects have been reviewed elsewhere.

Several force fields which are often used in the analysis of vibrational spectra have been compared[1]. The modified orbital valence force field gives the best overall agreement with observed frequencies. The effects of oxidation states are more important than those of mass, and for MF_6 and MCl_6^{2-} species the values of some of the force constants can be related to the number of non-bonding valence electrons and to crystal field stabilisation energies. Recent work on the electronic spectra of 3d elements with fluoride ligands in octahedral or tetragonal ligand fields has again emphasised the inadequacy of a purely electrostatic model to describe the spectra even of M^{II} species[2, 3]. Detailed assignments for many fluorides and their complexes are now available and phenomena such as nephelauxetic effects, optical electronegativities, vibrational fine structure and Jahn–Teller effects have been discussed.

1.2 HEPTAPOSITIVE ELEMENTS

Several preparative routes to TcO_3Cl have been described, the best being the reaction of $TcCl_4$ with O_2 in a sealed tube at 723 K [4]. Its vibrational spectrum, as expected, is very similar to those of ReO_3Cl, Re_2O_7 and Tc_2O_7[5]. The vibrational spectra of $ReOF_5$ and $OsOF_5$ in the gaseous state have been assigned in C_{4v} symmetry[6]; both $ReOF_5$ and ReF_7 behave as F^- ion acceptors towards FNO and FNO_2[7]. The products are formulated as ionic $A^+ReOF_6^-$ and $A^+ReF_8^-$ ($A = NO$ or NO_2); C_{5v} and D_{4d} symmetries, respectively, being suggested for the anions.

A study of the redox and halogen-exchange reactions of ReF_7 has yielded the rather surprising result that its properties in both these respects are identical to those of ReF_6. The only difference between the two compounds is that reactions of ReF_7 are more vigorous[8]. Some of the reactions of ReF_6 are discussed in Section 1.3.1.

1.3 HEXAPOSITIVE ELEMENTS

1.3.1 Hexahalides and their substituted derivatives

The classical route to high oxidation-state chlorides is the reaction of Cl_2 with high purity metal at elevated temperature. This is not a good method for $ReCl_6$ and $TcCl_6$ as mixtures of oxochlorides and lower chlorides are formed[4, 9]. However, $ReCl_6$ is readily formed from halogen-exchange reactions which occur below room temperature between ReF_6 and BCl_3 or CCl_4. A similar reaction using BBr_3 yields $ReBr_5$[8, 10], and it would be interesting to examine the reaction of TcF_6 with BCl_3. In passing, it should be noted that halogen-exchange reactions are of general use; for example, a wide variety of anhydrous bromides and iodides can be prepared from the corresponding chlorides using BBr_3 or BI_3[11].

Since previous reports on $ReCl_6$ and $ReBr_5$ were probably based on samples contaminated with $ReOX_4$, it is worth while summarising the properties of the compounds prepared by the halogen-exchange method. $ReCl_6$ is a dark green to black crystalline solid (m.p. 302 K) which decomposes on heating to give $ReCl_5$ and Cl_2. Its corrected magnetic moment (2.05 μ_B) is constant over the range 105–294 K. $ReBr_5$ is a dark brown solid which decomposes to $ReBr_3$ and Br_2. It does not show Curie–Weiss behaviour and magnetic interaction between Re^V centres is suggested[10]*.

ReF_6 is comparable with MoF_6 in its ability to undergo halogen-exchange and redox reactions. It oxidises PF_3 to PF_5, ReF_5 being the other product, and this represents an improved method for preparing ReF_5[8]. These types of reactions are common to many high oxidation-state transition metal fluorides and have recently been reviewed[12]. A complete rationalisation is not possible as the necessary thermodynamic data are lacking, but a start has been made to rectify this situation. The electron affinity of WF_6 has been estimated as 502 kJ mol^{-1} from the heat of hydrolysis of KWF_6[13]. WF_6 is the weakest oxidising agent in the series of 5d hexafluorides, and the derived value is compatible with estimated electron affinities for the other members made from preparative work.

The reactions of MoF_6 with methyl(methoxo)silanes are more vigorous than the corresponding reactions of WF_6, and the less substituted members of the series $MoF_{6-n}(OMe)_n$ are unstable thermally. $MoF(OMe)_5$ has been isolated, however, and appears to be monomeric[14]. The series of monomeric compounds $WCl_{6-n}(OMe)_n$ ($n = 1$–4) has been prepared from WCl_6 by a similar route, the last two members ($n = 5$ or 6) being formed from $WCl_4(OMe)_2$ and the OMe^- ion. The two series $WX_{6-n}(OMe)_n$ (X = F or Cl) may be interconverted using Me_3SiCl or SbF_3. Their stereochemistry differs in that both *cis* and *trans* isomers are observed where X = Cl and $n = 2$ or 4,

* Very recently the preparation of $ReCl_6$ from ReF_6 and BCl_3 has been questioned[11a] in view of an independent study of the same reaction in which the only Re-containing product isolated was $ReCl_5$. The authors suggest that although the conclusions of Ref. 10 are not disproved, the positive identification of $ReCl_6$ is still doubtful. In this study also, heats of formation of ReF_6(g) and $ReCl_5$(s) at 298 K were estimated as -1348 and -360 kJ mol^{-1}, respectively, from their heats of hydrolysis.

whereas the corresponding fluorides are exclusively *cis*[15]. One example of the resurgence of interest in alkyl and aryl derivatives of transition metal halides (see also Sections 1.4.1 and 1.5.2) is provided by the compounds $RWCl_5$ (R = Me or Ph) which are both green. They are formed from reactions of WCl_6 and R_2Zn in ether below room temperature. Little is known of their properties apart from their ability to form complexes with N-donor ligands. Apparently similar reactions occur with R_4Sn or R_3B[16].

A rather different derivative of WCl_6 is that obtained from the oxidation of WCl_5 by CCl_3CN. X-Ray work has shown that it is *trans*-$[(CCl_3CN)WCl_4(NCCl_2CCl_3)]$[17]. Recrystallisation of this compound from CH_2Cl_2 over several weeks gave a dimeric product $(WCl_6 \cdot CCl_3CN)_2$ whose x-ray structure[18] is shown in Figure 1.1. The presence of W—N triple bonds is suggested in the $W—N(C_2Cl_5)$ groups, but the W—N bond (237 pm) in the $W—N\vdots CCl_3$ group is weak.

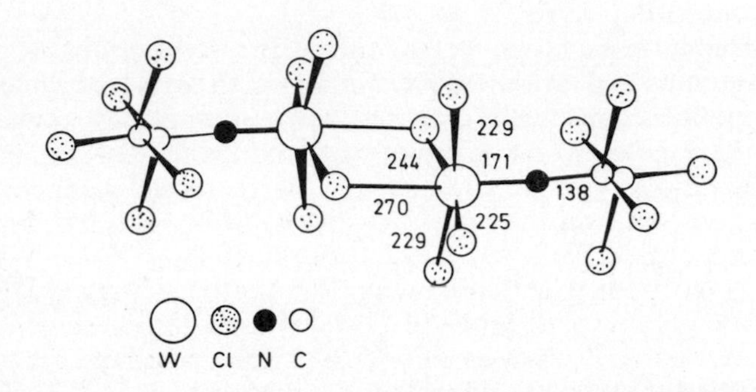

Figure 1.1 Structure of $(WCl_6 \cdot CCl_3CN)_2$. (From Fowles *et al.*[18] by courtesy of the Chemical Society.)

1.3.2 Oxo- and thio-halides

An earlier report that $OsOF_4$ is a yellow solid formed as a by-product in the fluorination of Os metal can now be discounted in view of the recent characterisation of the compound[19]. It is a grey-green solid, prepared by reduction of $OsOF_5$ on a hot W filament. X-Ray work suggests that the solid comprises tetrameric molecules, but that they either have a different structure or are packed differently from those in $(OsF_5)_4$. Unlike many pentafluorides, there is no evidence for polymeric species in its mass spectrum. The controversy over the identity of the bridging atoms in $(WOF_4)_4$ has been partially resolved. The x-ray data has been re-interpreted on the basis of a disordered structure with the tetramers having asymmetric W—F—W bridges and terminal O atoms[20]. This is consistent with the ^{19}F n.m.r. spectrum of polycrystalline WOF_4 and with the i.r. spectrum of $W^{18}OF_4$[21].

Solid $ReOCl_4$ contains square-pyramidal molecules (Re=O = 163 pm, Re—Cl = 226 pm) which are weakly associated through Re···Cl contacts *trans* to Re—O, to form dimers (Re···Cl = 355 pm) or endless chains

(Re···Cl = 365 pm)[22]. Its structure differs from that of $WOCl_4$ in that Re—O···Re linkages are absent, but the dimeric units are similar to those found in $WSCl_4$ and $WSBr_4$. An improved route to $ReOBr_4$, from Re metal and a 1:1 mixture of Br_2 and SO_2 at 673 K, has been described[23]. It is superior as other oxobromides are not formed, and blue-black $ReOBr_4$ appears to be isostructural with $ReOCl_4$.

To obtain unambiguous structural information on oxohalides in the gaseous state is not easy, and several experimental methods are normally required. A molecular beam mass spectrometric study of $MOCl_4$ (M = Mo, W, Re, or Os) and $WSCl_4$ vapours indicates that they are primarily monomeric[24]. The i.r. spectra of gaseous $ReOCl_4$ and $OsOCl_4$ favour C_{4v} symmetry[25]. The decision is not unambiguous but seems reasonable in view of the structure of solid $ReOCl_4$. Correlation of the data available on electron-impact ionisation potentials of Cr, Mo and W dioxo-dihalides suggests that the ionised electron originates mainly from oxygen in the fluorides, but from halogen in the other compounds[24].

The thermodynamic properties of Mo and W oxo-chlorides and -bromides have been studied in detail. Heats of formation (kJ mol^{-1}) at 298 K for some Mo^{VI} compounds, determined from heats of hydrolysis, are as follows[26, 27]:

$MoOCl_4(s)$	$MoO_2Cl_2(s)$	$MoO_2Br_2(s)$
−662	−727	−641

A thermodynamic study of the transport of MoO_2 by I_2 in the temperature gradient 1273–1073 K suggests that MoO_2I_2 exists in the gaseous state. The heat of formation of $MoO_2I_2(g)$ at 298 K is calculated as −416 kJ mol^{-1} [28].

1.3.3 Oxo- and thio-halide complexes

Oxygen-abstraction reactions of WCl_6 are well documented and similar behaviour has now been recognised for $WSCl_4$. This reacts with 1,2-dimethoxyethane (dme) to give a product whose structure is shown in Figure

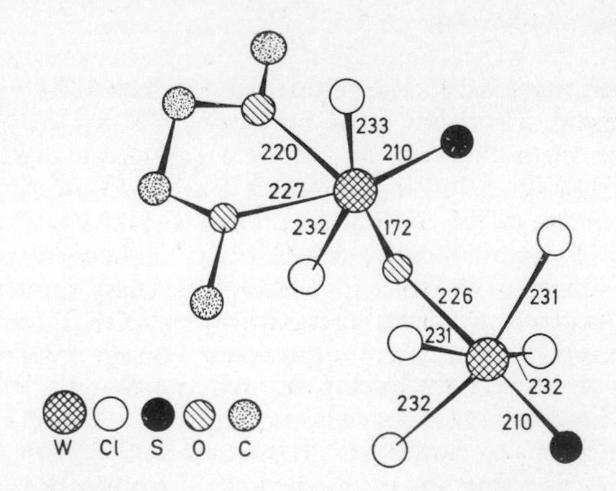

Figure 1.2 Structure of $WSCl_4 \cdot WOSCl_2$(dme). (From Fowles *et al.*[29] by courtesy of the Chemical Society.)

1.2. The bond distances suggest that the compound should be formulated as $S{:}W(Cl_4){\leftarrow}O{:}W({:}S)Cl_2(dme)$ [29]. The structures of two hydrates of oxohalides have been described. In $ReOCl_4 \cdot H_2O$, the O atom and H_2O molecule occupy *trans* positions in a very distorted octahedron. The Re—O and Re—Cl distances are similar to those in $ReOCl_4$, and $Re{-}OH_2 = 227$ pm [30]. The structure of $MoO_2Cl_2 \cdot H_2O$ is rather different from that of the anhydrous compound and is related to that of MoO_3. X-Ray work has been performed by two groups[31, 32], in one case[31] using vibrational and ^{1}H n.m.r. spectroscopy to provide additional evidence. The Mo coordination polyhedron and bond distances obtained by one group[31] are shown in Figure 1.3(a). A related environment for Mo^{VI} is found in $K[MoO_2Cl_2(H_2O)]_3Cl$ but here the trimmer unit is formed via *cis*-Mo—O—Mo bridges[33].

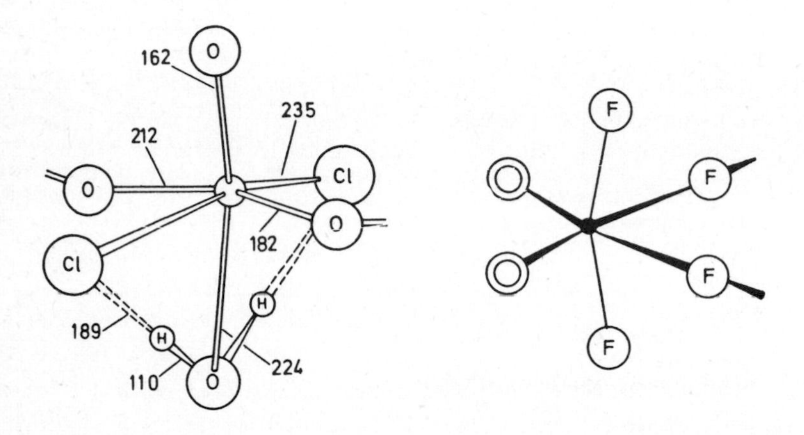

Figure 1.3 Coordination about Mo^{VI} (a) in $MoO_2Cl_2{\cdot}H_2O$ and (b) in $CsMoO_2F_3$. (From Schröder and Christensen[31] and from Mattes *et al.*[34] both by courtesy of J. A. Barth.) (Some bond distances in 3(a) taken from H. Schulz and F. A. Schröder (1973) *Acta Cryst.* *A29*, 322.)

Interest in polymeric oxofluoro anions has centred on the identification of the bridging ligand. The vibrational spectra of $A^IMO_2F_3$ (M = Mo or W) compounds are consistent with the presence of fluoro bridges[34], and x-ray work indicates that this is so for $NH_4MoO_2F_3$ [35] and $CsMoO_2F_3$ [34] (Figure 1.3(b)). The structure of the anion is identical to that found in $K_2VO_2F_3$, whose structure has now been reported in full[36]. Unpublished x-ray work, quoted in Ref. 35, that $CsMoO_2F_3$ contains a 5-coordinate anion would appear to be incorrect. A dimeric oxofluorotungstate anion $[W_2O_2F_9]^-$ (Figure 1.4(a)) has been identified in solution by ^{19}F and ^{183}W n.m.r. spectroscopy. It is one of the products from the reaction of WF_5OMe with $(MeO)_2SO$ [37]. To account for the observed spectrum, the W—F—W bridge must be nearly linear and must have considerable ionic character. The analogous Mo^{VI} anion is one of the products from the reaction of $MoOF_4$ with acetylacetone in MeCN [38], and has been isolated as a Rb^+ salt from the reaction of $MoOF_4$ with RbF in SO_2 [39]. The *trans* effect of a terminal oxygen

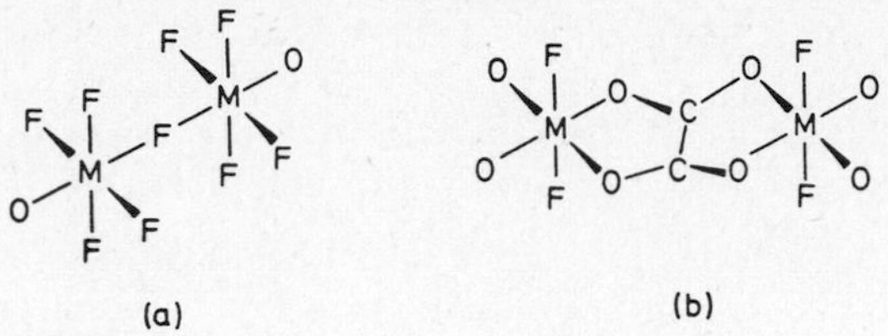

(a) (b)

Figure 1.4 Structures suggested for (a) $[M_2O_2F_9]^-$ and (b) $[M_2O_4F_4(C_2O_4)]^{2-}$, ($M = Mo^{VI}$ or W^{VI}).

ligand in oxohalides and their derivatives is well documented in the literature. Oxalic acid reacts with *cis*-$[MO_2F_4]^{2-}$ (M = Mo or W) to give $[M_2O_4F_4(C_2O_4)]^{2-}$, for which the structure shown in Figure 1.4(b) has been proposed from its spectra[40]. This suggests that fluorines *trans* to the terminal oxygens are preferentially replaced. However, a ^{19}F n.m.r. study of the reactions of WOF_5^- or $MoOF_4 \cdot NCMe$ with EtOH suggests that a fluorine *cis* to the terminal oxygen is preferentially replaced by OEt^- [38]; π(p—p) bonding in the *cis*-M(O)(OEt) moiety has been invoked to explain the latter observations but further work is required to establish this.

1.4 PENTAPOSITIVE ELEMENTS

1.4.1 Pentahalides, their substituted derivatives and oxohalides

Molecular beam mass spectrometry is a powerful tool for obtaining information about the degree of aggregation of molecules in the vapour state, and has been applied with great success to pentafluorides. Unlike VF_5 and CrF_5 whose vapours appear to be monomeric, polymeric species $(MF_5)_n$ have been observed in the vapour of Nb ($n \leqslant 4$), Ta ($n \leqslant 4$), Ir ($n \leqslant 5$) and Sb ($n \leqslant 5$) pentafluorides at temperatures in the range 340–390 K[41]. The spectrometer design rules out the possibility that the observed ions are the result of ion–molecule reactions, and the presence of pentameric species for Ir and Sb pentafluorides means that the polymers are not exclusively composed of the tetrameric rings found in the solid state. A similar method has been used to demonstrate the presence of $(VOF_3)_2$ species in the vapour of VOF_3 at ambient temperatures[42]. Temperature-dependent equilibria between monomeric and polymeric species are also indicated for gaseous NbF_5 and TaF_5 from their vibrational spectra and from vapour density measurements[43]. It is suggested that the polymers contain 6-coordinate metal atoms with *cis*-fluoro bridges, and that monomers are not the dominant species until *ca.* 673 K. This is a higher temperature than that previously proposed. The i.r. spectrum of NbF_5 isolated in an Ar matrix at liquid-hydrogen temperatures has been assigned on the basis of a C_{4v} monomer[44]. However the spectrum is similar to that obtained for 'polymeric' NbF_5[43], and the presence of such species cannot be ruled out.

The diethylamido-niobium and -tantalum fluorides, $MF_{5-n}(NEt_2)_n$ ($n=1$ or 2) are produced from reactions between MF_5OEt_2 or $(MF_5)_4$ and Me_3SiNEt_2 at room temperature. They are formulated as fluoro-bridged polymers from their i.r. spectra, and further substitution was not observed under the conditions used[45]. Further work on the reactions of $NbCl_5$ with Me_2Zn has characterised $NbCl_3Me_2$ and complexes of $NbCl_4Me$. These compounds are very labile and their solution equilibria are complex[46]. MeNC reacts with MCl_5 (M = Nb or Ta) in Et_2O at room temperature to give $[MCl_4\{C(Cl){:}NMe\}(CNMe)]_2$ and $[NbCl_3\{C(Cl){:}NMe\}_2(CNMe)]$ presumably via initial coordination of MeNC followed by its insertion into an M—Cl bond. Imino—N bridging is suggested for the dimeric products[47].

Many complexes of $ReOCl_3$ are known but the parent compound has so far proved elusive. A recent effort to prepare the compound has been partially successful however. A preliminary communication[48] claimed the isolation of purple α-$ReOCl_3$ in small yield from sealed tube reactions of $ReCl_5$ with ReO_3 or ReO_2 at 573 K. The same product was obtained by the photodecomposition of $ReOCl_4$ at 350 nm. On standing *in vacuo*, a second purple form is obtained by vapour-phase transport. This was originally designated β-$ReOCl_3$, as its x-ray powder pattern is distinct from the α form. Subsequently, however, single-crystal x-ray work has shown that β-$ReOCl_3$ is $(Re^{VI}_2O_3Cl_6)(Re^{VII}O_3Cl)_2$ (Figure 1.5) with approximately octahedral Re^{VI}

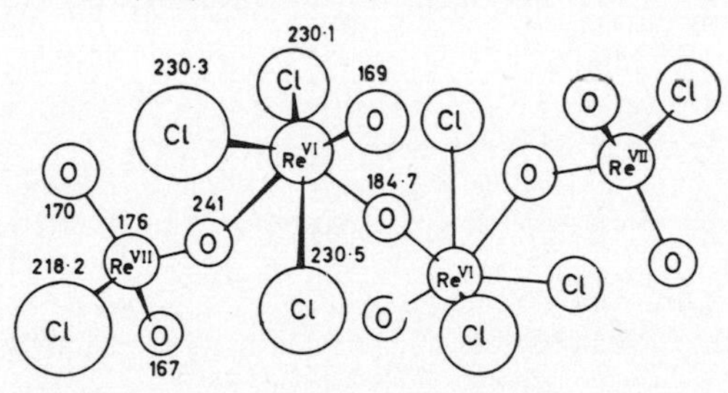

Figure 1.5 Structure of $(Re^{VI}{}_2O_3Cl_6)(Re^{VII}O_3Cl)_2$. (From Lock *et al.*[49] by courtesy of the National Research Council of Canada.)

and tetrahedral Re^{VII} atoms[49]. The compound can be reproducibly formed by the photochemical (350 nm) reaction between $ReOCl_4$, $ReOCl_4 \cdot H_2O$ and ReO_3Cl. In view of these results, the characterisation of $ReOCl_3$ must be questioned. The available data on α-$ReOCl_3$ is consistent with its existence but is not conclusive[49]. This work illustrates the fascinating (and difficult!) problems posed by Re chemistry.

The existence of MOX_3 (M = Mo or W, X = Cl or Br) compounds is well established and their heats of formation at 298 K have been determined from their heats of oxidative hydrolysis. Values obtained (in kJ mol^{-1}) are as follows[26, 50, 51]:

$MoOCl_3(s)$	$WOCl_3(s)$
−622	−728
$MoOBr_3(s)$	$WOBr_3(s)$
−458	−473

1.4.2 Hexahalometallates(V)

One of the most noteworthy developments in transition metal halide chemistry has been the characterisation of AuF_6^- salts[52, 53]. Oxidation of AuF_3 by F_2 in the presence of XeF_6 at high temperature and pressure leads to pale green $[F_5XeFXeF_5^+][AuF_6^-]$. The compounds $CsAuF_6$ and O_2AuF_6 have also been reported[52]. The AuF_6^- anion is essentially octahedral (Au—F = 185–190 pm), consistent with the expected t_{2g}^6 configuration of Au^V [53]. $CsAuF_6$ is isomorphous with other CsM^VF_6 (M = noble metal) compounds, and the frequency of the a_{1g} (totally symmetric) stretching mode in MF_6^- decreases along the series M = Os^V to Au^V. This may be indirect evidence for M←F π bonding in the anions where M = Os–Pt, as π bonding is not likely for a t_{2g}^6 configuration[52].

The $ReCl_6^-$ anion has been characterised for the first time. The compound $[PCl_4][ReCl_6]$, so formulated from i.r. conductance and magnetic studies, is formed either from $ReCl_5$ and PCl_5 at 573 K, or by the oxidation of Re metal by Cl_2 in the presence of PCl_5 at 773 K [54]. Metathetical reactions to prepare other salts were unsuccessful, either $ReCl_6^{2-}$ or decomposition products being obtained (cf. Section 1.5.2). Heats of formation of some TaF_6^- salts derived from a thermochemical study of acids and bases in BrF_3 are as follows: −2625 (K^+), −2594 (Na^+), −2220 (Ag^+) and −5232 (Ba^{2+}) kJ mol^{-1} [55]. They may be compared with the value of −2261 kJ mol^{-1} for KWF_6 calculated from a heat of hydrolysis study[13]. Mixed hexahalogenoniobates(V) and -tantalates(V) are readily generated in solution and may be identified using ^{19}F n.m.r. or ^{93}Nb n.m.r. spectroscopy[56–58]. For example, the series $[MF_{6-n}Cl_n]^-$ is formed in solutions of MCl_5 in aqueous HF [56] and the series $[NbCl_{6-n}Br_n]^-$ from mixtures of $NbCl_5$ and $NbBr_5$ in MeCN [58]. Structural isomers are usually observed and re-distribution among the species is random.

1.4.3 Oxohalometallates(V)

A re-investigation of the preparation of $ReOCl_5^{2-}$ salts has shown that the normally used method (reduction of ReO_4^- by I^- in HCl) leads to an equilibrium between $ReOCl_5^{2-}$, $ReCl_6^{2-}$ and ReO_4^- anions. The precise conditions necessary to obtain pure $A^I_2ReOCl_5$ have been established and it has been shown that for A = Cs or Rb, Re^V is spin-paired[59]. The crystal structure of $(NH_4)_2MoOBr_5$ shows the expected lengthening of the Mo—Br bond (283 pm) *trans* to Mo=O (186 pm) compared with the *cis*-Mo—Br bonds (255 pm)[60]. The *trans* ligand is labile in solution but the replacement of the *cis* ligands may be followed by e.p.r. spectroscopy. The four stages in the formation of $[MoOL_4]^{3+}$[L = $(MeO)_2P(O)H$] from $MoOBr_4^-$ in tetrahydrofuran have been followed in this way. The second, and subsequent steps, appear to be second-order displacement reactions, but the first step is first-order with respect to L and zero order with respect to $MoOBr_4^-$ although the rate depends on the initial concentration of $MoOBr_4^-$ [61]. There seems to be no simple mechanism that fits the latter observations.

$CsVOF_4$, which is orthorhombic, contains VOF_4 units of C_{4v} symmetry (V=O = 152.9 pm, V—F(average) = 178.8 pm). They are linked, via *cis* fluoro bridges involving a long V—F bond (231.2 pm) *trans* to oxygen, to form anionic chains[62]. The ^{19}F and ^{51}V n.m.r. data on VOF_4^- in anhydrous HF, are best interpreted in terms of a rapid rearrangement between C_{4v} and C_{2v} environments[63]. The anionic chains found in solid $CsVOF_4$ are related to those in $K_2VO_2F_3$[36], and the coordination about V is similar to that reported for orthorhombic Cs_2VF_6[64]. Bond distances in the isolated, very distorted octahedral VF_6^{2-} anion are $(V—F)_{eq.}$ = 192 pm and $(V—F)_{ax.}$ = 161 and 229 pm. The compound was prepared by the reaction of a 1:1 mixture of V_2O_5 and V_2O_3 with CsF in 40% HF. In view of this, and the very short V—F distance observed, further characterisation is desirable.

Several compounds containing the $VO_2F_2^-$ anion have been characterised and the structure of the anion appears to depend on the identity of the cation. Polymeric $[V(O)O_{2/2}F_2^-]_\infty$ anions are suggested for the Na^+, NH_4^+ and enH_2^{2+} salts from their i.r. spectra and x-ray powder data[65], and from the Raman spectrum of the NH_4^+ salt[66]. Monomeric anions having C_{2v} symmetry are indicated for the Ph_4M^+ (M = P or As) salts from their i.r. spectra and x-ray powder data[67]. $[Ph_4M][VO_2Cl_2]$ are similar.

1.5 TETRAPOSITIVE ELEMENTS

1.5.1 Tetrahalides and their substituted derivatives

IrF_4 cannot be prepared from Ir metal and fluorine but has been characterised for the first time as the product from the reduction of IrF_6 on a W or Ir filament at red heat[68]. The technique is similar to that used to prepare WF_5 and $OsOF_4$[19]. IrF_4 is a red-brown solid which sublimes at *ca.* 453 K and its mass spectrum shows no evidence for polymers in the vapour phase.

Although $PtCl_4$ has been known for many years, its structure is not known. Two reports have shown a previous suggestion that it comprises tetrahedral molecules like SnI_4 to be incorrect. The first[69] proposes that its structure is orthorhombic $[PtCl_{4/2}Cl_2]_\infty$ with the terminal Cl atoms *cis*. Its x-ray powder pattern is similar to those of $PtBr_4$ and α-PtI_4 which have this structure, and in its i.r. spectrum bands due to terminal and bridging Cl atoms can be distinguished. The second[70] compares the i.r. spectrum of $PtCl_4$ with those of $[PtMe_3Cl]_4$ and $[PtCl(C_3H_3)Cl]_4$, which have cubane structures comprising interlocking Pt_4 and Cl_4 tetrahedra. A similar cubane structure is suggested for $PtCl_4$, and the observation of $Pt_4Cl_{16}^+$ and fragment ions in its mass spectrum is used as supporting evidence.

$ReCl_4$ is conveniently prepared by the reduction of $ReCl_5$ with C_2Cl_4, either at 343–353 K for 42 h in CCl_4[71], or at 393–413 K for 6–24 h without a solvent[72]. In one report[72], it is claimed that the product is a new form of $ReCl_4$, designated γ (monoclinic but with different cell parameters from the β form), and that using longer reaction times or higher temperatures the products are β-$ReCl_4$ or $ReCl_3$. Reactions between C_3Cl_6 and $ReCl_5$, $ReOCl_4$ or Re_2O_7 at 433–453 K give $Re_2Cl_8(C_3Cl_6)$ which decomposes at 413 K to give γ-$ReCl_4$. $ReBr_4$ is formed from $ReCl_5$ and BBr_3 and is probably isostructural with γ-$ReCl_4$[72].

An exhaustive examination of the vibrational spectra of TiX_4 (X = Cl, Br, or I) compounds in solid, liquid and gaseous states and in solution shows that all the spectral features can be accounted for on the basis of tetrahedral molecular structures[73-77]. A previous suggestion that liquid $TiCl_4$ contains dimeric molecules is therefore not substantiated. The bond stretching force constants in MX_4 (M = Ti, Zr, or Hf; X = Cl, Br, or I) molecules are virtually independent of M, but show the expected variation with X, i.e. Cl > Br > I[78]. Similar behaviour has been observed previously in MX_6^{2-} anions.

Measurements of the thermal decomposition of $NbBr_4$ indicate that it is more stable in this respect than $NbCl_4$[79]. For example, the dissociation pressure at 573 K corresponding to

$$(5-x)NbBr_4(s) \rightleftharpoons NbBr_x(s) + (4-x)NbBr_5(g) (2.67 \leqslant x \leqslant 3.03),$$

is 4.3 Torr. The corresponding value for $NbCl_4$ is 230 Torr. MnF_4 has been examined as part of a thermochemical and molecular beam mass spectrometric study of Mn fluorides. Heats of formation of MnF_4(s) and MnF_4(g) at 298 K are -1108 and -966 kJ mol^{-1}, respectively, and $D(MnF_3-F)$ is 260 kJ mol^{-1} (Ref. 80).

The Raman spectra of $TiCl_4$–$TiBr_4$ liquid mixtures indicate that mixed halides $TiCl_{4-n}Br_n$ are formed, in agreement with previous i.r. work, and that ligand re-distribution is almost random[73]. The situation parallels that found for other mixed halides (cf. Section 1.4.2). A similar explanation has been invoked to explain the behaviour of ^{47}Ti and ^{49}Ti n.m.r. chemical shifts in $TiCl_4$–TiX_4 (X = Br or I) mixtures[81]. The crystal structure of $TiCl_3(NEt_2)$ confirms a previous suggestion made from a vibrational spectral study that the compound contains chloro rather than diethylamido bridges. Its structure is $[Ti(NEt_2)ClCl_{4/2}]_\infty$ with the terminal ligands in *cis* positions. The Ti—N distance (185.2 pm) is shorter than expected and may indicate some π(p→d) bonding[82]. The compounds $TiI_{4-n}(NR_2)_n$ (R = Me or Et; n = 2 or 3 but not 1) can be isolated from reactions between TiI_4 and $Ti(NR_2)_4$. Their spectra are similar to the analogous bromides, and the compounds are probably monomeric[83].

1.5.2 Halometallates(IV)

Two products from the $TiCl_4 \cdot PCl_5$ system have been characterised by x-ray crystallography. $[PCl_4]_2[Cl_4TiCl_2TiCl_4]$ is formed in $POCl_3$ solution, and the anion has an edge-sharing bi-octahedral structure (Figure 1.6(a)) with $(Ti—Cl)_{bridge}$ = 248.1–250.6 pm, $(Ti—Cl)_{terminal}$ = 224.7–230.0 pm. $[PCl_4]$–$[Cl_3TiCl_3TiCl_3]$ is formed in $SOCl_2$ solution and also from the reaction of PCl_3 with $TiCl_4$ in $SOCl_2$. The anion has a face-sharing bi-octahedral structure (Figure 1.6(b)) with $(Ti—Cl)_{bridge}$ = 247.7–254.6 pm, $(Ti—Cl)_{terminal}$ = 218.8–224.5 pm. The solvents' roles in these reactions are

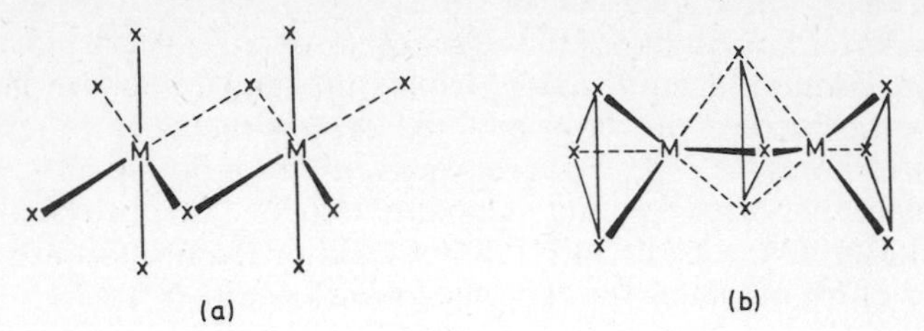

Figure 1.6 (a) M_2X_{10}: edge-sharing bi-octahedron found for $(Ti_2Cl_{10})^{2-}$ and $(Pt_2Cl_{10})^{2-}$ and suggested for $(Ti_2Me_2X_6Y_2)^{2-}$ (X, Y = Cl or Br).
(b) M_2X_9: face-sharing bi-octahedron found for $(Ti_2Cl_9)^-$, $(Cr_2X_9)^{3-}$, $(Mo_2X_9)^{3-}$ (X = Cl or Br), $(W_2Cl_9)^{3-}$ and $(Rh_2Cl_9)^{3-}$, and suggested for $(Ti_2Me_2X_6Y)^-$ (X, Y = Cl or Br)

not clear, but possibly the 2:1 ionic product is favoured by the more basic $POCl_3$ solvent[84]. Interestingly the 1:1 adduct formed between PCl_5 and VCl_4 contains a monomeric VCl_5^- anion. The electronic spectrum of the anion has been assigned on the basis of a square-pyramidal rather than a trigonal-bipyramidal structure[85]. Methyltitanium trichloride and tribromide show some similarities with the parent tetrahalides and they react with $R_4N^+Y^-$ (Y = Cl or Br) in CH_2Cl_2 at 273 K to form ionic compounds[86]. The products are $A_2[MeTiX_3Y_2]$, $A_2[Me_2Ti_2X_6Y_2]$, or $A[Me_2Ti_2X_6Y]$ (X, Y = Cl or Br; A = R_4N) depending on the stoichiometry of the reaction. Structures analogous to $Ti_2Cl_{10}^{2-}$ and $Ti_2Cl_9^-$ (Figure 1.6) are proposed for the dinuclear anions from their spectra. The dimeric species persist in solution, although fast intramolecular exchange is observed for $[Me_2Ti_2X_6Y]^-$. Oxygen is inserted readily into the Ti—C bonds of $[Me_2Ti_2Cl_7]^-$ to give the corresponding Ti—OMe species.

$A_2M^{IV}F_6$ compounds of 3d transition metals have attracted considerable attention in recent years (for a review of their electronic spectra see Ref. 3). New compounds of this type include Li_2CrF_6 which is isostructural with Na_2SnF_6. Other new complex fluorides of Cr^{IV} are $A^{II}CrF_6$ (A = Ba or Sr, $BaSiF_6$ type; A = Mg, Ca, Cd, or Hg, $LiSbF_6$ type; A = Ni or Zn, VF_3 type). Apart from $NiCrF_6$ which is brown, these compounds are yellow or pink in colour. They are prepared by fluorination of salt mixtures at high temperature and pressure[87]. Rb_2CoF_6 has been prepared for the first time. This compound like Cs_2CoF_6 has the K_2PtCl_6 structure and a $^2T_2$ ground state for Co^{IV}. The anomalously high magnetic moments of both compounds are attributed to ferromagnetic behaviour with Curie temperatures above ambient[88].

The structural chemistry of Zr^{IV} is very complex as coordination numbers of six, seven, or eight are commonly found. A very good illustration of this is $Rb_5Zr_4F_{21}$ whose anionic chain structure contains four crystallographically-independent Zr atoms. Their coordination polyhedra are a pentagonal-bipyramid (7-coordinate), an irregular antiprism (8-coordinate), an octahedron (6-coordinate) and an irregular antiprism with one vertex missing (7-coordinate)[89].

Although Cs_2NbCl_6 may be isolated from solutions of $NbCl_4$ in conc. HCl on the addition of CsCl, no e.s.r. signal due to $NbCl_6^{2-}$ is observed in solution.

The signals observed are attributed to the $NbOCl_5^{3-}$ and $[NbOCl_4(H_2O)]^{2-}$ anions. The latter ion is also one of the species in solution when Nb^V is reduced by Zn/HCl. $[NbOCl_4(EtOH)]^{2-}$ is observed if the reduction is performed in ethanol. Considerable delocalisation of the unpaired spin density into pπ orbitals of the equatorial Cl ligands is suggested[90]. The heats of formation of several A_2WCl_6(s) compounds at 298 K have been estimated from combustion calorimetry. The values are −1446 (Cs), −1430 (Rb) and −1359 (K) kJ mol^{-1}, numerically slightly smaller than those of the Mo^{IV} analogues[91]. These compounds may be used as organic reagents to accomplish the *cis* deoxygenation of vicinal diols to give olefins. Their utility is believed to be due to the ready formation of a *cis*-$W^{VI}O_2$ moiety; other reduced W species behave in a similar manner[92]. Attempts to prepare $ReCl_6^-$ salts (cf. Section 1.4.2) by the reactions of $ReCl_5$ with $R_4N^+Cl^-$ in CH_2Cl_2 under anhydrous, oxygen-free conditions, led to the formation of (R_4N)–$(ReCl_5)$[93]. The $ReCl_5^-$ anion is believed, from its spectra, to be polymeric with *cis*, non-linear Cl bridges. It reacts with N-donor ligands (L) to form $ReCl_5L^-$. These reactions clearly differentiate Re^V from Mo^V and W^V by the ease with which Re^V may be reduced to the tetrapositive state.

^{35}Cl n.q.r. studies have been made on several hexachlorometallates of 4d transition metals in order to make comparisons with previous work on 5d metals. In particular, n.q.r. frequencies may be used to discuss σ-covalent and Cl→M π-bonding contributions to the metal–chlorine bond. π-Bonding in MCl_6^{2-} (M = Tc–Pd) anions is considered to be more important than in the corresponding 5d complexes, particularly for d^3 and d^4 configurations. It is less important in $MnCl_6^{2-}$ but a high σ covalency is indicated[94].

Several new chloro anions containing Pt^{IV} have been characterised by x-ray crystallography. They were prepared from reactions of $PtCl_4$ with Ph_3CCl in various solvents and isolated as yellow, orange, or red Ph_3C^+ salts[70]. In CH_2Cl_2 solution, the monomeric anion $[PtCl_5(CH_2Cl_2)]^-$ appears to be the initial product, and this slowly dimerises to give $[Pt_2Cl_{10}]^{2-}$ which has an edge-sharing bi-octahedral structure, $(Pt{-}Cl)_{terminal}$ = 229.8 pm, $(Pt{-}Cl)_{bridge}$ = 237.4 pm, (Figure 1.6(a)). The third anion isolated from this system is $[Cl_4Pt^{IV}Cl_2Pt^{II}Cl_2Pt^{IV}Cl_4]^{2-}$. Its structure corresponds to three *trans* edge-sharing octahedra from which the two central terminal Cl atoms have been removed giving a square-planar environment for Pt^{II}. The Pt^{IV}—Cl—Pt^{II} bridges are unsymmetrical, Pt^{IV}—Cl = 238.4 pm (av.), Pt^{II}—Cl = 230.0 pm (av.). This anion may be formed from $[Pt_3^{IV}Cl_{14}]^{2-}$ by loss of Cl_2 with a corresponding $Pt^{IV} \rightarrow Pt^{II}$ reduction, as is known for other Pt^{IV} chloro complexes. $[Pt_3Cl_{12}]^{2-}$ is the only product isolated using $Cl_2HCCHCl_2$ as solvent, and no reaction occurs when more bulky, chlorohydrocarbon solvents are used. This emphasises the importance of solvent coordination in the initial stages of the reaction scheme. The reaction follows a different course when benzene is used as a solvent. A tetranuclear Pt^{IV} anion $[Pt_4(C_6H_4)_2Cl_{14}]^{2-}$, idealised symmetry D_{2h}, is formed whose structure (Figure 1.7) comprises two $[Pt_2(C_6H_4)Cl_6]$ groups, $(Pt{-}Cl)_{terminal}$ = 227.4 pm, $(Pt{-}Cl)_{bridge}$ = 235.2 pm, Pt—C = 197 pm, linked by two additional Cl bridges (253.3 pm).

The photochemistry of Pt^{IV} halide complexes is complex as photoreduction and photo-aquation are both possible reactions. These processes have been

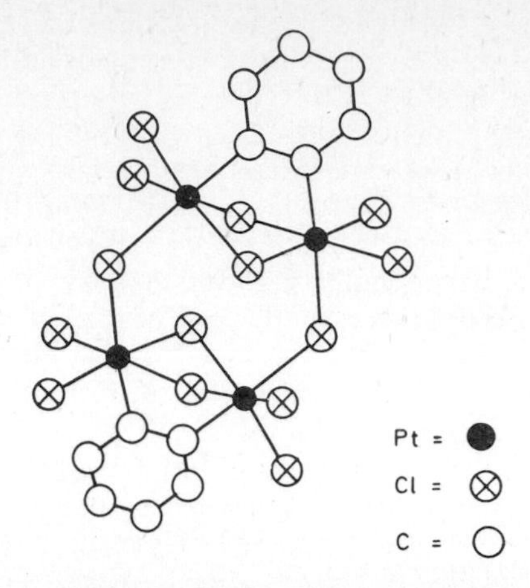

Figure 1.7 Structure of $[Pt_4(C_6H_4)_2Cl_{14}]^{2-}$. (From Dahl *et al.*[70] by courtesy of the American Chemical Society.)

observed in a recent flash-photolysis study of $PtCl_6^{2-}$. Energy absorption at the Cl→Pt charge-transfer band frequency produces both $[PtCl_5(OH)]^{2-}$ and $PtCl_4^-$ [95].

1.6 TRIPOSITIVE ELEMENTS

1.6.1 Halogeno compounds of 3d elements

TiOBr, which like TiOCl has the FeOCl structure, $[TiO_{2/2}O_{2/2}Br_{2/2}]_\infty$, has been prepared from Ti, TiO_2 and Br_2 using a temperature gradient of 923–823 K. It exhibits weak temperature-independent paramagnetism[96]. A re-investigation of compounds formed in the system VCl_3, A^+Cl^-, HCl, H_2O has characterised (by spectroscopic, magnetic and x-ray powder data) the V-containing complex ions that can exist[97]. $VCl_3 \cdot 6H_2O$, $RbVCl_4 \cdot 6H_2O$ and $Cs_2VCl_5 \cdot 4H_2O$ all contain the green, *trans*-$[VCl_2(H_2O)_4]^+$ cation. The analogous bromo cation is present in $VBr_3 \cdot 6H_2O$ and $Cs_2VBr_5 \cdot 5H_2O$. $A_2VCl_5 \cdot H_2O$ contain the red $[VCl_5(H_2O)]^{2-}$ anion and are of two structural types. Where A = K, NH_4, or Rb, they are isostructural with $K_2[FeCl_5(H_2O)]$, but the Cs^+ salt is of a different type. The magnetic properties of anhydrous A_2VCl_5 (A = Cs or Rb) suggest the presence of weak antiferromagnetic interactions in a polymeric structure.

There has been a vast amount of work dealing with the structures and spectroscopic properties of halometallates(III) of 3d elements and a selection only can be summarised here. The electronic spectrum of $PCl_5 \cdot CrCl_3$ indicates the presence of an octahedral $CrCl_6$ chromophore, not tetrahedral as had been proposed previously[98]. Octahedral Cr^{III} is proposed also in $[Cr(Cl_2AlCl_2)_3]$, which it is suggested is responsible for the transport of $CrCl_3$(s) by $AlCl_3$(g) in the temperature gradient 773–673 K [99]. Both Cr and

Ni are octahedrally coordinated in $RbNiCrF_6$ (Ni—F = Cr—F = 193 pm) and $KNiCrF_6 \cdot nH_2O$ ($n = 0$ or 1). The M_2X_6 structure found in these compounds is analogous to the pyrochlores[100]. Isolated MF_6^{3-} anions are commonly stabilised by large cations. Some examples are $[N_2H_5]_3[CrF_6]$, Cr—F = 190.5 pm (av.)[101], $[Cr(NH_3)_6][MnF_6]$, Mn—F = 192.2 pm, and $[Cr(NH_3)_6][FeF_6]$, Fe—F = 193.1 pm[102]. Hydrogen bonding is evident in all three cases and the regularity of the MnF_6^{3-} octahedron is ascribed to a dynamic Jahn–Teller effect. The i.r. spectrum of $[Cr(NH_3)_6][MnF_6]$ differs from those of K_3MnF_6 and K_2NaMnF_6 where static Jahn–Teller effects occur[103].

The x-ray structure of $[bipyH_2][MnCl_5]$ (bipy = bipyridyl) shows the presence of monomeric $MnCl_5^{2-}$, thus confirming previous magnetic work. The anion approximates to a square-pyramid with $(Mn—Cl)_{eq.} = 230.2$ pm (av.) and $(Mn—Cl)_{ax.} = 258.3$ pm[104]. In agreement with this, its single-crystal polarised electronic spectrum has been assigned in C_{2v} symmetry, and the spectrum of Mn^{III} doped in $(Et_4N)_2InCl_5$ in C_{4v} symmetry[105]. Anionic chains based on $[MF_4F_{2/2}]$ units with *trans*-F bridges are commonly found in (A_2^I or A^{II})MF_5 compounds, for example in $CaMF_5$ (M = Ti, V, Cr, or Co)[106], and in one of the two types of anionic chains present in $BaFeF_5$[107]. They are present also in $K_2MnF_5 \cdot H_2O$ where the H_2O molecule is not coordinated to Mn[108], but not in $K_2FeF_5 \cdot H_2O$ which contains discrete $[FeF_5(H_2O)]^{2-}$ anions[109]. A by-product in the preparation of $K_2MnF_5 \cdot H_2O$ is $K_2MnF_3(SO_4)$, which features anionic chains with Mn^{III} linked both through *trans*-F bridges and *trans*-SO_4 bridges. The distorted octahedral coordination,

$$[MnF_2F_{2/2}(SO_4)_{2/2}]$$

results in two short and four long bonds to Mn^{III}. This is the opposite to the normal situation for Mn^{III}, for example in $(NH_4)_2MnF_5$ and $K_2MnF_5 \cdot H_2O$, and probably results from the double bridge structure and not from Jahn–Teller distortion[110].

X-Ray work on $[MeNH_3]_2[FeBr_4]Br$ has provided the first crystallographic evidence for tetrahedral $FeBr_4^-$ [Fe—Br = 232 pm (av.)]. There is extensive $N—H \cdots Br^-$ bonding in the compound, and it is suggested that it may be ferro-electric[111]. The single-crystal electronic spectra of $AFeBr_4$ (A = Me_4N and related cations) compounds indicate Jahn–Teller distortion in the excited state. This effect is much less important in the isoelectronic $MnBr_4^{2-}$ anion possibly due to the larger nuclear charge of Fe^{III}. A similar situation exists for the chloro complexes[112].

The series of compounds A^ICoF_4 (A = Li, Na, K, Rb, or Cs) have been characterised and their abilities to fluorinate benzene investigated. The unit cells of the K, Rb and Cs salts are closely related to the corresponding $AFeF_4$ compounds and their magnetic moments (3.5 μ_B) are intermediate between those of CoF_3 and K_3CoF_6. Their behaviour towards benzene depends on the cation, the most interesting result being the direct conversion of C_6H_6 to C_6F_6, admittedly in small yield, by $CsCoF_4$ at 523 K. This is the first example of such a conversion by a high oxidation state metal fluoride[113]. Considerable oxidising ability is also apparent for other compounds in this area. For example, F_2 is liberated from solutions of K_3CuF_6, or Cs_2CoF_6 in anhydrous HF at 273 K[114].

1.6.2 Halogeno compounds of 4d and 5d elements

Hitherto, the existence of NbF_3 has been doubtful but recent work suggests that it can be prepared from NbF_5 and Nb metal at 1023 K and 3.5 kbar pressure. The product isolated has the ReO_3 structure being a semiconductor and weakly paramagnetic[115]. Refinement of the structure of $MoBr_3$ has indicated that it is of the orthorhombic $RuBr_3$ type, and not hexagonal as reported previously. The shorter Mo—Mo distance in the $(Mo_2Br_3)Br_{6/2}$ face-sharing octahedral chains is 292 pm [116].

The aquation of $CrCl_6^{3-}$ is well characterised but until recently little information about the corresponding process for $MoCl_6^{3-}$ was available. Treatment of this anion with 1 mol l^{-1} solutions of trifluoromethyl- or *p*-toluene-sulphuric acids gives a yellow solution which is considered to contain $Mo(H_2O)_6^{3+}$ [117]. Aquation is complete at equilibrium due to the extreme non-ligating character of the RSO_3^- anions. This work has been disputed on the basis of a kinetic study of the aquation of $MoCl_6^{3-}$ at pH 3.2–4.5 and 273 K. The first step occurs readily but the second step, giving $[MoCl_4(H_2O)_2]^-$, is very slow[118]. The yellow species obtained with the sulphuric acids is suggested to be an oxo- or hydroxo-bridged dimer[118], but its solution magnetic moment (3.69 μ_B) is consistent with monomeric Mo^{III} [119].

$[Me_4N][RhCl_4(H_2O)_2]$ contains slightly distorted octahedral anions with a *trans* configuration (Rh—Cl = 230.0–235.7 pm, Rh—O = 203.2 pm) [120], and distorted octahedral $RhCl_6^{3-}$ anions (Rh—Cl = 230.4–236.6 pm) are found in $K_3RhCl_6 \cdot H_2O$ [94]. The lack of symmetry in the latter anion is due to neighbouring H_2O molecules in the lattice. Low anion site symmetries are indicated also in $A_3IrCl_6 \cdot H_2O$ (A = K or NH_4) and K_3IrCl_6 from n.q.r. and i.r. studies. The compound of molecular formula $HK_8Rh_3Br_{18} \cdot 10H_2O$ contains three $RhBr_6^{3-}$ anions, (Rh—Br = 250 pm), of which two are crystallographically independent. The compound is prepared from $Rh(OH)_3 \cdot 3H_2O$, HBr and KBr[121].

1.6.3 Nonahalodimetallates(III) and octahalodimetallates(III) or (II)

Interest in $M_2X_9^{3-}$ anions, which have the face-sharing bi-octahedral structure (Figure 1.6(b)), stems from the possibility of direct M—M interaction in some cases. The two extreme situations for d^3 metals are $Cr_2Cl_9^{3-}$, in which there is a net repulsion between Cr^{III} centres, and $W_2Cl_9^{3-}$, where a direct interaction exists. The vibrational spectra of these anions have been analysed in D_{3h} symmetry and the calculated direct W—W force constant implies a bond order of between one and two[122]. Heats of formation for several $A^I_3W_2Cl_9(s)$ compounds at 298 K have been estimated from combustion and solution calorimetry. The values found are −2355 (K), −2477 (Rb) and −2507 (Cs) kJ mol^{-1} [123].

Some molecular parameters from recent single-crystal x-ray work on $M_2X_9^{3-}$ compounds[124, 125] are given below in Table 1.1 together with those of $Cr_2Cl_9^{3-}$ and $W_2Cl_9^{3-}$ for comparison.

Table 1.1

Compound	*M—X*$_{terminal}$ (*pm*)	*M—X*$_{bridge}$ (*pm*)		$M\hat{X}M_{bridge}$ (*degrees*)
$Cs_3Cr_2Br_9$ (Ref. 124)	241.7	257.7	331.7	80.0
$Cs_3Mo_2Cl_9$ (Ref. 124)	238.4	248.7	265.5	64.5
$Cs_3Mo_2Br_9$ (Ref. 124)	254.4	262.4	281.6	64.88
$(Me_3PhN)_3Rh_2Cl_9$ (Ref. 125)	229.6	239.7	312.1	81.3
$Cr_2Cl_9^{3-}$	234	252	312	76.4
$W_2Cl_9^{3-}$	240	250	241	58

Direct M—M interaction is indicated for the Mo anions but M—M repulsion is clearly present in $Cr_2Br_9^{3-}$, whose bond distances are consistent with its formulation as $[Br_3Cr(Br^-)_3CrBr_3]$[124], and in $Rh_2Cl_9^{3-}$, which is expected in view of the Rh^{III} t_{2g}^6 configuration[125]. Magnetic exchange in this type of anion may occur either directly, or via super-exchange through the bridging halogen atoms. In $A_3Mo_2X_9$ (A = Group I or R_4N cation, X = Cl or Br) compounds, direct exchange predominates and the (minor) contributions made by ferromagnetic and antiferromagnetic super-exchange are sensitive to structural changes. The total exchange in the Cr analogues is smaller by a factor $\geqslant 50$ and super-exchange and direct-exchange effects are comparable[126].

The vibrational spectra of $Re_2X_8^{2-}$ (X = Cl or Br), $Mo_2Cl_8^{4-}$ and related compounds have been examined in order to characterise the M—M stretching modes[127, 128]. Raman active bands in $Re_2X_8^{2-}$ at *ca.* 280 cm^{-1} are assigned to ^{1}(Re—Re), and the derived force constants are consistent with quadruple bonds[128]. Several reports of ligand-exchange reactions involving $Re_2Cl_8^{2-}$ or $Mo_2Cl_8^{4-}$ have appeared. One of the most interesting, deals with the solvolysis of $Mo_2Cl_8^{4-}$ under similar conditions to those described for $MoCl_6^{3-}$ (Ref. 117). The final product, obtained via $[Mo_2(SO_4)_4]^{4-}$ as an intermediate, is considered to be $Mo_2(aq)^{4+}$.

1.7 METAL ATOM CLUSTERS OF NIOBIUM, TANTALUM, MOLYBDENUM AND RHENIUM

Cluster compounds are commonly encountered in transition metal chemistry and the topic has been reviewed recently[129]. In this section, some developments in the chemistry of halide clusters will be summarised.

Halogen-exchange reactions of several cluster systems have been studied. The mixed halide $Re_3Br_6Cl_3$ was identified by spectroscopic methods from the reaction of Re_3Cl_9 with BBr_3. Re-distribution occurs on sublimation, and one of the products is thought to be Re_3Br_8Cl. Replacement of Cl by Br occurs preferentially in the terminal positions as would be expected[130]. Melt reactions at 673 K between HgY_2 and $[(Mo_6X_8)X_4]$ (X, Y = Cl, Br, or I) give $Hg[(Mo_6X_8)Y_6]$ or $Hg[(Mo_6Y_8)Y_6]$ depending on the identities of X and Y. A heavier face-bridging halogen is always substituted by a lighter one, presumably for thermodynamic reasons[131]. The anion in $Hg[(Mo_6Cl_8)Cl_6]$ appears to have an undistorted structure and this compound has a NaCl-type lattice[132]. The compounds $(pyH)_2[(M_6Br_{12})Cl_6]$ (M = Nb or Ta) undergo interesting intra-anionic exchange reactions at 473 K which are exothermic and irreversible[133]. The products are $(pyH)_2[(M_6Br_6Cl_6)Br_6]$ and x-ray work on the Nb compounds[134] has shown that the process involves six Cl and six Br atoms in two trigonal planes of the cluster (Figure 1.8). Metal exchange

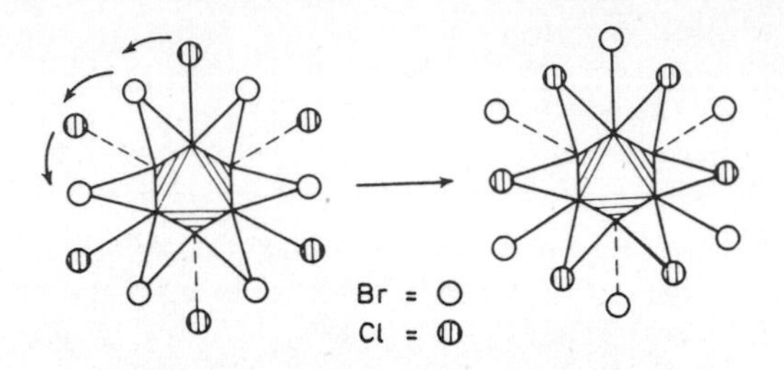

Figure 1.8 The halogen-exchange reaction $[(Nb_6Br_{12})Cl_6]^{2-} \rightarrow [(Nb_6Br_6Cl_6)Br_6]^{2-}$. (From Spreckelmeyer and von Schnering[134] by courtesy of J. A. Barth.)

has been little studied, but mixed clusters $[(Nb, Ta)_6Cl_{12}]^{2+}$ have been prepared from the reaction of $Na_4[(Nb_6Cl_{12})Cl_6]$ with its Ta analogue[135].

Several x-ray structural determinations on $(M_6Cl_{12})^{4+}$ (M = Nb or Ta) clusters are of interest in relation to previous work. The Ta_6 octahedron in $H_2[(Ta_6Cl_{12})Cl_6] \cdot 6H_2O$ is regular[136]. This contrasts with compounds having two or four negatively-charged ligands which have elongated and flattened Ta_6 octahedra respectively. The Nb—Nb bond distances (301.8–305.4 pm) in $[(Nb_6Cl_{12})Cl_6]^{2-}$ are longer, and the $(Nb—Cl)_{terminal}$ distances are shorter, than the corresponding distances in $[(Nb_6Cl_{12})Cl_6]^{4-}$ (Refs. 134, 137). This strongly implies that the electrons removed in the oxidation $(M_6X_{12})^{2+} \rightarrow (M_6X_{12})^{4+}$ are from a bonding molecular orbital centred primarily on the metal atoms[137]. Additional evidence comes from an n.q.r. study on $(Me_4N)_2[(Nb_6Cl_{12})Cl_6]$ which indicates that 14 electrons are available for Nb—Nb bonding in $(Nb_6Cl_{12})^{4+}$ [138]. Previous assignments of the electronic spectra of $(M_6X_{12})^{n+}$ (M = Nb or Ta; X = Cl or Br; n = 2, 3, or 4) clusters have been questioned as a result of a m.c.d. spectral study. A CF model with spin–orbit coupling taken into account was used to discuss the spectra and bonding[139].

The oxidation of $(Ta_6X_{12})^{2+}$ (X = Cl or Br) to the corresponding tripositive or tetrapositive cations by various transition metal species is well documented. This work has now been supplemented by an electrochemical study of $(Ta_6X_{12})^{n+}$ cations in aqueous solution[140]. Cyclic voltammetry, chronoamperometry and spectro-electrochemistry measurements were used to determine the following electrode potentials (in volts relative to standard calomel electrode):

$$(Ta_6X_{12})^{3+} + e = (Ta_6X_{12})^{2+};\ 0.25\ (Cl),\ 0.35\ (Br)$$
$$(Ta_6X_{12})^{4+} + e = (Ta_6X_{12})^{3+};\ 0.59\ (Cl),\ 0.65\ (Br)$$

Equilibrium quotients for the conproportionations

$$(Ta_6X_{12})^{2+} + (Ta_6X_{12})^{4+} \rightleftharpoons 2(Ta_6X_{12})^{3+}$$

are 5.7×10^4 (Cl) and 1.08×10^5 (Br), and for the reaction

$$(Ta_6Cl_{12})^{2+} + (Ta_6Br_{12})^{3+} \rightleftharpoons (Ta_6Br_{12})^{2+} + (Ta_6Cl_{12})^{3+}$$

the equilibrium quotient is 49. The study indicates that homogeneous electron transfer is rapid and that there is minimal structural change, particularly between the 2+ and 3+ ions. The series of hexamethylbenzene derivatives $[(Me_6C_6)_3M_3X_6]^+$ (M = Nb or Ta; X = Cl or Br) are readily oxidised by a variety of reagents to give dipositive cations. These may be isolated as PF_6^- or BPh_4^- salts and are diamagnetic. This suggests that they should be formulated as $[(Me_6C_6)_6M_6X_{12}]^{4+}$ and that they are structurally related to $[(M_6X_{12})Y_6]^{n-}$ ($n = 2$–4)[141]. The oxidations reactions would then be analogous to those discussed above.

1.8 DIPOSITIVE ELEMENTS

1.8.1 Halogeno compounds of 3d elements except copper(II) and zinc(II)

I.R. spectroscopy, either in the gaseous state at high temperatures or using matrix-isolation techniques, is often used to determine the structures of dihalides. The merits and pitfalls of these techniques have been reviewed in relation to recent work[142]. The i.r. spectra of matrix-isolated MCl_2 molecules (M = Ca–Fe inclusive and Ni) indicate that they are linear within experimental error (±10 degrees). Their geometries differ from the corresponding fluorides, for example in TiF_2, FTiF = 130 degrees, and a change in electronic configuration $3d4s \rightarrow 3d^2$ is suggested in going from TiF_2 to $TiCl_2$[143]. Electronic absorption and fluorescence spectroscopy are two other techniques that have been applied to the study of dihalides at high temperatures with great success[144]. Band assignments using ligand-field models are often subject to uncertainty, and several approaches have been developed. A recent treatment has emphasised the importance of X→M donor π interactions[145].

The dehydration of $CoCl_2 \cdot 6H_2O$ has been determined from dissociation pressure measurements in the range 292–393 K as $CoCl_2 \cdot nH_2O(s) \rightarrow CoCl_2 \cdot mH_2O(s) + (n-m)H_2O(g)$ ($n = 6$, 4, 2 and 1; $m = 4$, 2, 1 and 0). Derived thermodynamic parameters for these reactions and magnetic and spectroscopic measurements on the products, are taken to indicate that the major structural change occurs in forming $CoCl_2 \cdot H_2O$, although all the hydrates contain tetragonally-coordinated Co^{II}[146]. A re-examination of the $CoCl_2$-MeCN system suggests that the most important equilibrium is $3CoCl_2 + 6MeCN \rightleftharpoons [Co(NCMe)_6]^{2+} + [CoCl_3]_2^{2-}$. Formation of the dimeric anion is enhanced by the addition of Et_4NCl or LiCl, but is independent of the cation size[147]. This anion, $[Cl_2CoCl_2CoCl_2]^{2-}$, is present in the 1:2 adduct formed between the cyclohexaphosphazene $N_6P_6(NMe_2)_{12}$ and $CoCl_2$, which x-ray work[148] has shown to be $[\{N_6P_6(NMe_2)_{12}\}CoCl_2]$—

$[Co_2Cl_6]\cdot 2CHCl_3$; Co^{II} is approximately tetrahedrally-coordinated in the centrosymmetric anion (Figure 1.9) with $(Co-Cl)_{terminal} = 223.3$ pm (av.) and $(Co-Cl)_{bridge} = 233.6$ pm (av.). The cation is a trigonal-bipyramid isostructural with the Cu^{II} analogue, in which the counter-ion is linear $Cu^{I}Cl_2^-$.

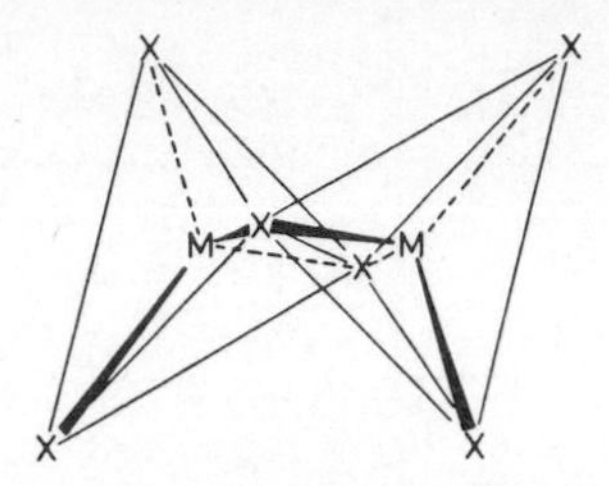

Figure 1.9 M_2X_6 group comprising two edge-sharing tetrahedra found for $(Co_2Cl_6)^{2-}$ and $(Cd_2I_6)^{2-}$

Far more information can be obtained from the electronic spectra of transition metal species when single crystals rather than powders are examined. Doping techniques enable the spectra of unusual isolated ions to be obtained; for example, that of Ti^{2+} isolated in NaCl [149]. The sample was prepared from the reaction of $CdCl_2$ with Ti in molten NaCl at 1223 K, NaCl being chosen as Na^+ and Ti^{2+} are of similar size and it is liquid at temperatures where Ti^{II} is expected to be thermodynamically stable. The electronic spectrum of Ti^{2+} is that expected for a d^2 ion in an octahedral ligand field.

Several papers have dealt with the effects of antiferromagnetic exchange between metal ions on their electronic spectra. The general spectral features are those expected for the isolated ions, but the intensities of spin-forbidden bands are greatly enhanced, and combination bands due to the simultaneous electronic excitation of pairs of ions are often observed. Among the compounds where these features have been recognised are VX_2 (X = Cl, Br, or I) [150], $CsCrCl_3$ [151] and $KCoF_3$ [152]. They are present to some extent also in $CsNiX_3$ (X = Cl or Br) [153]. $CsCrCl_3$ has a structure similar, but not identical, to $CsNiCl_3$. The face-sharing octahedral $[CrCl_{6/2}]_\infty$ chains do not have exact trigonal symmetry due to a small Jahn–Teller distortion associated with Cr^{II} [151].

The origin of the antiferromagnetism in $KMnCl_3\cdot 2H_2O$ has been investigated using ^{35}Cl and ^1H n.m.r. spectroscopy. The chemical unit cell contains two Mn^{2+} ions in an edge-sharing bi-octahedral unit $[Cl_2(H_2O)_2MnCl_2Mn(H_2O)_2Cl_2]$. Within this dimer the Mn^{2+} spins are parallel, and the overall antiferromagnetism results from a translation of this unit with spin reversal along the triclinic *a* axis. Octahedra of similar geometry are present in $MnCl_2\cdot 2H_2O$ but in this case nearest-neighbour spins are antiparallel. It appears therefore that the magnitude of nearest-neighbour exchange in such structures must be small [154]. Cation and magnetic ordering effects in phases derived from FeF_2 have been examined by Mössbauer spectroscopy and several different situations have been characterised[155]. The Néel temperatures of $Fe^{II}CoNiF_6$, $Mg_2Fe^{II}F_6$ and $MgFe_2^{II}F_6$ correspond to the weighted mean of those of the individual difluorides, indicating that these compounds are disordered rutile phases. Both the paramagnetic and the magnetically-ordered regions of the trirutile $LiFe_2F_6$ contain separate spectra due to Fe^{2+} and Fe^{3+}, and the signals have narrow line widths. Thus 'electron hopping' is absent and complete cation ordering is indicated. The repeating cation sequence in the *c* direction is $Li^+(Fe^{2+}\uparrow)$ — $(Fe^{3+}\downarrow)Li^+(Fe^{2+}\downarrow)(Fe^{3+}\uparrow)$ giving overall antiferromagnetic behaviour. However, in the trirutile $LiMgFe^{III}F_6$ there is ordering of Fe^{3+} into every third layer along the *c* axis, while Li^+ and Mg^{2+} are randomly disordered in the intervening two layers.

A new antiferromagnetic material $K_3Ni_2F_7$, which is the second member of the series $K_{n+1}Ni_nF_{3n+1}$, has been isolated as a by-product from the crystallisation of K_2NiF_4. Its magnetic properties are intermediate between the three-dimensional ordering in $KNiF_3$ and the two-dimensional ordering in K_2NiF_4[156].

1.8.2 Complex halides of palladium(II), platinum(II), copper(II) and silver(II)

Two classic examples of square-planar MX_4^{2-} ions are found in K_2PdCl_4 and K_2PtCl_4. The crystal structures of both compounds have been re-determined resulting in more accurate bond distances. The Pd—Cl and Pt—Cl distances are respectively 231.8 and 231.6 pm (both corrected for thermal motion), and in both compounds the K···Cl distances are 240 pm[157]. PdF_2 may be prepared by the reduction of 'PdF_3' or PdF_4 with Pd at *ca.* 1200 K in a gold-lined bomb, and the compounds $A^{II}PdF_4$ (A = Ca, Sr, Ba, or Pb) are formed from AF_2 and PdF_2 under similar conditions. They are formed also from the thermal decomposition of $A^{II}Pd^{IV}F_6$ salts at high temperatures. The compounds are brightly coloured, diamagnetic and have the tetragonal $KBrF_4$ structure[158].

Several ternary fluorides of Cu^{II} have been re-examined by x-ray crystallography as previous work had suggested the presence of tetrahedral Cu^{II}. This appears to be incorrect. Square-planar Cu^{II} is present in Sr_2CuF_6 (Cu—F = 195.4 pm) and in $A^{II}CuF_4$ (A^{II} = Ca or Sr). The latter compounds have the $KBrF_4$ structure (Cu—F = 188.0 and 185.8 pm, respectively), as does $SrCrF_4$ (Cr—F = 198.0 pm). Tetragonally-distorted Cu^{II} is found in $BaCuF_4$ the Cu—F bond distances being 185.2, 187.2, 190.9, 201.8 and 226.5 (× 2) pm[159]. This compound shows a deviation from Curie–Weiss behaviour below 212 K[160], and the electronic spectra of $SrMF_4$ (M = Cu or Cr) have been assigned on a square-planar basis[161]. Evidence for an elongated tetragonal $Ag^{II}F_6$ group comes from an electronic and e.p.r. spectral study[162] of some members of the series $AgM^{IV}F_6$ (M^{IV} = Ge, Sn, Pb, Ti, Zr, Hf, Rh, Pd, or Pt). These compounds are prepared by fluorination of Ag_2SO_4–A_2MX_6 mixtures at 753 K[163]. The distortions where M^{IV} = Zr or Hf are much greater than where M^{IV} = Sn or Pb, and substantial exchange between Ag^{II} sites is indicated for the former compounds[162,163]. In this respect the compounds' colours (M = Zr, Hf, blue-violet; M = Sn, Pb, light blue) are noteworthy. Other new Ag^{II} fluorides are $A^{I}AgF_3$ (A = K, Rb, or Cs), which have a similar structure to $KCuF_3$, blue-violet K_2AgF_4[164] and $CsAg^{II}M^{III}F_6$ (M^{III} = Tl, In, or Sc)[165]. The latter compounds have the cubic $RbNiCrF_6$ structure and their magnetic properties are complex.

Five-coordination in Cu^{II} halide complexes has aroused great interest recently, and several examples are now known. The $CuBr_5^{3-}$ anion in $[Cr(NH_3)_6][CuBr_5]$ has a trigonal-bipyramidal structure analogous to $CuCl_5^{3-}$ whose structure had been determined previously. Unlike many Main Group compounds of this structure, the axial bonds (Cu—Br = 245.00 pm) are shorter than the equatorial (Cu—Br = 251.91 pm). The difference in bond lengths however is much smaller in $CuBr_5^{3-}$ (6.91 pm) than in $CuCl_5^{3-}$ (9.8 pm). Axial compression in these species is assumed to be the result of the single occupation of the d_{z^2} orbital, and is less marked in $CuBr_5^{3-}$ as interligand repulsions are more important[166] (see also Section 1.8.3). $[CuCl_2Br_3]^{3-}$ is trigonal-bipyramidal also, with random disorder of Cl and Br atoms between axial and equatorial positions[166]. $[Co(NH_3)_6][CuCl_2Br_3]$ appears to be antiferromagnetic but the interaction is weaker than in the $CuCl_5^{3-}$ analogue. This is understandable if the interactions occur via the axial ligands[167]. The complex $Co(en)_3CuCl_5 \cdot H_2O$ has been known for many years but until recently structural information was lacking. X-Ray work has shown that Cu^{II} is 5-coordinate in this compound which is actually $[Co(en)_3] \times [Cu_2Cl_3]Cl_2 \cdot 2H_2O$. The complex anion corresponds to two trigonal-bipyramids fused along an equatorial–axial edge (Figure 1.10)[168]. Bond distances in $[Cl_3CuCl_2CuCl_3]^{4-}$ are $(Cu—Cl)_{ax.\ bridge} = 232.5$ pm, $(Cu—Cl)_{eq.\ bridge} = 270.3$ pm, $(Cu—Cl)_{ax.\ terminal} = 226.2$ pm and $(Cu—Cl)_{eq.\ terminal} = 227.3$ and 231.9 pm. The other interesting feature of this structure is the high energy configuration found for $Co(en)_3^{3+}$. The coordination chemistry of Cu^{II} halides is therefore complex, and this is illustrated further by the products isolated from reactions of CuX_2 (X = Cl or Br) with *N,N,N',N'*-tetraethylethylenediamine or its methyl analogue. Besides 1:1 adducts, the following anions were characterised as salts of the protonated ligands: distorted tetrahedral CuX_4^{2-}, monomeric $CuBr_3^-$ and $Cu_4OCl_{10}^{4-}$ in which each Cu has a trigonal-bipyramidal environment in a tetrahedral Cu_4O group[169].

1.8.3 Complex halides of zinc(II), cadmium(II) and mercury(II)

Of these three elements, cadmium has attracted the most interest. The mixed halide CdClBr has been identified by mass spectrometry in the vapour above solid $CdCl_2$–$CdBr_2$ mixtures at 720–770 K. The heat of formation of CdClBr(g) from CdX_2(g) is *ca.* 2 kJ mol^{-1} and the entropy increase associated with the decrease in molecular symmetry must be responsible for its formation[170].

The Raman spectra of molten CdX_2 (X = Cl or Br) indicate an average octahedral environment for Cd^{2+}, but addition of alkali metal halides increases the probability of tetrahedral coordination. The $CdCl_4^{2-}$ anion appears to have kinetic identity in such mixtures[171]. This may be compared

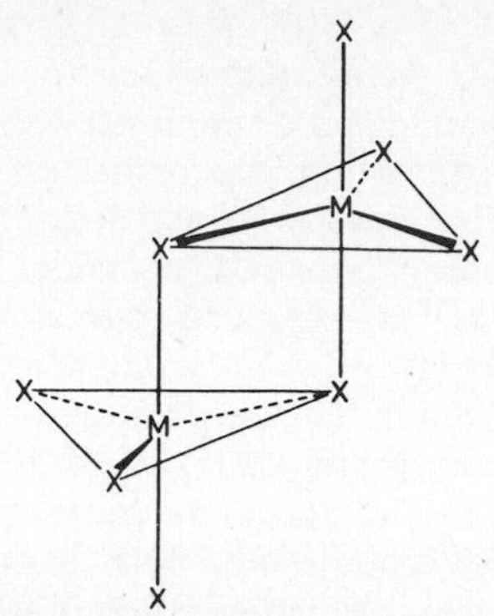

Figure 1.10 M_2X_8 group comprising two edge-sharing trigonal-bipyramids found for $(Cu_2Cl_8)^{4-}$

with the behaviour of CdX_2 and KX in nitrate melts where the predominant species are $CdCl_3^-$ and $CdBr_4^{2-}$ [172]. The compounds $(R_4N)_2CdX_4$ (X = Br or I) are conveniently prepared by mixing hot ethanolic solutions of the component halides in a 2:1 ratio. Their vibrational spectra indicate that CdX_4^{2-} are tetrahedral, and tetrahedral Cd^{II} is also indicated in $Bu_4^nNCdI_3$ which is prepared similarly using a 1:1 reactant ratio[173]. The anion is formulated as $Cd_2I_6^{2-}$ and is also present in the 1:2 adduct formed between tris(2-dimethylaminoethyl)amine(tda) and CdI_2. X-Ray work has shown this to be $[Cd(tda)I]_2[Cd_2I_6]$ the cation being a trigonal-bipyramid and the anion two tetrahedra fused along one edge (Figure 1.9) with $(Cd—I)_{terminal} = 270.4$ and 269.3 pm, and $(Cd—I)_{bridge} = 286.5$ and 284.9 pm[174].

A trigonal-bipyramidal $CdCl_5^{3-}$ anion has been found in $[Co(NH_3)_6]$—$[CdCl_5]$. Although the $(Cd—Cl)_{ax.}$ distances (252.6 pm) are shorter than the $(Cd—Cl)_{eq.}$ (256.1 pm), the difference is far smaller than in the analogous d^9 Cu^{II} species as expected[175]. Completing the structural comparison between Cd^{II} and Cu^{II} is $[Co(en)_3]_2[CdCl_6]Cl_2 \cdot 2H_2O$. This is prepared under similar conditions to the compound containing the $[Cu_2Cl_8]^{2-}$ anion discussed above, but the anion is a distorted octahedron (Cd—Cl = 258.8–276.5 pm)[176]. It is evident that the structural chemistry of Cd^{II} halogeno compounds is more varied than had been supposed.

X-Ray diffraction studies on solutions of $ZnCl_2$ in HCl (2.7–3.2 mol l^{-1}, $Zn^{2+}:Cl^- = 1:6$) together with previously obtained Raman spectra indicate that $ZnCl_4^{2-}$ with near tetrahedral symmetry is the predominant species. The mean Zn—Cl bond distance is 230 pm[177]. The Raman spectra of molten $ZnCl_2$–$AlCl_3$ mixtures provide no evidence for the existence of $ZnCl_4^{2-}$ nor for $AlCl_4^-$ or $Al_2Cl_7^-$. Evidently the Cl^--ion acceptor strengths of the two compounds are comparable. There is some evidence for interaction between the species however, and this results in a break-up of the polynuclear $ZnCl_2$ structure[178]. Reactions between HgO, HI and K_2MF_6 (M = Ti or Sn) in aqueous HF give products formulated as $[HgI]_2[MF_6]$. The Raman spectrum of HgI^+ is very similar to that of the isoelectronic AuI and a similar planar zig-zag chain structure is suggested for $(HgI)_n^{n+}$ [179].

1.9 MONOHALIDES

This review has contained many references to the application of mass spectrometry in transition metal halide chemistry, and the technique has been widely used to obtain thermodynamic information relating to the vaporisation of halides. Recent work in this area has included copper(I) chloride[180, 181] and bromide[182], silver(I) chloride[183], and silver(I) iodide in the presence of silver metal[184]. Polymeric species are observed for all these compounds in the vapour state.

The vapour above CuCl(s) in the temperature range 500–700 K is predominantly a mixture (approximately 1:1) of $(CuCl)_3$ and $(CuCl)_4$ species although a small amount of $(CuCl)_5$ is also present[180]. Trimer and tetramer species are observed above CuBr(s) at 700 K, but in this case the ratio is approximately 5:1[182]. There is no evidence that $(CuCl)_2$ and CuCl are present in the vapour above CuCl(s) below 900 K, but they have been observed in the superheated vapours above this temperature obtained using a double-oven Knudsen cell[181]. It has been suggested that $(CuCl)_2$ has a planar rhombohedral structure, $(CuCl)_3$ is a planar 6-membered ring and $(CuCl)_4$ is a distorted cube. Bands in the i.r. spectrum of matrix-isolated CuCl have been assigned to the latter two species[185].

The vapours above AgCl(l) at 800 K and an AgI(s) Ag metal mixture at 750 K both contain $(AgX)_3$ and AgX species[183, 184]. Contrary to previous work, no evidence was found for the sub-chloride Ag_3Cl_2[183].

It has been widely accepted that the Hg—Hg bond distances in mercury(I) halides depend on the electronegativity of the halogen. The correlation was based on bond distances determined from x-ray powder data, and single-crystal work has shown that the correlation is spurious. The Hg—Hg distances determined are 250.7 pm (Hg_2F_2), 252.6 pm (Hg_2Cl_2) and 249 pm (Hg_2Br_2); the difference between Hg_2F_2 and Hg_2Br_2 is not crystallographically significant[186]. These values are comparable with the Hg—Hg bond lengths in molecular $Hg_3(AlCl_4)_2$[187] and $[Hg_3^{2+}][AsF_6^-]_2$[188] which are 256 pm (av.) and 255 pm respectively. The —Cl—Hg—Hg—Hg—Cl— and Hg—Hg—Hg skeletons in these compounds are linear and symmetric, and are reminiscent of the linear X—Hg—Hg—X groups found in mercury(I) halides.

References

1. Labonville, P., Ferraro, J. R., Wall, M. C. and Basile, L. J. (1972). *Coord. Chem. Rev.*, **7,** 257
2. Oelkrug, D. (1971). *Struct. Bonding (Berlin)*, **9,** 1
3. Allen, G. C. and Warren, K. D. (1971). *Struct. Bonding (Berlin)*, **9,** 49
4. Guest, A. and Lock, C. J. L. (1972). *Can. J. Chem.*, **50,** 1807
5. Guest, A., Howard-Lock, H. E. and Lock, C. J. L. (1972). *J. Mol. Spectrosc.*, **43,** 273
6. Holloway, J. H., Selig, H. and Classen, H. H. (1971). *J. Chem. Phys.*, **54,** 4305
7. Selig, H. and Karpas, Z. (1971). *Israel J. Chem.*, **9,** 53
8. Canterford, J. H., O'Donnell, T. A. and Waugh, A. B. (1971). *Aust. J. Chem.*, **24,** 243
9. Guest, A. and Lock, C. J. L. (1971). *Can. J. Chem.*, **49,** 603
10. Canterford, J. H. and Waugh, A. B. (1971). *Inorg. Nucl. Chem. Lett.*, **7,** 395

11. Druce, P. M. and Lappert, M. F. (1971). *J. Chem. Soc.* (*A*), 3595
11a. Burgess, J., Fraser, C. J. W., Haigh, I. and Peacock, R. D. (1973). *J. Chem. Soc.* (*Dalton*), 501
12. O'Donnell, T. A. (1970). *Rev. Pure Appl. Chem.*, **20,** 159
13. Burgess, J., Haigh, I. and Peacock, R. D. (1971). *Chem. Commun.*, 977
14. Walker, D. W. and Winfield, J. M. (1971). *J. Fluorine Chem.*, **1,** 376
15. Handy, L. B., Sharp, K. G. and Brinckman, F. E. (1972). *Inorg. Chem.*, **11,** 523
16. Grahlert, W. and Thiele, K.-H. (1971). *Z. Anorg. Allgem. Chem.*, **383,** 144
17. Drew, M. G. B., Moss, K. C. and Rolfe, N. (1971). *Inorg. Nucl. Chem. Lett.*, **7,** 1219
18. Drew, M. G. B., Fowles, G. W. A., Rice, D. A. and Rolfe, N. (1971). *Chem. Commun.*, 231
19. Falconer, W. E., Burbank, R. D., Jones, G. R., Sunder, W. A. and Vasile, M. J. (1972). *Chem. Commun.*, 1080
20. Bennett, M. J., Haas, T. E. and Purdham, J. T. (1972). *Inorg. Chem.*, **11,** 207
21. Asprey, L. B., Ryan, R. R. and Fukushima, E. (1972). *Inorg. Chem.*, **11,** 3122
22. Edwards, A. J. (1972). *J. Chem. Soc.* (*Dalton*), 582
23. Edwards, D. A. and Ward, R. T. (1973). *Inorg. Nucl. Chem. Lett.*, **9,** 145
24. Singleton, D. L. and Stafford, F. E. (1972). *Inorg. Chem.*, **11,** 1208
25. Barraclough, C. G. and Kew, D. J. (1972). *Aust. J. Chem.*, **25,** 27
26. Oppermann, H., Stöver, G. and Kunze, G. (1972). *Z. Anorg. Allgem. Chem.*, **387,** 201
27. Oppermann, H. (1970). *Z. Anorg. Allgem. Chem.*, **379,** 262
28. Oppermann, H. (1971). *Z. Anorg. Allgem. Chem.*, **383,** 285
29. Britnell, D., Drew, M. G. B., Fowles, G. W. A. and Rice, D. A. (1972). *Chem. Commun.*, 462
30. Frais, P. W. and Lock, C. J. L. (1972). *Can. J. Chem.*, **50,** 1811
31. Schröder, F. A. and Christensen, A. N. (1972). *Z. Anorg. Allgem. Chem.*, **392,** 107
32. Atovmyan, L. O. and Aliev, Z. G. (1971). *J. Struct. Chem.*, (*U.S.S.R.*), **12,** 668
33. Atovmyan, L. O. and Krasochka, O. N. (1971). *Dokl. Akad. Nauk SSSR*, **196,** 91
34. Mattes, R., Müller, G. and Becher, H. J. (1972). *Z. Anorg. Allgem. Chem.*, **389,** 177
35. Atovmyan, L. O., Krasochka, O. N. and Rahlin, M. Ya. (1971). *Chem. Commun.*, 610
36. Ryan, R. R., Mastin, S. H. and Reisfeld, M. J. (1971). *Acta Crystallogr.*, **B27,** 1270
37. McFarlane, W., Noble, A. M. and Winfield, J. M. (1971). *J. Chem. Soc.* (*A*), 948
38. Buslayev, Yu. A., Kokunov, Yu. V., Bochkaryova, V. A. and Shustorovich, E. M. (1972). *J. Inorg. Nucl. Chem.*, **34,** 2861
39. Beuter, A. and Sawodny, W. (1972). *Angew. Chem. Int. Ed. Engl.*, **11,** 1020
40. Calves, J.-Y., Kergoat, R. and Guerchais, J.-E. (1972). *C. R. Acad. Sci. Ser. C*, **275,** 1423
41. Vasile, M. J., Jones, G. R. and Falconer, W. E. (1971). *Chem. Commun.*, 1355
42. Edwards, A. J. and Lloyd, D. R. (1972). *Chem. Commun.*, 719
43. Alexander, L. E., Beattie, I. R. and Jones, P. J. (1972). *J. Chem. Soc.* (*Dalton*), 210
44. Acquista, N. and Abramowitz, S. (1972). *J. Chem. Phys.*, **56,** 5221
45. Fuggle, J. C., Sharp, D. W. A. and Winfield, J. M., (1972). *J. Chem. Soc.* (*Dalton*), 1766
46. Fowles, G. W. A., Rice, D. A. and Wilkins, J. D. (1972). *J. Chem. Soc.* (*Dalton*), 2313
47. Crociani, B. and Richards, R. L. (1973). *Chem. Commun.*, 127
48. Frais, P. W., Lock, C. J. L. and Guest, A. (1971). *Chem. Commun.*, 75
49. Calvo, C., Frais, P. W. and Lock, C. J. L. (1972). *Can. J. Chem.*, **50,** 3607
50. Oppermann, H., Kunze, G. and Stöver, G. (1972). *Z. Anorg. Allgem. Chem.*, **387,** 339
51. Oppermann, H., Stöver, G. and Kunze, G. (1972). *Z. Anorg. Allgem. Chem.*, **387,** 317, 329
52. Leary, K. and Bartlett, N. (1972). *Chem. Commun.*, 903
53. Leary, K., Zalkin, A. and Bartlett, N. (1973). *Chem. Commun.*, 131
54. Frais, P. W., Lock, C. J. L. and Guest, A. (1970). *Chem. Commun.*, 1612
55. Richards, G. W. and Woolf, A. A. (1971). *J. Fluorine Chem.*, **1,** 129
56. Buslaev, Yu. A., Ilin, E. G., Bainova, S. V. and Krutkina, M. N. (1971). *Dokl. Akad. Nauk SSSR*, **196,** 374
57. Buslaev, Yu. A., Ilin, E. G. and Krutkina, M. N. (1971). *Dokl. Akad. Nauk SSSR*, **200,** 1345; **201,** 99
58. Buslaev, Yu. A., Kopanev, V. D. and Tarasov, V. P. (1971). *Chem. Commun.*, 1175
59. Fergusson, J. E. and Love, J. L. (1971). *Aust. J. Chem.*, **24,** 2689
60. Atovmyan, L. O., D'yachenko, O. A. and Lobkovskii, E. B. (1970). *J. Struct. Chem.*, (*U.S.S.R.*), **11,** 429
61. Lo, G. Y.-S. and Brubaker, C. H., Jr. (1972). *J. Coord. Chem.*, **2,** 5
62. Bushnell, G. W. and Moss, K. C. (1972). *Can. J. Chem.*, **50,** 3700
63. Howell, J. A. S. and Moss, K. C. (1971). *J. Chem. Soc.* (*A*), 270

64. Carpy, A. and Waltersson, K. (1972). *C. R. Acad. Sci., Ser. C*, **274,** 405
65. Sengupta, A. K. and Bhaumik, B. B. (1972). *Z. Anorg. Allgem. Chem.*, **384,** 255
66. Mattes, R. and Rieskamp, H. (1972). *Z. Naturforsch.*, **27b,** 1424
67. Ahlborn, E., Diemann, E. and Müller, A. (1972). *Z. Anorg. Allgem. Chem.*, **394,** 1
68. Sunder, W. A. and Falconer, W. E. (1972). *Inorg. Nucl. Chem. Lett.*, **8,** 537
69. Pilbrow, M. F. (1972). *Chem. Commun.*, 270
70. Cook, P. M., Dahl, L. F. and Dickerhoof, D. W. (1972). *J. Amer. Chem. Soc.*, **94,** 5511
71. Brignole, A. and Cotton, F. A. (1971). *Chem. Commun.*, 706
72. Muller, H. and Washinski, R. (1972). *Inorg. Nucl. Chem. Lett.*, **8,** 413
73. Clark, R. J. H. and Willis, C. J. (1971). *Inorg. Chem.*, **10,** 1118
74. Clark, R. J. H. and Hunter, B. K. (1971). *J. Chem. Soc. (A)*, 2999
75. Shurvell, H. F. (1971). *J. Mol. Spectrosc.*, **38,** 431
76. Kiefer, W. and Schrötter, H. W. (1970). *Z. Naturforsch.*, **25b,** 1374
77. Clark, R. J. H. and Willis, C. J. (1971). *J. Chem. Soc. (A)*, 828
78. Clark, R. J. H., Hunter, B. K. and Rippon, D. M. (1972). *Inorg. Chem.*, **11,** 56
79. Westland, A. D. and Lal, D. (1972). *Can. J. Chem.*, **50,** 1604
80. Ehlert, T. C. and Hsia, M. (1972). *J. Fluorine Chem.*, **2,** 33
81. Kidd, R. G., Matthews, R. W. and Spinney, H. G. (1972). *J. Amer. Chem. Soc.*, **94,** 6686
82. Fayos, J. and Mootz, D. (1971). *Z. Anorg. Allgem. Chem.*, **380,** 196
83. Bürger, H., Kluess, C. and Neese, H.-J. (1971). *Z. Anorg. Allgem. Chem.*, **381,** 198
84. Kistenmacher, T. J. and Stucky, G. D. (1971). *Inorg. Chem.*, **10,** 122
85. Russell, C. W. G. and Smith, D. W. (1972). *Inorg. Chem. Acta*, **6,** 677
86. Clark, R. J. H. and Coles, M. A. (1972). *J. Chem. Soc. (Dalton)*, 2454
87. Siebert, G. and Hoppe, R. (1972). *Z. Anorg. Allgem. Chem.*, **391,** 113, 126
88. Quail, J. W. and Rivett, G. A. (1972). *Can. J. Chem.*, **50,** 2447
89. Brunton, G. (1971). *Acta Crystallogr., Sect. B*, **27,** 1944
90. Johnson, D. P. and Bereman, R. D. (1972). *J. Inorg. Nucl. Chem.*, **34,** 679
91. Korol'kov, D. V. and Kudryashova, G. N. (1970). *Russ. J. Inorg. Chem.*, **15,** 1759
92. Sharpless, K. B. and Flood, T. C. (1972). *Chem. Commun.*, 370; see also Sharpless, K. B., Umbreit, M. A., Nieh, M. T. and Flood, T. C. (1972). *J. Amer. Chem. Soc.*, **94,** 6538
93. Tisley, D. G. and Walton, R. A. (1971). *J. Chem. Soc. (A)*, 3409
94. Cresswell, P. J., Fergusson, J. E., Penfold, B. R. and Scaife, D. E. (1972). *J. Chem. Soc. (Dalton)*, 254
95. Wright, R. C. and Laurence, G. S. (1972). *Chem. Commun.*, 132
96. von Schnering, H. G., Collin, M. and Hassheider, M. (1972). *Z. Anorg. Allgem. Chem.*, **387,** 137
97. Podmore, L. P. and Smith, P. W. (1972). *Aust. J. Chem.*, **25,** 2521
98. Dawson, J. H. J. and Smith, D. W. (1971). *Inorg. Nucl. Chem. Lett.*, **7,** 811
99. Lascelles, K. and Schäfer, H. (1971). *Z. Anorg. Allgem. Chem.*, **382,** 249
100. Babel, D. (1972). *Z. Anorg. Allgem. Chem.*, **387,** 161
101. Kojić-Prodić, B., Šcavnicar, S., Liminga, R. and Sljukić, M. (1972). *Acta Crystallogr., Sect. B*, **28,** 2028
102. Wieghardt, K. and Weiss, J. (1972). *Acta Crystallogr., Sect. B.*, **28,** 529
103. Wieghardt, K. and Siebert, H. (1971). *Z. Anorg. Allgem. Chem.*, **381,** 12
104. Bernal, I., Elliott, N. and Lalancette, R. (1971). *Chem. Commun.*, 803
105. Bellitto, C., Tomlinson, A. A. G. and Furlani, C. (1971). *J. Chem. Soc. (A)*, 3267
106. Dumora, D., Von der Mühll, R. and Ravez, J. (1971). *Mater. Res. Bull.*, **6,** 561
107. Von der Mühll, R., Andersson, S. and Galy, J. (1971). *Acta Crystallogr., Sect. B*, **27,** 2345
108. Edwards, A. J. (1971). *J. Chem. Soc. (A)*, 2653
109. Edwards, A. J. (1972). *J. Chem. Soc. (Dalton)*, 816
110. Edwards, A. J. (1971). *J. Chem. Soc. (A)*, 3074
111. Sproul, G. D. and Stucky, G. D. (1972). *Inorg. Chem.*, **11,** 1647
112. Vala, M., Jr., Mongan, P. and McCarthy, P. J. (1972). *J. Chem. Soc. (Dalton)*, 1870
113. Edwards, A. J., Plevey, R. G., Sallomi, I. J. and Tatlow, J. C. (1972). *Chem. Commun.*, 1028
114. Court, T. L. and Dove, M. F. A. (1971). *Chem. Commun.*, 726
115. Pouchard, M., Torki, M. R., Demazeau, G. and Hagenmuller, P. (1971). *C. R. Acad. Sci., Ser. C*, **273,** 1093
116. Babel, D. (1972). *J. Solid State Chem.*, **4,** 410
117. Bowen, A. R. and Taube, H. (1971). *J. Amer. Chem. Soc.*, **93,** 3287
118. Andruchow, W. Jr. and DiLiddo, J. (1972). *Inorg. Nucl. Chem. Lett.*, **8,** 689

119. Kustin, K. and Toppen, D. (1972). *Inorg. Chem.*, **11,** 2851
120. Thomas, C. K. and Stanko, J. A. (1973). *J. Coord. Chem.*, **2,** 211
121. Coetzer, J., Robb, W. Bekker, P. Z. (1972). *Acta Crystallogr., Sect. B*, **28,** 3587
122. Ziegler, R. J. and Risen, W. M., Jr. (1972). *Inorg. Chem.*, **11,** 2796
123. Kudryashova, G. N., Obozova, L. A. and Korol'kov, D. V. (1970). *Russ. J. Inorg. Chem.*, **15,** 1697
124. Saillant, R., Jackson, R. B., Streib, W. E., Folting, K. and Wentworth, R. A. D. (1971). *Inorg. Chem.*, **10,** 1453
125. Cotton, F. A. and Ucko, D. A. (1972). *Inorg. Chim. Acta*, **6,** 161
126. Grey, I. E. and Smith, P. W. (1971). *Aust. J. Chem.*, **24,** 73
127. Oldham, C., Davies, J. E. D. and Ketteringham, A. P. (1971). *Chem. Commun.*, 572
128. Bratton, W. K., Cotton, F. A., Debeau, M. and Walton, R. A. (1971). *J. Coord. Chem.*, **1,** 121
129. King, R. B. (1972). *Progr. Inorg. Chem.*, **15,** 287
130. Bush, M. A., Druce, P. M. and Lappert, M. F. (1972). *J. Chem. Soc. (Dalton)*, 500
131. Lesaar, H. and Schäfer, H. (1971). *Z. Anorg. Allgem. Chem.*, **385,** 65
132. von Schnering, H. G. (1971). *Z. Anorg. Allgem. Chem.*, **385,** 75
133. Spreckelmeyer, B., Brendel, C., Dartmann, M. and Schäfer, H. (1971). *Z. Anorg. Allgem. Chem.*, **386,** 15
134. Spreckelmeyer, B. and von Schnering, H. G. (1971). *Z. Anorg. Allgem. Chem.*, **386,** 27
135. Juza, D. and Schäfer, H. (1970). *Z. Anorg. Allgem. Chem.*, **379,** 122
136. Thaxton, C. B. and Jacobson, R. A. (1971). *Inorg. Chem.*, **10,** 1460
137. Koknat, F. W. and McCarley, R. E. (1972). *Inorg. Chem.*, **11,** 812
138. Edwards, P. A., McCarley, R. E. and Torgeson, D. R. (1972). *Inorg. Chem.*, **11,** 1185
139. Robbins, D. J. and Thomson, A. J. (1972). *J. Chem. Soc. (Dalton)*, 2350
140. Cooke, N. E., Kuwana, T. and Espenson, J. (1971). *Inorg. Chem.*, **10,** 1081
141. King, R. B., Braitsch, D. M. and Kapoor, P. N. (1972). *Chem. Commun.*, 1072
142. Eliezer, I. and Reger, A. (1972). *Coord. Chem. Rev.*, **9,** 189
143. Hastie, J. W., Hauge, R. H. and Margrave, J. L. (1971). *High Temp. Science*, **3,** 257
144. Gruen, D. M. (1971). *Progr. Inorg. Chem.*, **14,** 119
145. Lever, A. B. P. and Hollebone, B. R. (1972). *Inorg. Chem.*, **11,** 2183
146. Grindstaff, W. K. and Fogel, N. (1972). *J. Chem. Soc. (Dalton)*, 1476
147. Bobbitt, J. L. and Gladden, J. K. (1972). *Inorg. Chem.*, **11,** 2167
148. Harrison, W. and Trotter, J. (1973). *J. Chem. Soc. (Dalton)*, 61
149. Smith, W. E. (1972). *Chem. Commun.*, 1121
150. Smith, W. E. (1972). *J. Chem. Soc. (Dalton)*, 1634
151. McPherson, G. L., Kistenmacher, T. J., Folkers, J. B. and Stucky, G. D. (1972). *J. Chem. Phys.*, **57,** 3771
152. Ferguson, J., Wood, T. E., and Guggenheim, H. J. (1972). *Aust. J. Chem.*, **25,** 453
153. McPherson, G. L. and Stucky, G. D. (1972). *J. Chem. Phys.*, **57,** 3780
154. Spence, R. D., de Jonge, W. J. M. and Rama Rao, K. V. S. (1971). *J. Chem. Phys.*, **54,** 3438
155. Greenwood, N. N., Howe, A. T. and Ménil, F. (1971). *J. Chem. Soc. (A)*, 2218
156. Ferguson, J., Krausz, E. R., Robertson, G. B. and Guggenheim, H. J. (1972). *Chem. Phys. Lett.*, **17,** 551
157. Mais, R. H. B., Owston, P. G. and Wood, A. M. (1972). *Acta Crystallogr., Sect. B*, **28,** 393
158. Müller, B. and Hoppe, R. (1972). *Mater. Res. Bull.*, **7,** 1297
159. von Schnering, H. G., Kolloch, B. and Kolodziejczyk, A. (1971). *Angew. Chem. Int. Ed. Engl.*, **10,** 413
160. Chrétien, A. and Samouël, M. (1972). *Monatsh. Chem.*, **103,** 17
161. Dumora, D., Fouassier, C., Von der Mühll, R., Ravez, J. and Hagenmuller, P. (1971). *C. R. Acad. Sci., Ser. C*, **273,** 247
162. Allen, G. C., McMeeking, R. F., Hoppe, R. and Muller, B. (1972). *Chem. Commun.*, 291
163. Müller, B. and Hoppe, R. (1972). *Z. Anorg. Allgem. Chem.*, **392,** 37
164. Odenthal, R.-H. and Hoppe, R. (1971). *Monatsh. Chem.*, **102,** 1340
165. Müller, B. and Hoppe, R. (1973). *Z. Anorg. Allgem. Chem.*, **395,** 239
166. Goldfield, S. A. and Raymond, K. N. (1971). *Inorg. Chem.*, **10,** 2604
167. Jeter, D. Y. and Hatfield, W. E. (1972). *J. Coord. Chem.*, **2,** 39
168. Hodgson, D. J., Hale, P. K. and Hatfield, W. E. (1971). *Inorg. Chem.*, **10,** 1061
169. Belford, R., Fenton, D. E. and Truter, M. R. (1972). *J. Chem. Soc. (Dalton)*, 2345
170. Bloom, H. and Anthony, R. G. (1972). *Aust. J. Chem.*, **25,** 23
171. Clarke, J. H. R., Hartley, P. J. and Kuroda, Y. (1972). *J. Phys. Chem.*, **76,** 1831

172. Clarke, J. H. R., Hartley, P. J. and Kuroda, Y, (1972). *Inorg. Chem.*, **11,** 29
173. Ross, S. D., Siddiqi, I. W. and Tyrrell, H. J. V. (1972). *J. Chem. Soc.* (*Dalton*), 1611
174. Orioli, P. L. and Ciampolini, M. (1972). *Chem. Commun.*, 1280
175. Epstein, E. F. and Bernal, I. (1971). *J. Chem. Soc.* (*A*), 3628
176. Veal, J. T. and Hodgson, D. J. (1972). *Inorg. Chem.*, **11,** 597
177. Wertz, D. L. and Bell, J. R. (1973). *J. Inorg. Nucl. Chem.*, **35,** 137
178. Begun, G. M., Brynestad, J., Fung, K. W. and Mamantov, G. (1972). *Inorg. Nucl. Chem. Lett.*, **8,** 79
179. Breitinger, D. and Köhler, K. (1972). *Inorg. Nucl. Chem. Lett.*, **8,** 957
180. Guido, M., Balducci, G., Gigli, G. and Spoliti, M. (1971). *J. Chem. Phys.*, **55,** 4566
181. Guido, M., Gigli, G., and Balducci, G. (1972). *J. Chem. Phys.*, **57,** 3731
182. Schaaf, D. W. and Gregory, N. W. (1972). *J. Phys. Chem.*, **76,** 3271
183. Binnewies, M., Rinke, K. and Schäfer, H. (1973). *Z. Anorg. Allgem. Chem.*, **395,** 50
184. Binnewies, M. and Schäfer, H. (1973). *Z. Anorg. Allgem. Chem.*, **395,** 63
185. Cesaro, S. N., Coffari, E. and Spoliti, M. (1972). *Inorg. Chim. Acta*, **6,** 513
186. Dorm, E. (1971). *Chem. Commun.*, 466
187. Ellison, R. D., Levy, H. A. and Fung, K. W. (1972). *Inorg. Chem.*, **11,** 833
188. Davies, C. G., Dean, P. A. W., Gillespie, R. J. and Ummat, P. K. (1971). *Chem. Commun.*, 782

2
Thermodynamics of Binary and Ternary Transition Metal Oxides in the Solid State

ALEXANDRA NAVROTSKY
Arizona State University

2.1 INTRODUCTION

The last few years have seen great progress in the synthesis and characterisation, both structural and thermodynamic, of oxide materials. Out of this work is emerging the realisation that thermodynamics and structure are closely intertwined, especially when one begins to appreciate subtle structural changes on a microscopic level, such as substitutional disorder, point defects, extended defects and solid solution formation. On a somewhat more macroscopic level, the relation between crystal structure and stability is reflected in the thermodynamics of formation of isostructural series of compounds and in the thermodynamics of phase transitions. It is the purpose of this review to bring together selected thermodynamic and structural information about the oxides of the transition metals, to relate the structural features to their thermodynamic manifestations and to point out areas in which our knowledge is especially incomplete. The systems discussed specifically will be those for which both detailed structural data and thermodynamic data exist, but some thermodynamic data for other systems which are otherwise less well characterised will also be presented. The term 'transition metal' will be used rather loosely. In the many cases where, for example, Mg^{2+} behaves similarly to Co^{2+} and Fe^{2+}, or Al^{3+} substitutes for Cr^{3+} and Fe^{3+}, too much generality would be lost by arbitrarily confining the discussion to those ions containing a partially-filled d shell.

Specifically this article covers recent developments in the thermodynamics of the binary and ternary transition metal oxides of wide homogeneity ranges, of the binary and ternary line phases, of order–disorder phenomena within oxides of essentially constant composition, and of solid-solution formation in transition metal oxide systems. Solutions of oxygen in the metals and the suboxides closely related to them will not be included. The rare earth oxides will not be discussed. Measurements of thermal properties (heat capacities and heat contents) will be referred to only when these reveal some interesting structural change occurring in a given system.

2.2 BINARY METAL–OXYGEN SYSTEMS

2.2.1 Classification

A periodic classification of the transition metals and neighbouring elements, with a listing of their major known oxides, is shown in Figure 2.1. For the purposes of this review, these elements may be classified in four groups, as shown. These are (a) elements forming very stable oxides with the metal in only one oxidation state (Mg, Ca, Sr, Ba, Sc, Y, La, Zr, Hf); (b) elements of variable valence forming a very large number of oxides, often having wide homogeneity ranges and/or series of related structures (Ti, V, Cr, Nb, Mo, Ta, W); (c) elements of variable valence forming a smaller number of relatively stoichiometric oxides usually with rather ionic crystal structures (Mn, Fe, Co, Ni, Cu, Zn, Tc, Cd); and (d) the noble metals (Ru, Rh, Pd, Ag, Os, Ir, Pt, Au, Hg) which have oxides of limited stability. This review will emphasise the second and third groups, which account for the largest number of oxides and

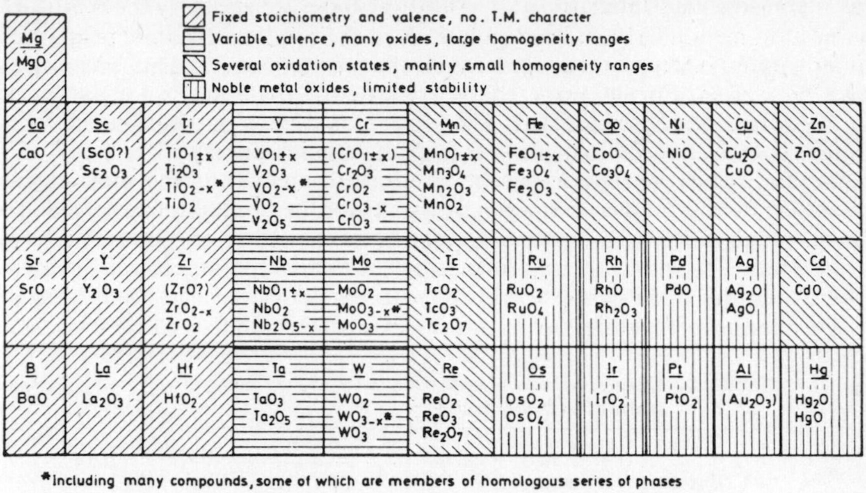

Fixed stoichiometry and valence, no. T.M. character
Variable valence, many oxides, large homogeneity ranges
Several oxidation states, mainly small homogeneity ranges
Noble metal oxides, limited stability

Mg MgO										
Ca CaO	Sc (ScO?) Sc_2O_3	Ti $TiO_{1\pm x}$ Ti_2O_3 TiO_{2-x}* TiO_2	V $VO_{1\pm x}$ V_2O_3 VO_{2-x}* VO_2 V_2O_5	Cr $(CrO_{1\pm x})$ Cr_2O_3 CrO_2 CrO_{3-x} CrO_3	Mn $MnO_{1\pm x}$ Mn_3O_4 Mn_2O_3 MnO_2	Fe $FeO_{1\pm x}$ Fe_3O_4 Fe_2O_3	Co CoO Co_3O_4	Ni NiO	Cu Cu_2O CuO	Zn ZnO
Sr SrO	Y Y_2O_3	Zr (ZrO?) ZrO_{2-x} ZrO_2	Nb $NbO_{1\pm x}$ NbO_2 Nb_2O_{5-x}	Mo MoO_2 MoO_{3-x}* MoO_3	Tc TcO_2 TcO_3 Tc_2O_7	Ru RuO_2 RuO_4	Rh RhO Rh_2O_3	Pd PdO	Ag Ag_2O AgO	Cd CdO
B BaO	La La_2O_3	Hf HfO_2	Ta TaO_3 Ta_2O_5	W WO_2 WO_{3-x}* WO_3	Re ReO_2 ReO_3 Re_2O_7	Os OsO_2 OsO_4	Ir IrO_2	Pt PtO_2	Al (Au_2O_3)	Hg Hg_2O HgO

*Including many compounds, some of which are members of homologous series of phases

Figure 2.1 Periodic classification of transition metal and related oxides

the most extensive chemistry. Since this classification is somewhat arbitrary it will be convenient to overstep its bounds on one or two occasions.

2.2.2 Stable oxides of fixed stoichiometry and no transition metal character

Thermodynamic data for the formation of these oxides from the elements at 1000 K are given in Table 2.1. More complete thermodynamic data are

Table 2.1 Thermodynamics of formation of some oxides of approximately fixed stoichiometry

Oxide	*Structure*	ΔH°_f *	*1000 K* ΔH°_f*	ΔS°_f†
MgO	rocksalt	−145.9 §	−117.7	−28.2
CaO	rocksalt	−151.5	−127.1	−24.4
SrO	rocksalt	−143.9	−121.1	−22.8
BaO	rocksalt	−134.0	−110.5	−23.5
$\frac{1}{3}Al_2O_3$ ‡	corundum	−134.9	−108.4	−26.5
$\frac{1}{3}Sc_2O_3$	bixbyite	−137.2	−112.0	−25.2
$\frac{1}{3}Y_2O_3$	bixbyite	−139.7	−115.5	−22.2
$\frac{1}{3}La_2O_3$	La_2O_3	−142.0	−119.3	−23.7
$\frac{1}{2}TiO_2$	rutile	−112.2	−91.3	−20.9
$\frac{1}{2}ZrO_2$	baddeleyite	−130.2	−108.0	−22.0
$\frac{1}{2}HfO_2$	baddeleyite	−131.8	−109.8	−22.0
$\frac{1}{2}SiO_2$	quartz	−69.4	−44.6	−24.8
$\frac{1}{2}GeO_2$	rutile	−63.9	−43.2	−20.7

* kcal/mol
† cal/deg mol
‡ all formula units given per mol oxygen atoms for direct comparison, thus $\frac{1}{3}M_2O_3$, $\frac{1}{2}MO_2$, etc.
§ sources of data, the standard tabulations, Refs. 1–6.

available in several tabulations[1–6]. Although these are, properly speaking, not transition metal (TM) oxides, regularities in their thermodynamic properties offer a useful background against which the properties of the true TM oxides can be viewed. Furthermore, these oxides participate in the formation of many mixed metal oxides with TM oxides, as well as in extensive isomorphous substitution for transition metal ions of similar size.

Several regularities are apparent. First, these are among the most stable oxides known; equilibrium oxygen pressures for their reduction at 1000 K lie in the range 10^{-40}–10^{-60} atm. Secondly, as data at other temperatures show, the enthalpies and entropies of formation are relatively temperature independent, thus ΔG_f° varies almost linearly with temperature.

Furthermore, regardless of variation in structure and stoichiometry, ΔS_f°

Table 2.2 Thermodynamics of formation of transition metal oxides of fixed stoichiometry

Oxide	*Structure*	1000 K		
		ΔH_f° *	ΔG_f° *	ΔS_f° †
TiO ‡	rocksalt	−123.1	−101.2	−21.9
$\frac{1}{3}Ti_2O_3$	corundum	−119.5	−99.1	−20.4
$\frac{1}{5}Ti_3O_5$	corundum	−116.1	−96.2	−19.9
$\frac{1}{2}TiO_2$	rutile	−112.2	−91.3	−21.0
VO ‡	rocksalt	−98.5	−79.0	−19.5
$\frac{1}{3}V_2O_3$	corundum	−97.3	−78.2	−19.2
$\frac{1}{2}VO_2$	rutile	−83.9	−65.3	−18.7
$\frac{1}{5}V_2O_5$	V_2O_5	−70.5	−54.4	−16.1
$\frac{1}{3}Cr_2O_3$	corundum	−90.2	−69.7	−20.4
$\frac{1}{2}CrO_2$	rutile	−69.7	−48.7	−21.0
$\frac{1}{3}CrO_3$	CrO_3	−42.3	−27.3	−15.0
MnO ‡	rocksalt	−92.3	−74.6	−17.7
$\frac{1}{4}Mn_3O_4$	tetrag. spinel	−82.9	−59.7	−23.2
$\frac{1}{3}Mn_2O_3$	bixbyite	−76.3	−55.7	−20.6
$\frac{1}{2}MnO_2$	SnO_2-type	−62.0	−40.8	−21.3
$Fe_{0.947}O$ ‡	rocksalt	−62.8	−42.5	−20.3
$\frac{1}{4}Fe_3O_4$	spinel	−64.9	−47.2	−17.7
$\frac{1}{3}Fe_2O_3$	corundum	−64.0	−44.6	−19.3
CoO	rocksalt	−56.9	−39.0	−17.9
$\frac{1}{4}Co_3O_4$	spinel	−52.4	−31.4	−21.0
NiO	rocksalt	−56.2	−35.9	−20.3
Cu_2O	cuprite	−41.9	−23.3	−18.6
CuO	tenorite	−37.5	−16.2	−21.3
ZnO	zincite	−84.4	−58.9	−25.5
$\frac{1}{2}NbO_2$	rutile	−94.5	−73.7	−20.8
$\frac{1}{5}Nb_2O_5$	Nb_2O_5	−90.2	−69.9	−20.2
$\frac{1}{5}Ta_2O_5$	Nb_2O_5	−96.7	−76.7	−20.0
$\frac{1}{2}MoO_2$	SnO_2-type	−66.5	−48.5	−18.1
$\frac{1}{3}MoO_3$	MoO_3	−58.2	−39.4	−18.8
$\frac{1}{2}WO_2$	rutile	−67.3	−48.0	−19.4
$\frac{1}{3}WO_3$	ReO_3	−66.1	−47.0	−19.1
CdO	rocksalt	−63.4	−36.7	−26.7
$\frac{1}{3}ReO_3$	ReO_3	−44.4	−31.5	−13.0
$\frac{1}{7}Re_2O_7$	Re_2O_7	−42.8 §	−32.0 §	−21.6 ¶

* kcal mol^{-1}
† cal deg^{-1} mol^{-1}
‡ compound has wide stoichiometry range, data are for composition shown
§ sources of data, the standard tabulations, Refs. 1–6
¶ at 500 K

remains fairly constant, *ca.* 22 ± 4 cal deg^{-1} mol^{-1} of oxygen atoms. This results principally from the large negative volume change in the gas–solid reaction and is the factor responsible for the decreasing stability with temperature of all oxides, and especially of those with the metal in high oxidation states. Since this entropy of formation is comparable, per mol of oxygen atoms, for TM oxides as well (see Table 2.2), the variation in stability among different oxides reflects predominantly a change in their enthalpies of formation. This in turn, depends on the variation of the difference in lattice energies between the metallic and oxide phases, rather than just the variation of the lattice energy of the oxide. Thus, although lattice energies decrease monotonically in the series MgO, CaO, BaO, SrO, the standard enthalpy of formation of the oxides passes through a minimum at CaO.

Structurally, most of these oxides are relatively simple; their structures are determined by stoichiometry and cation size. The rocksalt structure is typical of the divalent oxides. Al_2O_3 (as well as Fe_2O_3, Cr_2O_3, and others) crystallises in the corundum structure, while the M_2O_3 oxides with larger cations prefer the cubic bixbyite and/or hexagonal La_2O_3 structure types. In the MO_2 oxides, increasing cation size brings one from the quartz structure (SiO_2, high

Table 2.3 Phase transitions in stoichiometric binary oxides

Oxide	*Transition, structure types*	*Conditions*	ΔH °/kcal mol^{-1}	ΔS °/cal mol^{-1} deg^{-1}
TiO_2	rutile→anatase	metastable	+1.6 (968)[6]	
TiO_2	rutile→brookite	metastable		
TiO_2	rutile→α-PbO_2	$P > 40$ kb[4]	+0.76 (298)[5]	
TiO_2	rutile→fluorite	$P > 330$ kb[11]		
VO_2	dist. rutile→rutile	345 K, 1 atm	+1.0[10]	+2.8
ZrO_2	tetrag→cubic	1478 K, 1 atm	+1.45[10]	+1.0
SiO_2	quartz→rutile	rutile form stable at $P > 100$ kb	+11.8 (298)[9]	+3.2
GeO_2	rutile→quartz	1306 K, 1 atm	+5.2 (968)[8]	+4.0
NbO_2	dist. rutile→rutile	1068 K, 1 atm		
MgO	rocksalt→zincite	metastable	$\Delta G^\circ_{1478} = 9.6$[2]	
CoO	rocksalt→zincite	metastable	$\Delta G^\circ_{1323} = +4.3$[3]	
NiO	rocksalt→zincite	metastable	$\Delta G^\circ_{1323} = +8.7$[3]	
CuO	tenorite→rocksalt	?	$\Delta G^\circ_{1273} = 2.3$[7]	
ZnO	zincite→rocksalt	673 K, 105 kb[4]	$\Delta G^\circ_{1323} = +5.8$[3]	
Mn_3O_4	tetrag. spinel→ cubic spinel	1443 K, 1 atm[1]	+4.96[1]	+3.93[1]
Mn_2O_3	bixbyite→ corundum		$\Delta G^\circ_{1148} = 2.7$[1]	
Fe_2O_3	corundum→ bixbyite	metastable	$\Delta G^\circ_{1148} = 0.8$[1]	

References for Table 2.3
1. Hennings, D. (1967). *Ph.D. Thesis*, Saarbrücken
2. Kenny, D. S. and Navrotsky, A. (1972). *J. Inorg. Nucl. Chem.*, **34**, 2115
3. Navrotsky, A. and Muan, A. (1971). *J. Inorg. Nucl. Chem.*, **33**, 35
4. Bates, C. H. White, W. B. and Roy, R. (1962). *Science*, **137**, 993
5. Navrotsky, A. Jamieson, J. C. and Kleppa, O. J. (1967). *Science*, **158**, 3799
6. Navrotsky, A. and Kleppa, O. J. (1967). *J. Amer. Ceram. Soc.*, **50**, 626
7. Landolt, C. and Muan, A. (1969). *J. Inorg. Nucl. Chem.*, **31**, 1319
8. Navrotsky, A. (1971). *J. Inorg. Nucl. Chem.*, **33**, 1119
9. Holm, J. L. Kleppa, O. J., and Westrum, E. F. Jr., (1967). *Geochim. Cosmochim. Acta*, **31**, 2289
10. Elliot, J. F. and Gleiser, M. (1960). *Thermochemistry for Steelmaking*, Vol. 1, (Reading; Mass: Addison Wesley)
11. McQueen, R. G. *et al.*, (1967). *Science*, **155**, 1404

GeO_2), to rutile (low GeO_2, TiO_2), through distortions of the fluorite structure (baddleyite form of ZrO_2, tetragonal ZrO_2) to the fluorite structure (cubic ZrO_2). Similar trends are evident in the TM oxides.

The application of pressure to oxides effectively increases the metal/oxygen ionic radius ratio, and transitions to the more dense structures are observed: quartz to rutile (GeO_2 and SiO_2), rutile to distorted or true fluorite (TiO_2). Some thermodynamic parameters for such transitions have been summarised in Table 2.3.

2.2.3 Metal–oxygen systems showing numerous compounds, wide homogeneity ranges, and homologous series

The systems Ti—O, V—O, Cr—O, Nb—O, Mo—O, Ta—O, and W—O show a bewildering number of discrete phases. Structurally, these phases fall into three groups: (a) compounds of relatively simple structures and small homogeneity ranges (corundum-type Cr_2O_3, Ti_2O_3, V_2O_3; rutile-type TiO_2, VO_2; V_2O_5; Nb_2O_5; Ta_2O_5; WO_2; WO_3; etc.); (b) homologous series of shear structures or Magnéli phases (Ti_nO_{2n-1}, Mo_nO_{3n-1}, W_nO_{3n-2}, and many other binary and ternary series), and (c) oxides $M_{1\pm x}O$, with x as large as 0.25, based on a parent rocksalt structure with a large concentration of ordered and/or disordered defects.

2.2.3.1 Oxides of fixed stoichiometry

This group comprises the 'classical' chemistry of these TM oxides, with oxidation states and structures easily understood in terms of experience gained from aqueous solutions. Their thermodynamics of formation presents no special problems. Heats, free energies, and entropies of formation of these and some other TM compounds are summarised in Table 2.2. The oxides of V, W and Mo are low melting and volatile, while those of Ti, Cr, Nb and Ta (in oxidation states stable at high temperature) are more refractory. Higher oxidation states become increasingly unstable as temperature increases; thus CrO_2 and CrO_3 (and ternary chromates) need not be considered above 500–600 °C, while V_2O_5 (m.p. 690 °C) loses oxygen from the melt beginning at *ca.* 800 °C.

A number of equilibrium phase transitions and several metastable crystallographic modifications occur in these materials. VO_2 transforms reversibly at 345 K from a monoclinic (α) to the tetragonal rutile (β) structures, with an enthalpy of transition of 1.025 kcal mol^{-1} [3]. This transformation has been described as a simple semiconductor-to-metal transition[7] but recent work favours a more complex description[8]. Rutile appears to be the only stable polymorph of TiO_2 at atmospheric pressure[9]; the thermodynamic data for anatase indicate its metastability, whereas brookite may not exist as a totally pure and anhydrous material. At high pressure, TiO_2 transforms to the α-PbO_2 structure, which can be quenched back to ambient conditions and has been studied by calorimetry[10] and also to a fluorite type phase, which apparently cannot be quenched[11]. Thermodynamic data for these transformations are included in Table 2.3.

2.2.3.2 *Homologous series of Magnéli phases*

These present unique problems both structurally and thermodynamically. These phases accommodate reduction of the parent composition (MO_2, MO_3 or other), not by the creation of random defects over a range of stoichiometry, but by the existence of metal rich regions at regularly-spaced crystallographic shear planes in the structure. These shear planes become integral building blocks of the structure, their orientation and spacing determining (or being determined by) the stoichiometry. This results in families of compounds, based on a parent structure (rutile, ReO_3, α-PbO_2, or other), with a common orientation but increasing frequency of shear planes. These structural families can be represented by formulae such as Ti_nO_{2n-1} or W_nO_{3n-2}—a series of essentially stoichiometric compounds whose spacing, in terms of metal/oxygen ratio becomes very close as n becomes large. Families of 'block' structures, in which two sets of crystallographic shear planes of different orientation dissect the structure into blocks (rather than slabs) of parent structure bounded by metal-rich regions have also been characterised for a number of systems. Furthermore, new structures can be created as coherent intergrowths of two or more shear or block structures. Thus, at least formally, one is faced with an essentially infinite number of possible compounds, each of fixed stoichiometry, though with an unwieldly formula, between the compositions M_2O_3 and MO_2 or MO_2 and MO_2. The structural principles determining these phases have been reviewed quite extensively recently[12–14]

The existence of Magnéli phases immediately raises three basic questions about their formation and thermodynamic stability. First, is the concept of a shear plane also a mechanistic one—do such materials actually form and react by the creation and movement of shear planes through the structure? Secondly, are these shear phases thermodynamically stable (rather than 'artifacts' of a specific mechanism of oxidation or reduction), and do thermodynamic considerations impose any limits to the number or spacing of possible phases? Third, what are the detailed energetics of shear plane formation and ordering—how much difference, in terms of ΔH, ΔG, ΔS is there between different possible structures and different phase assemblages?

Electron microscopy has answered the first question mainly in the affirmative. In several systems (TiO_{2-x}[15], TiO_2–Cr_2O_3[16,17], $[W^{6+}M^{5+}]O_x$[18], Nb_2O_5[19], TiO_2–Fe_2O_3[20], TiO_2–Ga_2O_3[20]), reduction is first accommodated by point defects, followed by disordered shear planes, followed by lamellae with closely-spaced shear planes, followed finally by families of shear structures in which both the spacing and orientation of shear planes changes with increasing reduction. These observations permit some hypotheses to be made about the energetics of shear plane formation, interaction, and migration. A recent summary of possible shear plane formation mechanisms is given by van Landuyt and Amelinckx[17].

The second and third questions admit of no complete answers at present. Neither empirical thermodynamic data nor statistical thermodynamic theory are sufficiently detailed and precise to predict unequivocally the stability of any given phase under given conditions of T, P and f_{O2}. Several general conclusions can be drawn, however. Anderson and Burch[21] have calculated electrostatic lattice energies of phases in the series Ti_nO_{2n-1} ($n = 4$–9). They found that the shear structures are stabilised by 10–15% relative to structures of the same stoichiometry containing random oxygen vacancies, but that ordering of the vacancies would produce a similar stabilisation.* Calculation of actual heats of formation depends on an appropriate choice of repulsive and other non-electrostatic terms, making the final numerical values subject to considerable uncertainty. The number of possible stable phases will be limited, in principle, when the difference in free energy between successive pairs becomes small enough to be attainable by probable thermal fluctuations. In practice, there is growing evidence, see below, that the higher Magnéli phases, at least in V–O and W–O, become unstable long before any such limit is approached. These theoretical considerations have been discussed at some length by Anderson[22].

Experimentally, the free energies of formation of these oxides have been obtained by equilibrium methods in which the oxygen fugacity over a two-phase assemblage, or over a given composition in a single phase of variable stoichiometry, is measured either electrochemically or by using buffered gas mixtures. Because of the large number of closely-spaced phases, extreme care is needed in these measurements. Such data are available for the systems Ti–O[23], V–O[24,25] and W–O[26,27]. From the temperature dependence of these equilibria, attempts have been made to extract enthalpy and entropy of formation values. More recently, direct calorimetric measurements at 600–800 °C of the enthalpies of oxidation of some Magnéli phases in the systems V–O[28] and W–O[29] have been carried out.

The heats and free energies of formation, per g atom of compounds from V_2O_3 to V_2O_5 and from WO_2 to WO_3 are shown in Figures 2.2 and 2.3 respectively. There is still some controversy concerning the homogeneity ranges of V_2O_3, V_3O_5 and V_6O_{11} and imperfect agreement between the results obtained by Anderson and Khan[24] and by Wakihara and Katsura[25] but several points are obvious from the graphs. In the V–O system, the oxides V_3O_5, V_4O_7, have free energies of formation 0.3–0.5 kcal (g atom)$^{-1}$ more negative and enthalpies of formation 0.7–1.1 kcal (g atom)$^{-1}$ more negative than do the corresponding mixture of V_2O_3 and VO_2. The oxides V_6O_{11} and V_8O_{15}, on the other hand, fall on or above the line joining V_2O_3 and VO_2 on the graphs and the data further suggest that these are unstable relative to V_5O_9 and VO_2. Similarly, the Magnéli phases in the W–O system are clearly stable with respect to decomposition into WO_2 and WO_3 but, within experimental error, the free energy of reactions such as $WO_{2.90} \rightarrow WO_{2.72} + WO_{2.96}$ is zero. Both these sets of data suggest that, as n increases, the Magnéli phases become unstable with respect to decomposition into a Magnéli phase of lower n value and unreduced parent oxide. Some of the higher phases that have been synthesised may indeed be metastable. Whether such trends apply generally remains to be seen.

The fact that different possible shear structures or intergrowths thereof lie extremely close to each other in energy is further indicated by the wealth of structures shown by Nb_2O_5. Recent electron microscopy[19] shows that many of these structures result from twinning, stacking faults and the incorporation of 'wrong' blocks into a parent structure. This behaviour presumably bears

* Recent calculations by von Dreele (to be published) for the niobium oxides yield similar results.

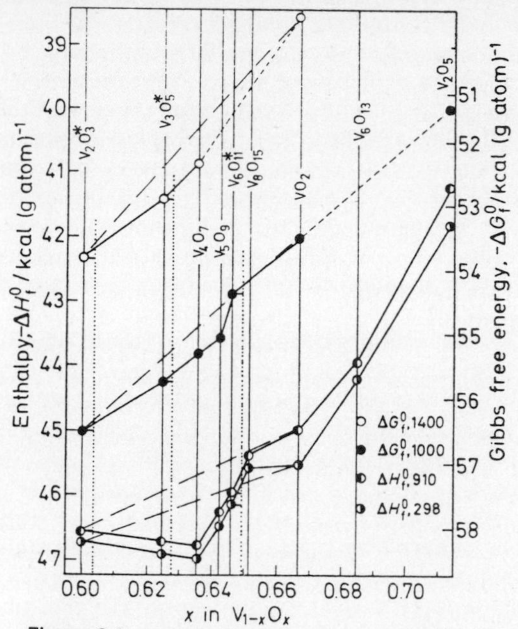

Figure 2.2 Thermodynamics of the vanadium–oxygen system. (Enthalpies of formation from Charlu and Kleppa, in press. Free energies from Anderson and Khan[24].)

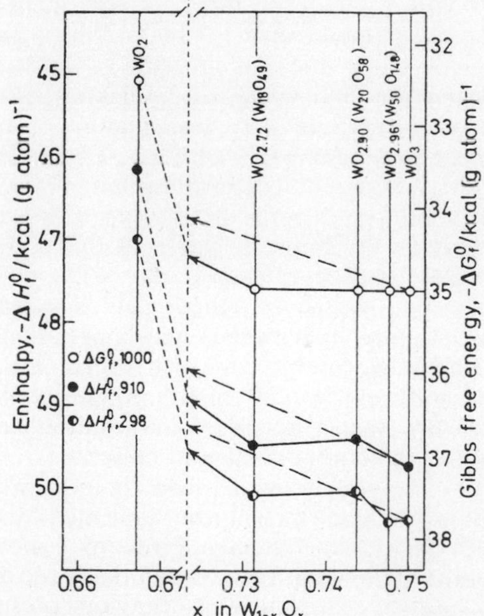

Figure 2.3 Thermodynamics of the tungsten–oxygen system. (Enthalpies of formation from Charlu and Kleppa[29]. Free energies from Elliot and Gleiser[3].)

some analogy to the polytypism observed in, for example silicon carbide and zinc sulphide.

The experimentally-determined oxygen pressure isotherms in the systems V–O and Ti–O appear reversible on oxidation and reduction, in marked contrast to the hysteresis observed by Eyring and his co-workers[30] in rare earth oxide systems, e.g., Pr–O, where homologous series of structures may also be involved. Furthermore, the Magnéli phases in Ti–O, V–O, Mo–O and W–O either melt congruently or decompose incongruently to liquid plus another shear phase[23, 24]. There is thus no evidence for the merging of these shear phases into a single phase (solid) region of wide homogeneity range at high temperature, again in contrast to the wide region of α-phase found in the rare earth oxide systems at high temperature.

Lastly, it should be noted that the problem of coherent intergrowths, of microdomains, of possible stabilisation due to coherent interfaces between regions of 10–100 Å linear extent present new thermodynamic problems. Obviously, the usual definition of a *phase* and of *bulk* thermodynamic properties no longer suffice, and the formalisms of small systems thermodynamics or some similar approach must be brought to bear. Some first steps in this direction, with applications also to partially ordered phases of wide homogeneity ranges are being made[22]. A rather more empirical approach to the calculation of thermodynamic properties has been taken by Kimura[118], and may prove useful.

2.2.3.3 *Oxides with the rocksalt structure and wide homogeneity ranges*

This group includes $NbO_{1 \pm x}$, $TiO_{1 \pm x}$, $VO_{1 \pm x}$ and, in violation of the classification scheme adopted above, MnO_{1+x} and FeO_{1+x} as well. Pertinent phase diagrams at atmospheric pressure are shown in Figure 2.4. The transition from a nearly stoichiometric oxide having essentially non-interacting point defects (see Section 2.2.4.2) to a highly non-stoichiometric oxide, in which the defects are present at such high concentrations that they clearly must interact, is a gradual one, depending on P and T as well as on composition. Thus the homogeneity range of MnO_{1+x} 'opens up' only above 1300 K, while stoichiometric and almost defect free FeO[31] and TiO[32] can be made under high pressure.

Thermodynamically, three related questions can be posed about such non-stoichiometric oxides. First, under what conditions can a phase of wide homogeneity range be stable? Secondly, how does the chemical potential and partial molar entropy and enthalpy of each component vary across the composition range? Thirdly, how is this observed variation of thermochemical parameters related to the structure details—the kinds and number of defects present and the interactions which lead to their ordering and/or clustering? Thus three types of work are most relevant to these questions: direct measurement of thermodynamic quantities (oxygen fugacities and enthalpies of oxidation), structural studies characterising the defects present (diffraction of x-rays, electrons, and neutrons, low-angle x-ray scattering, Mössbauer resonance, etc.), and statistical mechanical calculations based on specific structural models, the results of which can be compared with or fitted

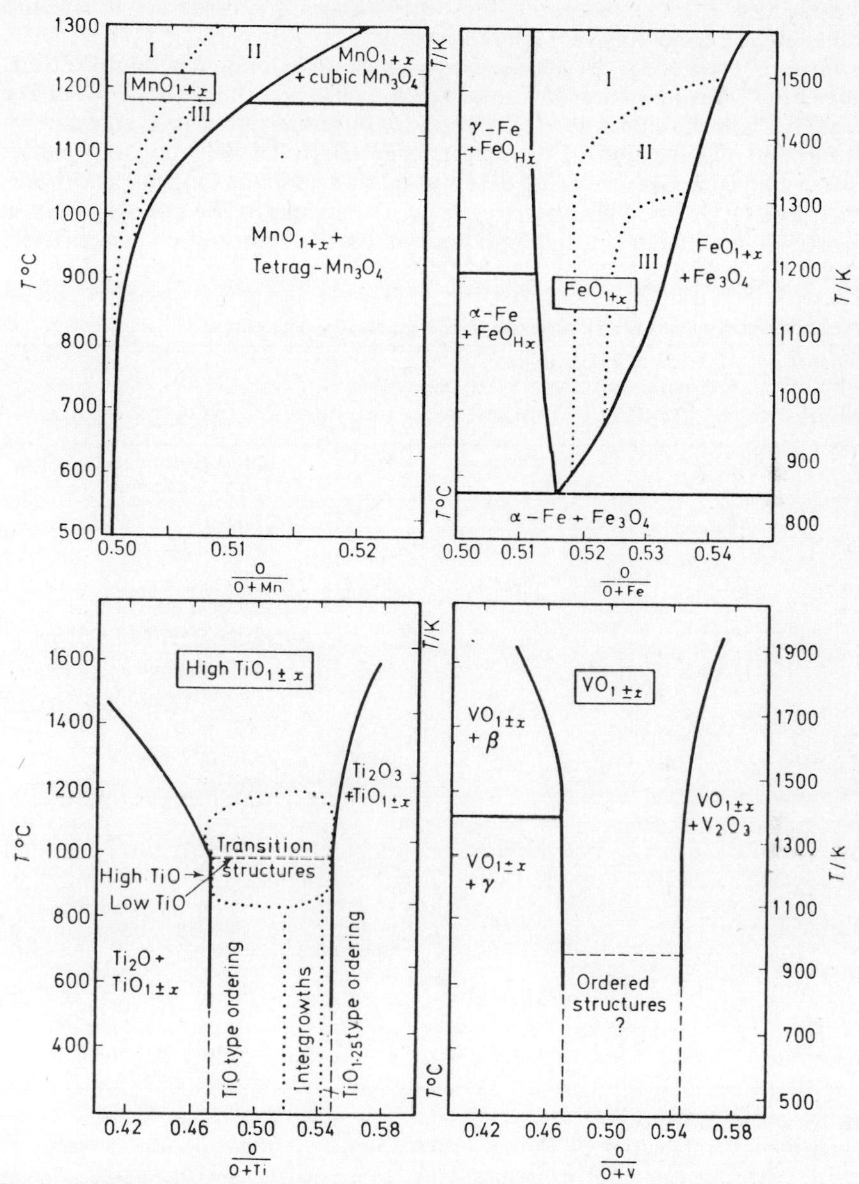

Figure 2.4 Phase diagrams of the Mn–O, Fe–O, Ti–O, and V–O systems near $X_0 = 0.50$. (Data combined from standard tabulations and from Fender and Riley[37] Fender and Riley[38], D. Watanabe[46].)

to experimental data. The present state of the art, in the reviewer's opinion, shows considerable progress in each of these three areas separately, but claims little success as yet in integrating these approaches to provide a complete and coherent picture of any real oxide systems.

General criteria for the stability of a phase over a range of compositions have been given by Anderson[22] and by Libowitz[33]; both the shape of the free energy *v.* composition curve for that phase (a broad rather than sharp minimum) and the location of the free energy curves of neighbouring phases (coexisting compositions being determined by a common tangent) are important. The larger the homogeneity range, the smaller is the energy of defect formation and the less steeply f_{O2} varies across the composition range. These

Table 2.4 Thermodynamics of oxides having significant homogeneity ranges

		Composition range				
		Reduced limit		*Oxidised limit*		
Oxide	T(K)	*Composition*	$-\log_{10} p_{O_2}$	*Composition*	$-\log_{10} p_{O_2}$	*Predominant defect type*
TiO	1673	$TiO_{0.80}$	44.1	$TiO_{1.30}$	41.5[1,2]	Strongly interacting cation and anion vacancies
VO	1673	$VO_{0.80}$	34.5	$VO_{1.30}$	33.2[1,2]	Strongly interacting cation and anion vacancies
MnO	1773	MnO	14.7	$MnO_{1.13}$	1.4[1,3]	Cation vacancies and Mn^{3+} ions, complex interactions at higher defect concentrations
FeO	1348	$FeO_{1.045}$	13.6	$FeO_{1.145}$	11.3[1,4]	Cation vacancies, octahedral and tetrahedral Fe^{3+} ions, defect clusters, at least below 1273 °K
CoO	1473	~CoO	9.2	$CoO_{1.01}$	in air[1,5] single phase	Cation vacancies and Co^{3+} ions
NiO	1273	~NiO	10.3	$NiO_{1.0002}$	in air[1,6] single phase	Cation vacancies and Ni^{3+} ions
CuO	1273	~CuO	0.9	~CuO	in air[1,7] single phase	Interstitial copper atoms (?)
ZnO	~$Zn_{1.001}O$ at 1273 K when $p_{Zn}=1$ atm[1,8]					Interstitial zinc, probably unipositive ions
NbO	1300	$NbO_{0.95}$	24.6	$NbO_{1.02}$	22.0[1,9]	Strongly interacting cation and anion vacancies

References for Table 2.4

1. Kofstad, P. (1972). *Nonstoichiometry, Diffusion, and Electrical Conductivity in Binary Metal Oxides*, (New York: Wiley)
2. Reed, T. (1970). *The Chemistry of Extended Defects in Nonmetallic Solids*, (L. Eyring and M. O'Keefe, editors) (Amsterdam: North Holland Pub.)
3. Hed, A. Z. and Tannhauser, D. (1967). *J. Electrochem. Soc.*, **114**, 314
4. Marucco, J. *et al.* (1970). *J. Chim. Phys.*, **67**, 914
5. Sockel, H. and Schmalzried, H. (1968). *Ber. Bunsenges. Phys. Chem.*, **72**, 745
6. Tretyakov, Yu. D. and Rapp, R. A. (1969). *Trans AIME*, **245**, 1235
7. Tretyakov, Yu. D. *et al.* (1972). *J. Solid State Chem.*, **5**, 157
8. Moore, W. J. and Williams, E. L. (1959). *Discuss. Faraday Soc.*, **28**, 86
9. calc. from M. Hoch *et al.*, (1962). *J. Phys. Chem. Sol.*, **23**, 1463

trends are borne out by the data shown in Table 2.4, which also contains data for some almost stoichiometric oxides.

Recently, the equilibrium oxygen fugacity within the homogeneous wüstite phase has been measured by several investigators[34–37, 40] and similar measurements have been made on the single-phase region in MnO[38]. Partial molar enthalpies and entropies for the solution of oxygen in the non-stoichiometric oxide have been calculated from the temperature dependence of these data. In addition, the partial molar enthalpy of solution of oxygen in wüstite has been measured directly by high temperature calorimetry[39]. In Figure 2.5, the

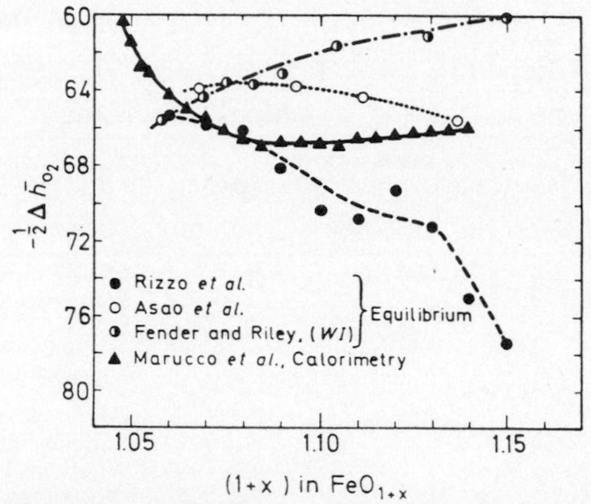

Figure 2.5 Partial molar enthalpy of oxygen in wüstite. (Data from Fender and Riley[37] Rizzo *et al.*[36] Asao *et al.*[35] Marucco *et al.*[39].)

variation with composition of the partial molar enthalpy of solution of $\frac{1}{2}O_2$ in wüstite is shown. The several sets of values derived from the temperature dependence of equilibrium data differ among themselves and from the calorimetric values. The latter, being the most directly obtained, are probably the most reliable. They show that $\Delta\bar{h}_{1/2O_2}$ first decreases fairly sharply with increasing oxygen content and then levels off or increases slightly as the magnetite boundary is approached. The results of Fender and Riley for both FeO[37] and MnO[38] and of Rizzo *et al.*[36] and Raccah and Vallet[40] for FeO have led to the hypothesis that the single-phase region consists of three distinct parts, possibly separated by second-order phase transitions (see Figure 2.4) and perhaps distinguished by different patterns of defect ordering.

On the basis of structural studies on quenched samples of wüstite, several models have been proposed for the defect structure of FeO. The simple Schottky–Wagner picture of a complete anion sublattice with cation vacancies and Fe^{3+} ions randomly present on normal cation sites is insufficient. Clusters of associated defects appear to be present. Roth[41] has proposed islands of a magnetite-like structure, with Fe^{3+} on both octahedral and tetrahedral sites. Koch[42] modified this model to clusters containing four interstitials and 13 vacancies, other work[43] favours even more complex clusters. Recent Mössbauer work[44] suggests, on the other hand, that Fe^{2+} and Fe^{3+} ions are not distinguishable in $Fe_{1-x}O$ and (FeMg)O solid solutions. Furthermore, one cannot reliably assess whether the structural features characteristic of quenched samples persist at high temperature. The data of Bransky[113] suggest that above 1273 K the formation of defect clusters is not significant, at least in its affect on electrical and transport properties.

It seems unlikely that thermodynamic measurements alone will be able to distinguish between closely-related structural models, since these structures will be very close in energy. Nevertheless, some comparisons, using the tools of statistical mechanics, between the fairly detailed thermodynamic data and the proposed defect models should prove valuable.

The oxides $TiO_{1\pm x}$, $VO_{1\pm x}$, and $NbO_{1\pm x}$ present an even more complex picture. As seen from Figure 2.4, these oxides exist in the composition range $M_{0.8}O$ to $M_{1.2}O$. Even at the stoichiometric composition, *ca.* 13% of both cation and anion sites are apparently vacant in TiO. The concentrations of defects in the high-temperature non-stoichiometric TiO and VO phases as a function of composition have been deduced from comparisons of x-ray and picnometric densities[45]; some recent results are shown in Figure 2.6. Samples

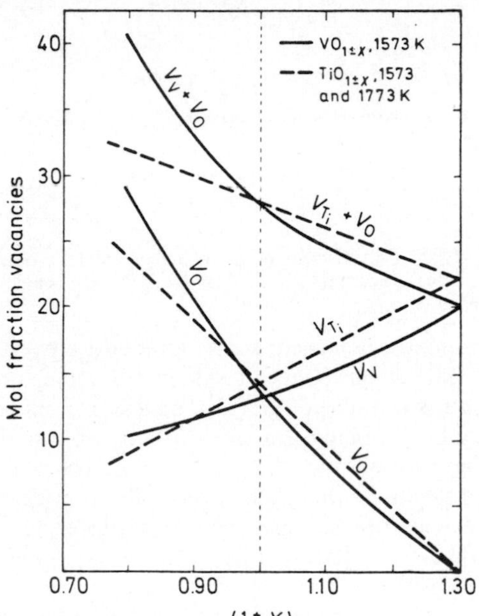

Figure 2.6 Vacancy concentration in vanadium and titanium oxides. (From Banus and Reed[45], by courtesy of the authors and the American Institute of Physics.)

quenched from 1200–1400 °C show no detectable ordering and the cubic NaCl structure. Annealing $TiO_{1\pm x}$ at temperatures below *ca.* 950 °C results in a wealth of ordered and partially-ordered structures[46], both for stoichiometric TiO and for other compositions. The Russian school[47], on the other hand, has proposed microdomain structures for these oxides, in which, rather than a uniform ordering of cations and vacancies throughout the structure, the sample consists of areas, intergrown on a very small scale, of structures corresponding to the neighbouring higher and lower oxides.

The thermodynamics of $TiO_{1\pm x}$ and $VO_{1\pm x}$ are imperfectly known; indeed there are serious inconsistencies in the thermodynamics of the titanium oxygen system[48]. Hoch[49] had measured the equilibrium oxygen pressure for both two-phase regions bounding $TiO_{1\pm x}$, $VO_{1\pm x}$, and $NbO_{1\pm x}$ and from these has calculated defect interaction energies based on a statistical model. No detailed thermodynamic studies within the single-phase regions exist.

Solid solubility between the high temperature forms of VO and TiO has been found to be very limited[50], which is suggestive of different defect structures of the two end members. The oxide CrO has not been made, but *ca.* 12 mol % CrO can be substituted into TiO or VO[51]. Obviously, oxygen fugacity plays a decisive role in determining the solid solubility limits in these ternary (M—M′—O) systems, and since most high temperature preparations are carried out in vacuum or in sealed tubes with ill-defined fo_2, no thermodynamical analysis of the results is possible.

Lastly, reference should be made to several rather extensive theoretical treatments of interacting defects[52–54]. These employ various means to calculate the partition function for the non-stoichiometric crystal and to develop a self-consistent model of defect interactions.

2.2.4 Metal oxides of narrow homogeneity ranges with the transition metal in well-defined oxidation states

These involve chiefly the transition metals in their +2 and +3 oxidation states. The +2 state is seen in the oxides of rocksalt structure; FeO and MnO (non-stoichiometric as discussed above); CoO, NiO, and CdO; in zincite, ZnO; and in tenorite, CuO. A mixture of +2 and +3 oxidation states is observed in the spinels Mn_3O_4, Fe_3O_4 and Co_3O_4, which may also be viewed as interoxidic compounds between MO and M_2O_3. The +3 state occurs in the corundum oxides Ti_2O_3, V_2O_3, Cr_2O_3 and Fe_2O_3, and in Mn_2O_3 for which the Jahn–Teller effect stabilises a distorted structure. In addition, Cu_2O is a relatively stable oxide with the +1 oxidation state, while oxidation states higher than +3 are seen in TiO_2, VO_2, CrO_2 and MnO_2 (+4), and CrO_3 (+6). The last three compounds are stable at low temperatures only. The synthesis of Co_2O_3 (all +3 cobalt) and nickel oxide approaching Ni_3O_4 in composition may be possible, but these do not appear to be stable under ordinary conditions.

Thermodynamic data at 1000 K are given in Table 2.2 for these oxides. The ease of reduction to the metal increases in the sequence Mn, Fe, Co, Ni. The higher oxidation states become unstable at high temperature; thus in air the reactions $MnO_2 \rightarrow Mn_2O_3$, $Mn_2O_3 \rightarrow Mn_3O_4$, $Fe_2O_3 \rightarrow Fe_3O_4$, $Co_3O_4 \rightarrow CoO$ and $CuO \rightarrow Cu_2O$ occur at 850, 1150, 1663, 1193 and 1308 K respectively.

Solid solutions among oxide pairs of the same structure form extensively, those among the rocksalt oxides often deviate little from thermodynamic ideality (see Section 2.3.6), while those of corundum and rutile structure deviate more from ideality and may decompose by a spinodal mechanism at temperatures of 700–1200 K[114].

2.2.5 Thermodynamics of point defects in nearly stoichiometric oxides

In dealing with reaction rates and electronic properties, the defect structure of transition metal oxides is all-important. In terms of bulk thermodynamic properties, however, the presence of defects becomes influential only when their concentration approaches 0.1–1%, as in $Fe_{1-x}O$ and in silver halides near their melting points. For the nearly stoichiometric oxides, defect concentrations range from 0.1% down. The defect equilibria can often be treated by a simple equilibrium approach, in which a defect formation reaction is written for which an equilibrium constant can be formulated, with the thermodynamic activities of defect species replaced by their concentrations. From the temperature dependence of the equilibrium constant, the enthalpies and entropies of defect formation can be calculated. From the dependence of defect concentration on oxygen partial pressure, the nature of the defect species can be deduced. The defect structures of a number of transition metal oxides have been rather well characterised, and some recent work is summarised in Table 2.4. For a more detailed discussion, the reader is referred to some of the review articles available[55–58]

2.2.6 Oxides of the noble metals

Pertinent thermodynamic data for these oxides are summarised in Table 2.5. On the whole, the stability of such oxides is very limited, and their interest, especially to high temperature chemists, lies more in the undesirable aspects of crucible and thermocouple corrosion than in their chemistry *per se*. Many of these oxides are not well characterised with respect to composition and crystal structure.

Table 2.5 Thermodynamics of formation of oxides of the noble metals

Oxide	ΔH/at K	ΔG/at K	*Upper stability limit*/at K	*Ref.*
$\frac{1}{2}RuO_2$	−27.3 (1000)	−10.0 (1000)	*ca.* 1420	1
RhO	−19.5 (1000)	−4.0 (1000)	*ca.* 1300	1
Rh_2O_3	−21.3 (1000)	−3.0 (1000)	*ca.* 1150	1
PdO	−21.9 (1000)	−2.9 (1000)	*ca.* 1140	1
Ag_2O	−7.2 (298)	−2.5 (298)	*ca.* 450	1
AgO	−6.2 (298)	+6.6 (298)	metastable	1
$\frac{1}{2}OsO_2$	−35.2 (1000)	−16.7 (1000)	—	2
$\frac{1}{4}OsO_4$	−23.5 (298)	−17.9 (298)	melts 329, boils 403	1
$\frac{1}{2}IrO_2$	−18.3 (1000)	−0.5 (1000)	*ca.* 1115	1
$\frac{1}{3}Au_2O_3$	−0.3 (298)	+6.1 (298)	metastable	1
Hg_2O	−22.0 (298)	−13.0 (298)	*ca.* 670	1
HgO	−21.7 (298)	−14.0 (298)	*ca.* 750	1

References for Table 2.5

1. Coughlin, J. P. (1954). *U.S. Bur. Mines Bull.*, 542
2. Franco, J. I. and Kleykamp, H. (1972). *Ber. Bunsenges. Phys. Chem.*, **76**, 691

2.3 TERNARY OXIDES

2.3.1 Introduction

Transition metal ions take part in a vast number of ternary oxides. Chief among these are compounds of the stoichiometries ABO_3 (perovskite, ilmenite, and pyroxene structures), AB_2O_4 (spinel, olivine, and phenacite structures), ABO_2, AB_2O_7.

Recent thermodynamic and structural work has focused on (a) using refined values of ionic radii and crystallographic parameters and improved lattice energy calculations to determine and predict the stability fields of various structure types and (b) using high temperature experimental techniques to obtain precise and systematic thermodynamic data. Sufficient data on the heats, free energies, and entropies of formation of mixed metal oxides from their constituent binary oxides now exist to warrant some generalisations and to hazard some predictions.

Before discussing specifics, a general comment on magnitudes is appropriate. Lattice energies, which measure the total cohesive energy of a solid, for typical oxides are in the region of 400–1000 kcal mol^{-1}. Standard heats and free energies of formation of oxides from a metal and oxygen usually range from 50–300 kcal mol^{-1}. Heats and free energies of reactions involving solids only (formation of ternary compounds, phase transitions, etc.) rarely exceed 50 kcal mol^{-1}, and often are 0–15 kcal mol^{-1} in magnitude. Thus the heats of formation of ternary compounds from their binary constituents represent only a minute fraction of their lattice energies.

The majority of thermodynamic data discussed below will refer to high temperatures ($T \approx 1000$ K) and to formation reactions from the oxides. It appears to the reviewer that uniform reference states for oxide thermodynamics may far more conveniently be taken to be the binary oxides in their most stable modifications and, when possible, at 1000 K and 1 atm. This has been done for the binary oxides above. Conversion to a traditional standard state which refers to the elements at 298 K introduces additional uncertainty and makes most thermodynamics of formation parameters unnecessarily large in magnitude, often swamping the small thermodynamic effects of real interest.

2.3.2 Crystal chemistry—ionic radii, stability fields, lattice energies, etc.

The structure adopted by a ternary oxide of given metal : oxygen ratio depends on temperature, pressure and the cations present. Qualitatively, concepts such as radius ratio and tolerance factor have been very useful in predicting crystal structures[59, 60, 116]. The possible structures for a given stoichiometry, e.g. AB_2O_4, can be arranged in qualitative stability fields, plotted as functions of r_A and r_B. Since pressure effectively compresses an oxide ion more than a cation, and increasing temperature favours less dense structures of higher entropy, the stability-field diagrams often bear some relation to P–T phase diagrams. For the stoichiometry AB_2O_4, both types of diagrams are shown schematically in Figure 2.7, with some known phase transitions indicated.

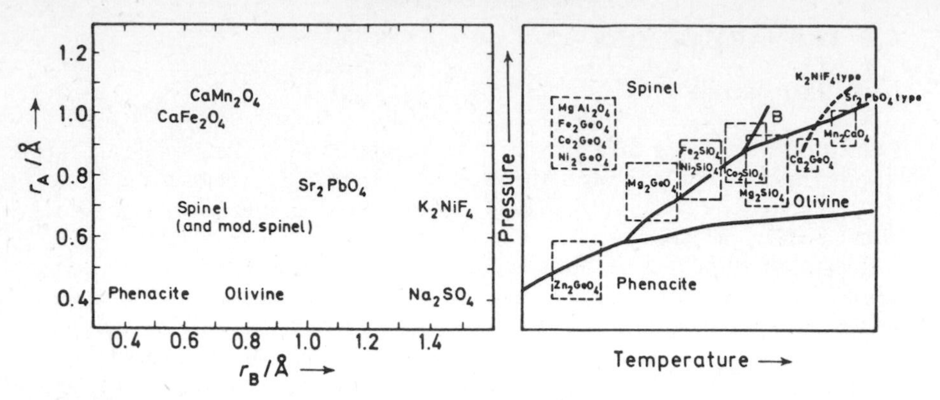

Figure 2.7 AB_2O_4 structures: schematic stability fields and pressure temperature relations

Quantitatively, numerous attempts have been made at *a priori* lattice energy calculations of oxides. The reviews of Ladd and Lee[61, 62] summarise the situation for the simpler structures. For the less symmetric structures (e.g. chain silicates) the refinement of crystal structures has led to much more precise knowledge of bond angles and bond lengths, and computer techniques have made feasible the calculation of electrostatic lattice energies. Rather than discuss such calculations in detail, we present a tabular summary of some recent studies (Table 2.6). In this connection, the derivation by Shannon and Prewitt[63] of a set of ionic radii best consistent with crystallographic interatomic distances is a valuable contribution. These values stress the fact that the effective ionic radius depends on coordination number as well as valence.

At present, the calculation of lattice energies with precision sufficient to predict phase transitions and to ascertain which of a series of closely-related structures is stable is seldom possible. The calculation of the non-ionic contribution to the lattice energy remains a major problem, especially in low symmetry structures where such contributions are more important.

2.3.3 Systematics of formation of series of ternary oxides of fixed stoichiometry: AB_2O_4, ABO_3, etc.

Recent advances in high temperature solution calorimetry[64] and in equilibrium measurements using gas mixtures of controlled oxygen fugacity[65] or high temperature electrochemical cells with solid electrolytes[66] have made possible the accurate determination of the enthalpies and free energies of formation of ternary mixed metal oxides from their binary oxide components. From $\Delta G°$ and $\Delta H°$, values of $\Delta S°$ can be calculated and compared to values obtained from the third law. Certain general conclusions can now be drawn about the thermodynamics of ternary oxide formation.

Thermodynamic data, arranged by stoichiometry and structure type, are presented in Table 2.7. Data are most complete for the spinels, but the trends observed seem to apply to many groups of oxides. Except in cases where substitutional disorder plays a major role, the entropies of formation are small enough so that the enthalpies and free energies of formation lie within

Table 2.6 Summary of recent lattice energy calculations and related topics

Subject	*Ref.*
Self-consistent ionic radii from crystallographic data.	Shannon and Prewitt, (1969). *Acta Crystallogr.*, **B25**, 925
Summary of lattice energies of inorganic compounds.	Ladd and Lee, (1964). *Progr. Solid State Chem.*, **1**, 37; (1967). **3**, 265
Crystal chemistry of olivine, spinel, and modified spinel structures in silicates.	Moore and Smith, (1970). *Phys. Earth. Planet. Interiors*, **3**, 166
Lattice energy and thermodynamic properties of $\gamma.Al_2O_3$.	Borer and Gunthard, (1970). *Helv. Chim. Acta*, **53**, 1043
Lattice energies of shear phases in Ti_nO_{2n-1} series.	Anderson and Burch, (1971). *J. Phys. Chem. Solids*, **37**, 923
Crystal chemistry of AB_2O_4 structures.	Reid and Ringwood, (1970). *J. Solid State Chem.*, **1**, 557
Lattice energies of some cubic oxides.	Hanna, (1965). *J. Phys. Chem.*, **69**, 2971
Classification, representation, and prediction of crystal structures of ionic compounds.	Gorter, (1970). *J. Solid State Chem.*, **1**, 279
The structon theory applied to inorganic crystals.	Huggins, (1968). *Macromolecules*, **1**, 184
Systematic survey of cubic crystal structures.	Loeb, (1970). *J. Solid State Chem.*, **1**, 237
Cohesive energy of MnO.	Nagai, (1965). *J. Phys. Soc. Jap.*, **20**, 1366
Electric field gradients in the spinel lattice.	Rosenberg, *et al.* (1970). *J. Appl. Phys.*, **41**, 1114
Structural basis of the olivine-spinel transition.	Kamb, (1968). *Amer. Mineral.* **53**, 1439
Lattice self-potentials, Madelung constants of oxides and fluorides.	van Gool and Piken, (1968). *J. Mat. Sci.*, **4**, 95; (1968). **4**, 105
Madelung energies and site preference in amphiboles.	Whittaker, (1971). *Amer. Mineral.*, **56**, 980
Madelung energy of ordered and disordered TiO.	O'Keefe and Valigi, (1969). *J. Chem. Phys.*, **50**, 1490
Computer simulation of Mg_2SiO_4 polymorphs.	Baur, (1972). *Amer. Mineral.* **57**, 709

Table 2.7 Thermodynamics of formation of ternary oxides from their binary oxide compounds§

Compound	ΔG_f°/kcal mol^{-1} (at K)	ΔH_f°/kcal mol^{-1} (at K)	ΔS_f°/cal deg^{-1} mol^{-1}
Stoichiometry AB_2O_4			
SPINELS			
$MgAl_2O_4$ (N)*	−8.4 (1273)[2], −8.4 (1673)[3]	−5.9 (965)[15]	+2.0
$MnAl_2O_4$ (N)	est. −10 (1273)[1]		
$FeAl_2O_4$ (N)	−5.7 (1273)[4]		
$CoAl_2O_4$ (N)	−7.2 (1273)[2], −6.9 (1273)[5]	−8.9 (970)[1]	−1.3
$NiAl_2O_4$ (Int)	−5.2 (1273)[2]	−0.7 (970)[1]	+3.5
$CuAl_2O_4$ (Int)	*ca.* −0.5 (1000)[33]	+5.2 (970)[1]	> +5
$ZnAl_2O_4$ (N)	−13.0 (1000)[6]	−10.6 (970)[1]	+2.6
$MgFe_2O_4$ (Int)	−5.3 (1273)[2]	−1.6 (970)[14]	+2.9
$MnFe_2O_4$ (Int)	−6.8 (1273)[2]	[−4.8 (298)[35]]	(+1.6)
Fe_3O_4 (I)	−7.5 (1000)[7]	−5.1 (1000)[7]	+2.4
$CoFe_2O_4$ (I)	−9.4 (1273)[2]	−5.9 (970)[1]	+2.8
$NiFe_2O_4$ (I)	−6.0 (1273)[2]	−1.2 (970)[1]	+3.8
$CuFe_2O_4$ (I)	~ −0.5 (1000)[33]	+5.1 (970)[1]	> +5
$ZnFe_2O_4$ (N)	−6.5 (1400)[6]	−2.7 (970)[1]	+2.7
$CdFe_2O_4$ (N)		+1.8 (970)[1]	
$MgCr_2O_4$ (N)	−8.1 (1273)[2]	−11.1 (1173)[12]	−2.4
$MnCr_2O_4$ (N)	est. −10 (1273)		
$FeCr_2O_4$ (N)	−8.8 (1273)[2], −9.9 (1573)[10]	[−13.8][2]	(−3.9)
$CoCr_2O_4$ (N)	−12.0 (1273)[2]	[−14.0][2]	(−1.6)
$NiCr_2O_4$ (Int, T)	−6.4 (1273)[2]	−1.3 (1173)[12]	(+4.0)
$CuCr_2O_4$ (Int, T)	−4.5 (1273)[11]	+2.5 (1173)[12]	(+5.5)
$ZnCr_2O_4$ (N)		−15.0 (1173)[12]	
$CdCr_2O_4$ (N)		−10.1 (1173)[12]	
$MgGa_2O_4$ (Int)		−6.8 (970)[14]	
$CoGa_2O_4$ (I)		−7.5 (970)[1]	
$NiGa_2O_4$ (I)		−0.8 (970)[1]	
$CuGa_2O_4$ (I)		+4.0 (970)[1]	
$ZnGa_2O_4$ (N)		−8.5 (970)[1]	
$CdGa_2O_4$ (N)		−1.8 (970)[1]	
$MgMn_2O_4$ (N, T)		−2.7 (970)[1]	
Mn_3O_4 (N, T)	−5.7 (1000)[8]		
$FeMn_2O_4$ (N, T)			
$CoMn_2O_4$ (N, T)	−8.6 (1473)[9]	−3.8 (970)[1]	+3.2
$NiMn_2O_4$ (Int)		+0.5 (970)[1]	
$CuMn_2O_4$ (Int)		+3.0 (970)[1]	
$ZnMn_2O_4$ (N)		−6.6 (970)[1]	
$CdMn_2O_4$ (N)		−7.5 (970)[1]	
Mg_2TiO_4 (I)	−5.8 (1673)[27], −5.1 (1573)[26]	−6.0 (970)[1]	
Mn_2TiO_4 (I)	−7.6 (1523)[28]		
Fe_2TiO_4 (I)	−8.2 (1573)[29]		
Co_2TiO_4 (I)	−6.0 (970)[1]		
[Ni_2TiO_4]‡	[−1.8 (1673)[27]], [−1.0 (1523)[28]]		
Zn_2TiO_4 (I)	−4.7 (1323)[31]	−0.8 (970)[1]	+2.9
Mg_2GeO_4 (N)		−19.3 (1058)[37]	
Co_2GeO_4 (N)		−14.1 (970)[1]	
Ni_2GeO_4 (N)		−9.5 (970)[1]	
OLIVINES			
Mg_2SiO_4	−14.8 (1000)[16]	−15.7 (1000)[16]	−1.0
		−15.4 (964)[14]	
Mn_2SiO_4	−10.1 (1000)[17]	−11.8 (298)[16]	*ca.* −1
Fe_2SiO_4	−4.9 (1000)[18], −4.2 (1423)[18]	−5.9 (298)[16]	−1.0
Co_2SiO_4	−3.2 (1273)[5, 20]	−4.3 (965)[13]	−0.9
Ni_2SiO_4	−1.2 (1273)[20]	−3.3 (965)[13]	−1.6
Mg_2GeO_4	stable	−16.3 (965)[14]	
Mn_2GeO_4	stable		
Ca_2SiO_4	−31.7 (1000)[16]	−29.7 (1000)[16]	−1.8
Ca_2GeO_4		−41.6 (298)[34]	

Table 2.7 (continued)

PHENACITES			
Zn_2SiO_4	−6.7 (1173)[22]	−7.8 (965)[13]	−0.9
Zn_2GeO_4		−10.8 (965)[13]	
OTHER			
$CdAl_2O_4$		−4.5 (970)[1]	
Ca_2SnO_4	−16.2 (1400)[36]		
Stoichiometry ABO_3			
ILMENITES			
$MgTiO_3$	−5.1 (1673)[27], −4.8 (1573)[26]		
$MnTiO_3$	−7.0 (1523)[28]		
$FeTiO_3$	−4.3 (1573)[29], −6.5 (298)[32]	[−6.8 (298)[32]]	
$CoTiO_3$	−3.6 (1373)[30]		
$NiTiO_3$	−2.3 (1673)[27]		
$ZnTiO_3$	[+1.7 (1323)][31]		
$CdTiO_3$		−6.6 (965)[23]	
PEROVSKITES			
$CdTiO_3$		−3.0 (965)[23]	
$CaTiO_3$	−20.5 (1000)[16]	−18.9 (1000)[16]	−1.6
$CaSnO_3$	−14.9 (1400)[36]		
PYROXENES AND RELATED STRUCTURES			
$MgSiO_3$	−8.1 (1000)[16]	−9.5 (1000)[16], −8.7 (965)[14]	−1.4
$MnSiO_3$	−6.1 (1000)[17]		−0.3
[$FeSiO_3$]	[−1.3 (1423)][18], [−0.9 (1523)][19]		
[$CoSiO_3$]	[−0.6 (1523)][17]		
[$NiSiO_3$]	[*ca.* 0.0 (1523)][21]		
$MgGeO_3$		−10.9 (965)[14]	
$MnGeO_3$	stable		
$FeGeO_3$	stable		
$CoGeO_3$		−7.4 (965)[13]	
[$NiGeO_3$]	unstable		
$CaSiO_3$	−21.1 (1000)[16]	−21.6 (1000)[16]	−0.5
$CaGeO_3$		−25.2 (298)[34]	
OTHER			
$CuGeO_3$	−4.0 (1273)[37]	−5.3 (973)[37]	
Stoichiometry AB_2O_5			
PSEUDOBROOKITES AND RELATED COMPOUNDS			
$MgTi_2O_5$	−7.3 (1673)[27], −6.8 (1573)[26]		
[$MnTi_2O_5$]	[−6.3 (1523)[28]]		
$FeTi_2O_5$	−4.3 (1573)[29]		
$CoTi_2O_5$	−3.5 (1373)[30]		
[$NiTi_2O_5$]	[−2.0 (1573)][27]		
Stoichiometry ABO_4			
TUNGSTATES			
$MgWO_4$		−17.7 (965)[24]	
$MnWO_4$		−19.2 (298)[25]	
$FeWO_4$	−10.3 (1173)[15]		
$CoWO_4$		−14.4 (965)[24]	
$NiWO_4$		−10.6 (965)[24], −12.9 (298)[25]	
$CuWO_4$		−6.6 (965)[24]	
$ZnWO_4$		−10.1 (965)[24]	
$CdWO_4$		−18.9 (965)[24]	

* N = normal spinel, I = inverse spinel, Int = intermediate cation distribution, T = tetragonal.
† Values in brackets are estimated or have large uncertainties.
‡ Compounds in brackets not stable under normal $P_1 T$ conditions.
§ References for Table 2.7 on p. 50.

0–2 kcal mol^{-1} of each other at 1000 K. Typical data for the aluminates, MAl_2O_4; orthosilicates, M_2SiO_4; metatitanates, $MTiO_3$; and tungstates, MWO_4; are shown in Figure 2.8.

From Figure 2.8 and Table 2.7, we see that variation of the divalent cation in the series Mg, Mn, Fe, Co, Ni, Cu, Zn, Cd gives fairly uniform variations in ΔH° and ΔG° in all these groups of compounds, although the overall magnitude of ΔG° and ΔH° is different for each series (that is, displaced vertically in Figure 2.8). The calcium compounds (not shown) are considerably more stable but these often have crystal structures different from the other members of a series. This stability is a reflection of cation size and oxide basicity, and is further shown in the series of perovskites $CaTiO_3$, $SrTiO_3$, $BaTiO_3$. Among the dipositive ions of similar size, the compounds of magnesium and manganese generally have the greatest stability, followed, more or less closely, by those of iron and cobalt. The decrease in ΔH_f° of iron and cobalt ternary oxides appears to be greater for the tungstates and olivine orthosilicates than for the spinels and ilmenites, possibly because of the greater distortion of the octahedral sites in the former. The nickel compounds have even smaller heats and free energies of formation in all cases; and Ni_2SiO_4, in particular, either melts incongruently[67] or decomposes to the oxides before melting[68]. The copper compounds are destabilised by another 3–5 kcal mol^{-1} compared to the nickel ternary oxides. This makes the copper silicates and Cu_2GeO_4 unstable. The copper spinels have positive enthalpies of formation and are apparently stabilised at high temperatures by the configurational entropy due to an almost random cation distribution. The 3–5 kcal difference between copper and nickel spinels corresponds in magnitude to, and probably has its origin in, the enthalpy and free energy of the hypothetical transformation of CuO from the tenorite to the rocksalt structure and the analogous change in

---

References for Table 2.7

1. Navrotsky, A. and Kleppa, O. J. (1968). *J. Inorg. Nucl. Chem.*, **30**, 479
2. Tretjakow, J. D. and Schmalzried, H. (1965). *Ber. Bunsenges. Phys. Chem.*, **69**, 396
3. Rosen, E. and Muan, A. (1966). *J. Amer. Ceram. Soc.*, **49**, 107
4. Rezukhina, T. N. *et al.* (1963). *Russ. J. Phys. Chem.*, **37**, 358
5. Aukrust, E. and Muan, A. (1963). *J. Amer. Ceram. Soc.*, **46**, 358
6. Gilbert, I. and Kitchener, (1965). *J. Chem. Soc.*, 3922
7. Coughlin, J. P. (1954). *U.S. Bur. Mines Bull.*, 542
8. Mah, A. D. (1960). *U.S. Bur. Mines Rept. Invest.*, 5600
9. Aukrust, E. and Muan, A. (1964). *Trans. Met. Soc. AIME*, **230**, 378
10. Rezukhina, T. N. (1965). *Elektrokhimiya*, **1**, 467
11. Schmahl, N. and Minzl, E. (1965). *Z. Physik. Chemie N.F.*, **47**, 358
12. Müller, F. and Kleppa, O. J. (1973). *J. Inorg. Nucl. Chem.*, **25**, 2673
13. Navrotsky, A. (1971). *J. Inorg. Nucl. Chem.*, **33**, 4035
14. Shearer, J. A. and Kleppa, O. J. (1973). *J. Inorg. Nucl. Chem.*, **35**, 1073
15. Schmahl, N. and Dillenburg, H. (1972). *Z. Phys. Chem.*, **77**, 113
16. Robie, R. and Waldbaum, D. (1968). *U.S. Geol. Survey Bull.*, 1259
17. Biggers, J. V. and Muan, A. (1967). *J. Amer. Ceram. Soc.*, **50**, 230
18. Schwerdtfeger, K. and Muan, A. (1966). *Trans. AIME*, **236**, 201
19. Nafziger, R. and Muan, A. (1967). *Amer. Mineral.*, **52**, 1364
20. Taylor, R. W. and Schmalzried, H. (1964). *J. Phys. Chem.*, **68**, 2444
21. Estimated from data of Campbell, F. E. and Roeder, P. (1968). *Amer. Mineral.* **53**, 257
22. Kitchener, J. A. and Ignatowicz, (1951). *Trans. Faraday Soc.*, **47**, 1278
23. Neil, J. M. Navrotsky, A. and Kleppa, O. J. (1971). *Inorg. Chem.*, **10**, 2076
24. Navrotsky, A. and Kleppa, O. J. (1969). *Inorg. Chem.*, **8**, 756
25. Proshina, Z. V. and Rezukhina, T. N. (1960). *Russ. J. Inorg. Chem.*, **5**, 488
26. Brezny, B. and Muan, A. (1971). *Thermochim. Acta*, **2**, 107
27. Evans, L. G. and Muan, A. (1971). *Thermochim. Acta*, **2**, 121
28. Evans, L. G. and Muan, A. (1971). *Thermochim. Acta*, **2**, 277
29. Johnson, R. E. Woermann, E. and Muan, A. (1971). *Amer. J. Sci.*, **271**, 278
30. Brezny, B. and Muan, A. (1969). *J. Inorg. Nucl. Chem.*, **31**, 649
31. Navrotsky, A. and Muan, A. (1970). *J. Inorg. Nucl. Chem.*, **32**, 3471
32. Levitskii, V. A., Popov, S. G. and Ratiani, D. D. (1971). *Russ. J. Phys. Chem.*, **45**, 294
33. Estimated from reduction equilibria and range of stability of Cu^{2+} – containing spinel.
34. Grebenschikov, R. G. and Pasechnova, R. A. (1967). *Russ. J. Inorg. Chem.*, **12**, 598
35. Fischer, M. (1966). *Z. Anorg. Allgem. Chem.*, **345**, 134
36. Möller, B. (1970). *Z. Anorg. Allgem. Chem.*, **376**, 144
37. Navrotsky, A. *J. Solid State Chem.* (in the press)

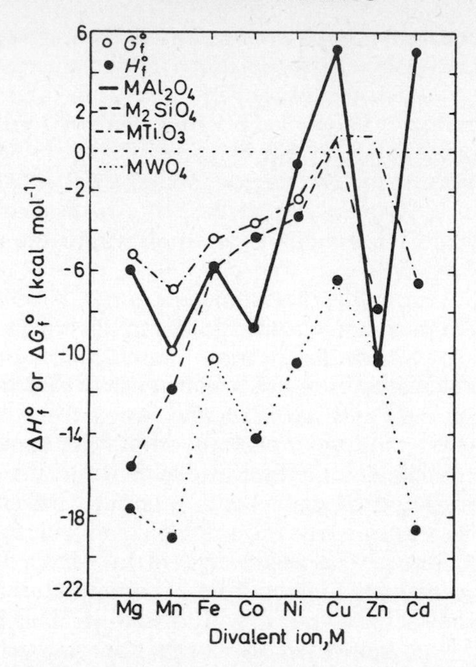

Figure 2.8 Enthalpies of formation of ternary oxides from their binary oxide components. Data from sources listed in Table 2.7.

coordination from square planar to octahedral upon spinel formation. This explanation is further substantiated by the observation that $CuGeO_3$, which has a unique chain structure in which the environment of Cu^{2+} is much like that in tenorite, has an enthalpy of formation of -5.3 kcal mol^{-1} [69]. The ternary oxides of zinc have stabilities similar to those of magnesium when the zinc retains tetrahedral coordination as in its binary oxide (e.g. $ZnAl_2O_4$, Zn_2SiO_4); but are considerably destabilised (e.g., Zn_2TiO_4 and the non-existence at atmospheric pressure of $ZnTiO_3$ and $ZnSiO_3$) when Zn^{2+} is forced to assume octahedral coordination. Again, this destabilising effect is comparable in magnitude to ΔG° for the reaction ZnO(zincite)→ZnO (rocksalt)[72, 175]. The Cd^{2+} ion is intermediate between Mg^{2+} and Ca^{2+} in size and crystal-chemical behaviour. The stability of cadmium compounds appears to depend strongly on the size of the octahedral or tetrahedral site available to Cd^{2+}, and the enthalpy of formation of the spinel increases with increasing lattice parameter in the series $CdAl_2O_4$, $CdFe_2O_4$, $CdGa_2O_4$, $CdMn_2O_4$, $CdCr_2O_4$. $CdTiO_3$ and $CdSnO_3$ exhibit dimorphism between the ilmenite and perovskite structure; the enthalpy of formation of $CdTiO_3$ (ilmenite) is comparable to that of the other TM ilmenites and the enthalpy of the ilmenite–perovskite transition is 3.6 kcal mol^{-1} [70].

The variation of the tripositive or tetrapositive cation in the ternary compound has the following effects on stability. Among the spinels, the 2–3 aluminate, ferrite, gallate and manganite spinel series and the 2–4 titanate series have generally comparable stabilities for the same dipositive cation.

The germanate and chromite spinels are on the whole more stable than other spinels, and the germanates are in general more stable than the corresponding silicates. The tungstates show the most negative enthalpies and free energies of formation of the compounds shown here. This reflects the greater acidity of the oxide WO_3 compared to Al_2O_3, Fe_2O_3, SiO_2, TiO_2, etc. The variation in acidity could be extended to include the carbonates ($\Delta H^\circ = -15$ to -30 kcal mol^{-1}) from MO and CO_2 and the sulphates ($\Delta H^\circ = -50$ to -70 kcal mol^{-1} from MO and SO_3) and the trends observed above for the dipositive ions would still apply[71].

The entropies of formation deserve some comment. When no configurational disorder exists in the ternary oxide, ΔS_f° from the binary oxides is quite small, generally -2 to $+2$ cal deg^{-1} mol^{-1}, and good agreement is seen between ΔS_f° values calculated from heat capacity data (Third law) and from the difference between ΔH_f° and ΔG_f° (Second law). However, the effect of cation disorder in inverse, random or partially disordered spinels (see Section 2.3.5.2 below) is to contribute a configurational term to the entropy of the spinel. Since this substitutional disorder is generally 'frozen in' below *ca.* 1000 K, heat capacity measurements alone are not a reliable means of obtaining the true entropies. Because the initial stages of disorder provide the largest entropy increments, a small amount, *ca.* 10%, of configurational disorder, can substantially affect the thermodynamics of formation. This point has been discussed for spinels[72] and more recently for other phases, such as sillimanite, Al_2SiO_5 [73, 74], as well. A corrolary of substitutional disorder is the 'entropy stabilisation' of such disordered phases at high temperature. With positive enthalpies of formation, the copper spinels such as $CuAl_2O_4$, $CuFe_2O_4$, $CuGa_2O_4$ and $CuCr_2O_4$ are stabilised at high temperature by the large $T\Delta S$ term, and persist metastably to room temperature. Positive enthalpies of formation from the oxides have also been found for a beryllium aluminate, $BeAl_6O_{10}$ [75], and for mullite, $2Al_2O_3 \cdot 3SiO_2$ [76], in both of which equivalent tetrahedral sites are occupied by more than one type of ion. In addition, the pseudobrookite structure is apparently stable only at high temperatures; Fe_2TiO_5 [77], $FeTi_2O_5$ [77], $CoTi_2O_5$ [79], Al_2TiO_5 [80] and Ga_2TiO_5 [80] decompose to other phase assemblages below 858, 1413, 1408, *ca.* 1600 and *ca.* 1500 K, respectively. The last two compounds are claimed to decompose to oxides with strongly exothermic heat effects (endothermic heats of formation). In these materials there may well be considerable randomisation of titanium and the other cation over two or perhaps three non-equivalent octahedral sites which would lead to their entropy stabilisation at high temperature. A recent Mossbauer study of the $FeTi_2O_5$–Ti_3O_5 series confirms that considerable configurational disorder exists[119].

2.3.4 Stabilities of structure types and phase transitions

A number of ternary oxides show phase transformations at high pressures to denser polymorphs. In addition, extensive solid solubility exists between many pairs of ternary oxides of similar stoichiometry but different structure which suggests that the energies of transformation are not prohibitively large. Figure 2.7 implies that such transformation should be fairly general phenomena for compounds which lie near the limits of a particular stability field. Phase transitions in silicates and germanates have been examined rather extensively because of their possible application to the mineralogy of the earth's mantle[81].

Experimentally, several methods are available for obtaining thermodynamic data for these phase transformations. For high-pressure transitions, the phase boundary can be determined as a function of P and T and the Clausius–Clapeyron equation used to obtain ΔS if ΔV is known[81]. In systems showing extensive terminal solid solubility at atmospheric pressure, the activity–composition relations in these solid solutions can be determined and free energies of transformation estimated therefrom[82]. The second method has the advantage of being equally applicable to the study of metastable cryptomodifications not attainable by high pressure methods, for example, phases less dense than the atmospheric pressure structures. Thirdly, thermochemical measurements at atmospheric pressure can be performed on quenched high pressure samples[83]. Recent advances in high temperature solution and reaction calorimetry permit such measurements to be made on samples of a size (200–300 mg total) that can reasonably be made in a present-day high pressure apparatus.

Thermodynamic data for phase transitions at the stoichiometries AB_2O_4 and ABO_3 are summarised in Table 2.8. A number of recent discussions stress the crystal chemical regularities of these transitions[81, 83, 84, 88]. Several generalities stand out. Firstly, a transition such as olivine–spinel occurs with a fairly constant ΔS and ΔV, although Ni_2SiO_4 may be an exception to this[83, 85, 86]. Secondly, the germanates generally act as accessible high pressure analogues to the silicates (possess the high pressure structures at atmospheric pressure or transform to them at much lower pressures) but the analogy breaks down in several instances[87, 81]. Thirdly, there is a greater complexity of possible transitions than initially imagined. For the AB_2O_4 compounds, modified spinel, magnetoplumbite, and K_2NiF_4 structures must be considered as well as spinel[89]. For the stoichiometry ABO_3, the ilmenite, perovskite, and garnet structures occur at high pressure. This complexity of possible structures renders predictions of transitions and extrapolations to higher pressure ranges somewhat uncertain. Further complications result because decomposition to dense polymorphs of the binary oxides may be energetically competitive reactions as well[85]. Lastly, the dependence of the standard free energy of the transformation on the dipositive cation can be rationalised on the basis of ligand field effects[85, 88]; the structure of greatest octahedral site symmetry (spinel) is increasingly stabilised in the series Mg, Co $\approx$ Fe, Ni, and this is true in both silicates and germanates. The behaviour of zinc ternary oxides at high pressure is somewhat unpredictable because Zn^{2+} can apparently compete with Si^{4+} or Ge^{4+} for tetrahedral coordination, leading to new structures and/or spinels of randomised or inverse cation distribution[87].

Another important manifestation of pressure is the destabilisation of Fe^{3+} relative to Fe^{2+}. This has been confirmed for many inorganic and organometallic compounds by Mössbauer resonance techniques[90] and applies also to ternary iron oxides including silicates[91].

Table 2.8 Thermodynamics of phase transitions in ternary oxides†

Compound	*Transition*	*Conditions*	*Thermodynamic parameters* ΔH°/kcal mol^{-1} ΔG°/kcal mol^{-1}	ΔS°/cal deg^{-1} mol^{-1}	ΔV/cm^3 mol^{-1}
STOICHIOMETRY AB_2O_4					
Mg_2SiO_4	olivine→modified spinel	1273 K, 120 kb[1]			−3.17[1]
	olivine→spinel	metastable with respect to modified spinel (1273 K, 125 kb)[1]			−4.21[2]
	olivine→phenacite	less dense cryptomodification			+8.65[2]
Fe_2SiO_4	olivine→spinel	1273 K,52 kb[7]	$\Delta H^\circ \approx 0$[7]	−4.6[7]	−4.35[2]
Co_2SiO_4	olivine→spinel	1073 K, 63 kb[10]	$\Delta H^\circ = 2.6$[10]	−3.2[10]	−3.98[2]
	olivine→modified spinel	1473 K, 73 kb[10]			−3.0[9]
Ni_2SiO_4	olivine→spinel	1273 K, 30 kb[7,8]	$\Delta H^\circ = 1.2$[7,8] $\Delta H^\circ_{986} = 1.2$[9]	−1.0	−3.33[2]
Zn_2SiO_4	phenacite→modified spinel	several intervening transitions but modified spinel stable above 1173 K, 130 kb[5]			−10.6[5]
	phenacite→olivine	metastable			−7.71[2]
Mg_2GeO_4	olivine→spinel	1083 K, 1 atm[4]	* $\Delta H^\circ_{1083} = -3.7$[4]	−3.5[4]	−3.52[2]
	olivine→phenacite	less dense cryptomodification	$\Delta H^\circ_{1058} = -3.0$[25]		−9.44[2]
Mn_2GeO_4	olivine→Sr_2PbO_4 type	1173 K, 90 kb[11]			−7.79[11]
Co_2GeO_4	spinel→olivine	less dense cryptomodification	$\Delta G^\circ_{1473} = +3.5$[2]		+3.9[2]
	spinel→phenacite	less dense cryptomodification	$\Delta G^\circ_{1323} = +12.7$[3]		+12.9[2]
Ni_2GeO_4	spinel→olivine	less dense cryptomodification	$\Delta G^\circ_{1473} = +8.2$[2]		+3.1[2]
	spinel→phenacite	less dense cryptomodification	$\Delta G^\circ_{1323} = +23.0$[3]		
Zn_2GeO_4	phenacite→spinel	773 K, *ca.* 40 kb[6]	$\Delta G^\circ_{1323} = +12.6$[3]		−12.1[2]
	phenacite→olivine	metastable			−8.6[2]
Ca_2GeO_4	olivine→K_2NiF_4 type	1173 K, 110 kb[13]			−12.52[13]
Cd_2SiO_4	olivine→spinel				
$FeMnGeO_4$	olivine→spinel	1373 K, 235 kb[12]			−4.13[12]
	olivine→Sr_2PbO_4 type				−7.55[13]
$CoMnCeO_4$	olivine→spinel	1373 K, <35 kb[12]			−4.24[12]
$MgMnGeO_4$	olivine→spinel	1373 K, <35 kb[12]			−4.35[12]
Mn_3O_4	tetrag. spinel→$CaMn_2O_4$ type	1173 K, 120 kb[23]			
$CaAl_2O_4$	$CaAl_2O_4$ type→$CaFe_2O_4$ type	1173 K, 100 kb[23]			

STOICHIOMETRY ABO_3					
$MgSiO_3$	pyroxene→ilmenite or pyroxene→spinel or mod. sp. + stishovite	est. > 250 kb.			
$FeSiO_3$	olivine + quartz→pyroxene	1273 K, 15 kb[18]	$\Delta H° = -0.04$[18]	-0.45[18]	-1.34[18]
	pyroxene→spinel + stishovite	1173 K, 125 kb[19]			
$CoSiO_3$	olivine + quartz→pyroxene	1273 K, 20 kb[18]	$\Delta H° = +0.60$[18]	-0.35[18]	-1.63[18]
	pyroxene→spinel + stishovite	1173 K, 125 kb[19]			
$NiSiO_3$	apparently not stable under any conditions, extensive Ni^{2+} substitution possible, in pyroxene				
$ZnSiO_3$	$Zn_2SiO_4 + SiO_2 \rightarrow 2ZuSiO_3$ (pyroxene)	1473 K, 31 kb[5]			
$CaSiO_3$	wollastonite→perovskite	estimate 200–250 kb			
$MgGeO_3$	pyroxene→ilmenite	973 K, 28 kb[17]			
$MnGeO_3$	pyroxene→ilmenite	973 K, 25 kb[17]			
$FeGeO_3$	pyroxene→spinel + rutile	973 K, 10 kb[17]			
$CoGeO_3$	pyroxene→spinel + rutile	973 K, 10 kb[17]			
$NiGeO_3$	apparently not stable under any conditions, extensive Ni^{2+} substitution possible, in pyroxene				
$ZnGeO_3$	ilmenite	synthesis at high P[5]			
$CdTiO_3$	ilmenite→perovskite	1103 K, 1 atm[15]	$\Delta H° = 3.74$[15] $\Delta H° = 3.58$[16]	3.4[15]	-2.94[15]
$CdSnO_3$	ilmenite→perovskite	*ca.* 1223 K, 1 atm[15]			-2.06[15]
$CaGeO_3$	wollastonite→garnet	973 K, 40 kb[17]			-4.2[21]
	garnet→perovskite				
$MnVO_3$	distorted ilmenite→ distorted perovskite	1300 K, 45 kb[21]			
$MnTiO_3$	ilmenite→corundum (disordered ilmenite)	1273 K, 66 kb[22]		+0.5	-1.6%
OTHER STOICHIOMETRIES					
$ZrSiO_4$	zircon→scheelite	1173 K, 120 kb[14]			

* Both these values belong to Mg_2GeO_4 ol→sp
† References for Table 2.8 on p. 56

2.3.5 Defect equilibria in ternary oxides

2.3.5.1 Point defects present at small concentrations

These are, in general, analogous to the point defects in binary oxides. When the ternary oxide contains one TM oxide (NiO, CoO, etc.) and one oxide of essentially fixed stoichiometry (Al_2O_3, SiO_2, etc.), the predominant defect type usually consists of electron holes (tripositive TM ions) and cation vacancies, much as in the parent TM oxides. However, the concentration of defects is much smaller in the ternary oxides, thus, the order of magnitude of maximum deviation from stoichiometry is: FeO, 10%; Fe_2SiO_4, 1%; CoO, 1%; Co_2SiO_4, 0.01%; $CoAl_2O_4$, 0.1%[92]. Furthermore the defect concentration both in the binary TM oxide and in the ternary compound decreases markedly in solid solutions when the second component contributes essentially no defects. A roughly exponential decrease in defect concentration with mol fraction of the other component has been observed in systems such as CoO–MgO[93], $CoAl_2O_4$–$MgAl_2O_4$[92] and Fe_2SiO_4–Mg_2SiO_4[94]. This has been interpreted to mean that the enthalpy of defect formation increases linearly with the mol fraction of other component[92, 93].

Point defect equilibria have been studied extensively in spinels and less extensively in other ternary systems. Pertinent work dealing with the defect type and the energetics of defect formation is summarised in Table 2.9. A review of point defects in ternary ionic crystals has been given by Schmalzried[95].

References for Table 2.8

1. Ringwood, A. E. and Major, A. (1970). *Phys. Earth Planetary Interiors*, **3**, 89
2. Navrotsky, A. (1973). *J. Solid State Chem.*, **6**, 21
3. Navrotsky, A. (1973). *J. Solid State Chem.*, **6**, 42
4. Dachille, F. and Roy, R. (1960). *Amer. J. Science*, **258**, 225
5. Syono, Y. Akimoto, S. and Matsui, T. (1971). *J. Solid State Chem.*, **3**, 369
6. Rooymans, C. J. M. (1966). *Phillips Res. Rept. Suppl.*, No. 5
7. Akimoto, S. and Fujisawa, H. (1965). *J. Geol. Res.*, **70**, 1969
8. Ma, C.-B. (1972). *Ph.D. Thesis*, Geology, Harvard University
9. Navrotsky, A. (1973). Earth Planet Sci. Lett., **19**, 471
10. Akimoto, S. and Sato, Y. (1968). *Phys. Earth Planet. Interiors*, **1**, 498
11. Wadsley, A. D. and Reid, A. F. (1968). *Acta Crystallogr.*, **B24**, 740
12. Ringwood, A. E. and Reid, A. F. (1970). *J. Phys. Chem. Solids*, **31**, 2791
13. Reid, A. F. and Ringwood, A. E. (1970). *J. Solid State Chem.*, **1**, 557
14. Reid, A. F. and Ringwood, A. E. (1969). *Earth Planet. Sci. Lett.*, **6**, 205
15. Liebertz, J. and Rooymans, C. J. M. (1965). *Z. Phys. Chem.*, **44**, 242
16. Neil, J. M. Navrotsky, A. and Kleppa, O. J. (1971). *Inorg. Chem.*, **10**, 2076
17. Ringwood, A. E. (1966). *Advances in Earth Science*, (P. M. Hurley, editor) (Cambridge, Mass: M.I.T. Press) p. 381.
18. Akimoto, S. *et al.*, (1965). *J. Geophys. Res.*, **70**, 5269
19. Ringwood, A. E. and Major, A. (1965). *Earth Planet. Sci. Lett.*, **1**, 209
20. Hayashi, H. *et al.*, (1964). *Chem. Abstr.*, **61**, 12 720c
21. Syono, Y. Akimoto, S. and Endoh, Y. (1971). *J. Phys. Chem. Solids*, **32**, 243
22. Syono, Y. *et al.*, (1969). *J. Phys. Chem. Sol.*, **30**, 1665
23. Navrotsky, A. Mat. Res. Bull. in the press

Table 2.9 Studies of defect equilibria in ternary oxides

System	*Measurements*	*Comments*	*Refs.*
$Co_xFe_{3-x}O_4$	Diffusion, oxygen pressure	Equilibrium constants for defect reactions given	1–3
$Ni_xFe_{3-x}O_4$	Oxygen pressure	Equilibrium constants for defect reactions given	1, 3
$Mg_xFe_{3-x}O_4$	Diffusion, oxygen pressure	Equilibrium constants for defect reactions given	1, 33
$Mn_xFe_{3-x}O_4$	Diffusion, oxygen pressure	Equilibrium constants for defect reactions given	1
Co_2SiO_4	Diffusion, oxygen pressure, formation of ternary oxide	Equilibrium constants for defect reactions given	4, 5
$(CoMg)Al_2O_4$	Oxygen pressures	Equilibrium constants for defect reactions given	4
Zn_2SiO_4	Diffusion, formation of ternary oxide		5
Mg_2SiO_4	Diffusion, formation of ternary oxide		5
$(FeMg)_2SiO_4$	Diffusion, conductivity, as $f(T, X, f_{O_2})$		6, 7
$CoCr_2O_4$	Diffusion, oxygen pressure	Equilibrium constants for defect equilibrium given	8
$CoAl_2O_4$	Diffusion, oxygen pressure		
$(MnMgFe)_3O_{4+x}$	Defect structure		9
$CaWO_4$	Diffusion	Energy of Schottky defect pair formation given	10
$LiFe_5O_8$	Oxygen pressures	Defect structure proposed	11
$(CuFe)_2O_{4+x}$	Oxygen pressures	Enthalpies of defect formation calculated	12
$(CuMgMnFe)_3O_4$	Physical properties	Defect type proposed	9

References for Table 2.9

1. Hohmann, H. H. *et al.*, (1967). *Proc. Brit. Ceram. Soc.*, **8**, 91
2. Müller, W. and Schmalzried, H. (1964). *Ber. Bunsenges. Phys. Chem.*, **68**, 270
3. Schmalzried, H. and Tretjakow, J. D. (1966). *Ber. Bunsenges. Phys. Chem.*, **70**, 180
4. Greskovich, C. and Schmalzried, H. (1970). *J. Phys. Chem. Solids*, **31**, 639
5. Borchardt, G. and Schmalzried, H. (1971). *Z. Phys. Chem.* (*N.F.*), **74**, 265
6. Buening, D. and Buseck, P. R. (1973). *J. Geophys. Res.* **78**, 6852
7. Duba, A. (1972). *J. Geophys. Res.*, **77**, 2483
8. Schmalzried, H. (1963). *Ber. Bunsenges. Phys. Chem.*, **67**, 93
9. Amemiya, M. (1970). *J. Inorg. Nucl. Chem.*, **32**, 2187
10. Gupta, Y. P. and Weirick, L. J. (1967). *J. Phys. Chem. Solids*, **28**, 2545
11. Rapp, R. A. and Tretyakov, Yu. D. (1969). *Trans AIME*, **245**, 1235
12. Tretyakov, Yu. D. *et al.*, (1972). *J. Solid State Chem.*, **5**, 157

2.3.5.2 Substitutional disorder: cation distributions among non-equivalent sites

The existence of substitutional disorder in spinels and other ternary oxides has been mentioned in connection with their thermodynamics of formation

(Section 2.3.3). This section will focus on the thermodynamics of the disordering process itself. The interchange of ions A and B between sites a and b can be written as:

$$A_a + B_b = A_b + B_a$$

For this reaction an equilibrium constant K may be written, with which free energy, enthalpy and entropy terns may be associated. A number of models can then be applied to calculate the thermodynamics of the disordering process from experimental data. The simplest model assumes that the free energy of disordering consists of two terms, a molar interchange enthalpy, ΔH_{int} independent of temperature, pressure, and the degree of disorder, and a purely configurational entropy term arising from the random mixing of cations on each sublattice. Thus, once the number of sites participating in the disordering process is known, the only parameter to be determined is the interchange enthalpy. For spinels, AB_2O_4, in which one tetrahedral and two octahedral sites are involved, the equilibrium constant is[96]:

$$K = \frac{x^2}{(1-x)(2-x)} = \exp^{(-\Delta H_{int}/RT)}$$

where x is the mol fraction of B ions on a(tetrahedral) sites. For pyroxenes[97], in which one each of two distinct octahedral sites are involved, and for sillimanite[73, 74], with two tetrahedral sites participating, the equilibrium constant is:

$$K = \frac{x^2}{(1-x)^2} = \exp^{(-\Delta H_{int}/RT)}$$

where x is the degree of disorder. The data used to test these equations and to derive values of ΔH_{int} are of two kinds. The first consists of experimental values of the disorder parameter, x, of samples equilibrated at, and quenched from, a known temperature or series of temperatures. These site-occupancy parameters can be determined by refinements of x-ray structure data[98] (when the two ions differ sufficiently in atomic number), by neutron diffraction[99], by Mössbauer spectroscopy[97, 98] (mainly for iron-containing samples), or by visible or infrared spectroscopy[100, 101]. The second type of experiment is calorimetric; the heat of solution in an appropriate solvent of samples equilibrated at different temperatures is determined[74, 96]. Since these samples will have different degrees of disorder, the difference in heat of solution between different samples can be related to a change in the order parameter and, using the model above or some other model, molar interchange enthalpy can be calculated. In Table 2.10 are listed some of the systems that have been studied by these methods and the molar interchange enthalpies derived for their disordering reactions.

In the spinels, disordering involves sites of different symmetry, octahedral and tetrahedral, and numerous attempts have been made to relate the interchange enthalpies and the observed cation distributions to octahedral site reference energies for individual ions. Several sets of such site preference energies are available; the crystal field stabilisation energies of Dunitz and Orgel[102] and of McClure[103] and site preference energies calculated

Table 2.10 Cation disordering reactions and molar interchange enthalpies in ternary oxides

Oxide	*Reaction*	ΔH_{int}/kcal mol^{-1}	*Comments*
SPINELS			
$MgAl_2O_4$	$Mg_A^{2+} + Al_B^{3+} \rightarrow Mg_B^{2+} + Al_A^{3+}$	+11[1], 9±2[2]	
$CoAl_2O_4$	$Co_A^{2+} + Al_B^{3+} \rightarrow Co_B^{2+} + Al_A^{3+}$	+13[1], 14±3[3]	
$MgGa_2O_4$	$Mg_A^{2+} + Ga_B^{3+} \rightarrow Mg_B^{2+} + Ga_A^{3+}$	−2.5[1]	
$MgFe_2O_4$	$Mg_A^{2+} + Fe_B^{3+} \rightarrow Mg_B^{2+} + Fe_A^{3+}$	−2.7[1], −2.5[4]	Ordering thermodynamics may be more complicated[12]
$CuAl_2O_4$	$Cu_A^{2+} + Al_B^{3+} \rightarrow Cu_B^{2+} + Al_A^{3+}$	+1[1]	Ordering thermodynamics may be more complicated[11]
$NiAl_2O_4$	$Ni_A^{2+} + Al_B^{3+} \rightarrow Ni_B^{2+} + Al_A^{3+}$	−2[1], −2.4[4]	Ordering thermodynamics may be more complicated[11]
$CuCr_2O_4$	$Cu_A^{2+} + Cr_B^{3+} \rightarrow Cu_B^{2+} + Cr_A^{3+}$	+12[1]	
$CuMn_2O_4$	$Cu_A^{2+} + Mn_B^{3+} \rightarrow Cu_B^{2+} + Mn_A^{3+}$	+4.7[1]	May be complicated by valence changes
$MnAl_2O_4$	$Mn_A^{2+} + Al_B^{3+} \rightarrow Mn_B^{2+} + Al_A^{3+}$	+10[1]	
$FeAl_2O_4$	$Fe_A^{2+} + Al_B^{3+} \rightarrow Fe_B^{2+} + Al_A^{3+}$	+9.2[1]	
$MnGa_2O_4$	$Mn_A^{2+} + Ga_B^{3+} \rightarrow Mn_B^{2+} + Ga_A^{3+}$	+5.4[1]	
$NiFe_2O_4$	$Ni_A^{2+} + Fe_B^{3+} = Ni_B^{2+} + Fe_A^{3+}$	−17[1], −17.2[4]	
	$Ni_B^{2+} + Fe_B^{3+} \rightarrow Ni_B^{3+} + Fe_B^{2+}$	*ca.* 200[8]	
$CoGa_2O_4$	$Co_A^{2+} + Ga_B^{3+} \rightarrow Co_B^{2+} + Ga_A^{3+}$	*ca.* −5[1]	
$Co_xFe_{3-x}O_4$	$Co_B^{3+} + Fe_B^{2+} \rightarrow Co_B^{2+} + Fe_B^{3+}$	−16[5]	
		$\Delta G_{1453} = -16.7$[8]	
	$Co_A^{2+} + Fe_B^{3+} \rightarrow Co_B^{2+} + Fe_A^{3+}$	−1[5], −2.6[7]	
$Mn_xCo_{3-x}O_4$	$Mn_B^{2+} + Co_B^{3+} \rightarrow Mn_B^{3+} + Co_B^{2+}$	*ca.* −10[5]	
	$Co_A^{2+} + Mn_B^{3+} \rightarrow Co_B^{2+} + Mn_A^{3+}$	*ca.* +16[15]	
$Fe_xMn_{3-x}O_4$	$Mn_B^{3+} + Fe_B^{2+} \rightarrow Mn_B^{2+} + Fe_B^{3+}$	−6.2[5], −6.9[6]	
	$Mn_B^{2+} + Fe_A^{3+} \rightarrow Mn_A^{2+} + Fe_B^{3+}$	*ca.* 0[5]	
OTHER			
$Mg_xFe_{1-x}SiO_3$	$Mg_{M1}^{2+} + Fe_{M2}^{2+} \rightarrow Mg_{M2}^{2+} + Fe_{M1}^{2+}$	+3.6[9]	
Al_2SiO_5	$Al_{Al}^{3+} + Si_{Si}^{4+} \rightarrow Al_{Si}^{3+} + Si_{Al}^{4+}$	+16[10]	

References for Table 2.10

1. Navrotsky, A. and Kleppa, O. J. (1967). *J. Inorg. Nucl. Chem.*, **29,** 2701 from cation distribution at one temperature.
2. Ref. 1, by solution calorimetry of heated and unheated natural spinels.
3. Schmalzried, H. (1961). *Z. Phys. Chem.* (*Frankfurt*), **28,** 203 cation dist. as $f(T) \rightarrow K$.
4. Ref. 1, cation dist. as $f(T) \rightarrow K$.
5. Navrotsky, A. (1969). *J. Inorg. Nucl. Chem.*, **31,** 59, from redox thermodynamics of these systems.
6. Lotgering, F. K. (1964). *J. Phys. Chem. Solids*, **25,** 95
7. Müller, W. and Schmalzried, H. (1964). *Ber. Bunsenges. Phys. Chem.*, **68,** 270, studies defect equilibria.
8. Hohmann, H. H. *et al.*, (1967). *Proc. Brit. Ceram. Soc.*, **8,** 91, studies of defect equilibria.
9. Virgo, D. and Hafner, S. (1969). *Mineral Soc. Amer. Spec. Paper*, **2,** 67, cation dist. as $f(T)$.
10. Navrotsky, A. Newton, R. C. and Kleppa, O. J. (1973). *Geochim. Cosmachim. Acta* **37,** 2497
11. Cooley, R. F. and Reed, J. S. (1972). *J. Amer. Ceram. Soc.*, **55,** 395
12. Kriessman, C. J. and Harrison, S. F. (1956). *Phys. Rev.*, **103,** 857

specifically for the spinel structure by Miller[104] and by Navrotsky and Kleppa[96]. These are shown in Figure 2.9 and show a fairly consistent pattern, although the choice of the zero of energy differs among them. Semiquantitatively, such site preference energies offer a satisfactory systemisation of cation distributions in most spinels.

When the two sites in question are more similar (such as the M_1 and M_2 sites of pyroxenes, amphiboles or olivines), the site preference among ions of similar size and charge is less pronounced, and it must be rationalised on more subtle grounds, such as the distortions and average metal–oxygen distances in each coordination polyhedron. A good deal of work is currently in progress in characterising such cation distributions, especially in complex silicates. Of particular interest is the possible use of cation distributions in selected minerals as reliable geothermometers[97].

The simple thermodynamic model outlined above is clearly a vast oversimplification, and numerous attempts, both empirical and theoretical, have been made to use equations with several parameters to fit experimental cation distribution data. Several types of modifications to the simple model have been proposed. These include (a) allowing the molar interchange enthalpy to vary linearly with the degree of disorder[105]; (b) allowing the site preference

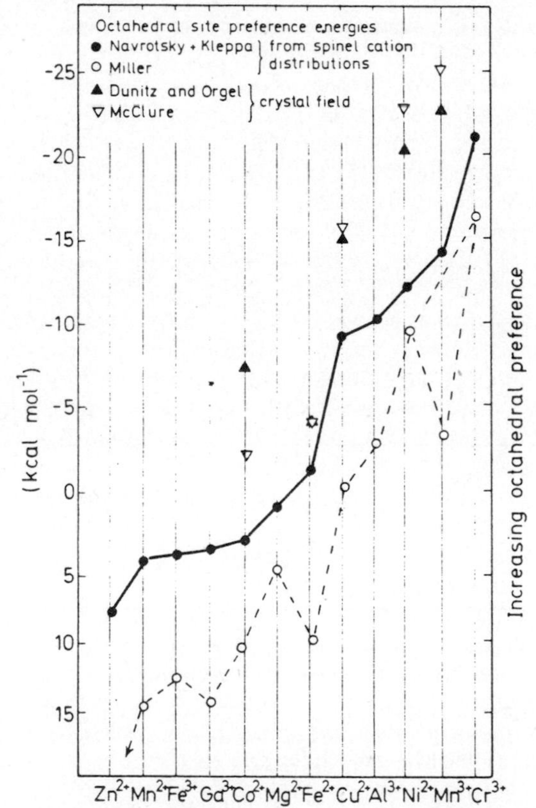

Figure 2.9 Octahedral site preference energies. (From Navrotsky and Kleppa[72], by courtesy of Pergamon Press.)

energies and therefore the interchange enthalpy to change with temperature[106]; (c) adding a non-configurational (lattice) term to the entropy of mixing[107]; and (d) taking into account non-ideal mixing on each sublattice[108]. These modifications are not conceptually independent of each other, and which approach is most fruitful is not clear at present. The major difficulty, in the reviewer's opinion, is to obtain sufficiently accurate site occupancy numbers as functions of temperature for a sufficient variety of systems to be able to rigorously test the applicability of different equations, each of which contains two to five adjustable parameters. Until more much detailed experimental data become available, the simple model, which provides one parameter, an interchange enthalpy that is generally found with a standard deviation of $\pm 10\%$, may be the most useful general first approximation.

The effect of pressure on cation distribution equilibria is generally negligible because the volume changes on disordering are of the order of 0.5% or less[74].

2.3.6 Oxide solid solutions

This section deals with the thermodynamics of formation of solid solutions of the type AO–A'O, AB_2O_4–$A'B_2O_4$, AB_2O_4–AB'_2O_4, AB_2O_4–$A'B'_2O_4$, and ABO_3–$A'BO_3$, where the end members may have the same or different structures. Generally, three phenomena can be distinguished in these solid solutions: (a) the influence on the lattice energy and on the vibrational and configurational entropies of the mixing of ions of different sizes, polarisabilities, and electronic properties; (b) possible changes in cation distributions among non-equivalent sites in the solid solutions compared to the end members; and (c) the necessary transformations in structure when the two end members are of different structure types and show mutual terminal solubilities. Solutions in which only the first effect occurs are the simplest, but even for these a totally satisfactory theoretical treatment is not available at present. The rigorous calculation of a partition function for such ionic crystals in which the nearest neighbours of a cation remain the same (oxide ions) but the next nearest neighbours change presents conceptual difficulties in terms of what is really meant by a pair interaction energy. Several theoretical approaches have met with some success, mainly in alloys and alkali halide solutions[109–112]. These each calculate some form of a strain energy, sometimes allowing for local relaxation about a given ion. To a first approximation the *positive* enthalpy of mixing should increase roughly with the square of the difference in lattice parameters or in molar volumes of end members. The simplest one-parameter fit of thermodynamic data is formally equivalent to a regular solution approximation. Clearly this model does not adequately represent activity–composition relations for oxides, but it at least allows the comparison of a characteristic parameter for a wide variety of systems. Again, as was the case for cation distribution data, the experimental activity–composition relations are seldom accurate enough to warrant fitting them to equations with several free parameters.

To illustrate the magnitude of deviations from ideality in solid solutions

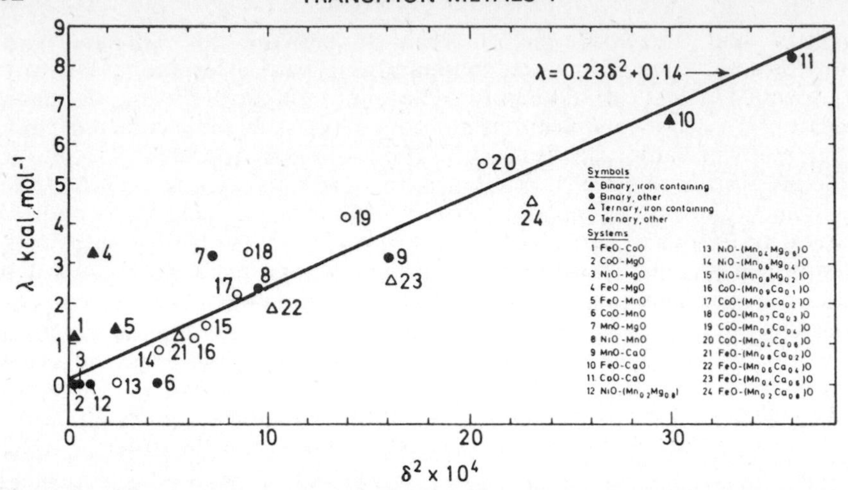

Figure 2.10 Binary oxide solid solutions with rocksalt structure—thermodynamic interaction parameter, λ, versus size parameter, δ^2, see text

with the rocksalt structure, Figure 2.10 shows the interaction parameter, λ, for a regular solution model, plotted against the square of a normalised size difference parameter, δ^2, where

$$\delta = \frac{a_{0,AO} - a_{0,A'O}}{a_{0,AO} + a_{0,A'O}}$$

where a_0 is the lattice constant.

This interaction parameter is related to the heat of mixing at a mol fraction, X, of 0.5 by the relation $\Delta H^M_{0.5} = X_{AO}X_{A'O}\lambda = 0.25\ \lambda$, and is related to the activity coefficients by the symmetric equations:

$$\ln \gamma_{AO} = \frac{\lambda}{RT} X^2_{A'O} \text{ and } \ln \gamma_{A'O} = \frac{\lambda}{RT} X^2_{AO}$$

The points on the graph refer to different temperatures and to a somewhat arbitrary fit of the regular solution model to all systems. Some data from ternary systems are also included, in which one uses the activity coefficient of a binary oxide which is being mixed with a solid solution of fixed composition. Although there is considerable scatter, the points confirm an approximately linear dependence of λ on δ^2. Several of the iron containing systems, especially FeO–MgO, lie significantly above the line. This may be ascribed to the change in defect concentrations in FeO upon solid-solution formation, but the exact nature of the energetics is not clear. The system NiO–MgO may show a small negative deviation from ideality, which cannot be explained in terms of 'strain' in the mixed crystal. Further work on such solid solutions, especially the computation of entropies of mixing by combining calorimetric enthalpies with the free energies (when the former become available) would be useful.

Ternary solid solutions in which two ions of similar size are mixed on one sublattice only usually show small deviations from ideality. Data are summarised in Table 2.11. When cation redistributions occur, as in solid solutions between spinels of different cation distributions containing ions of marked site preference, fairly large and sometimes negative deviations from ideality can occur. When, in addition, a redistribution of oxidation state occurs, as in the Co_3O_4–Mn_3O_4, Co_3O_4–Fe_3O_4, and Mn_3O_4–Fe_3O_4 series, very large negative deviations from ideality are seen. The activity–composition relations in such systems can be rather complicated, and a rigorous analysis of enthalpy terms arising from the cation redistribution and from the lattice readjustments is not feasible. Several such activity–composition relations are shown in Figure 2.11. It is important to note that these should be calculated such that one mol of ions is being mixed[78], thus the appropriate components to be used in the system Co_2TiO_4–Zn_2TiO_4 are $CoTi_{0.5}O_2$ and $ZnTi_{0.5}O_2$ and in the system Fe_3O_4–Co_3O_4 are $FeO_{1.33}$ and $CoO_{1.33}$.

When the two end-members have different structures, a temperature dependent miscibility gap between the two terminal solid solutions must exist. The activity–composition relations for such a case, Co_2GeO_4–Zn_2GeO_4, further complicated by the existence of another spinel phase near $ZnCoGeO_4$ are also shown in Figure 2.11. Table 2.11 also summarises the systems of this kind for

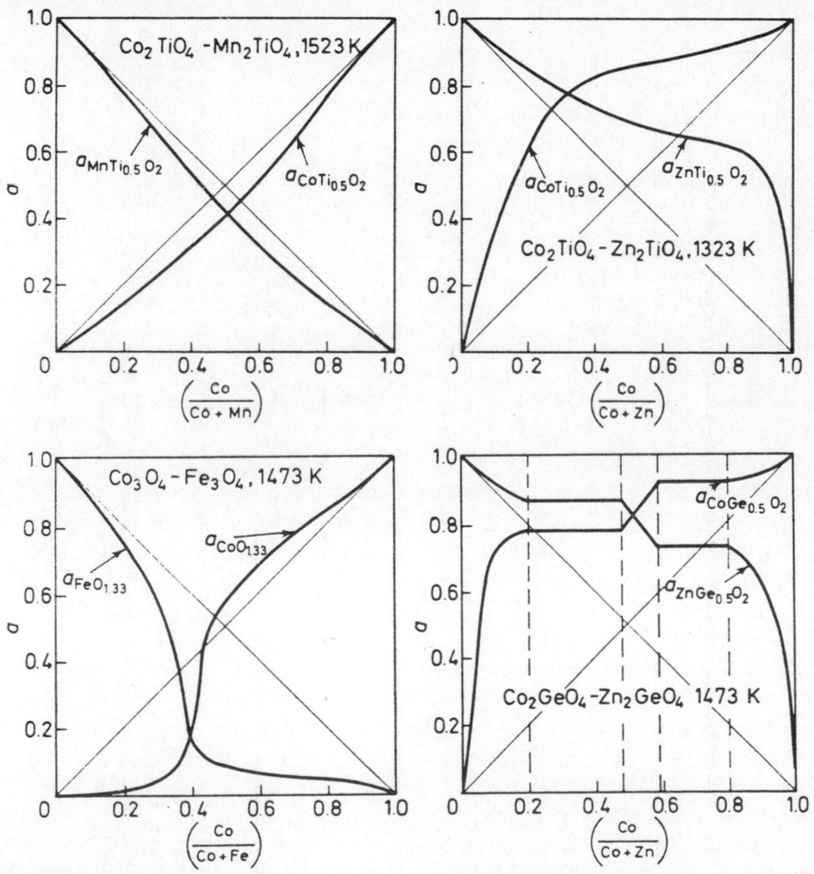

Figure 2.11 Activity–composition relatons in solid solutions of some ternary oxides. (Data from Evans and Muan[120]. Aukrust and Muan[121], Navrotsky[122], A. Navrotsky and A. Muan[115].)

Table 2.11 Thermodynamics of solid solution formation between ternary oxides†

System	*Thermodynamics*	*Comments*	*Ref.*
A–B			
	Essentially no cation ordering		
$MnTiO_3$–$CoTiO_3$	small pos. dev. at 1523 K		1
$MnTiO_3$–$NiTiO_3$	small pos. dev. at 1523 K		1
$MnTi_2O_5$–$CoTi_2O_5$	very small pos. dev. at 1523 K		1
$MgTiO_3$–$NiTiO_3$	pos. dev., $\Delta G^{ex}_{1673} = 0.67$ (0.5)*		2
$MgTi_2O_5$–$NiTi_2O_5$	ideal at 1673 K	s.s. not complete to $NiTi_2O_5$	2
Co_2SiO_4–Fe_2SiO_4	pos. dev., $\Delta G^{ex}_{1463} = 0.331$ (0.5)		3
$CoAl_2O_4$–$MgAl_2O_4$	ideal at 1673 K		4
Fe_2SiO_4–Mn_2SiO_4	ideal at 1423 K		5
Fe_2SiO_4–Mg_2SiO_4	pos. dev., $\Delta G^{ex}_{1473} = 0.59$ (0.5)		7
	pos. dev., $\Delta G^{ex}_{1477} = 0.4$ (0.5)		22, 23
	analysis of available data		27
Mn_2SiO_4–Co_2SiO_4	small neg. dev., $\Delta G^{ex}_{1523} = -0.25$ (0.5)		8
Mg_2TiO_4–Co_2TiO_4	small pos. dev., $\Delta G^{ex}_{1573} = +0.24$ (0.5)		9
$MgTiO_3$–$CoTiO_3$	very small neg. dev., $\Delta G^{ex}_{1573} = -0.13$ (0.5)		9
$MgTi_2O_5$–$CoTi_2O_5$	ideal at 1573 K		9
Mg_2TiO_4–Fe_2TiO_4	pos. dev., $\Delta G^{\circ}_{1573} = +0.30$ (0.5)	complicated by large homogeneity ranges	10
$MgTiO_3$–$FeTiO_3$	pos. dev.,	complicated by large homogeneity ranges	10
$MgTi_2O_5$–$FeTi_2O_5$	ideal at 1573 K	complicated by large homogeneity ranges	10
FeO–$Li_{0.5}Fe_{0.5}O$	neg. dev., $\Delta G^{ex}_{1273} = -0.9$ (0.5)		17
$FeAl_2O_4$–$FeCr_2O_4$	miscibility gap below *ca.* 1173 K		20

Some re-adjustments of cation distributions probable

Mn_2TiO_4–Co_2TiO_4	neg. dev., $\Delta G^{ex}_{1523} = -0.32$ (0.5)		1
Mn_2TiO_4–Ni_2TiO_4	neg. dev., $\Delta G^{ex}_{1523} = -0.60$ (0.5)	s.s. not complete to Ni_2TiO_4	1
Mg_2TiO_4–Ni_2TiO_4	ideal at 1673 K	s.s. not complete to Ni_2TiO_4	2
$FeSiO_3$–$MnSiO_3$	ideal at 1423 K	s.s. not complete to $FeSiO_3$	5
$FeSiO_3$–$CaSiO_3$	ideal at 1353 K	s.s. not complete to $FeSiO_3$	6
$FeSiO_3$–$MgSiO_3$	ideal at 1473 K	s.s. not complete to $FeSiO_3$	7
	pos. dev., $\Delta G^{ex}_{1477} = +0.3$ (0.5)	s.s. not complete to $FeSiO_3$	22, 23
$MnSiO_3$–$CoSiO_3$	ideal at 1473 K	s.s. not complete to $CoSiO_3$	8
Co_2TiO_4–Zn_2TiO_4	large pos. dev. in activities but neg. enthalpies of mixing, thus $\frac{1}{2}Co_2TiO_4 + \frac{1}{2}Zn_2TiO_4 \rightarrow CoZnTiO_4$ $\Delta G_{1323} = -1.2$ $\Delta G^{ex}_{1323} = +2.4$ $\Delta H_{970} = -2.30$	$Zn[ZnTi]O_4 \rightarrow Zn[CoTi]O_4 \rightarrow Co[CoTi]O_4$	16
Ni_2TiO_4–Zn_2TiO_4	complex dev. in activities, for $\frac{1}{2}Ni_2TiO_4 + \frac{1}{2}Zn_2TiO_4 \rightarrow NiZnTiO_4$ $\Delta G_{1323} = -3.4$ $\Delta G^{ex}_{1323} = +0.2$	$Zn[ZnTi]O_4 \rightarrow Zn[NiTi]O_4 \rightarrow$ not complete s.s. series	16
$Li_{1/2}Fe_{5/2}O_4$–Fe_3O_4	complex dev., depend on oxygen pressure		17
FeV_2O_4–Fe_3O_4	complex deviations		19
$FeCr_2O_4$–Fe_3O_4	miscibility gap below *ca.* 1173 K		20
$FeAl_2O_4$–Fe_3O_4	miscibility gap below *ca.* 1073 K		20
$CuFe_2O_4$–Mn_3O_4–Fe_3O_4	thermo. analysis along quasibinary lines		24
$MgFe_2O_4$–Fe_3O_4	pos. dev., $\Delta G^{ex}_{1133} =$ *ca.* $+0.5$ (0.5)		26
$NiFe_2O_4$–Fe_3O_4	neg. dev. 873–1173 K		30

Large re-adjustments of cation and valence distribution			
Co_3O_4–Mn_3O_4 (cubic phase)	large neg. deviations for $\frac{1}{3}Co_3O_4+\frac{2}{3}Mn_3O_4 \rightarrow CoMn_2O_4$ $\Delta G_{1473} = -9.3$, $\Delta G^{ex}_{1473} = -3.7$ for $\frac{2}{3}Co_3O_4+\frac{1}{3}Mn_3O_4 \rightarrow Co_2MnO_4$ $\Delta G_{1473} = -9.3$, $\Delta G^{ex}_{1473} = -3.72$	$Co^{2+}[Co_2^{3+}]O_4 \rightarrow Co^{2+}[Mn^{3+}Co^{3+}]O_4$ $\rightarrow Co^{2+}[Mn_2^{3+}]O_4 \rightarrow Mn^{2+}[Mn_2^{3+}]O_4$	11, 15
Co_3O_4–Fe_3O_4	complex deviations, overall negative excess free energies for $\frac{1}{3}Co_3O_4+\frac{2}{3}Fe_3O_4 \rightarrow CoFe_3O_4$ $\Delta H_{970} = -11.6$, $\Delta G_{1473} \rightarrow -13.5$, $\Delta G^{ex}_{1473} = -7.9$ for $\frac{2}{3}Co_3O_4+\frac{1}{3}Fe_3O_4 \rightarrow Co_2FeO_4$ $\Delta G_{1473} = -9.2$, $\Delta G^{ex}_{1473} \rightarrow -3.6$	$Co^{2+}[Co_2^{3+}]O_4 \rightarrow Fe^{3+}[Co^{2+}Co^{3+}]O_4$ $\rightarrow Fe^{3+}[Co^{2+}Fe^{3+}]O_4 \rightarrow Fe^{3+}[Fe^{2+}Fe^{3+}]O_4$	12, 15 18, 25
Mn_3O_4–Fe_3O_4 (cubic phase)	complex deviations, overall negative excess free energies for $\frac{1}{3}Mn_3O_4+\frac{2}{3}Fe_3O_4 \rightarrow MnFe_2O_4$ $\Delta G_{1473} = -8.0$, $\Delta G^{ex}_{1473} = -2.4$ for $\frac{2}{3}Mn_3O_4+\frac{1}{3}Fe_3O_4 \rightarrow Mn_2FeO_4$ $\Delta G_{1473} = -11.2$, $\Delta G^{ex}_{1473} = -5.6$	$Mn^{2+}[Mn_2^{3+}]O_4 \rightarrow Mn^{2+}_{1/2}Fe^{3+}_{1/2}[Mn^{2+}_{1/2}Fe^{3+}_{1/2}Mn^{3+}]O_4$ $\rightarrow Mn^{2+}_{1/3}Fe^{3+}_{2/3}[Mn^{2+}_{2/3}Fe^{3+}_{4/3}]O_4$ $\rightarrow Fe^{3+}[Fe^{2+}Fe^{3+}]O_4$	13–15
Terminal solid solutions only			
Fe_2SiO_4–Ca_2SiO_4	complex pos. deviations at ends neg. in middle for $\frac{1}{2}Ca_2SiO_4+\frac{1}{2}Fe_2SiO_4 \rightarrow FeCaSiO_4$ $\Delta G = -0.3$ kcal	miscibility gap at Ca_2SiO_4 rich end	6
Be_2SiO_4–Zn_2SiO_4	limited solubility		21
Be_2GeO_4–Zn_2GeO_4	limited solubility		21
Co_2GeO_4–Mg_2GeO_4	activity–comp. rel. 1473 K		28
Ni_2GeO_4–Mg_2GeO_4	activity–comp. rel. 1473 K		28
Co_2GeO_4–Zn_2GeO_4	activity–comp. rel. 1323 K	second spinel phase near $CoZnGeO_4$	29
Ni_2GeO_4–Zn_2GeO_4	activity–comp. rel. 1323 K	second spinel phase near $NiZnGeO_4$	29

* (0.5) = at mol fraction of 0.5, ΔG^{ex} in kcal mol^{-1}

† References for Table 2.11 on p. 67

which activity–composition relations have been measured. These relations permit estimates to be made of the free energies of structural transformations in the two end members[82]

2.3.7 Ternary oxides of variable stoichiometry

Thermodynamic data for such materials are almost completely lacking. Structurally, several classes of compounds can be recognised (a) shear structures or Magnéli phases in ternary systems such as VO_2–WO_3 and WO_3–Nb_2O_5; (b) defect spinels and other defect structures with partially vacant cation sublattices; (c) the tungsten, molybdenum and vanadium bronzes, and (d) oxyfluorides in which the substitution of F^- for O^{2-} forces the adjustment of cation valence and/or composition. Some systems which have been partially characterised by phase equilibrium studies are listed in Table 2.12, together with any available indication of their thermodynamic stability. Clearly this remains a largely unexplored area, especially in terms of thermodynamics.

References for Table 2.11

1. Evans, L. G. and Muan, A. (1971). *Thermochim. Acta*, **2**, 277
2. Evans, L. G. and Muan, A. (1971). *Thermochim. Acta*, **2**, 121
3. Masse, D. P. Rosén, E. and Muan, A. (1966). *J. Amer. Ceram. Soc.*, **49**, 328
4. Rosén, E. and Muan, A. (1966). *J. Amer. Ceram. Soc.*, **49**, 107
5. Schwertfeger, K. and Muan, A. (1964). *Trans. AIME*, **236**, 201
6. Johnson, R. E. and Muan, A. (1967). *Trans. AIME*, **239**, 1931
7. Nafziger, R. and Muan, A. (1967). *Amer. Mineral.*, **52**, 1364
8. Beggers, J. V. and Muan, A. (1967). *J. Amer. Ceram. Soc.*, **50**, 230
9. Brezny, B. and Muan, A. (1971). *Thermochim. Acta*, **2**, 107
10. Johnson, R. E. Woermann, E. and Muan, A. (1971). *Amer. J. Sci.*, **271**, 278
11. Aukrust, E. and Muan, A. (1964). *Trans. AIME*, **230**, 378
12. Aukrust, E. and Muan, A. (1964). *Trans. AIME*, **230**, 1395
13. Schwerdtfeger, K. and Muan, A. (1967). *Trans. AIME*, **239**, 1119
14. Tretyakov, Yu. D. (1964). *Zh. Neorg. Khim.*, **9**, 161
15. Navrotsky, A. (1969). *J. Inorg. Nucl. Chem.*, **31**, 59
16. Navrotsky, A. and Muan, A. (1970). *J. Inorg. Nucl. Chem.*, **32**, 3471
17. Tretyakov, Yu. D. *et al.*, (1972). *J. Solid State Chem.*, **5**, 191
18. Carter, R. E. (1960). *J. Amer. Ceram. Soc.*, **43**, 448
19. Vorobyev, Yu. P. *et al.*, (1968). *Dokl. Akad. Nauk SSSR-Khim.*, **54**, 3
20. Cremer, V. (1969). *N. Jb. Miner Abh.*, 184
21. Hahn, Th. and Eysel, W. (1970). *N. Jb. Miner Abh.*, 263
22. Kitayama, K. and Katsura, T. (1968). *Bull. Chem. Soc. Jap.*, **41**, 1146
23. Kitayama, K. (1970). *Bull. Chem. Soc. Jap.*, **43**, 1390
24. Schepetkin, A. A. *et al.*, (1971). *Russ. J. Phys. Chem.*, **45**, 929
25. Tretyakov, Yu. D. and Khomyakov, G. (1963). *Russ. J. Inorg. Chem.*, **8**, 1345
26. Gordeev, I. V. and Tretyakov, Yu. D. (1963). *Russ. J. Inorg. Chem.*, **8**, 944
27. Williams, R. J. (1972). *Earth Planet. Science Lett.*, **15**, 296
28. Navrotsky, A. (1973). *J. Solid State Chem.*, **6**, 21
29. Navrotsky, A. (1973). *J. Solid State Chem.*, **6**, 42
30. Popov, G. P. and Chufarov, G. I. (1963). *Russ. J. Phys. Chem.*, **37**, 301

Table 2.12 Ternary oxides of variable stoichiometry

System	*Characteristics*	*Thermodynamics*	*Ref.*
Li–Mn–O	Oxygen deficient spinels		1
Al_2O_3–ZnO	Non-stoichiometric spinels	Metastable?	2
Mo–W–O and many others	Magneli phases		3
various	Non-stoichiometric ternary oxides with ordered vacancies, especially spinels	Metastable	4
ZrO_2–MO_n	Extensive solid solutions with defective anion lattices		5
various, e.g. Na_xWO_3, $NaV_2O_{5-x}F_x$, Na_x–MnO_2	Bronzes—metallic oxides with TM in several oxidation states, stable over a range of compositions	Oxygen pressures for various assemblages given	6–8
		Enthalpies of formation given	9, 10

References for Table 2.12

1. Bergstein, (1965). *J. Phys. Chem. Solids*, **26**, 1181
2. Colin and Thery, (1965); *C. R. Acad. Sci. Paris*, **261**, 3141, (1965). **261**, 3826
3. Ekstrom, (1972). *Mater. Res. Bull.*, **7**, 19
4. Joubert, *et al.*, (1970). *'Problems of Nonstoichiometry'*, (A. Rabenau, editor), (Amsterdam: North Holland)
5. Leonov, (1971). *Russ. J. Inorg. Chem.*, **16**, 755
6. Volkov *et al.*, (1970). *Russ. J. Phys. Chem.*, **44**, 340
7. Volkov *et al.*, (1970). *Russ. J. Phys. Chem.*, **44**, 1163
8. Hagenmuller, (1971). *Progr. Solid State Chem.*, **5**, 71
9. Dickens *et al.*, (1973). *J. C. S., Dalton Trans.*, **1**, 30
10. Dickens and Neild, (1973). *J. Solid State Chem.*, **7**, 474

References

1. Stull, D. R. and Prophet, H. (1971). *JANAF Thermochemical Tables*, 2nd edn., National Standard Data Reference Service, U.S. Govt. Printing Office
2. Robie, R. A. and Waldbaum, D. R. (1968). *U.S. Geol. Survey Bull.*, 1259
3. Elliot, J. F. and Gleiser, M. (1960). *'Thermochemistry for Steelmaking,'* (Reading, Mass: Addison Wesley)
4. Kelley, K. K. (1960). *U.S. Bur. Mines Bull.* 584
5. Coughlin, J. P. (1954). *U.S. Bur. Min. Bull.* 542
6. Kubaschewski, O., Evans, E. and Alcock, E. B. (1967). *Metallurgical Thermochemistry*, 3rd edn. (Oxford: Pergamon Press)
7. Morin, F. J. (1959). *Phys. Rev. Letts.*, **3**, 34
8. Goodenough, J. B. (1971). *J. Solid State Chem.*, **3**, 490
9. Navrotsky, A. and Kleppa, O. J. (1967). *J. Amer. Ceram. Soc.*, **50**, 626
10. Navrotsky, A. Jamieson, J. C. and Kleppa, O. J. (1967). *Science*, **158**, 3799
11. McQueen, R. G. Jamieson, J. C. and Marsh, S. P. (1967). *Science*, **155**, 1404
12. Tilley, R. J. D. MTP Inorganic Chemistry, Series 2, Volume 10 in preparation
13. Magnéli, A. (1970). *The Chemistry on Extended Defects in Nonmetallic Solids*, (L. Eyring and M. O'Keeffe, editors) pp. 148–163. (Amsterdam: North Holland)
14. Wadsley, D. (1963). *Nonstoichiometric Compounds*, (R. Ward, editor), (Washington D.C: American Chemical Society)
15. Bursill, L. A. and Hyde, B. G. (1971). *Phil. Mag.*, **23**, 3
16. Gibb, R. M. and Anderson, J. S. (1972). *J. Solid State Chem.*, **4**, 379
17. Bursill, L. A. Hyde, B. G. and Philp, K. D. (1971). *Phil. Mag.*, **23**, 1501
18. Bursill, L. A. and Hyde, B. G. (1972). *J. Solid State Chem.*, **4**, 430
19. Anderson, J. S. Browne, J. M. and Hutchinson, J. L. (1972). *J. Solid State Chem.*, **4**, 419
20. Gibb, R. M. and Anderson, J. S. (1972). *J. Solid State Chem.*, **5**, 212
21. Anderson, J. S. and Burch, R. (1971). *J. Phys. Chem. Solids*, **32**, 923

22. Anderson, J. S. (1969). *Problems in Non-stoichiometry*, (H. Rabenau, editor) (Amsterdam: North Holland)
23. Anderson, J. S. and Khan, A. S. (1970). *J. Less Common Met.*, **22**, 219
24. Anderson, J. S. and Khan, A. S. (1970). *J. Less. Common Met.*, **22**, 209
25. Wakihara, M. and Katsura, T. (1970). *Met. Trans.*, **1**, 363
26. St. Pierre, G. R. Ebihara, W. T., Pool, M. J. and Speiser, R. (1962). *Trans. AIME*, **224**, 259
27. Bosquet, J. and Perachon, G. (1963). *C. R. Acad. Sci. (Paris)*, **256**, 694
28. Charlu, T. V. and Kleppa, O. J. (1973). *High Temp. Science*, **5**, 260
29. Charlu, T. V. and Kleppa, O. J. (1973). *J. Chem. Thermodynamics*, **5**, 325
30. Eyring, L. *Lanthanide and Actinide Oxides. A Case History in Solid State Chemistry*, (C. N. R. Rao, editor) (New York: Marcel Dekker Publishers)
31. Katsura, T. Iwasaki, B. Kimura, S. and Akimoto, S. (1967). *J. Chem. Phys.*, **47**, 4559
32. Taylor, A. and Doyle, N. J. (1969). *High Temp. High Press.*, **1**, 679
33. Libowitz, G. (1969). *Energetics in Metallurgical Phenomena*, Vol. 4, (W. M. Mueller, editor) (New York: Gordon and Breach).
34. Marucco, J. Picard, C. Gerdanian, P. and Dodé, M. (1970). *J. Chim. Phys.*, **67**, 914
35. Asao, H. Ono, K. Yamaguchi, Y. and Moriyama, J. (1970). *Mem. Fac. Eng. Kyoto Univ.*, **32**, 66
36. Rizzo, H. F. Gordon, R. S. and Cutler, I. B. (1968). *U.S. National Bureau of Standards Special Publication* 296, p. 129
37. Fender, B. E. F. and Riley, F. D. (1969). *J. Phys. Chem. Sol.*, **30**, 793
38. Fender, B. E. F. and Riley, F. D. Ref. 13, p. 54
39. Marucco, J. Gerdanian, P. and Dodé, M. (1970). *J. Chim. Phys.*, **67**, 906
40. Raccah, P. and Vallet, P. (1965). *Mem. Sci. Rev. Met.*, **62**, 1
41. Roth, W. L. (1960). *Acta Crystallogr.*, **13**, 140
42. Koch, F. and Cohen, J. B. (1969). *Acta Crystallogr.*, **B25**, 275
43. Childs, P. E. (1967). *Ph.D. Thesis*, University of Oxford
44. Bokshtein, B. S. Zhukhovitskii, A. A. Kozheurov, V. A. Lykasov, A. A. and Nikolskii, G. S. (1972). *Russ. J. Phys. Chem.*, **46**, 507
45. Banus, M. D. and Reed, T. B. (1972). *Phys. Rev. B*, **5**, 2775
46. Watanabe, D. Terasaki, O. Jostsons, A. and Castles, J. R. Ref. 13, p. 238
47. Ariya, S. M. Morozova, M. P. and Vol'f, E. (1957). *Russ. J. Inorg. Chem.*, **2**, 16
48. Gilles, P. Ref. 13, p. 75
49. Hoch, M. Iyer, A. S. and Nelken, J. (1962). *J. Phys. Chem. Solids*, **23**, 1463
50. Lochman, R. E. Rao, C. N. R. and Honig, J. M. (1969). *J. Phys. Chem.*, **73**, 1781
51. Brauer, G. Reuther, H. Walz, H. and Zapp, K. H. (1969). *Z. Anorg. Allgem. Chem.*, **369**, 144
52. Libowitz, G. G. (1967). *J. Phys. Chem. Solids*, **28**, 1145
53. Alex, K. and McLellan, R. (1971). *J. Phys. Chem. Solids*, **32**, 449
54. Atlas, L. (1968). *J. Phys. Chem. Solids*, **29**, 91
55. Kröger, F. A. (1964). *The Chemistry of Imperfect Crystals*, (Amsterdam: North Holland)
56. Kofstad, P. (1972). *Nonstoichiometry, Diffusion, and Electrical Conductivity in Binary Metal Oxides*, (New York: Wiley)
57. Kröger, F. A. and Vink, H. S. (1956). *Solid State Physics*, **3**, 307
58. Sockel, H. G. and Schmalzried, H. (1962). *Ber. Bunsenges. Phys. Chem.*, **72**, 745
59. Roy, R. (1963). *Physics and Chemistry of Ceramics*, (C. Klingsberg, editor) (New York: Gordon and Breach)
60. Gattow, G. (1964). *Anorg. Z. Allgem. Chem.*, **333**, 134
61. Ladd, M. F. C. and Lee, W. H. (1964). *Progr. Solid State Chem.*, **1**, 37
62. Lee, W. H. and Ladd, M. F. C. (1967). *Progr. Solid State Chem.*, **3**, 265
63. Shannon, R. D. and Prewitt, C. T. (1969). *Acta Crystallogr.*, **B25**, 925
64. Kleppa, O. J. (1967). *Proc. Brit. Ceram. Soc.*, **8**, 31
65. Muan, A. (1967). *Amer. Min.*, **52**, 797
66. Tretjakow, J. D. and Schmalzried, H. (1965). *Ber. Bunsen. Phys. Chem.*, **69**, 396
67. Ringwood, A. E. (1956). *Geochim. Cosmochim. Acta*, **10**, 297
68. Phillips, B., Hutta, J. J. and Warshaw, I. (1963). *J. Amer. Ceram. Soc.*, **46**, 579
69. Navrotsky, A. *J. Solid State Chem.* (in the press)
70. Neil, J. M. Navrotsky, A. and Kleppa, O. J. (1971). *Inorg. Chem.*, **10**, 2076
71. Navrotsky, A. and Kleppa, O. J. (1969). *Inorg. Chem.*, **8**, 756
72. Navrotsky, A. and Kleppa, O. J. (1968). *J. Inorg. Nucl. Chem.*, **30**, 479
73. Holdaway, M. J. (1971). *Amer. J. Sci.*, **271**, 97

74. Navrotsky, A., Newton, R. C. and Kleppa, O. (1973). *J. Geochim. Cosmochim. Acta*, **37**, 2497
75. Holm, J. L. and Kleppa, O. J. (1966). *Acta Chem. Scand.*, **20**, 2568
76. Holm, J. L. and Kleppa, O. J. (1966). *Amer. Min.*, **51**, 1608
77. Haggerty, S. E. and Lindsley, D. H. (1969). *Carnegie Inst. Year Book*, **68**, 247
78. Schwertfeger, K. and Muan, A. (1966). *Trans AIME*, **236**, 201
79. Brézny, B. and Muan, A. (1969). *J. Inorg. Nucl. Chem.*, **31**, 649
80. Boyer, G. (1971). *J. Less Common Met.*, **24**, 129
81. Ringwood, A. E. (1970). *Phys. Earth Planet. Interiors*, **3**, 109
82. Navrotsky, A. (1973). *J. Solid State Chem.*, **6**, 21
83. Navrotsky, A. in press.
84. Syono, Y. Tokonami, M. and Matsui, Y. (1971). *Phys. Earth Planet. Int.*, **4**, 347
85. Akimoto, S. Fujisawa, H. and Katsura, T. (1965). *J. Geophys. Res.*, **70**, 1969
86. Ma, C. B. (1972). *Ph.D. Thesis*, Geology, Harvard University
87. Syono, Y. Akimoto, S. and Matsui, Y. (1971). *J. Solid State Chem.*, **3**, 369
88. Kamb, B. (1968). *Amer. Min.*, **53**, 1439
89. Reid, A. F. and Ringwood, A. E. (1970). *J. Solid State Chem.*, **1**, 557
90. Drickamer, H. G. *et al.*, (1970). *J. Solid State Chem.*, **2**, 94
91. Burns, R. G. Huggins, F. E. and Drickamer, H. G. (1972). *Proc. 24th Int. Geol. Congr.*, Sect. 14, 113
92. Greskovich, C. and Schmalzried, H. (1970). *J. Phys. Chem. Solids*, **31**, 639
93. Zintl, G. (1966). *Z. Phys. Chem.*, **48**, 340
94. Buseck, P. and Buening, O. (1973). *J. Geophys. Res.*, 78, 6852
95. Schmalzried, H. *Progr. Solid State Chem.*, **2**,
96. Navrotsky, A. and Kleppa, O. J. (1967). *J. Inorg. Nucl. Chem.*, **29**, 2701
97. Hafner, S. and Virgo, D. (1969). *Miner. Soc. Amer. Spec. Paper*, **2**, 67
98. Burnham, C. W. Ohashi, Y., Hafner, S. S. and Virgo, D. (1971). *Amer. Mineral.*, **56**, 850
99. Fischer, P. (1967). *Zeit. J. Kristall.*, **124**, 275
100. Reinen, D. (1968). *Z. Anorg. Allgem. Chem.*, **356**, 172
101. Burns, R. G. and Huggins, F. E. (1972). *Amer. Mineral.*, **57**, 967
102. Dunitz, J. D. and Orgel, L. E. (1957). *J. Phys. Chem. Solids*, **3**, 318
103. McClure, D. S. (1957). *J. Phys. Chem. Solids*, **3**, 311
104. Miller, A. (1959). *J. Appl. Phys.*, **30**, 248
105. Callen, H. B. Harrison, S. E. and Kriessman, C. J. (1956). *Phys. Rev.*, **103**, 851
106. Datta, R. K. and Roy, R. (1965). *Z: Kristall*, **121**, 410
107. Cooley, R. F. and Reed, J. S. (1972). *J. Amer. Ceram. Soc.*, **55**, 395
108. Grover, J. E. and Orville, P. M. (1969). *Geochim. Cosmochim. Acta*, **33**, 205
109. Fancher, D. L. and Barsch, G. B. (1969). *J. Phys. Chem. Solids*, **30**, 2503
110. Wasastjerna, J. A. (1948). *Soc. Scient. Fennicae, Comment. Phys. Math.*, **14**, 5
111. Lawson, A. W. (1947). *J. Chem. Phys.*, **15**, 831
112. Driessens, F. C. M. (1968). *Ber. Bunsenges. Phys. Chem.*, **72**, 764
113. Bransky, I. and Hed, A. Z. (1964). *J. Amer. Ceram. Soc.*, **51**, 231
114. Schultz, A. H. and Stubican, V. S. (1970). *J. Amer. Ceram. Soc.*, **53**, 613
115. Navrotsky, A. and Muan, A. (1971). *J. Inorg. Nucl. Chem.*, **33**, 35
116. Reid, A. F. and Ringwood, A. E. (1970). *J. Solid State Chem.*, **1**, 557
117. Van Landuyt, I. and Amelinckx, S. (1973). *J. Solid State Chem.*, **6**, 222
118. Kimura, S. (1973). *J. Solid State Chem.*, **6**, 438
119. Grey, I. E. and Ward, J. (1973). *J. Solid State Chem.*, **7**, 300
120. Evans, G. and Muan, A. (1971). *Thermochem. Acta*, **2**, 277
121. Aukrust, E. and Muan, A. (1964). *Trans AIME*, **230**, 1395
122. Navrotsky, J. (1973). *J. Solid State Chem.*, **6**, 42

3
Transition Metal Oxyanions

A. MÜLLER AND E. DIEMANN
University of Dortmund

3.1 INTRODUCTION

This chapter is concerned with recent developments in the chemistry of transition metal ions involving the oxo(oxide) ligand, O^{2-}, and covers most of the literature published between 1968 and 1972. In order to keep this review at a reasonable length the coverage has arbitrarily been limited to compounds which contain discrete molecular units of the type $[MO_xY_y]_z^{n-}$ with $Z \leqslant 2$ and $n \neq 0$. This will exclude a large number of oxo-compounds, i.e. almost all uncharged metal oxocompounds, most isopoly- and hetero-poly-species and ternary oxides which contain polymeric units.

Some older reviews on this topic should be listed here. Literature surveys of orthoxo salts[4], group V and VI isopolyanions[2,3], binuclear complexes involving oxo bridges[4], peroxo species[5], oxohalides[6], oxides[7,8], alkoxides[9] and oxocations[10] have been published. Carrington and Symons[11] have reviewed the structures and reactivities of tetraoxoanions and recently Griffith[12] has published a general survey of transition metal oxo-complexes, which covers the most important literature up to 1968. Müller *et al.* have reviewed the chemistry[13], structural chemistry[14] vibrational spectra[15] and electronic structures[16] of sulphur and selenium substituted transition metal oxo-compounds.

This subject is divided in two parts. The first reports in a general survey physical measurements of transition metal oxo-complexes and some new aspects of their bonding. The second part contains the more descriptive chemistry, such as new preparations, and also includes some papers published on classical physico-chemical measurements, such as thermochemistry, kinetics and electrochemistry.

3.2. GENERAL SURVEY

3.2.1 Vibrational spectra and normal coordinate analysis

3.2.1.1 Tetraoxo species

Many transition metal oxides and oxometallates of the type MX_4 (T_d) present difficulties for the assignments of their vibrational spectra[17–22]. Even in a new and very useful book on vibrational spectra of inorganic compounds the correct vibrational frequencies for several oxoanions such as VO_4^{3-} or MoO_4^{2-} have not been reported[23]. Only three lines appear in the Raman spectra of the aqueous solutions of VO_4^{3-}, MoO_4^{2-}, WO_4^{2-}, ReO_4^- and OsO_4, ν_1 (A_1) being observed at higher wave numbers than ν_3 (F_2) (but for CrO_4^{2-}, MnO_4^- and RuO_4 the ratio ν_1/ν_3 is less than 1). An explanation for the differing ratios ν_1/ν_3 has recently been given[25–27, 549]. One of the present authors and his co-workers have developed a method to assign the frequencies ν_2 (E) and ν_4 (F_2) in tetrahedral molecules and ions[28]. The method employs the calculation of the relative Raman intensities. Two different conclusions with respect to the assignment of ν_2 and ν_4 were made in earlier papers. Whereas some authors suppose that ν_2 cannot be observed due to its low intensity others pointed out that both vibrational frequencies lie so close together, that they cannot be resolved (for references see Ref. 28).

The calculations show that ν_2 should have a higher intensity than ν_4. With the help of the L-matrix approximation method of Müller[29, 30] the intensitv ratio[28] is,

$$\frac{I\nu_2}{I\nu_4}=\frac{2(\mu_x+\frac{4}{3}\mu_M)}{2\mu_x+4\mu_M}$$

The frequencies and relative Raman intensities of the fundamentals of oxoanions containing closed shells have been summarised in Tables 3.1a[28] and Table 3.1b where the frequencies of other anions, which have mainly been measured in solids, are given.

Table 3.1a Vibrational frequencies and relative Raman intensities[28] of oxoions with T_d symmetry (in cm^{-1})

	Raman				*Infrared*		I_{ν_2}/I_{ν_4}	
	$\nu_1(A_1)$	$\nu_3(F_2)$	$\nu_2(E)$	$\nu_4(F_2)$	$\nu_4(F_2)$	*Ref.*	*calc.*	*exp.*
VO_4^{3-}	826	804	336‡		—	28	1.26	—
CrO_4^{2-}	846	890	349,	378	—	28	1.26	1.4
MoO_4^{2-}	897	837	317‡		325	28	1.47	—
WO_4^{2-}	931	838	325‡		—	28	1.66	—
MnO_4^-	839	914	360,	430	—	28	1.28	1
TcO_4^-	912	912	325,	336	334	28	1.48	1.6
ReO_4^-	971	920	332†		322	28	1.66	—
RuO_4	882	914	323,	334	330	*	1.48	1.4
OsO_4	965	960	333,	323	329	†	1.66	1.5

* Dodd, R. E. (1959). *Trans. Faraday Soc.* **55**, 1480 and Ref. 41
† Ref. 39 and: Huston, J. L. and Claassen, H. H. (1970). *J. Chem. Phys.* **52**, 5646
‡ ν_2 and ν_4 coincide

In an earlier paper the positions of ν_2 (E) and ν_4 (F_2) in the isoelectronic ions VO_4^{3-}, CrO_4^{2-}, MnO_4^- have been discussed empirically[112]. Recently two groups have also measured the Raman spectra of CrO_4^{2-} ions in molten fluorides[30] and aqueous solution[31] (the data[31] are not completely correct). One of the present authors and co-workers has measured the spectra of $^{92}MoO_4^{2-}$, $^{100}MoO_4^{2-}$, $^{50}CrO_4^{2-}$, $^{53}CrO_4^{2-}$, $Cr^{18}O_4^{2-}$ and $Re^{18}O_4^-$ [32–34].

Pinchas has studied the vibrational spectra of ^{18}O-labelled vanadates and permanganates in solids and in aqueous solution and has discussed the isotope effect (^{16}O—^{18}O) on hydrogen bonding[35, 36] (see also Ref. 37). The

Table 3.1b Vibrational frequencies* of oxoions with T_d symmetry (in cm^{-1})[599]

	$\nu_1(A_1)$	$\nu_3(F_2)$	$\nu_2(E)$	$\nu_4(F_2)$
$KRuO_4$	830	845	339	312
MnO_4^{2-}	812	820	325	332
FeO_4^{2-}	832	790	340	322
RuO_4^{2-}	840	804	331	336
CrO_4^{3-}	834	860	260	324
K_3MnO_4	810	838	324	349
Li_3ReO_4	808	853	264	319
K_3FeO_4	776	805	265	335
Ba_2TiO_4	761	770	306	371
Li_4ZrO_4	792	846	332	387
Li_4HfO_4	796	800	325	379
Ba_2VO_4	818	780	319	368
Ba_2CrO_4	806	855	353	404
Ba_2MoO_4	792	808	328	373
Ba_2WO_4	821	840	323	367
Ba_2FeO_4	762	857	257	314
Ba_2CoO_4	790	855	300	340

* The data and assignments for the species given in this table are not as sure as for Table 3.1a. (The force constants f_r of HfO_4^{4-} should for instance be larger than for ZrO_4^{4-}, see Table 3.4)

vibrational spectra of RuO_4 and OsO_4 (using ^{18}O labelled compounds[38, 39]) have been measured[38–42] (wrong results in Ref. 42). Raman studies on species in aqueous solution (vanadates, molybdates and tungstates[43]) have been made by Griffith in a series of papers (for earlier investigations see Ref. 43). The laser Raman spectrum (He–Ne excitation) of the MnO_4^- ion has been measured[44]. The results agree with an earlier study[45].

The resonance Raman spectra of MnO_4^- and CrO_4^{2-} have been scanned[46–48]. Four overtones of the ν_1 (A_1) fundamental of MnO_4^- could be observed[47]. The dissociation energy of MnO_4^- into Mn + 4O was estimated[47]. From the Raman intensity of ν_1 (A_1) of different oxoanions approximate bond orders were estimated[49]. The i.r. reflection spectra of fused salts A_2^I MO_4 (A = Li, Na, K, Rb, Cs; M = Cr, Mo, W) have been measured[50].

3.2.1.2 *Oxospecies with C_{3v} and C_{2v} symmetry*

In this chapter the results of studies of those species where one or more oxygen atoms have been replaced by other atoms such as X = F, Cl, Br, S, Se and N will be given (type MO_xX_{4-x} (M = Cr, Mo, W, Mn, Tc, Re, Os))[51–57].

Müller *et al.* have measured the spectra of the mixed chalcogeno anions[43], CrO_3Br^- [57] and of ReO_3N^{2-} [58]. The problem of assigning the ν (M–halogen) stretching frequency and $\delta(MO_3)$ vibration has been discussed for compounds of the type MO_3X^{n-} (M = Cr, Mn, Tc, Re; X = F, Cl, Br, S) by the same group in detail[57]. For the ions $MoOS_3^{2-}$ and WOS_3^{2-} a similar problem occurs as in the case of the tetraoxoanions, as only one Raman line is observed in aqueous solution instead of the expected two for the vibrations $\delta_s(MS_3) \hat{=} \nu_3(A_1)$ and $\delta_{as}(MS_3) \hat{=} \nu_6(E)$[56]. The results are summarised in Tables 3.2 and 3.3. The related compounds ($O_3ReOReO_3$ and $(HO)ReO_3$[59]) and vibrational spectra of the ions $^{92}MoOS_3^{2-}$, $^{100}MoOS_3^{2-}$ [60], $^{92}MoO_2S_2^{2-}$ and $^{100}MoO_2S_2^{2-}$ [61] have been recorded.

Table 3.2 Vibrational frequencies of oxoions with C_{3v} symmetry (in cm^{-1})

		$\nu_1(A_1)$ $\nu_s(MX)$	$\nu_2(A_1)$ $\nu_s(MY)$	$\nu_3(A_1)$ $\delta_s(MY_3)$	$\nu_4(E)$ $\nu_{as}(MY)$	$\nu_5(E)$ ρ_r	$\nu_6(E)$ $\delta_{as}(MY_3)$	*Ref.*
$SMoO_3^{2-}$	K-salt; i.r., Ra solid	475	882	331	833	239	314	cf. 13
SWO_3^{2-}	K-salt; i.r., Ra solid	461	914	333	855	239	317.5	cf. 13
$OMoS_3^{2-}$	K-salt; i.r., DMSO-soln. Ra, aq.soln.	862	461	183	470	263	183	55, 56
OWS_3^{2-}	K-salt; i.r., DMSO-soln. Ra, aq.soln.	878	474	182	451	264	182	55, 56
$OMoSe_3^{2-}$	Cs-salt; i.r., Ra solid	858	293	120	355	188	120	55
$OWSe_3^{2-}$	Cs-salt; i.r., Ra solid	878	292	(120)	312	194	(120)	55
$SMoSe_3^{2-}$	Cs-salt; i.r.	471	276	121	342	130	121	cf. 13, 15
$SWSe_3^{2-}$	Cs-salt; i.r.	471	284	108	310		108	cf. 13
$SeMoS_3^{2-}$	Cs-salt; i.r. solid	345	457		469			cf. 13
$SReO_3^-$	K-salt; Ra, aq.soln.	528	948	322	906	(240)*	322	cf. 57
$FCrO_3^-$	K-salt; i.r. solid	635	912	338	952	257	370	cf. 57
$ClCrO_3^-$	Ra, soln.	438	907	295	954	209	365	cf. 57
$BrCrO_3^-$	Cs-salt; i.r., Ra, solid	395	906	242	948	200	364	57
$NOsO_3^-$	Ra, aq.soln.	1021	897	309	871	309	372	cf. 22
$NReO_3^{2-}$	K-salt; i.r. solid	1022	878	315	830	273	380	58

* Tl-salt

3.2.1.3 Normal coordinate analysis

For the calculation of diagonal force constants for the ions mentioned in the last two sections approximation methods such as the Fadini[22], the L matrix[29, 30] and the Extended L-Matrix Methods[62] can be used because the mass coupling is relatively small[29] (the elements G_{ij} $(i \neq j)$ are in general small). This is especially valid for the ions with T_d symmetry, where the frequencies of the two F_2 fundamentals differ approximately by a factor of two or more.

Some more rigorous calculations of exact force constants using, e.g. isotope shifts or especially Coriolis coupling constants have been published (CrO_4^{2-}, MoO_4^{2-}, MoS_4^{2-} [32–34], RuO_4 [38, 42, 63, 64] and OsO_4 [39, 65]). The reported results[42] for RuO_4 are incorrect (harmonic force constants for RuO_4 and OsO_4 are given in Refs. 38, 39).

Force constant calculations for species with T_d symmetry using approximation methods and physical models (U.B.F.F., O.V.F.F.) have been reported[66–70] for several oxoanions. (for mean amplitudes of vibration see Ref. 71–74). According to the present authors the U.B.F.F. and O.V.F.F. models do not give physically reasonable values for the stretching force constant K or the symmetry force constants for the tetrahedral oxoanions, though in isoelectronic series such as VO_4^{3-}, CrO_4^{2-} and MnO_4^- the repulsion between the non-bonded atoms can be explained[75, 76]. This repulsion cannot be interpreted by van der Waals forces alone as has been assumed earlier. The force constants of the tetrahedral oxyanions are summarised in Table 3.4.

Table 3.3 Vibrational frequencies of oxoions with C_{2v} symmetry (in cm^{-1})

		$\nu_1(A_1)$ $\nu_s(MX)$	$\nu_2(A_1)$ $\nu_s(MY)$	$\nu_3(A_1)$ $\delta_s(XMX)$	$\nu_4(A_1)$ $\delta_s(YMY)$	$\nu_5(A_2)$ τ	$\nu_6(B_1)$ $\nu_{as}(MX)$	$\nu_7(B_2)$ $\delta(XMY)$	$\nu_8(B_1)$ $\nu_{as}(MY)$	$\nu_9(B_2)$ $\delta(XMY)$	*Ref.*
$O_2MoS_2^{2-}$	K-salt; i.r.(DMSO-soln.)	867	451	310	200	270	842	248	471	270	55
	NH_4-salt; i.r., Ra solid										cf. 13
$O_2WS_2^{2-}$	K-salt; i.r.(DMSO-soln.)	886	454	310	196	280	848	235	442	280	55
	NH_4-salt; i.r., Ra solid										cf. 13
$O_2MoSe_2^{2-}$	$[Ni(NH_3)_6]$-salt; i.r. solid	864	283	339	114		834		353	251	55
	NH_4-salt; i.r., Ra solid										
$O_2WSe_2^{2-}$	Cs-salt; i.r., Ra solid	888	282	319	116	235	845	156	329	235	55
	NH_4-salt; i.r., Ra solid										

Table 3.4 Symmetry and internal force constants (in mdyn Å^{-1})[600] calculated with the 'Extended L Matrix Approximation Method' of Müller *et al.*[62] ($L_{34} = -0.075$) and from frequency data of Table 3.1‡

	*Solution II**						*Solution I**		
	$F_{11}(A_1)$	$F_{33}(F_2)$	$F_{34}(F_2)$	$F_{44}(F_2)$	f_r	f_{rr}	F_{33}	F_{34}	F_{44}
VO_4^{3-}	6.43	4.26	0.08	0.34	4.80	0.54	4.52	0.29	0.34
CrO_4^{2-}	6.75	5.28	0.11	0.43	5.65	0.37	5.58	0.36	0.43
MoO_4^{2-}	7.58	5.38	0.03	0.35	5.93	0.55	5.50	0.20	0.35
WO_4^{2-}	8.17	5.92	0.03	0.41	6.48	0.56	5.97	0.13	0.41
MnO_4^-	6.64	5.68	0.18	0.56	5.92	0.24	5.97	0.43	0.56
TcO_4^-	7.84	6.42	0.03	0.40	6.78	0.36	6.56	0.23	0.39
ReO_4^-	8.90	7.11	0.02	0.43	7.56	0.45	7.17	0.15	0.43
RuO_4†	7.50	6.78	0.03	0.41	6.96	0.18	6.92	0.24	0.41
OsO_4†	8.95	8.07	0.01	0.47	8.29	0.22	8.13	0.16	0.47
$KRuO_4$	6.49	5.53	0.02	0.34	5.77	0.24	5.65	0.19	0.34
MnO_4^{2-}	6.22	4.53	0.07	0.34	4.95	0.42	4.77	0.28	0.34
FeO_4^{2-}	6.53	4.22	0.07	0.32	4.80	0.58	4.45	0.26	0.32
RuO_4^{2-}	6.65	5.02	0.06	0.40	5.43	0.41	5.12	0.20	0.40
CrO_4^{3-}	6.56	4.88	0.05	0.32	5.30	0.42	5.17	0.29	0.32
K_3MnO_4	6.18	4.74	0.08	0.37	5.10	0.36	4.99	0.31	0.37
Li_3ReO_4	6.15	6.14	0.02	0.40	6.14	0.00	6.19	0.13	0.40
K_3FeO_4	5.68	4.39	0.08	0.34	4.71	0.32	4.62	0.28	0.34
Ba_2TiO_4	5.46	3.89	0.15	0.40	4.28	0.39	4.12	0.33	0.40
Li_4ZrO_4	5.91	5.47	0.10	0.51	5.58	0.11	5.59	0.27	0.51
Li_4HfO_4	5.97	5.39	0.07	0.56	5.54	0.15	5.43	0.16	0.56
Ba_2VO_4	6.31	4.05	0.13	0.40	4.61	0.56	4.28	0.32	0.40
Ba_2CrO_4	6.12	4.90	0.16	0.49	5.21	0.31	5.16	0.38	0.49
Ba_2MoO_4	5.91	5.04	0.09	0.48	5.26	0.22	5.14	0.24	0.48
Ba_2WO_4	6.35	5.96	0.05	0.53	6.06	0.10	6.00	0.16	0.53
Ba_2FeO_4	5.47	4.94	0.03	0.30	5.07	0.13	5.21	0.27	0.30
Ba_2CoO_4	5.88	5.02	0.06	0.36	5.24	0.22	5.26	0.29	0.36

* See ref. 34, 62
† Calculated from harmonic frequencies
‡ Some inconsistencies in the force constants indicate that some frequencies from Ref. 599 are probably wrong

Data from normal coordinate analysis for the species MO_3X with C_{3v} and C_{2v} symmetries (M = Cr, Mo, W, Mn, Tc, Re, Os; X = F, Cl, Br, S, Se, N) and MOS_3 (M = Mo, W, Re) have generally been published with the spectral data (see Section 3.2.1.3.) (for OsO_3N^- see Ref. 77 and for several other oxospecies with C_{3v} symmetry see Ref. 78).

3.2.1.4 *Spectra of solids*

The vibrational spectra of solids with isolated groups give additional information. Internal vibrations are in general not identical with those in solution as can be explained by the selection rules of the site or factor group (inactive vibrations can become active and degenerate vibrations can lose their degeneracy). Especially interesting is the measurement of Raman spectra of single crystals (e.g. A_2CrO_4 (A = K, Rb, Cs)) because this allows the direct assignment of the factor group fundamentals which is in general difficult in the case of measurements on powders. The interpretation of the spectra of solids gives information on cation-anion interactions.

Spectra of the following metallates (with different cations) have been reported: vanadates[79–82], Na_3NbO_4[83], chromates[84, 85, 155], molybdates and tungstates[86–92], permanganates[93–95], perrhenates[96, 97] and OsO_4[98].

The single crystal Raman and infrared spectra have been reported for

K_2CrO_4[99]. The influence of several factors on the halfwidth of the spectra (internal vibration ν_3 (F_2) of oxometallates) have been investigated[100]. The halfwidth of ν_3 decreases markedly with decreasing charge on the anion, e.g. in the series TiO_4^{4-}, VO_4^{3-}, CrO_4^{2-}, MnO_4^- [100]. The effect of impurities on the splitting of degenerate modes has been investigated[101]. The asymmetric stretching frequency ν_3 of different barium tetraoxometallates has been correlated with the electronic structure of the anions[102].

The Raman spectra of $KCrO_3F$, $KCrO_3Cl$ and $KOsO_3N$ show that there is a statistical orientation of the CrO_3F^- and OsO_3N^- ions in the solids[103]. Some new low frequency bands have been found in the i.r. spectrum of $K_2Cr_2O_7$[104]. There has been some recent interest in the spectra of anions in host lattices. There is one important consequence. If the anion can be regarded as an impurity, the original correlation field splitting can be excluded (VO_4^{3-} in KBr[105], CrO_4^{2-} in KCl and KBr[106–108], MnO_4^- in KCl, RbCl, KBr, RbBr[109] and MnO_4^- in $KReO_4$[110]).

Some results are available for isomorphous substitution (MnO_4^- in $KClO_4$[110], $^{92}MoO_4^{2-}$, $^{92}MoS_4^{2-}$, $^{100}MoO_4^{2-}$ and $^{100}MoS_4^{2-}$ in K_2SO_4[34]). In all these cases the triply degenerate fundamentals of species F_2 are split into three components according to the selection rules of the site symmetry (site C_s). One of the present authors has used the isotope shift $\Delta\nu_3$ (F_2) of the above labelled ions for the calculation of force constants of the anions, as the accuracy of these measurements is better than for normal solids (very sharp bands!)[111] (partially incorrect data[15]).

3.2.1.5 Complexes of the type MO_nX_{6-n} *(n=2,3,4 and 6) and other oxycomplexes*

The structure of complexes of these types can be determined by i.r. and Raman spectroscopy. The ν(MO) bands for mono-oxy-species are found in the i.r. at *ca.* 950 cm^{-1}, near 820 cm^{-1} for *trans*-dioxo- and at approximately 930 and 890 cm^{-1} for *cis*-dioxo- species. A paper including many new data[113] and a systematic study has been published by Griffith.

The vibrational spectra of solid hexoxometallates of rhenium containing the isolated group ReO_6 have been measured by Hauck[116] and the force constants for the ReO_6^{6-} as well as for WO_6^{6-} and TeO_6^{6-} ions have been calculated[117].

New measurements on MO_nX_{6-n} type oxo complexes have been published[118–125, 161], these investigations contain partial assignments. Bent oxoanions containing the system M_2O may be mentioned here. Until now the δ (M_2O) vibration was wrongly assigned to the band at 220 cm^{-1} in $Cr_2O_7^{2-}$. We could conclude from the spectra of $^{50}Cr_2O_7^{2-}$ and $^{53}CrO_4^{2-}$ that this vibration should lie at a much lower frequency (see also Refs. 114, 115).

Spectral data on other oxohalogenometallates have been reported: $TiOCl_4^{2-}$ [126], $TiOCl_5^{3-}$ [126], $ReOCl_4$[127], $ReOCl_5^-$ [127–130], $[Re_2O_3Cl_8]^{2-}$ [127], $ReOBr_5^{2-}$ [128–130]. The compounds $AMoO_2F_3$(A = K, NH_4, Rb, Cs, Tl) and AWO_2F_3(A = Rb, Cs) do not contain isolated ions, but the octahedrally coordinated metal atoms are linked by *cis*-bridging F atoms into infinite chains[131].

The vibrational spectra of transition metal peroxy-compounds[158], and of compounds with rhenium–, ruthenium–, and osmium–oxygen bonds have been reported[134, 159]. A characterisation of the vibrational properties of the metal–oxygen bridge system has been given[115].

The infrared absorption intensity of the ν(VO) vibrational frequency of the vanadyl(IV) ion and its thiocyanato complexes has been measured and discussed with respect to bond properties[135]. The trends in the VO stretching frequencies has been shown to be parallel to the nephelauxetic series for some monomeric complexes of VO^{2+}[136]. Muller *et al.* have reported the i.r. spectra of oxalato-oxoniobates and tantalates[137].

3.2.2 Electronic spectra

3.2.2.1 Tetraoxo species

Many groups have investigated the electronic spectra of tetraoxoanions of the type MO_4^{n-} with a d^0 configuration (for a review see Müller *et al.*[16]) and a very large number of MO calculations have been published (see Section 3.2.2.4). Not so much interest has been shown for species with d^n (e.g. MnO_4^{2-}) configurations but up to now the problem of assigning the spectra has not been solved completely. Though it is established that Δ is negative, the numerical value has been the subject for intensive discussions. Two extremes are $\Delta = -25\,000$ cm^{-1} and $\Delta \approx 0$ (see Ref. 16). But there is now general agreement that the order of MO energies is[16]

$$d(t_2) > d(e) > (\pi)t_1 > (\pi)t_2 > [(\pi)e] > [(\sigma)t_2 > [(\sigma)a_1]$$

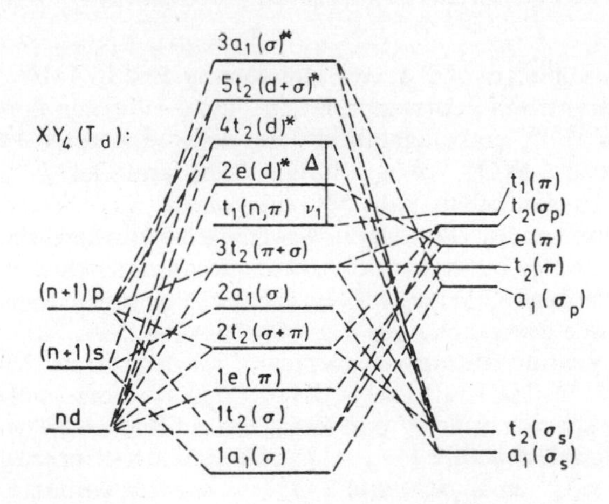

Figure 3.1 Simplified MO diagram for ions and molecules with d^0 configuration (T_d symmetry)

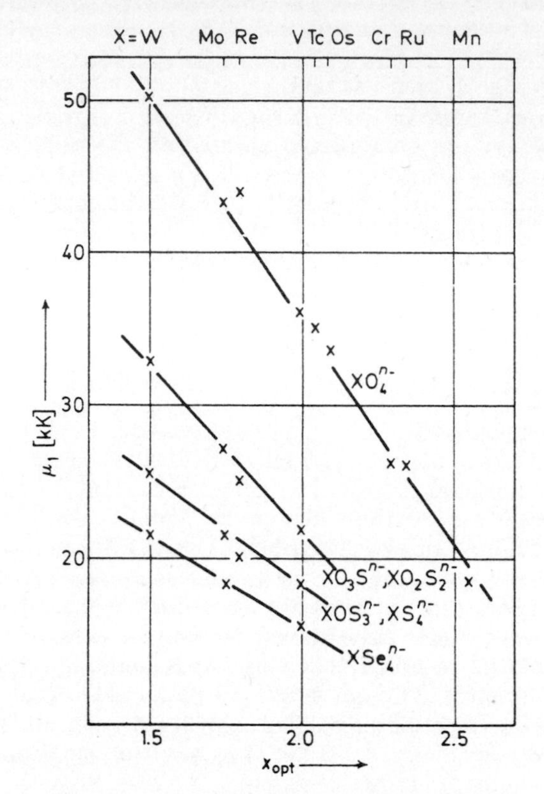

Figure 3.2 The energy of the longest wavelength transition in the electronic spectra of some transition metal oxo ions as a function of the optical electronegativity

The electronic spectra of d^0 species are summarised in Table 3.5. Empirically the first transition corresponds to $(\pi)t_1 \rightarrow d(e)$ $(^1A_1 \rightarrow {}^1T_2)$[138]. MO calculations[139, 140] and magnetic circular dichroism spectra of VO_4^{3-}, VS_4^{3-}, MoO_4^{2-}, MoS_4^{2-}, WO_4^{2-}, WS_4^{2-}, ReO_4^-, ReS_4^- and OsO_4[553] agree with this assignment. In spite of this many MO calculations have been published with a wrong assignment for the longest wavelength band and the assignment of some higher energy bands of thio- and seleno-anions (which are presented for comparison in Table 3.5) are problematical[16]. For some recent studies on the spectra of these compounds see Ref. 141–153, 164, 165.

Low temperature absorption spectra of single crystals (M_2CrO_4(M = K, NH_4, Na)[166, 167] $LiClO_4(MnO_4) \cdot 3H_2O$[168]) have been published. The low-temperature spectra of TcO_4^- and ReO_4^- (in $KClO_4$) show two band systems with vibrational structure ($^1A_1 \rightarrow {}^1T_2$[169]) (for low-temperature absorption spectra of CrO_4^{2-} in K_2SO_4 and VO_4^{3-} in sodium sulphate dodecahydrate see Ref. 170).

The electronic spectra of the open-shell anions CrO_4^{3-}, MnO_4^{2-} have been reported[171–176]. The observation of an absorption maximum at

Table 3.5 Electronic spectra frequencies in kK (1 kK = 1000 cm^{-1}) of maxima (or baricentres of vibrational structures) with shoulders in parentheses and molar extinction coefficients of complexes of point-group T_d. Identifications of M.O. symmetry type one-electron excitations are given for the first transitions

	$(\pi)t_1 \rightarrow (d)e$	$(\pi)t_2 \rightarrow (d)e$	$(\pi)t_1 \rightarrow (d)t_2$	*Other transitions*			*Ref.*
VO_4^{3-}	36.9(8000)	45.0(5700)	—	—			
VS_4^{3-}	18.6(5800)	—	—	25.4(3500)	28.5(4800)	37.5(11000)	142
VSe_4^{3-}	15.6	—	—	21.5			
CrO_4^{2-}	(24),26.8(4800)	36.6(3700)	39.2(3000)	48.8(4700)			138
MnO_4^-	(16),18.6(2400)	32.3(1700)	28.5(1200)	44.0(1000)	52.9(20200)		143
MoO_4^{2-}	43.2(2800)	48.0(8400)	—	—			11, 138
MoS_4^{2-}	(19),21.4(13000)	—	—	31.5(17000)	41.3(24000)	48.3(40000)	138
$MoSe_4^{2-}$	(16),18.0	—	—	27.8	37.2		138
TcO_4^-	34.6(2100)	40.5(5700)	—	(53.2)(1500)			143
RuO_4	26.0	32.3	—	40 (weak)			138
WO_4^{2-}	50.3(8300)	—	—	—			11
WS_4^{2-}	25.5(18500)	—	—	36.1(28500)	46.2(30000)		138
WSe_4^{2-}	(18.9),21.6(16000)	—	—	31.6(24000)	41.3(30000)		138
ReO_4^-	43.7(3600)	48.3(6100)	—	next above 54			143, 138
ReS_4^-	19.8(9600)	—	—	32.0(18000)	44.0(30000)		25
OsO_4	33.8	40.5	—	48 (weak)			138

10 000–12 000 cm^{-1} in solutions containing MnO_4^{2-} confirms the earlier assignment to the $e \rightarrow t_2$ (d→d) transition. This band was also detected in single crystals of $Ba(MnO_4, SO_4)$[175] (see also Refs. 174, 176). The polarised spectrum of $Ca_2[PO_4, CrO_4]Cl$ has been recorded[176]. According to Banks *et al.*[171] the spectra of CrO_4^{3-} show that the band at 17 kK should be assigned to a charge transfer band, whereas other workers assign this to a d→d transition.

The electronic spectra of the compounds Li_6MO_6 (M = Tc, W, Re, Os, Te), Li_7MO_6 (M = Nb, Ta, Sb, Bi, Ru, Os) Li_8MO_6 (M = Zr, Hf, Sn, Pb, Ir, Pt) and Li_5MO_6 (M = Re, I, Tc, Os) have been reported by Hauck[177].

3.2.2.2 *Substituted tetraoxo compounds*

(a) *Complexes with three oxide and a fourth ligand* — Müller *et al.* have investigated the spectra of several substituted species, e.g. CrO_3X^- (X = F, Cl, Br) and OsO_3N^-. The a_2 component of the $(\pi)t_2$ orbitals in MO_3X is concentrated on the three O ligands and cannot involve π orbitals of X (for discussion see also Refs. 16, 178, 179). The assignment of the bands is especially difficult if that electronegativity of X is similar to that of O.

(b) *Complexes with four chalcogenide ligands* — Several new complexes with four chalcogenide ligands were prepared and the spectra investigated although in some cases the existence of the species could only be deduced from the electronic spectra ($VO_2S_2^{3-}$, $VO_2Se_2^{3-}$, WO_3Se^{2-}, MoO_3Se^{2-}, $ReOS_3^-$). Some of the corresponding acids could be prepared in solution and their existence proved by the electronic spectra. Data are summarised in Tables 3.6 and Table 3.7.

In the case of the ions MY_3X^{n-} with C_{3v} symmetry the first two bands could be assigned in cases where the electronegativity of X is much higher than that of Y because the a_2 MO is located on the three ligands Y. The transitions $a_2 \rightarrow e$ and $t_1 \rightarrow e$ almost coincide for species VOS_3^{3-}–VS_4^{3-}, $MoOS_3^{2-}$–MoS_4^{2-} or WOS_3^{2-}–WS_4^{2-} [178]. The electronic spectra of the ions MoO_3S^{2-},

Table 3.6 Absorption bands of complexes of point-group C_{3v}. Notation as in Table 3.5

	$(\pi)a_2 \rightarrow (d)e$	$(\pi)e \rightarrow (d)e$	*Other transitions*			*Ref.*
VOS_3^{3-}	(19.2)	21.8	30.8	33.9		178
$VOSe_3^{3-}$	15.6	19.2	—			376
MoO_3S^{2-}	—	25.4	34.7(strong)	44.7	52.0	181, 16
$MoOS_3^{2-}$	21.5(2300)	25.5(8700)	32.0(6600)(38,5)	44.1(14000)		178
MoO_3Se^{2-}	—	22.0	32.0			180
$MoOSe_3^{2-}$	17.9	22.0	28.45	(35.5)	40	179
MoS_3Se^{2-}	(18)(weak) and	20.6	30.6	40.4		179
$MoSSe_3^{2-}$	(16)(weak) and	18.55	28.4	38.0		179
WO_3S^{2-}	—	30.6	41.0(strong)			181
WOS_3^{2-}	(26.7)(3000)	29.9(11300)	37.0(7200)	41.1(9700)		178
WO_3Se^{2-}	—	27.0	38.0			180
$WOSe_3^{2-}$	22.1	26.0(strong)	34.1	38.2		447
ReO_3S^-	—	28.6	33.6(strong)	46.5		375
$ReOS_3^-$	19.8	25.5	32.3			375
OsO_3N^-	28	31.5(1000)	35.8	44.0(2500)		137
CrO_3F^-	22.2	28.9	36.4			*
CrO_3Cl^-	22.75	28.1	35.1	41.4		*
CrO_3Br^-	22.3	27.6	34.3	38.4		457

* unpublished data

Table 3.7 Absorption bands of complexes of point groups C_{2v} (and C_s and C_1). Notion as in Table 3.5

$VO_2S_2^{3-}$	21.8	27.8	32.8			*Ref.* 375
$VO_2Se_2^{3-}$	19.2?					376
$MoO_2S_2^{2-}$	25.4(3000)	31.4(6000)	34.7(3000)			181
$MoO_2Se_2^{2-}$	22.0	28.5	32.0			cf. 16
$MoOS_2Se^{2-}$	(20)	24.3	31	42.5		179
$MoOSSe_2^{2-}$	18.5	23.15	30.7			179
$WO_2S_2^{2-}$	30.6(4000)	36.6(6900)	41.0(3900)			181
$WO_2Se_2^{2-}$	27.0	34.0	38.0			cf. 16
WOS_2Se^{2-}	(25)	29.1	36	41	48.4	181
$WOSSe_2^{2-}$	23.2	26.7	33.5			181
H_2WS_4	(23.2)	26.1	36.0	39.1		
H_2WSe_4	(20.4)	23.0	32.1	35.6		
$H_2WO_2S_2$	(19.4)	23.7	33.8(strong)	37	43.7	13
$H_2WO_2Se_2$	(16.4)	(20.8)	30.5	35.7	40.5	
H_2WOS_3	21.7	27.4(strong)	32.1	38.8(strong)		
H_2WOSe_3	17.6	24.5(strong)	27.9	35.3(strong)		

WO_3S^{2-} [91], MoO_3Se^{2-} and WO_3Se^{2-} [180] have been measured and discussed by comparison with the spectra of the ions $MO_xX_{4-x}^{2-}$ (M = Mo, W; X = S, Se)[179–181]. In a paper on the assignment of electron transfer bands of molybdates(VI), tungstates(VI) and rhenates(VII) of type $MXY_3(C_{3v})$ (X, Y = O, S, Se) an attempt was made to assign the higher-energy bands[179], but according to our present knowledge only the first two bands of species such as MOX_3^{n-} can be assigned[16] unambiguously.

For a detailed discussion of the spectra of the species with four chalcogen and species containing ligands such as halogens or nitrogen see Ref. 16.

3.2.2.3 *Other oxo complexes*

The electronic spectra of complexes containing the VO group[182-189] and different oxo-complexes of Ti[126], Fe[191], Tc[192], Co[193] and Mo[191] with different other ligands have been investigated. The electronic spectra of MOX_5^{2-} (X = Cl, Br; M = Nb, Ta)[128, 194, 195] and $NbOCl_4^-$ [194, 195] have been measured by Horner *et al.* and Miranda *et al.*[195].

3.2.2.4 *MO calculations*

Many MO and *ab initio* calculations have been published for closed shell tetrachalcogeno anions (TcO_4^-, ReO_4^-, ReS_4^- [140], VO_4^{3-}, VS_4^{3-}, VSe_4^{3-}, MoO_4^{2-}, MoS_4^{2-}, $MoSe_4^{2-}$, WO_4^{2-}, WS_4^{2-} and WSe_4^{2-} [139], MnO_4^- [196–208] CrO_4^{2-} [197–200, 205–206, 209, 210], VO_4^{3-} [197, 210], TiO_4^{4-} [211], RuO_4[197] however the results are contradictory and it is the opinion of the present authors that the most of the calculations do not help to solve the problems of the assignment of the higher-energy bands in the electronic spectra (see also Ref. 139).

Dipole transition moments for the oxoanions have been calculated by Tandon *et al.*[211–216]. For a molecular orbital interpretation of x-ray absorption edges of chromates, manganates and CrO_3Cl^- see Ref. 217. With the help of extended Hückel MO calculations on some titanium–oxygen systems an explanation of the electronic spectra of TiO_2 and $BaTiO_3$ was attempted[218]. A review on the different molecular orbital approaches of inorganic complexes has been published[219] and in other papers the electronic structures of $[VO(OH_2)_5]^{2+}$ [220], $[ReOCl_5]^{2-}$ [221], $[MoOF_5]^{2-}$ [222] and an oxygen-bridged dimer molybdate[223], have been investigated.

3.2.3 Photo-electron, x-ray and magnetic circular dichroism spectra

The above mentioned methods often give useful information on the electronic structure of complex compounds. The He(I) photo-electron spectra of RuO_4 and OsO_4 have been measured and five molecular orbitals identified[224]. There is evidence for a Jahn–Teller effect in the $^2T_2$ state. The photo-electron spectra (ESCA) of many chalcogenometallates of the transition metals have been reported[225] and Fischer[226] has examined the soft x-ray spectra of Cr_2O_3, CrO_3, CrO_4^{2-} and $Cr_2O_7^{2-}$. The conclusion that the highest filled orbital in CrO_4^{2-} is $3t_2$ instead of t_1 seems to be dubious. A similar study (*ab initio* calculations and XPS spectra) of MnO_4^-, CrO_4^{2-} and VO_4^{3-} has been published[227]. The $L_{\alpha 1,2}$ and $L_{\beta 1}$ emission profile of $KMnO_4$ was investigated and the peak energies were related to the energy levels in the MO diagram[228]. The results from valence region XPS spectra of MnO_4^- and CrO_4^{2-} have been compared with various MO calculations[229] (see also Refs. 230, 231) mcd spectra have been little studied (MnO_4^- and CrO_4^{2-})[232-234]. The present authors and co-workers[553, 554] have measured the spectra of VO_4^{3-}, VS_4^{3-}, MoO_4^{2-}, MoS_4^{2-}, WO_4^{2-}, WS_4^{2-}, ReO_4^-, ReS_4^- and OsO_4. According to the A/D value of the first transition the corresponding bands in the electronic spectrum must be assigned to $t_1 \rightarrow 2e$ transitions for all species.

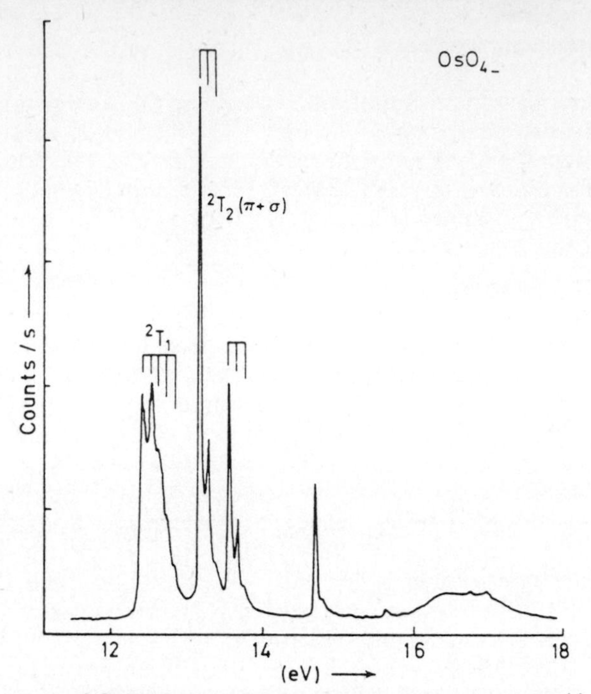

Figure 3.3 He (I) photo-electron spectrum of osmium tetroxide

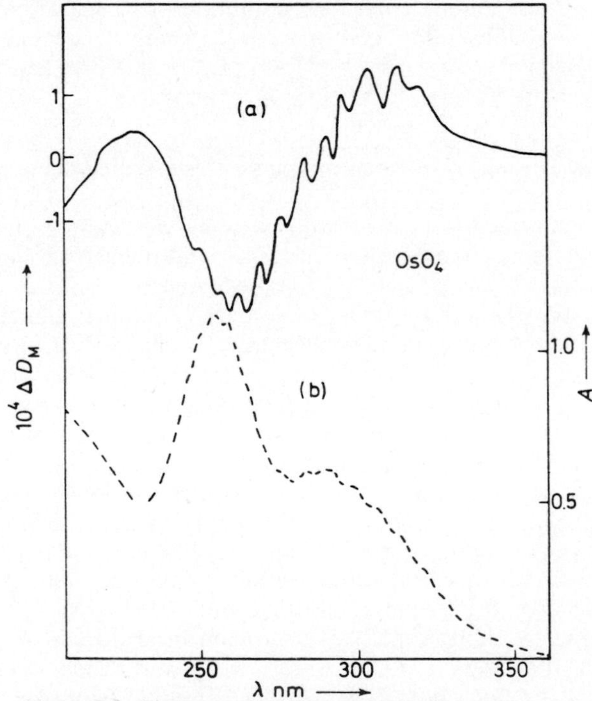

Figure 3.4 (a) Magnetic circular dichroism spectrum
(b) electronic absorption spectrum of osmium tetroxide at room temperature

3.2.4 Electron spin resonance

E.S.R. studies are an important method to obtain knowledge of the electronic structure of open shell oxyanions. An investigation of single crystals of K_2CrO_4 containing 1% K_2MnO_4 showed a well resolved spectrum at 20 K with ^{55}Mn hyperfine structure[235]. The following conclusions were obtained:

(a) the highest occupied orbital is 2*e*, containing the unpaired electron;
(b) the *e* orbital is mainly concentrated on the central atom;
(c) the first unoccupied orbital of t_2 symmetry is spread over the whole ion.

These studies were also of help in the interpretation of the electronic structure of closed shell ions (from the above results a negative value of the crystal field parameter Δ as defined by *Jørgensen* follows).

Different groups have investigated the e.s.r. spectra of oxyanions (partly radicals formed by γ-irradiation) VO_4^{2-} [236, 237], CrO_4^{3-} [238, 239], MnO_4^{3-} [239], CrO_4^{-} [238, 240], CrO_2^{-} [238], MnO_4^{2-} [241, 242], NbO_4^{3-} [236, 243], (for e.s.r. measurements of irradiated anion-doped potassium chromate see Ref. 244).

A self-consistent charge calculation has been performed on the e.s.r. spin-Hamiltonian parameters of CrO_4^{3-} and MnO_4^{3-}, in which values of the spin–orbit constants and $\langle r^{-3} \rangle$ for different charges on the ion are varied until the calculated charge is consistent with the assumed one. For both ions the charge on the central metal ion is $\leqslant +2$[239]. The e.s.r. and optical spectra of MnO_4^{2-} in $BaSO_4$ showed that the single unpaired electron is essentially in a d_{z^2} orbital. An electron paramagnetic resonance study of some iron group salts has been made by Rahman *et al.*[245].

E.S.R. spectra of the oxo-halogeno metallates VOF_5^{3-} [246], $CrOF_5^{2-}$ [247–249], $CrOF_4^{-}$ [256], $CrOClF_3^{-}$ [250], $CrOCl_2F_2^{-}$ [250], $CrOCl_3F^{-}$ [250], $CrOCl_4^{-}$ [250], $CrOCl_5^{2-}$ [251], $NbOF_4^{2-}$ [252], $NbOCl_4^{2-}$ [252], $MoOF_5^{2-}$ [247–249], $MoOCl_5^{2-}$ [248], $MoOBr_5^{2-}$ [251], $MoOI_5^{2-}$ [251], WOF_5^{2-} [247, 249] and molybdenum thiourea complexes[306] (see also Ref. 342) have been described.

The interpretation of the spectrum of VOF_5^{3-} showed that the strongest interaction results from in-plane π-bonding between the $3d_{xy}$ orbital of V and the $2p_y$ orbitals of F[246]. The degree of metal–ligand bond covalency decreases in the order Cr > Mo > W in the case of the ions MOF_5^{2-} (M = Cr, Mo, W)[247] and this can be generalised for all corresponding transition metal compounds[415]. E.S.R. spectra show that in solutions of $KCrOF_4$ in aqueous HF, $CrOF_4^{-}$ and $CrOF_5^{2-}$ are present[248].

E.S.R. spectra are reported for vanadyl compounds (tartrates[133, 253–256] and others[257–261]). In the vanadyl(IV) aquo-complex the effective charge of the VO^{2+} ion with respect to the ligands is greater than that of complexes of dipositive transition metal ions[260].

A Russian group studied the e.s.r. spectral linewidth of liquid solutions of an ethylene glycol complex of CrV for even and odd isotopes of chromium[262].

3.2.5 Nuclear magnetic resonance spectra

As ^{16}O has no nuclear magnetic moment, n.m.r. studies of transition metal oxygen compounds are only possible with, e.g. ^{17}O- labelled compounds or in the case of compounds with transition metals, where the metal resonance spectra can be measured (e.g. ^{53}Cr, ^{55}Mn). Sometimes these studies allow an interpretation of the metal oxygen bonding, such as for the compounds CrO_4^{2-}, CrO_2Cl_2 and $Cr_2O_7^{2-}$[263].

Very interesting results[264] were obtained by measuring the ^{17}O resonances of VO_4^{3-}, CrO_4^{2-}, MoO_4^{2-}, WO_4^{2-}, TcO_4^-, ReO_4^-, RuO_4 and OsO_4, as there is a linear relation between the energy of the $t_1 \rightarrow 2e$ transition and $\delta(^{18}O)$. This result can easily be interpreted from the formula for the chemical shift tensor. The ^1H magnetic resonance line broadening of non-aqueous solvents by vanadyl ions and complexes has been investigated by Angerman *et al.*[265, 266]. ^1H n.m.r. spectra give evidence for π-bonding between the VO_2^+ and water in oxovanadium(IV) complexes[267]. Several groups report ^{19}F n.m.r. studies on fluoro complexes[268–274]. Buslaev has studied spectra of aqueous solutions of fluoro complexes of niobium and tungsten[271, 275] such as $(NH_4)_3Nb_2O_2F_9$ and $(NH_4)_2WO_2F_2$. The exchange of F atoms by H_2O is discussed (for the interaction of $(NH_4)_2WO_2F_4$ with H_2O_2 and WOF_5^- with EtOH and acac see Ref. 270, 271). Evidence has been obtained for stereochemical non-rigidity in 5- and 7-coordinate structures such as WOF_4 and $[Pr_4N]WOF_5$[274]. ^{19}F n.m.r. studies showed that $NbOF_5^{2-}$ is present in solutions of Nb^V containing more than 30% HF[273]. ^{19}F and ^{51}V n.m.r. data suggest that there is rapid F rearrangement between a square-pyramidal and trigonal-bipyramidal geometry for VOF_4^- in HF solution[272]. Other studies use ^{17}O, ^{35}Cl, ^{23}Na, ^{51}V, ^{53}Cr, ^{55}Mn, $^{95, 97}Mo$, ^{93}Nb n.m.r. spectra. Ligand substitution processes of complex oxo-vanadium(IV) species in aqueous HCl solution have been studied by Zeltmann *et al.* by ^{17}O and ^{35}Cl n.m.r.[548] and the ^{23}Na n.m.r. of powder samples of Na_2MoO_4 and Na_2WO_4 (with and without crystal water) has been investigated[276], the ^{23}Na quadrupole couplings are the largest yet observed. The $^{95, 97}Mo$ studies of the molybdate showed no quadrupole effects[276]. The broadening of ^{53}Cr n.m.r. in aqueous solutions of Na_2CrO_4 was interpreted in terms of the reaction:

$$CrO_4^{2-} + Cr_2O_7^{2-} \rightleftharpoons Cr_2O_7^{2-} + CrO_4^{2-}$$ [277].

A ^{55}Mn n.m.r. study of $KMnO_4$ in D_2O has been reported by Lutz *et al.*[278] and the ^{51}V n.m.r. spectra Na_3VO_4 and KVO_3–H_2O studied[279].

In two other papers the n.m.r. of aqueous solutions of sodium perrhenate (^{185}Re and ^{187}Re)[280] and solid state n.m.r. in the presence of completely asymmetric quadrupole and chemical shift effects for $(NH_4)VO_3$, KVO_3, $NaVO_3$ have been investigated[281].

3.2.6 Mössbauer and nuclear quadrupole resonance spectra

The internal magnetic field of hexapositive ^{57}Fe (K_2FeO_4, $BaFeO_4$, $SrFeO_4$) has been studied[282]. The isomer shifts of K_2FeO_4 at 293 and 1.6 K are −0.88 and −0.9 mm s^{-1}, the quadrupole splitting is zero[283] and Fe is in a tetrahedral site. The compound is antiferromagnetic at 1.6 K[283] (for the Mössbauer effect using the 46.5 keV line of ^{183}W in $LiWO_4$ giving the Debye-temperature $\theta_M = 172 \pm 9$ K, see Ref. 284).

The n.q.r. spectrum of ^{185}Re and ^{187}Re in $KReO_4$ shows that the electric field gradient is non-zero because the ReO_4^- ions have symmetry lower than T_d[285]. Large quadrupole coupling constants (0.95 and 1.00 mm s^{-1}) and small isomer shifts (0.13 mm s^{-1} relative to Fe-oxide) indicate covalent character for the Fe—O bond in both the low temperature and the high-temperature phases of Li_5FeO_4[286].

3.2.7 Magnetochemistry

One of the most interesting phenomena in the magnetochemistry of transition metaloxoanions is the temperature independent paramagnetism (TIP) of the closed shell anions MnO_4^- and CrO_4^{2-}. The effect is caused by non-vanishing matrix elements of the orbital angular momentum operator between the ground and excited states. The TIP can only be explained for the oxoanions if electrons are delocalised from 2p orbitals of oxygen into d orbitals of the central metal and in the case of vanishing covalent bonding the TIP would become negligible. Oxometallates with open shell anions, such as K_3MnO_4 and $LaCrO_4$ are paramagnetic. These compounds have been investigated extensively over many years. Complexes of the type $M_2VOCl_4' \, nH_2O$ ($M = NH_4$, K, Rb, Cs; $n = 0, 1$) and M_3VOCl_4 (M = K, Rb, Cs) have been investigated by Kalinikov *et al.*[287, 288]. The effective magnetic moments of the compounds M_3VOCl_4 imply possible V—V interaction along with a delocalisation along the V—O—V chains[287]. Cs_2ReOCl_5 (0.43 B.M.) and Rb_2ReOCl_5 (0.52 B.M.) exhibit spin pairing at room temperature[289] (see also Ref. 290). The magnetic susceptibility of M_2ReOX_5 (X = Cl, Br) was measured between 80–380 K and the effect of the cation on the magnetic properties discussed[291]. O and OH bridges in transition metal and non-transition element complexes have been discussed in respect of the magnetic properties[4] (see also Refs. 292, 293). The temperature-dependent magnetic susceptibilities of A_2FeO_4 (A = K, Rb, Cs) and $BaFeO_4$ obey the Curie–Weiss law, effective magnetic moments have been reported[294]. A large number of bi- and tetranuclear Mo^V complexes has been prepared and their magnetic properties discussed[160].

3.2.8 Structural chemistry

Several groups have determined or redetermined the crystal structures of simple oxometallates. The most important results are summarised in Table 3.8. Interesting investigations include the electron diffraction measurements of gaseous Tl_2MoO_4[295] and $K_2CrO_4$296.

Other studies reported include x-ray investigations of $Na_3VO_4 \cdot 12H_2O$ and some hydrates of fluorinated or hydroxylated salts of general formula $M_3XO_4 \cdot MY \cdot (10-x)H_2O$ (M = Na, K; X = P, As, V; Y = F, OH)[297], and the hexaoxometallates Na_5TcO_6, Na_5ReO_6, Na_5OsO_6, Li_8MO_6 (M = Zr, Hf), Li_7MO_6 (M = Ru, Os, Nb), Li_6OsO_6, β-Li_6MO_6 (M = Re, Tc) and Li_5MO_6 (M = Tc, Re, Os)[298–300].

The preparations, structure and properties of some alkali metal mono-thioperrhenates $MReO_3S$ (M = K, Rb, Cs) have been reported[301]. The compounds $[A(NH_3)_4]B_2$ (A = Zn, Cd; B = MnO_4^-, ReO_4^- and OsO_3N^-) and

Table 3.8 Structural data

	Ref.	*Lattice parameters Å space group*	*Bond length Å*	*Additional remarks*
K_2CrO_4	601	*Pnma*	$d(CrO) = 1.636, 1.651, 1.643$	Redetermination
K_2CrO_4	296			electron diffraction in the gas phase
Ag_2CrO_4	602	*Pnma*	$d(CrO) = 1.69, 1.67, 1.63$ $d(AgO) = 2.34$ (shortest)	AgO interactions disturb CrO_4 tetrahedrons
Tl_2CrO_4	603	$a = 5.910$, $b = 10.727$ $c = 7.910$, *Pmcn*		
$Tl^{I}Tl^{III}(CrO_4)_2$	604	$a = 14.668$, $b = 5.718$ $c = 8.639$, *Pnma*		O atoms form tetrahedra around Cr and octahedra around Tl^{III}
$(NH_4)_2CrO_4$	605	$a = 12.300$, $b = 6.294$, $c = 7.664$, $\beta = 115.6°$, $C2/m$	$d(CrO) = 1.662$	
$(NH_4)_2CrO_4$	606	$a = 12.21$, $b = 6.258$ $c = 7.630$, $\beta = 115.17°$	$d(CrO) = 1.658 \pm 0.004$	H atoms located
$MgCrO_4 \cdot 5H_2O$	607	$a = 6.363$, $b = 10.750$ $c = 6.156$, $\alpha = 97° 17'$, $\beta = 108° 14'$, $\gamma = 76° 09'$		nearly the same structure as $CuSO_4 \cdot 5H_2O$
$Ba_3(CrO_4)_2$	608	$R\bar{3}m$	$d(CrO) = 1.88$ $(3x)$; 1.73	contains tetrahedral CrO_4^{3-} ions
β-$Na_2Cr_2O_7$	609	$P\bar{1}$	$d(CrO)_{br} = 1.78$, $d(CrO)_{term} = 1.61$ CrOCr = 131	
$K_2Cr_2O_7$	610	$a = 13.367$, $b = 7.376$ $c = 7.445$, $\alpha = 90.75°$, $\beta = 96.21°$, $\gamma = 97.96°$	$d(CrO)_{br} = 1.79$, $d(CrO)_{term} = 1.63$ CrOCr = 124°, 128°	

Tl_2MoO_4	295		$d(MoO) = 1.81$, $d(TlO) = 2.30$	by electron diffr. in the gas phase: a two ring lying in mutually perpendicular planes
K_2MoO_4	611	$a = 12.348$, $b = 6.081$ $c = 7.538$, $\beta = 115.74°$, $C2/m$	$d(MoO) = 1.76 \pm 0.01$	hexagonal variety see Ref. 438
K_2WO_4	612	$a = 12.39$, $b = 6.105$ $c = 7.560$, $\beta = 115.96°$	$d(WO) = 1.79$	see also Ref. 162, hexagonal variety see Ref. 438
$(NH_4)_2WO_2S_2$	613	$a = 11.400$, $b = 7.334$ $c = 9.385$, $\beta = 117.05°$, $C2/m$	$d(WO) = 1.775$, $d(WS) = 2.193$	
$K_3(WOS_3)Cl$	614 615	$a = 12.507$, $b = 6.317$ $c = 12.371$, $Pca\ 2_1$	$d(WO) = 1.760$ $d(WS) = 2.208, 2.197, 2.196$	
Cs_2MoOS_3	616	$a = 9.771$, $b = 7.242$ $b = 12.223$, $Pnma$	$d(MoO) = 1.785$ $d(MoS) = 2.179, 2.176$	
$Na_2MoO_4 \cdot 2H_2O$	617	$a = 13.823$, $b = 10.562$ $c = 8.514$, $Pcab$, $Z = 8$	$d(MoO) = 1.76$	infinite chains $MoO_4–H_2O–MoO_4$
K_2MnO_4	618	$a = 7.667$, $b = 5.895$ $c = 10.359$, $Pnma$	$d(MnO) = 1.659$	
$Ba(MnO_4)_2$	619		$d(MnO) = 1.67$	Refinement, $R = 0.17$
$Ba_2(CoO_4)$	620	$a = 7.650$, $b = 5.850$ $c = 10.337$, $Pnma$	$d(CoO) = 1.76–1.80$	
Li_5ReO_6	298	$a = 5.034$, $c = 14.141$ hex., $P3_112$	$d(ReO) = 2.05$	isolated ReO_6^{5-}-octahedra
Li_6WO_6	290	$a = 8.902$, $b = 2.879$ $c = 4.090$, $Immm$	$d(WO) = 2.04$	isolated WO_6^{6-}-octahedra

$[Co(NH_3)_4](ReO_4)_2$ are isostructural and crystallise in the space group $F43m$[4, 302–305]. X-Ray data have been determined for the cubic compounds: $[Co(NH_3)_6](MnO_4)_3$, $[Cr(NH_3)_6](MnO_4)_3$, $[Ni(NH_3)_6]MoO_4$, $[Ni(NH_3)_6]WO_4$, $[Zn(NH_3)_4]WO_4$[303, 550]. Müller *et al.* have studied the structural chemistry of the compounds A_2MOX_3 (A = K, Rb, Cs; M = Mo, W; X = S, Se), Cs_2MOSSe_2 and Cs_2MOS_2Se (M = Mo, W), which all crystallise with the β-K_2SO_4 type (space group *Pnma*) lattice[307]. Due to hydrogen bonding the ammonium dithiosalts $(NH_4)_2MO_2X_2$ (M = Mo, W; X = S, Se) have a monoclinic (space group *C*2/*c*) structure[313]. X-Ray data have also been published for $KOsO_3N$ and $Ba(OsO_3N)_2$[308–310].

The lattice of $K_2Mo_2O_7$ is built up from chains of MoO_4 tetrahedra and MoO_6 octahedra running parallel to the *b*-axis[311]. Thermal and structural properties of Tl_2CrO_4 and Tl_2MoO_4 have been reported[312] and the Madelung contribution to the lattice energy and Madelung constants calculated for Li_5ReO_6, Li_6WO_6 and some other hexaoxometallates[314]. Other papers dealing with the crystal or molecular structures of oxocompounds which do not contain simple units will only be listed here:
$(NH_4)_3[VO_2(C_2O_4)_2]\cdot 2H_2O$[315], $Cs_2VOCl_4\cdot H_2O$[196], $(NH_4)_3VO_2F_4$[316], $Na_4[VO\text{-}d,l\text{-}C_4H_2O_6]_2\cdot 12H_2O$[317], $K_2VO_2F_3$[318], $N_2H_6[NbOF_5]\cdot H_2O$[319], $[N(Et)_4]_2MOCl_4$ (M=Ti, V)[320], $NaNbO_2F_2$[321], Cs_2NbOBr_5[321], Cs_2NbOBr_5[322], Rb_2NbOBr_5[322], Cs_2WOBr_5[322], $(NH_4)_3NbO(C_2O_4)_3\cdot H_2O$[323, 324], $(NH_4)_3VOF_5$[325], $(NH_4)_2VO(C_2O_4)_2\cdot 2H_2O$[326], *cis*-$[VO(oxal)_2H_2O]^{2-}$[327], $Mo[(CH_3)_2NCHO]_2O_2Cl_2$[328], $NaK_3[MoO_2(CN)_4]\cdot 6H_2O$[329], $[Mo_2O_5L_2]^{2-}$(L = pyrocatechol)[330], $(NH_4)_2MoOBr_5$[331], $Na_2Mo_2O_4$(cystein)$\cdot 5H_2O$[332, 333], $K_4[Cl_5WOWCl_5]$[334], $(NH_4)_2MoO_2F_3$[335], $2Na_4(O_3Mo(EDTA)MoO_3)\cdot 8H_2O$[336], $Rb_2MoO_2F_4$[337], $K_4[MoO_2(CN)_4]\cdot 6H_2O$[338], $(pyH)_2[O(H_2O)(O_2)_2MoOMo(O_2)_2(H_2O)O]$[545], $[ORe(CN)_4]_2O^{4-}$[339], $K_3[ReO_2(CN)_4]$[340, 341], tris-(μ-isobutyrato)di[chlororhenium(III)] perrhenate[344], $Re_2O_3en_2Cl_4$[345], $(NH_4)_2[Zn(NH_3)_2(CrO_4)_2]$[346], di-$\mu$-oxo-tetrakis(2,2'-bipy)dimanganese(III,IV) perchlorate[347]. *Cis-* and *trans-arrangements* of the MO_2 grouping in dioxo-compounds of groups V, VI, VII and VIII transition metals have been discussed[348].

3.2.9 Chemical bonding

3.2.9.1 Tetraoxospecies

It is evident from much evidence that there is a considerable degree of M—O double bonding in oxyanions of the type MO_4^{n-}.

(a) There are high values of the stretching force constants and the estimated bond orders according to Siebert[13, 22, 55] (see Table 3.9).

Table 3.9 Frequencies of ν_1 (A_1), the $t_1 \rightarrow 2e$ transition ν_e, bond distances r, the mean polarisability derivatives with respect to Q_1 $\bar{\alpha}'$, calculated bond orders N and stretching force constants f_r

	$\nu_1(A_1)$ (cm^{-1})	$\nu_e(t_1 \rightarrow 2e)$ (kK) (Ref. 16)	$\bar{\alpha}'$ ($Å)^2$	N†	N‡	r (Å)	f_1 (mdyn $Å^{-1}$)¶
(Ba_2TiO_4)[28]	761[599]	—			1.3	1.63, 1.78 [12]	4.28
VO_4^{3-}	826[28]	36.9	2.67	2.1	1.3	1.71[621]	4.80
CrO_4^{2-}	846[28]	26.8	2.80	2.2	1.5	1.66[621]	5.30
MnO_4^-	839[28]	18.6			1.5	1.63[621]	5.92
NbO_4^{3-}	—	—			—	1.89[12]	—
MoO_4^{2-}	897[28]	43.2	2.51	2.4	1.8	1.77[621]	5.93
TcO_4^-	912[28]	34.6			1.9	1.71§	6.78
RuO_4	882[28]	26.0			2.0	1.706‖	6.96
WO_4^{2-}	931[28]	50.3	2.31	2.7	1.9	1.79[621]	6.48
ReO_4^-	972[28]	43.7	2.51	2.6	2.2	1.76[621]	7.56
OsO_4	965[28]	33.8	2.55	2.2	2.2	1.75[621]	8.29
CrO_4^{3-}	834[599]	28.2*			1.4	1.84[608]	5.30
MnO_4^{2-}	812[599]	22.9*			1.3	1.66[621]	4.95
(K_3MnO_4)	810[599]	30.8*			1.3	1.71[621]	5.10
FeO_4^{2-}	832[599]	—			1.2	(1.62)[621]	4.80
$(KRuO_4)$	830[599]	31.7			1.7	1.79[12]	5.77
(Ba_2CoO_4)	790[599]	—			1.3	1.78[620]	5.24

* Viste, A. and Gray, H. B. (1964). *Inorg. Chem.*, **3**, 1113
† bond orders Ref. **49** calculated from $\bar{\alpha}'$ and optical electronegativity data from Jørgensen
‡ dto. calculated with Siebert's formula (see Ref. 590)
§ Krebs, B. (private communication)
‖ Schäfer, L. and Seip, H. M. (1967). *Acta Chem. Scand.*, **21**, 737
¶ see Table 3.4

(b) The sum of the covalent single-bond radii is larger than the M—O bond distance[13].

(c) According to a rule of Ballhausen and Liehr it is only possible to explain the high values of the oscillator strength of the electronic transitions if one assumes a stabilisation of the π-bonding molecular orbitals[13].

(d) MO calculations indicate that the bonding e MO is strongly stabilised[139, 140].

(e) Raman intensity data indicate double bonding[49] (see Table 3.9).

The calculated values of force constants f_{MO} of MO_4^{n-} species with the d^0 configuration have been evaluated as empirical functions of n as well as of the lowest energy electronic transitions $t_1 \rightarrow 2e$. There are nearly linear relationships between the force constants and each of these parameters for isoelectronic species and for species with central atoms from the same group of the periodic table[415].

(a) Force constants f_{MO} increase for isoelectronic series $f(MO_4^{n-}) > f(MO_4^{n-1}) > f(MO_4^{n-2})$.

(b) Force constants for species of the same group show the sequence $f(M^{3d}O_4^{n-}) < f(M^{4d}O_4^{n-}) < f(M^{5d}O_4^{n-})$.

3.2.9.2 *Substituted tetraoxo species*

In the case of substituted oxospecies of the type MO_3X, MO_2X_2 and MOX_3 the bond properties of MO depend on the nature of the other ligand (X: isoelectronic with O^{2-}, e.g. N^{3-} or F^- or pseudo-isoelectronic, e.g. Cl^-, Br^-, S^{2-}, Se^{2-}).

(a) If X is a chalcogen atom, S or Se, the metal ligand π-bonding is large for all four metal–ligand bonds[55] (e.g. $MoOS_3^{2-}$ and ReO_3S^-; for a MO diagram see for example[178]). In the series MO_4^{2-}, MO_3S^{2-}, $MO_2S_2^{2-}$, MO_3S^{2-}, MS_4^{2-} (M = Mo, W) the stretching force constants f_{MO} and f_{MS} are nearly the same[13, 55, 349].

(b) If X is a halogen atom as in CrO_3F^-, CrO_3Cl^- and CrO_3Br^- the double bond order of M—O increases compared with that in the corresponding MO_4^{n-} mainly due to the fact that π-bonding is not so important for the metal halogen bond. ($f_{MO}(MO_3X) > f_{MO}(MO_4)$; f_{MO} is approximately the same for all species MO_3X)[51, 57, 349].

(c) For species with X=N, e.g. OsO_3N^-, ReO_3N^{2-} ($f_{MO}(MO_3N) < f_{MO}(MO_4)$) because N^{3-} is a very strong π-donor[58, 349].

3.2.9.3 *Other oxospecies*

The bond properties for the other species mentioned in Section 3.2.1.5 will not be given in detail here. Generally terminal metal–oxygen bonds are appreciably double bonded whereas in M—O—M bridges there is much less multiple character (compare the Cr—O distances in $Cr_2O_7^{2-}$; see Section 3.2.8). If the coordination number increases (e.g. from $MO_2X_2 \rightarrow MO_2X_4$) the force constants decrease.

3.3 DESCRIPTIVE CHEMISTRY

3.3.1 Group IV

The older literature on the chemistry of titanium has been reviewed by Clark[350]. More recent work includes the paper of Muto *et al.*, who have reported the hydrothermal syntheses and structures of alkali metal titanates[351]. Oxofluorides of the type $M^IM^{II} \leqslant OF_3 \cdot nH_2O$ (M^I=K, Rb, Cs; M^{II}=Zr, Hf) and $TiOF_2 \cdot nH_2O$ have been prepared[352] and $KTiO_2F$ obtained by the reaction of KF with TiO_2 under high pressure[353]. Raman spectroscopic studies of $TiCl_4$ solutions in aqueous HCl show that the ions $TiO_2Cl_4^{4-}$ or $TiOCl_5^{3-}$ are probably present[354]. The system MF_2–$TiOF_2$ has been investigated[355] and the reaction of potassium fluorotitanate with metallic Ti has been studied in aqueous solution[356]. M_2TiCl_5 (M = Rb, Cs) reacts with Sb_2O_3 or As_2O_3 giving $M_2[TiOCl_3]$[35]. $KZrOF_3 \cdot 2H_2O$ and $KHfOF_3 \cdot 2H_2O$ were prepared from the corresponding hexafluorometallate, $KF \cdot 2H_2O$ and KOH. The reaction products were characterised by thermogravimetry, x-ray and i.r. spectra[358]. The zirconium (and hafnium) oxychloride–HCl–H_2O systems have been studied at room temperature[359]. The preparation and properties of tetraethylammonium-decachloro-oxodizirconate has been reported by Feltz[360]. On the basis of PMR studies it has been claimed that the solid oxychloride and oxysulphate of hafnium do not contain OH group but have the formulae $HfOSO_4 \cdot nH_2O$ and $HfOCl_2 \cdot 8H_2O$ respectively[361]. Investigations on oxocompounds of Ti and Zr which do not contain halogens have been published[362–367].

A cryometric determination of the solvation and of the actual nature of oxymetallic species dissolved in a molten halide solvent medium has been reported[368]. Thermoanalytical investigations on $K_2TiO(C_2O_4)_2 \cdot nH_2O$[369], $H_2HfO(SO_4)_2 \cdot 3H_2O$[370] and on $M_2TiF_6(M_2TiO_2F_2)$[371] have been described. The enthalpy of the reaction of Ti^{V} with hydrogen peroxide in aqueous solution has been determined[372].

Gutmann *et al.* have carried out a polarographic investigation on $TiOCl_2$ and titanium(IV) oxyperchlorate in DMSO and DMF[373].

3.3.2 Group V

Reference to the older literature of group V oxyanions is made in the monographs of Clark[350] and Fairbrother[190].

Cubic and monoclinic forms of Na_3TaO_4 were prepared by the reaction of Ta_2O_5 with sodium carbonate or NaOH at 320–950 °C. Orthorhombic Na_3MO_4 (M = Nb, Ta) could be obtained by the reaction of Na_2O with Nb and Ta oxides at 550 °C *in vacuo*[374]. Evidence for the existence of the ions $VO_2S_2^{3-}$, $VO_2Se_2^{3-}$ and $VOSe_3^{3-}$ was obtained[375, 376].

Salts containing isolated $VO_2F_2^-$ and $VO_2Cl_2^-$ entities have been prepared[377], $[PCl_4][VOCl_4]$ is obtained from phosphorus pentachloride and $VOCl_3$[378, 379]. VCl_4 in liquid sulphur dioxide gives $VOCl_3$ and $VOCl_5^{2-}$[380]. In the system metal fluoride–V_2O_5–HF–H_2O the formation of $K_3V_2O_4F_5$ and $K_2VO_2F_3$ is claimed. The latter should contain non-linear VO_2^+ groups[381]. Several metal oxofluorovanadates have been studied by Davidovich and his group[382]. The reaction of VCl_4 with alkali metal chlorides in hydrochloric acid gives the compounds $M_2VOCl_4 \cdot 2H_2O$ (M = NH_4, K, Rb, Cs)[383]. $M_2VOCl_5 \cdot 2H_2O$ is formed from $VOCl_3$ and the corresponding alkali metal chlorides[384]. $VOCl_3$ reacts with NH_3 to give $[VO(NH_3)_5]Cl_2$[385]. $M_3[VOCl_4]$ (M = K, Rb, Cs) are formed by heating a 3:1 mixture of M_3VCl_6 and Sb_2O_3[386]. Two new methods for the preparation of oxotetrachlorovanadates(V) have been reported by Nicholls *et al.*[387]. The oxofluorovanadates Na_3VOF_5, $Cs_3V_2O_2F_7$, K_2VOF_4 and Rb_2VOF_4 have been obtained from MF–VO_2–HF–H_2O systems[388]. The same group has reported the structure of several oxofluorovanadates[388] and Sengupta *et al.* have claimed the preparation of several oxofluorovanadates(V)[389–393], which should, according to their physical data, contain polymeric units. High symmetry VO^{2+} complexes with halide and NCS—ligands have been prepared and their electronic spectra examined[394].

The synthesis and crystal structure of Li_2NbOF_5 have been reported[395]. The corresponding $NbOF_5^{2-}$ ion and $TaOF_4^-$ could be extracted with di-n-octylamino-alcohols from hydrofluoric acid solutions of the metals[396]. The reaction of potassium heptafluorotantalate with ammonia and alkali hydroxides in aqueous solution has been studied[397]. $MTaO_2Cl_2 \cdot H_2O$ (M = K, Rb, Cs, NH_4) were prepared by the reaction of $TaCl_5$ and MCl in saturated HCl solution[386]. The formation of the polymeric species $NbO(OH)F_2 \cdot nH_2O$ and $KNbO_2F_2 \cdot 0.75H_2O$ has been described[398]. Other oxofluorometallates of Nb and Ta could be obtained from aqueous NH_4F and the metal fluorides[399].

The synthesis and spectroscopic properties of $K_3[VO(CN)_5]$[400] and of oxo-isothiocyanatotantalates(V)[401] have been described. Several groups have worked on vanadium[402–407], niobium[408–411] and tantalum[412, 413] oxo complexes containing more complicated organic ligands.

The kinetics of the fission and formation of $[VO(OH)_2]^{2+}$ has been investigated[414]. The thermochemistry, thermoanalytical studies and measurements of chemical equilibria are the subject of several papers[156, 416–430].

Voltametric, potentiometric and conductometric studies have been made on vanadium oxo species[431–434] and potentials of the VO_2^+–VO^{2+} systems in concentrated solutions in sulphuric, phosphoric and acetic acid tabulated[435].

3.3.3 Group VI

The highly hygroscopic alkali metal chromates(V) M_3CrO_4 (M = K, Rb, Cs) have been prepared[436] and two new phases K_4MO_5 (M = Mo, W) observed in the K_2MoO_4–K_2O system at 400 °C[437]. Kessler *et al.* have characterised a hexagonal variety of potassium molybdate and tungstate[438]. $K_6M_2O_9$ (M = Mo, W) were prepared by heating a stoichiometric mixture of K_2MO_4 and K_2O at 420 °C[439]. It is suggested that orthomolybdic acid exists as $Mo(OH)_6$ and not as $(OH)_2MoO_2$; the corresponding dissociated anions would be $MoO(OH)_5^-$ and MoO_4^{2-} [440]. The existence of the ions $[CrO_3Cr]^{4+}$ and $[CrO_2CrO_2Cr]^{5+}$ has been claimed[154]. Salts of the ions $MoOS_3^{2-}$ [441, 442] WOS_3^{2-} [442], $MoOSe_3^{2-}$ [443], $WOSe_3^{2-}$ [444], $MoO_2Se_2^{2-}$ [445], $WO_2Se_2^{2-}$ [445], $MoOS_2Se^{2-}$ [446, 447], WOS_2Se^{2-} [446, 447], $MoOSSe_2^{2-}$ and $WOSSe_2^{2-}$ [448] have been characterised by the present writers. The corresponding acids H_2WOS_3, H_2WOSe_3, $H_2WO_2S_2$ and $H_2WO_2Se_2$ along with H_2WS_4 and H_2WSe_4 have been obtained in Et_2O solutions[13]. Complexes with thio- and seleno-anions acting as ligands, such as, e.g. $[Ni(WOS_3)_2]^{2-}$ [449] and $[Ni(WO_2S_2)_2]^{2-}$ [450] and complexes of MoS_4^{2-} [451], WS_4^{2-} [452, 453] and WSe_4^{2-} [451] are known, for oxoanions as ligands see Ref. 157. Evidence for the ions MO_3X^{2-} (M = Mo, W; X = S, Se)[180, 454], has been obtained from the electronic spectra of the MO_4^{2-}–H_2X systems measured as a function of time. The results and the relative reaction rates of molybdates and tungstates have been extensively discussed[343, 455]. Compounds of the type $K_3(MOS_3)X(M = Mo, W; X = Cl, Br)$ have been reported[456].

Several papers deal with oxohalogenometallates. Bromochromates(VI) have been obtained in a pure form for the first time. The alkali-salts contain the isolated CrO_3Br^- entity[457]. Fluoro- and chloro-chromates(VI) with organic cations have been studied[458]. A procedure has been described for a gravimetric determination of chromium as $Ph_4As[CrO_3Cl]$[458]. A study of the complex formation of $(NH_4)_2MoO_4$ with NaF in aqueous solution indicates the formation of MoO_3F^- and another ion at high NaF concentrations[459]. $Cs_3MoO_3F_3$ and $Cs_3(MoO_3)_2F_3$ were prepared from the $CsF–MoO_3–H_2O$ system[460]. Mixed crystals of $Cs_2MoOCl_{5-x}Br_x$ and $Cs_2MoO_xBr_{5-x}$ have been reported[461], and pure oxopentabromomolybdates prepared[462]. $K_2W(OH)Cl_5$ has been shown to be an oxo bridged dimer $K_4(W_2OCl_{10}]$[463]. The hexapositive tungsten derivatives $M_2W(O–O)OCl_4$ ($M = K, NH_4, Rb, Cs$) have been studied by x-ray diffraction and i.r. spectroscopy showing that the compounds are not hydrates[464]. The reaction between $H_2WO_2F_4$ and CaF_2 in acid media has been investigated[465]. ^{19}F N.M.R. measurements on tungsten oxofluoride solutions have shown the presence of $[WO_2F_6]^{2-}$ and $[WO_2F_3(H_2O]^-$ [466]. The reaction of MoO_3 with $(NH_4)HF_2$ at 110 and 200 °C gives $(NH_4)_2MoO_2F_4$ and $(NH_4)MoO_2F_3$ respectively[467]. The alkali metal oxofluorometallates $A_3MO_xF_{6-x}(x = 1–3;\ M = Mo^{VI},\ W^{VI},\ Ti^{IV},\ V^{V\ and\ IV},\ Nb^{V})$[468] and $Rb_2MoO_2F_4$, $Cs_2MoO_2F_4$, Rb_2NbOF_5 and Cs_2NbOF_5 have been prepared[469]. Saha and co-workers claim the preparation of $H_2[MoOCl_5]\cdot 2H_2O$, $H[MoOCl_4]$, $H_2[MoOBr_5]$ and $H[MoOBr_4]$[24, 470, 471] and the preparation and properties of oxohalogenometallates containing other ligands (e.g. thiourea, pyridine, CN^-)[472–483] and oxometallates with organic ligands (e.g. $(CF_3)_2C(OH)C(OH)(CF_3)_2$ cysteine, 2,2′-bipyridyl phenanthroline, picoline, $C_2O_4^{2-}$ [484–490] and SCN^-[491] have been reported.

Several groups have investigated the thermochemistry (including TG and DTA), kinetics and the chemical equilibria of reactions involving group VI oxometallates[492–507]. The thermodynamic properties of the molybdate ion have been studied[508] and Karov *et al.*[509] have reported on the thermal decomposition of $(NH_4)_2MoO_4$. The equilibrium $HCrO_4^- \rightleftharpoons CrO_4^{2-} + H^+$ has been investigated by potentiometric and spectroscopic methods in the temperature range 5–60 °C[510]. The heats of formation of some alkali chromates(VI) were calculated[511] and the equilibrium constant of the reaction:

$$2CrO_4^{2-} + 2H^+ \rightleftharpoons Cr_2O_7^{2-} + H_2O$$

determined[512]. The thermal decompositions of $(NH_4)_2MoO_2S_2$ and $(NH_4)_2WO_2S_2$ have been studied by t.g. and d.t.a. methods[513]. The stability constants for the formation of citrate complexes $[C_6H_5O_7MoO_2]^-$ have been reported[505].

Electrochemical (polarographic, voltametric, conductivity, potentiometric and electroreduction) studies on oxometallates and substituted species have been reported[397, 514–519]. The limiting molar conductance of CrO_4^{2-} in H_2O (25–60 °C) and the extent of hydrolysis of the ion has been determined by Jones[514].

3.3.4 Group VII

The solid-state synthesis of M_3ReO_5 (M = K, Rb, Cs) by using the corresponding alkaline metal peroxides has been described[520, 521] and manganates M_2MnO_4 (M = K, Rb, Cs) were obtained similarly[522]. The preparations and properties of Na_3VO_4, Na_3CrO_4 and Na_3MnO_4 have been reported[523]. $KMMnO_4$ and $KMCrO_4$ (M = Sr, Ba) prepared by heating K_2MnO_4 or K_2CrO_4, MO and MnO_2[524]. Ternary oxides of rhenium(VII), such as $Ti_2Re_2O_{11}$, $Nb_4Re_2O_{17}$ and $NbReO_6$[525] and of rhenium(IV) and lanthanum are described[526]. $Ba_5(ReO_6)_2$ and $Ba_3(ReO_5)_2$ were investigated by vibrational and electronic absorption spectroscopy, indicating that isolated ReO_6 octahedra are present in $Ba_5(ReO_6)_2$[527]. A mass spectroscopic study of the vaporisation of barium and magnesium perrhenates showed the presence of $Ba(ReO_4)_2$ and Re_2O_7 in the gas phase above the solid. The mass spectrum also shows $HReO_4^+$ formed by secondary reaction of Re_2O_7 with traces of water in the vacuum system[528]. In the opinion of the present authors an investigation of Petrov *et al.*[529] on the vibrational spectrum of solid perrhenic acid $HOReO_3$ is incorrect and the solid does not contain isolated $HOReO_3$ groups. Synthesis, i.r. and u.v. spectra of substituted ammonium perrhenates[530, 531] and x-ray studies and the reduction of compounds of the type $MMnO_4$ and M_2MnO_4 (M = Rb, Cs)[532] have been reported. Spectrophotometric measurements show that ReO_4^- is converted into a meso form in strongly alkaline solution[533], a reaction which was not observed with TcO_4^-. Permanganic acid, $HOMnO_3$[534], is obtained by the reaction of equimolar amounts of $Ba(MnO_4)_2$ and H_2SO_4. Measurements of the ^1H chemical shift and the Raman intensity of the $\nu_1(A_1)$ line show that $HOReO_3$ is completely ionised in concentrations up to 7 mol l^{-1}[535]. $[Co(NH_3)_5OReO_3]^{2+}$, containing coordinated ReO_4^-, has been reported[536].

Although salts of the tetrahedral anions $ReO_2S_2^-$ and $ReOS_3^-$ are unknown, evidence has been obtained for their existence in aqueous solution[375]. The chemistry and properties of salts containing the anions $TcOCl_5^{2-}$, $ReOCl_5^-$, $ReOBr_5^-$, $ReOCl_5^{2-}$ and $ReOBr_5^{2-}$[537–542] and the preparation of the compounds $M[ReOF(H_2O)(CN)_3]$ (M = Na, K, NH_4)[543], $NOReOF_6$ and NO_2ReOF_6[544] and $K_4[Re_2O_3(CN)_8]$[555] have been reported. NN'-ethylenebis(salicylideneiminato)manganese(II), MnL, decomposes in DMSO to give $(MnOL)_2O_2$. Magnetic, i.r. and thermogravimetric data indicate that $(MnOL)_2O_2$ contains both oxo- and μ-peroxo-bonding[556].

During the cathodic reduction of TcO_4^- solutions, half wave potentials −0.81, −0.90 and −1.10 V, corresponding to the reactions $Tc^{VII} \rightarrow Tc^{V}$; $Tc^{V} \rightarrow Tc^{IV}$ and $Tc^{IV} \rightarrow Tc^{III}$ are observed respectively[557]. Several other electrochemical investigations on group VII metallates have been published[558–567].

The kinetics and mechanism of the reduction of MnO_4^- in aqueous solution has been reviewed[568]. Isotopic ligand exchange studies have been made on $[Re(amine)_4O_2]^+$ ions[569] and thermochemical thermoanalytical (d.t.a., t.g.), mass spectroscopic and chemical equilibrium measurements on group VII oxometallates[576–583].

3.3.5 Group VIII

Few studies have been made of group VIII oxoions. The decomposition of ferrates(VI) in strongly alkaline solutions has been investigated and there is a decreasing rate of decomposition with increasing purity of the reaction solution[584].

γ-Irradiation of sodium ruthenate solution causes the precipitation of ruthenium(IV) hydroxide; the reaction occurs through the intermediate formation of RuO_4^- [585]. The chemical behaviour of a binuclear oxygen-bridged nitratonitrosylruthenium complex in an aqueous environment has been studied. $[RuO_2(NO)_2NO_3]_2O$ decomposes in neutral aqueous solution to give the mononuclear $(RuO_2(NO_2)(NO_3)_2H_2O]$ ion[586]. *Trans*-$Cs_2[RuO_2X_4]$ (X = Br, CN, 0.5 oxalate) were prepared by the reaction of RuO_4 in CCl_4 with the appropriate acid and Cs^+ ions; *trans*-$[RuO_2(NH_3)_4]Cl$ and $[RuO_2py_2Cl_2]$ were obtained similarly[587]. Bol'shakov and co-workers[588] have studied the chemical states of ruthenium in sulphate–chloride solutions. At high HCl concentrations the Ru is present only as the $[Ru_2OCl_{10}]^{4-}$ ion[588]. A checked procedure for the preparation of the corresponding potassium salt $K_4[Ru_2OCl_{10}]\cdot H_2O$ has been reported[589]. Other species characterised include the μ_3-oxotriruthenium carboxylate complexes $[Ru_3O(OAc)_6(H_2O)_3]OAc$ and $[Ru_3O[OAc)_6$-$py_3]^+$ [590] and oxocomplexes of Ir, Ru and Os with the ligands py, and $Mc_4C_2O_2$ resulting from the action of olefines[132]. Bardin and Goncharenko[591] report the formation of red $[OsO_2(OH)_4]^{2-}$ from alkali hydroxide solutions of osmiumtetroxide, via the intermediate $[OsO_4(OH)_2]^{2-}$. From polarography osmium species in sodium hydroxide solutions apparently include: $OsO(OH)_5^-$, $OsO(OH)^{3+}$, $OsO(OH)_3^+$, $OsO(OH_3^-$, $Os_2O_2(OH)_{10}^{6-}$, $OsO(OH)_5^{3-}$ and $Os_2O_2(OH)_6$[592]. One of the present authors has made extensive studies of compounds containing the nitrido-osmate (VIII) ion $[OsO_3N]^-$ by means of vibrational and electronic spectroscopy and x-ray diffraction[309, 310, 593]. The tetraphenylarsonium salt has been prepared and its use for the analytical determination of osmium described[594]. The properties of K_2OsO_4 have been investigated[595]. Studies on the complex compounds $[OsO_2L](OH)_2\cdot H_2O$ and $[OsO_2L]Cl_2\cdot 5H_2O$ containing the piperazine-dibiguanide ligand ($L=C_8H_{18}N_{10}$) have been reported[596].

Ferrate(VI) ion can be determined voltametrically in 10.5 M KOH at a rotating Fe electrode[597]. The FeO_4^{2-} reduction occurs at *ca.* −0.3 V *v.* the Hg–HgO–KOH electrode. The thermodynamic properties of ruthenium oxo compounds have been studied[508, 598]. Heats of formation ΔH_f reported were (in kcal mol^{-1})[598]: Na_2RuO_4(aq.), -224.1 ± 0.8; RuO_4^{2-}(aq.), -109.4 ± 1.0; RuO_4(aq.), -57.5 ± 1.1; and RuO_4(liq.), -57.0 ± 1.1.

Acknowledgement

Grants of the Deutsche Forschungsgemeinschaft, Fonds der Chemischen Industrie, NATO scientific affairs division, and Wissenschaftsministerium (Landesamt für Forschung) NRW for our own work included in this review are gratefully acknowledged. Rhenium, ruthenium and osmium compounds were kindly supplied by Degussa, Hanau (Germany). We also thank Prof. Dr. O. Glemser for discussions and Dipl. chem. H. H. Heinsen for kind assistance.

References

1. Scholder, R. (1958). *Angew. Chem.*, **70,** 583
2. Jahr, K. F. and Fuchs, J. (1966). *Angew. Chem. Int. Ed.*, **5,** 689
3. Pope, M. T. and Dale, B. W. (1968). *Quart. Rev. Chem. Soc.*, **22,** 527
4. Jezowska-Trzebiatowska, B. (1968). *Coord. Chem. Rev.*, **3,** 255
5. Connor, J. A. and Ebsworth, E. A. V. (1964). *Advan. Inorg. Chem. Radiochem.*, **6,** 279
6. Canterford, J. H. and Colton, R. J. (1968). *Halides of First Row Transition Elements* and *Halides of Second and Third Row Transition Elements* (New York: J. Wiley)
7. Brewer, L. and Rosenblatt, G. M. (1961). *Chem. Rev.*, **61,** 257
8. McConnell, J. D. M. (1972). *MTP Int. Rev. Sci.*, Inorganic, Ser. 1, Vol. 5, p. 33
9. Bradley, D. C. (1967). *Coord. Chem. Rev.*, **2,** 299; Bradley, D. C. and Fisher, K. J. (1972). *MTP Int. Rev. Sci.*, Inorganic Chemistry, Ser. 1, Vol. 5, p. 65
10. Selbin, J. (1964). *J. Chem. Educ.*, **41,** 86; (1965). *Angew. Chem. Int. Ed.*, **5,** 712
11. Carrington, A. and Symons, M. C. R. (1963). *Chem. Rev.*, **63,** 443
12. Griffith, W. P. (1970). *Coord. Chem. Rev.*, **5,** 459
13. Diemann, E. and Müller, A. (1974). *Coord. Chem. Rev.*, **10,** 79
14. Sievert, W. and Müller, A. (1974). *Z. Chem.*, to be published.
15. Schmidt, K. H. and Müller, A. (1973). *Coord. Chem. Rev.*, to be published
16. Müller, A., Diemann, E. and Jørgensen, C. K. (1973). *Structure and Bonding*, **14,** 23
17. Wu, T. Y. (1946). *Vibrational Spectra and Structure of Polyatomic Molecules* (Ann Arbor Mich.: I. W. Edwards)
18. Landolt-Börnstein (1951). *Zahlenwerte und Funktionen* (Berlin: Springer), Band I, 2. Teil
19. Kohlrausch, K. W. F. (1943). *Raman-Spektren* (Leipzig: Akad. Verlagsgesellschaft)
20. Lecomte, J. (1958). in: *Handbuch der Physik* (S. Flügge, editor), Vol. 26, p. 825 (Berlin: Springer)
21. Nakamoto, K. (1970). *Infrared Spectra of Inorganic and Coordination Compounds*, 2nd edn. (New York: Wiley-Interscience)
22. Siebert, H. (1966). *Anwendungen der Schwingungsspektroskopie in der anorganischen Chemie* (Berlin: Springer)
23. Ross, S. D. (1972). *Inorganic Infrared and Raman Spectra* (London: McGraw Hill)
24. Saha, H. K. and Halder, M. C. (1971). *J. Inorg. Nucl. Chem.*, **33,** 705
25. Müller, A., Diemann, E. and Rao, V. V. K. (1970). *Chem. Ber.*, **103,** 2961
26. Müller, A., Schmidt, K. H., Tytko, K. H., Bouwma, J. and Jellinek, F. (1972). *Spectrochim. Acta*, **A28,** 381
27. Müller, A., Baran, E. J. and Aymonino, P. J. (1968). *An. Asoc. Quim. Argent.*, **56,** 85
28. Weinstock, N., Schulze, H. and Müller, A. (1973). *J. Chem. Phys.*, **59,** 5063
29. Müller, A. (1968). *Z. Phys. Chem. (Leipzig)*, **238,** 116
30. Peacock, C. J. and Müller, A. (1968). *J. Molec. Spectrosc.*, **26,** 454
31. Aono, S., Sado, A., Chihara, K. and Kanzin, E. (1970). *Sci. Rep. Kanazawa Univ.*, **15,** 11
32. Müller, A., Weinstock, N., Mohan, N., Schläpfer, C. W. and Nakamoto, K. (1972). *Z. Naturforsch.*, **A27,** 542
33. Müller, A., Weinstock, N., Mohan, N., Schläpfer, C. W. and Nakamoto, K. (1973). *Appl. Spectrosc.*, **27,** 257
34. Müller, A., Königer, F. and Weinstock, N. (1974). *Spectrochim. Acta*, in the press
35. Pinchas, S. (1969). *Israel J. Chem.*, **7,** 5
36. Pinchas, S. (1969). *J. Chem. Phys.*, **51,** 2284
37. Pinchas, S. and Shamir, J. (1969). *Israel J. Chem.*, **7,** 805
38. McDowell, R. S., Asprey, L. B. and Hoskins, L. C. (1972). *J. Chem. Phys.*, **56,** 5712
39. McDowell, R. S. and Goldblatt, M. (1971). *Inorg. Chem.*, **10,** 625
40. Davidson, G., Logan, N. and Morris, A. (1968). *Chem. Commun.*, 1044
41. Griffith, W. P. (1968). *J. Chem. Soc. A*, 1663
42. Levin, I. W. and Abramowitz, S. (1969). *J. Chem. Phys.*, **50,** 4860
43. Griffith, W. P. and Lesniak, P. J. B. (1969). *J. Chem. Soc. A*, 1066
44. Hendra, P. J. (1968). *Spectrochim. Acta*, **A24,** 125

45. Müller, A. and Krebs, B. (1966). *Z. Naturforsch.*, **B21**, 3
46. Kiefer, W. and Bernstein, H. J. (1972). *Molec. Phys.*, **23**, 835
47. Kiefer, W. and Bernstein, H. J. (1971). *Chem. Phys. Lett.*, **8**, 381
48. Krasser, W. (1969). *Ber. Kernforschungsanlage Jülich, JUEL-628-CA*, 157
49. Müller, A. and Schulze, H., *Advances in Raman Spectroscopy*, Vol. 1, p. 546 (London: Heyden)
50. Ereshko, N. A. and Malt'sev, A. A. (1971). *Kolebatel'nye Spektry Neorg. Khim.*, 93
51. Aymonino, P. J., Schulze, H. and Müller, A. (1969). *Z. Naturforsch.*, **B24**, 1508
52. Leroy, M. J. F., Burgard, M. and Müller, A. (1971). *Bull. Soc. Chim. Fr.*, 1183
53. Müller, A., Krebs, B., Kebabcioglu, R., Stockburger, M. and Glemser, O. (1968). *Spectrochim. Acta*, **A24**, 1831
54. Müller, A., Böschen, I. and Sievert, W. (1970). *Z. Naturforsch.*, **B25**, 311
55. Schmidt, K. H. and Müller, A. (1972). *Spectrochim. Acta*, **A28**, 1829
56. Müller, A., Weinstock, N. and Schulze, H. (1972). *Spectrochim. Acta*, **A28**, 1075
57. Müller, A., Schmidt, K. H., Ahlborn, E. and Lock, C. J. L. (1973). *Spectrochim. Acta*, **A29**, 1773
58. Krebs, B. and Müller, A. (1968). *J. Inorg. Nucl. Chem.*, **30**, 463
59. Spoliti, M. and Stafford, F. E. (1968). *Inorg. Chim. Acta*, **2**, 301
60. Müller, A., Schmidt, K. H. and Mohan, N. (1972). *J. Chem. Phys.*, **57**, 1752
61. Müller, A., Weinstock, N., Schmidt, K. H., Nakamoto, K. and Schläpfer, C. W. (1972). *Spectrochim. Acta*, **A28**, 2289
62. Müller, A., Mohan, N. and Heidborn, U. (1972). *Z. Naturforsch.*, **A27**, 129
63. Müller, A. and Krebs, B. (1968). *J. Molec. Spectroscopy*, **26**, 136
64. Müller, A., Krebs, B., Cyvin, S. J. and Diemann, E. (1968). *Z. Anorg. Allg. Chem.*, **359**, 194
65. Müller, A., Krebs, B. and Cyvin, S. J. (1967). *Acta Chem. Scand.*, **21**, 2399
66. Müller, A., Kebabcioglu, R., Leroy, M. J. F. and Kaufmann, G. (1968). *Z. Naturforsch.*, **B23**, 738
67. Ram, R. S. and Thakur, S. N. (1971). *Ind. J. Pure Appl. Phys.*, **9**, 34
68. Sanyal, N. K., Singh, H. S., Pandey, A. N. and Singh, B. P. (1970). *Ind. J. Pure Appl. Phys.*, **8**, 72
69. Müller, A. and Krebs, B. (1967). *J. Molec. Spectrosc.*, **24**, 180
70. Krebs, B., Müller, A. and Fadini, A. (1967). *J. Molec. Spectrosc.*, **24**, 198
71. Narayanan, V. A. and Nagarajan, G. (1971). *Acta Phys. Pol. A*, **40**, 401
72. Sanyal, N. K., Pandey, A. N. and Singh, H. S. (1969). *J. Quant. Spectrosc. Radiat. Transfer*, **9**, 1035
73. Sanyal, N. K., Pandey, A. N., Singh, H. S. and Singh, B. P. (1970). *J. Quant. Spectrosc. Radiat. Transfer*, **10**, 1343
74. Müller, A. and Cyvin, S. J. (1968). *J. Molec. Spectrosc.*, **26**, 315
75. Müller, A., Ahlrichs, R. and Krebs, B. (1966). *Z. Naturforsch.*, **B21**, 719
76. Krebs, B. and Müller, A. (1967). *J. Molec. Spectrosc.*, **22**, 290
77. Müller, A., Baran, E. J., Bollmann, F. and Aymonino, P. J. (1969). *Z. Naturforsch.*, **B24**, 960
78. Müller, A., Krebs, B., Fadini, A. and Glemser, O. (1968). *Z. Naturforsch.*, **A23**, 1656
79. Baran, E. J. and Aymonino, P. J. (1969). *Z. Anorg. Allg. Chem.*, **365**, 211
80. Baran, E. J. and Aymonino, P. J. (1972). *Z. Naturforsch.*, **B27**, 76
81. Olivier, D. (1969). *Rev. Chim. Miner.*, **6**, 1033
82. Baran, E., Aymonino, P. J. and Müller, A. (1972). *J. Molec. Structure*, **11**, 453
83. Blasse, G. and van den Heuvel, G. P. M. (1972). *Mater. Res. Bull.*, **7**, 1041
84. Baran, E. J. and Aymonino, P. J. (1968). *An. Asoc. Quim. Argent.*, **56**, 91
85. Müller, A., Stockburger, M. and Baran, E. J. (1969). *An. Asoc. Quim. Argent.*, **57**, 65
86. Balashov, V. A., Karetnikov, G. S. and Maier, A. A. (1970). *Tr. Mosk. Khim.-Tekhnol. Inst.*, No. **67**, 151, 154
87. Caillet, P. and Saumagne, P. (1969). *J. Molec. Struct.*, **4**, 191
88. Petrov, K. I., Veronskaya, G. N., Shakhno, I. V. and Savel'eva, A. V. (1970). *Izv. Akad. Nauk SSSR, Neorg. Mater.*, **6**, 515
89. Plyusnina, I. I. and Zaitseva, L. A. (1969). *Vestn. Mosk. Univ. Geol.*, **24**, 110
90. Schwing-Weill, M. J. and Arnaud-Neu, F. (1970). *Bull. Soc. Chim. Fr.*, 853
91. Zaitsev, B. E., Zakharikova, E. I., Ivanov-Emin, B. N. and Cherenkova, G. I. (1969). *Zh. Neorg. Khim.*, **14**, 1493
92. Zaitsev, B. E., Ivanov-Emin, B. N., Korotaeva, L. G. and Remizov, V. G. (1971). *Kolebatel'nye Spektry Neorg. Khim.*, 300

93. Balicheva, T. G. and Lavrov, B. B. (1968). *Probl. Sovrem. Khim. Koord. Soedin. Leningrad. Gos. Univ.*, **2,** 190
94. Baran, E. J. and Aymonino, P. J. (1968). *Monatsh. Chem.*, **99,** 1584
95. Doyle, W. P. and Kirkpatrick, I. (1968). *Spectrochim. Acta*, **A24,** 1495
96. Johnson, L. W., Rogers, M. T. and Leroi, G. E. (1972). *J. Chem. Phys.*, **56,** 789
97. Ulbricht, K. and Kriegsmann, H. (1968). *Z. Anorg. Allg. Chem.*, **358,** 193
98. Levin, I. W. (1969). *Inorg. Chem.*, **8,** 1018
99. Adams, D. M., Hooper, M. A. and Lloyd, M. H. (1971). *J. Chem. Soc. A*, 946
100. Baran, E. J. and Müller, A. (1971). *Spectrochim. Acta*, **A27,** 517
101. Verlan, E. M., Dem'yanenko, V. P. and Tsyashenko, Yu. P. (1971). *Fiz. Tverd. Tela (Leningrad)*, **13,** 2459
102. Baran, E. J. and Aymonino, P. J. (1968). *Spectrochim. Acta*, **A24,** 291
103. Whiting, F. L., Mamantov, G., Begun, G. M. and Young, J. P. (1972). *Inorg. Chim. Acta*, **5,** 260
104. Mathur, M. S., Frenzel, C. A. and Bradley, E. B. (1968). *J. Molec. Struct.*, **2,** 429
105. Kondilenko, I. I., Tsyashchenko, Yu. P. and Pasechnyi, V. A. (1970). *Ukr. Fiz. Zh.*, **15,** 1917
106. Berezhinskii, L. I., Krulikovskii, B. K. and Bereza, V. F. (1971). *Ukr. Fiz. Zh.*, **16,** 261
107. Dem'yanenko, V. P., Tsyashenko, Yu. P. and Verlan, E. M. (1970). *Fiz. Tverd. Tela*, **12,** 545
108. Lisitsa, M. P. and Berezhinskii, L. I. (1971). *Ukr. Fiz. Zh.*, **16,** 1347
109. Manzelli, P. and Taddei, G. (1969). *J. Chem. Phys.*, **51,** 1484
110. Barraclough, C. G. and Bilander, I. (1970). *Aust. J. Chem.*, 00, 000
111. Müller, A., Königer, F. and Weinstock, N. (1973). *Spectrochim. Acta*, in the press
112. Müller, A., Baran, E. J. and Hendra, P. J. (1969). *Spectrochim. Acta*, **A25,** 1654
113. Griffith, W. P. (1969). *J. Chem. Soc. A*, 211
114. Königer, F., Müller, A. and Mattes, R. (1973). *Spectrochim. Acta*, 00, 000
115. Wing, R. M. and Callahan, K. P. (1969). *Inorg. Chem.*, **8,** 871
116. Hauck, J. (1970). *Z. Naturforsch.*, **B25,** 468
117. Hauck, J. and Fadini, A. (1970). *Z. Naturforsch.*, **B25,** 422
118. Beuter, A. and Sawodny, W. (1971). *Z. Anorg. Allg. Chem.*, **381,** 1
119. Buslaev, Yu. A. and Davidovich, R. L. (1968). *Zh. Neorg. Khim.*, **13,** 1254
120. Dehnicke, K. Pausewang, G. and Rüdorff, W. (1969). *Z. Anorg. Allg. Chem.*, **366,** 64
121. Griffith, W. P. and Wickins, T. D. (1968). *J. Chem. Soc. A*, 400
122. Petrov, K. I., Kravchenko, V. V. and Sinitryn, N. M. (1970). *Zh. Neorg. Khim.*, **15,** 2732
123. Soptrajanov, B., Nikolovski, A. and Petrov, I. (1968). *Spectrosc. Lett.*, **1,** 117
124. Van den Poel, J. and Neumann, H. M. (1968). *Inorg. Chem.*, **7,** 2086
125. Wasson, J. R. (1968). *J. Inorg. Nucl. Chem.*, **30,** 171
126. Fowles, G. W. A., Lewis, D. F. and Walton, R. A. (1968). *J. Chem. Soc. A*, 1468
127. Brisdon, B. J. and Edwards, D. A. (1968). *Inorg. Chem.*, **7,** 1898
128. Edwards, D. A. and Ward, R. T. (1970). *J. Molec. Struct.*, **6,** 421
129. Petrov, K. I. and Kravchenko, V. V. (1970). *Zh. Neorg. Khim.*, **15,** 2216
130. Petrov, K. I. and Kravchenko, V. V. (1969). *Zh. Neorg. Khim.*, **14,** 000
131. Mattes, R., Müller, G. and Becher, H. J. (1972). *Z. Anorg. Allg. Chem.*, **389,** 177
132. Griffith, W. P. and Rossetti, R. (1972). *J. C. S. Dalton Trans.*, 1449
133. James, P. G. and Luckhurst, G. R. (1970). *Molec. Phys.*, **18,** 141
134. Jezowska-Trzebiatowska, B., Hanuza, J. and Baluka, M. (1970). *Acta Phys. Pol. A*, **38,** 563
135. Larsson, R. and Nunziata, G. (1970). *Acta Chem. Scand.*, **24,** 1
136. Sathyanarayana, D. N. and Patel, C. C. (1968). *J. Inorg. Nucl. Chem.*, **30,** 207
137. Muller, M. and Dehand, J. (1971). *Bull. Soc. Chim. Fr.*, 2837, 2843
138. Müller, A. and Diemann, E. (1971). *Chem. Phys. Lett.*, **9,** 369
139. Kebabcioglu, R. and Müller, A. (1971). *Chem. Phys. Lett.*, **8,** 59
140. Kebabcioglu, R., Müller, A. and Rittner, W. (1971). *J. Molec. Struct.*, **9,** 207
141. Radhakrishna, S. and Pande, K. P. (1972). *Chem. Phys. Lett.*, **13,** 62
142. Müller, A., Diemann, E. and Ranade, A. C. (1969). *Chem. Phys. Lett.*, **3,** 467
143. Mullen, P., Schwochau, K. and Jørgensen, C. K. (1969). *Chem. Phys. Lett.*, **3,** 49
144. Mattoo, B. N. (1968). *Ind. J. Chem.*, **6,** 364
145. Maria, H. J., Srinivasan, B. N. and McGlynn, S. P. (1968). *Molec. Lumin. Int. Conf.*, 787
146. Marculescu, L. and Ghita, C. (1972). *Krist. Tech.*, **7,** K 39
147. Johnson, L. W., Hughes, E. and McGlynn, S. P. (1971). *J. Chem. Phys.*, **55,** 4476
148. Demuynck, J., Kaufmann, G. and Brunette, J. P. (1969). *Bull. Soc. Chim. Fr.*, 3840

149. Dembicka, D. and Bartecki, A. (1970). *Rocz. Chem.*, **44**, 1011
150. Canit, J. C., Billardon, M. and Badoz, J. P. (1969). *J. Opt. Soc. Amer.*, **59**, 1000
151. Butowiez, B. (1971). *C. R. Acad. Sci. Ser. B.*, **272**, 534
152. Butowiez, B. (1970). *C. R. Acad. Sci. Ser. B.*, **271**, 1141
153. Boulon, G., Gaume-Mahn, F. and Curie, D. (1970). *C. R. Acad. Sci. Ser. B.*, **270**, 111
154. Adams, A. C., Crook, J. R., Bockhoff, F. and King, E. L. (1968). *J. Amer. Chem. Soc.*, **90**, 5761
155. Carter, R. L. and Bricker, C. E. (1971). *Spectrochim. Acta*, **A27**, 569
156. Churikova, M. B., Smirnov, E. K. and Vasil'kova, I. V. (1971). *Zh. Neorg. Khim.*, **16**, 547
157. Coomber, R. and Griffith, W. P. (1968). *J. Chem. Soc. A*, 1128
158. Griffith, W. P. and Wickins, T. D. (1968). *J. Chem. Soc. A*, 397
159. Jezowska-Trzebiatowska, B., Hanuza, J. and Baluka, M. (1971). *Spectrochim. Acta*, **A27**, 1753
160. Jezowska-Trzebiatowska, B. and Rudolf, M. (1970). *Rocz. Chem.*, **44**, 745
161. Kharitonov, Yu. Ya. and Buslaev, Yu. A. (1969). *Spektrosk. Tverd. Tela*, 197
162. Kools, F. X. N. M., Koster, A. S. and Rieck, G. D. (1970). *Acta Crystallogr.*, **B26**, 1974
163. Carter, R. L. and Bricker, C. E. (1971). *Spectrochim. Acta*, **A27**, 825
164. Whiting, F. L., Mamantov, G. and Young, J. P. (1972). *J. Inorg. Nucl. Chem.*, **34**, 2475
165. Hush, N. S. and Hobbs, R. J. M. (1968). *Progr. Inorg. Chem.*, **10**, 259
166. Butowiez, B. (1968). *C. R. Acad. Sci. Ser. B*, **267**, 1234
167. Butowiez, B. (1970). *J. Phys. (Paris)*, **31**, 477
168. Johnson, L. W. and McGlynn, S. P. (1971). *J. Chem. Phys.*, **55**, 2985
169. Guedel, H. U. and Ballhausen, C. J. (1972). *Theoret. Chim. Acta*, **25**, 331
170. Duinker, J. C. and Ballhausen, C. J. (1969). *Theoret. Chim. Acta*, **12**, 325
171. Banks, E., Greenblatt, M. and Holt, S. (1968). *J. Chem. Phys.*, **49**, 1431
172. Banks, E., Greenblatt, M. and McGarvey, B. R. (1971). *J. Solid State Chem.*, **3**, 308
173. Day, P., DiSipio, L. and Oleari, L. (1970). *Chem. Phys. Lett.*, **5**, 533
174. DiSipio, L., Oleari, L. and Day, P. (1972). *J. C. S. Faraday Trans. II*, **68**, 776
175. Kosky, C. A. and Holt, S. L. (1970). *Chem. Commun.*, 668
176. Simo, C., Banks, E. and Holt, S. (1970). *Inorg. Chem.*, **9**, 183
177. Hauck, J. (1969). *Z. Naturforsch.*, **B24**, 1349
178. Diemann, E. and Müller, A. (1970). *Spectrochim. Acta*, **A26**, 215
179. Müller, A., Diemann, E., Neumann, F. and Menge, R. (1972). *Chem. Phys. Lett.*, **16**, 521
180. Müller, A., Ranade, A. C. and Rittner, W. (1971). *Z. Anorg. Allg. Chem.*, **380**, 76
181. Müller, A., Diemann, E., Ranade, A. C. and Aymonino, P. J. (1969). *Z. Naturforsch.*, **B24**, 1247
182. Bartecki, A. and Kaminski, J. (1970). *Rocz. Chem.*, **44**, 1839
183. Selbin, J. and Vigee, A. (1968). *J. Inorg. Nucl. Chem.*, **30**, 1644
184. Schmidtke, H. H. and Garthoff, D. (1969). *Z. Naturforsch.*, **A24**, 126
185. Vanquickenborne, L. G. and McGlynn, S. P. (1968). *Theoret. Chim. Acta*, **9**, 390
186. Wendling, E. and Lavillandre, de J. (1968). *Bull. Soc. Chim. Fr.*, 35
187. Feltz, A. and Langbein, H. (1970). *J. Inorg. Nucl. Chem.*, **32**, 2951
188. Bartecki, A. and Kaminski, J. (1971). *Rocz. Chem.*, **45**, 315
189. Drake, J. E., Vekris, J. E. and Wood, J. S. (1969). *J. Chem. Soc. A*, 345
190. Fairbrother, F. (1967). *The Chemistry of Niobium and Tantalum* (Amsterdam: Elsevier)
191. Schugar, H. J., Rossman, R. G., Barraclough, C. G. and Gray, H. B. (1972). *J. Amer. Chem. Soc.*, **94**, 2683
192. Baluka, M., Hanuza, J. and Jezowska-Trzebiatowska (1972). *Bull. Acad. Pol. Sci. Ser. Sci. Chim.*, **20**, 271
193. Barratt, J. (1968). *Chem. Commun.*, 874
194. Horner, S. M., Clark, R. J. H., Crociani, B., Copley, D. B., Horner, W. W., Collier, F. N. and Tyree, S. Y. (1968). *Inorg. Chem.*, **7**, 1859
195. Miranda, C., DaSilveira, M. and Vernois, J. (1970). *J. Inorg. Nucl. Chem.*, **32**, 839
196. Atovmyan, L. O. and Aliev, Z. G. (1970). *Zh. Strukt. Khim.*, **11**, 782
197. Baranovskii, V. I. and Sizova, O. V. (1971). *Dokl. Akad. Nauk SSSR*, **200**, 871
198. Brown, R. D., James, B. H. and O'Dwyer, M. F. (1970). *Theoret. Chim. Acta*, **17**, 362
199. Brown, R. D., James, B. H., McQuade, T. J. V. and O'Dwyer, M. F. (1970). *Theoret. Chim. Acta*, **17**, 279
200. Brown, R. D., James, B. H. and O'Dwyer, M. F. (1970). *Theoret. Chim. Acta*, **19**, 45
201. Clack, D. W. and Farrimond, M. S. (1970). *Theoret. Chim. Acta*, **19**, 373
202. Dacre, P. D. and Elder, M. (1971). *Chem. Phys. Lett.*, **11**, 377

203. Dahl, J. P. and Johansen, H. (1968). *Theoret. Chim. Acta*, **11**, 13
204. De Alti, G. and Galasso, V. (1971). *Chem. Phys. Lett.*, **8**, 223
205. Hillier, I. H. and Saunders, V. R. (1970). *Proc. Roy. Soc.* (*London*) *A*, **320**, 161
206. Hillier, I. H. and Saunders, V. R. (1971). *Chem. Phys. Lett.*, **9**, 219
207. Hillier, I. H. and Saunders, V. R. (1969). *Chem. Commun.*, 1275
208. Johnson, K. H. and Smith, F. C. jun. (1971). *Chem. Phys. Lett.*, **10**, 219
209. Astier, M. (1970). *Molec. Phys.*, **19**, 209
210. Dahl, J. P. and Johansen, H. (1968). *Theoret. Chim. Acta*, **11**, 8
211. Surana, S. S. L. and Tandon, S. P. (1971). *Ind. J. Pure Appl. Phys.*, **9**, 300
212. Surana, S. S. L. and Tandon, S. P. (1972). *Can. J. Spectrosc.*, **17**, 10
213. Surana, S. S. L. and Tandon, S. P. (1970). *Spectrosc. Lett.*, **3**, 329
214. Tandon, S. P., Surana, S. S. L. and Tandon, K. (1969). *Spectrosc. Lett.*, **2**, 295
215. Tandon, S. P. and Surana, S. S. L. (1970). *J. Chem. Phys.*, **52**, 3521
216. Tandon, S. P. and Surana, S. S. L. (1972). *J. Inorg. Nucl. Chem.*, **34**, 3089
217. Seka, W. and Hanson, H. P. (1969). *J. Chem. Phys.*, **50**, 344
218. Hsia, Y. P. (1968). *Can. J. Chem.*, **46**, 2667
219. Dahl, J. P. and Ballhausen, C. J. (1968). *Advan. Quantum Chem.*, **4**, 170
220. Nikolov, G. S. and Yatsimirskii, K. B. (1969). *Teorec. Eksp. Khim.*, **5**, 592
221. Natkaniec, L. and Jezowska-Trzebiatkowska, B. (1971). *Bull. Acad. Pol. Sci., Ser. Sci. Chem.*, **19**, 129
222. Wendling, E. and Lavillandre, de J. (1968). *Bull. Soc. Chim. Fr.*, 2743
223. Brown, D. H., Perkins, P. G. and Stewart, J. J. (1972). *J. C. S. Dalton Trans.*, 1105
224. Diemann, E. and Müller, A. (1973). *Chem. Phys. Lett.*, **19**, 538
225. Müller, A., Jørgensen, C. K. and Diemann, E. (1972). *Z. Anorg. Allg. Chem.*, **391**, 38
226. Fischer, D. W. (1971). *J. Phys. Chem. Solids*, **32**, 2455
227. Connor, J. A., Hillier, I. H., Saunders, V. R., Wood, M. H. and Barber, M. (1972). *Molec. Phys.*, **24**, 497
228. Koster, A. S. (1971). *Proc. Kon. Ned. Akad. Wetenschap. Ser. B*, **74**, 332
229. Prins, R. and Novakov, T. (1972). *Chem. Phys. Lett.*, **16**, 86
230. Clack, D. W. (1972). *J. C. S. Faraday Trans. II*, 1672
231. Singh, R. B., Handel, S. K. and Stenerhag, B. (1972). *Z. Phys.*, **249**, 241
232. Collingwood, J. C., Day, P., Denning, R. G., Robbins, D. J., DiSipio, L. and Olearie, L. (1972). *Chem. Phys. Lett.*, **13**, 567
233. Stephens, P. J. (1968). *Chem. Phys. Lett.*, **2**, 241
234. Stephens, P. J., Mowery, R. L. and Schatz, P. N. (1971). *J. Chem. Phys.*, **55**, 224
235. Carrington, A., Ingram, D. J. E., Schonland, D. and Symons, M. C. R. (1956). *J. Chem. Soc.*, 4710
236. Edwards, P. R., Subramanian, S. and Symons, M. C. R. (1968). *Chem. Commun.*, 799
237. Subramanian, S. and Symons, M. C. R. (1970). *J. Chem. Soc. A*, 2367
238. Debuyst, R., Apers, D. J. and Capron, P. C. (1972). *J. Inorg. Nucl. Chem.*, **34**, 1541
239. McGarvey, B. R. (1968). *Electron Spin Resonance Metal Complexes Proc. Symp.*, 1
240. Lister, D. H. and Symons, M. C. R. (1970). *J. Chem. Soc. A*, 782
241. Kosky, C. A., McGarvey, B. R. and Holt, S. L. (1972). *J. Chem. Phys.*, **56**, 5904
242. Subramanian, S. and Rogers, M. T. (1972). *J. Chem. Phys.*, **57**, 2192
243. Edwards, P. R., Subramanian, S. and Symons, M. C. R. (1968). *J. Chem. Soc. A*, 2985
244. Schara, M., Sentjure, M., Milenkovic, S. M. and Veljkovic, S. R. (1970). *J. Inorg. Nucl. Chem.*, **32**, 369
245. Rahman, S. M. F., Imamuddin, M. and Choudhury, S. H. (1968). *Sci. Ind.* (*Karachi*), 139
246. Manoharan, P. T. and Rogers, M. T. (1968). *J. Chem. Phys.*, **49**, 3912
247. Abdrakhmanov, R. S., Garif'yanov, N. S. and Semenova, E. I. (1968). *Zh. Strukt. Khim.*, **9**, 530
248. Manoharan, P. T. and Rogers, M. T. (1968). *J. Chem. Phys.*, **49**, 5510
249. Van Kemenade, J. T. C. (1970). *Rec. Trav. Chim. Pays-Bas*, **89**, 1100
250. Marov, I. N., Abramova, T. I. and Ermarkov, A. N. (1970). *Zh. Neorg. Khim.*, **15**, 2383
251. Dalton, L. A., Bereman, R. D. and Brubaker, C. H. (1969). *Inorg. Chem.*, **8**, 2477
252. Gainullin, I. F., Garif'yanov, N. S. and Kozyrev, B. M. (1968). *Dokl. Akad. Nauk SSSR*, **180**, 858
253. Belford, R. L., Chasteen, N. D., So, H. and Tapscott, R. E. (1969). *J. Amer. Chem. Soc.*, **91**, 4675
254. Dunnhill, R. H. and Smith, T. D. (1968). *J. Chem. Soc. A*, 2189

255. Dunnhill, R. H. and Symons, M. C. R. (1968). *Molec. Phys.*, **15,** 105
256. Hasegawa, A., Yamada, Y. and Miura, M. (1971). *Bull. Chem. Soc. Jap.*, **44,** 3335
257. Hasegawa, A., Yamada, Y. and Miura, M. (1969). *Bull. Chem. Soc. Jap.*, **42,** 846
258. Jordan, R. B. and Angerman, N. S. (1968). *J. Chem. Phys.*, **48,** 3983
259. Rao, K. V. S., Sastry, M. D. and Venkateswarlu, P. (1968). *J. Chem. Phys.*, **49,** 4984
260. Reuben, J. and Fiat, D. (1969). *Inorg. Chem.*, **8,** 1821
261. Srinivasan, R. and Subramanian, C. K. (1971). *Ind. J. Pure Appl. Phys.*, **9,** 24
262. Garif'yanov, N. S., Kozyrev, B. M. and Fedotov, V. N. (1968). *Dokl. Akad. Nauk SSSR*, **178,** 808
263. Kidd, R. G. (1967). *Can. J. Chem.*, **45,** 605
264. Figgis, B. N., Kidd, R. G. and Nyholm, R. S. (1962). *Proc. Roy. Soc. (London)*, **A269,** 469
265. Angerman, N. S. and Jordan, R. B. (1969). *Inorg. Chem.*, **8,** 65
266. Angerman, N. S. and Jordan, R. B. (1969). *Inorg. Chem.*, **8,** 1824
267. Vigee, G. and Selbin, J. (1968). *J. Inorg. Nucl. Chem.*, **30,** 2273
268. Buslaev, Yu. A. and Petrosyants, S. P. (1969). *Zh. Strukt. Khim.*, **10,** 1105
269. Buslaev, Yu. A., Petrosyants, S. P. and Tarasov, V. P. (1970). *Dokl. Akad. Nauk SSSR*, **193,** 611
270. Buslaev, Yu. A., Kokunov, Yu. V., Bochkareva, V. A. and Shustorovich, E. M. (1972). *J. Inorg. Nucl. Chem.*, **34,** 2861
271. Buslaev, Yu. A., Petrosyant, S. P. and Tarasov, V. P. (1970). *Zh. Strukt. Khim.*, **11,** 1023
272. Moss, K. C. and Howell, J. A. S. (1971). *J. Chem. Soc. A*, 270
273. Moss, K. C. and Howell, J. A. S. (1971). *J. Chem. Soc. A*, 2481
274. Tebbe, F. N. and Muetterties, E. L. (1968). *Inorg. Chem.*, **7,** 172
275. Buslaev, Yu. A., Il'in, E. G. and Kopanev, V. D. (1971). *Dokl. Akad. Nauk SSSR*, **196,** 829
276. Lynch, G. F. and Segel, S. L. (1972). *Can. J. Phys.*, **50,** 567
277. Egozy, Y. and Loewenstein, A. (1969). *J. Magn. Resonance*, **1,** 494
278. Lutz, O. and Steinkilberg, W. (1969). *Phys. Lett. A*, **30,** 183
279. Segel, S. and Creel, R. B. (1970). *Can. J. Phys.*, **48,** 2673
280. Dwek, R. A., Luz, Z. and Shporer, M. (1970). *J. Phys. Chem.*, **74,** 2232
281. Baugher, J. F., Taylor, P. C., Oja, T. and Bray, P. J. (1969). *J. Chem. Phys.*, **50,** 4914
282. Shinjo, T., Ichida, T. and Takada, T. (1969). *J. Phys. Soc. Jap.*, **26,** 1547
283. Ito, A. and Ono, K. (1969). *J. Phys. Soc. Jap.*, **26,** 1548
284. Kaltseis, J., Posch, H. A. and Vogel, W. (1972). *J. Phys., Ser. C*, **5,** 2523
285. Rogers, M. T. and Rama Rao, K. V. S. (1968). *J. Chem. Phys.*, **49,** 1229
286. Le Corre, C., Malve, A., Gleitzer, C. and Foct, J. (1972). *C. R. Acad. Sci. Ser. C*, **274,** 466
287. Kalinikov, V. T., Morozov, A. I., Lebedev, V. G. and Ubozhenko, O. D. (1971). *Zh. Neorg. Khim.*, **16,** 2034
288. Kalinikov, V. T., Morozov, A. I., Lebedev, V. G., Ubozhenko, O. D. and Volkov, M. N. (1972). *Zh. Neorg. Khim.*, **17,** 675
289. Fergusson, J. E. and Love, J. L. (1971). *Aust. J. Chem.*, **24,** 2689
290. Belova, V. I., Syrkin, Ya. K., Chakrabarti, D. K. and Ivanov-Emin, B. N. (1968). *Zh. Strukt. Khim.*, **9,** 73
291. Hauck, J. (1969). *Z. Naturforsch.*, **B24,** 251
292. Jezowska-Trzebiatowska, B. and Rudolf, M. (1970). *Rocz. Chem.*, **44,** 1031
293. Jezowska-Trzebiatowska, B. and Rudolf, M. (1969). *Bull. Acad. Pol. Sci., Ser. Sci. Chim.*, **17,** 419
294. Audette, R. J. and Quail, J. W. (1972). *Inorg. Chem.*, **11,** 1904
295. Tolmachev, S. M. and Rambidi, N. G. (1972). *Zh. Strukt. Khim.*, **13,** 3
296. Spiridonov, V. P. and Lutoshkin, B. I. (1970). *Vestn. Mosk. Univ. Khim.*, **11,** 509
297. Remy, F. and Guerin, H. (1970). *Bull. Soc. Chim. Fr.*, 2073
298. Hauck, J. (1968). *Z. Naturforsch.*, **B23,** 1603
299. Hauck, J. (1969). *Z. Naturforsch.*, **B24,** 1067
300. Hauck, J. (1969). *Z. Naturforsch.*, **B24,** 1064
301. Krebs, B., Müller, A. and Beyer, H. H. (1968). *Z. Anorg. Allg. Chem.*, **362,** 44
302. Müller, A., Böschen, I. and Sievert, W. (1970). *Z. Naturforsch.*, **B25,** 311
303. Müller, A., Christophliemk, P. and Tossidis, I. (1973). *J. Molec. Structure*, **15,** 289
304. Müller, A, Böschen, I., Baran, E. J. and Aymonino, P. J. (1973). *Monatsh. Chem.*, **104,** 836
305. Müller, A. and Böschen, I. (1971). *Z. Naturforsch.*, **B26,** 843
306. Marov, I. N., Dubrov, Yu. N., Belyaeva, V. K., Ermarkov, A. N. and Avdonina, S. I. (1969). *Zh. Neorg. Khim.*, **14,** 3019

307. Müller, A. and Sievert, W. (1974). *Z. Anorg. Allg. Chem.*, in the press
308. Dyachenko, O. A., Golyshev, V. M. and Atovmyan, L. O. (1968). *Zh. Strukt. Khim.*, **9,** 329
309. Müller, A., Bollmann, F. and Krebs, B. (1969). *Z. Anorg. Allg. Chem.*, **368,** 155
310. Müller, A., Bollmann, F. and Baran, E. J. (1969). *Z. Anorg. Allg. Chem.*, **370,** 238
311. Magarill, S. A. and Klevtsova, R. F. (1971). *Kristallografiya*, **16,** 742
312. Gaultier, M. and Pannetier, G. (1972). *Rev. Chim. Miner.*, **9,** 271
313. Müller, A., Diemann, E. and Baran, E. J. (1970). *Z. Anorg. Allg. Chem.*, **375,** 87
314. Hauck, J. (1970). *Z. Naturforsch.*, **B25,** 749
315. Atovmyan, L. O. and Sokolava, Yu. A. (1971). *Zh. Strukt. Khim.*, **12,** 934
316. Markin, V. N. (1968). *Vestn. Leningrad Univ. Fiz. Khim.*, **23,** 151
317. Tapscott, R. E., Belford, R. L. and Paul, L. C. (1968). *Inorg. Chem.*, **7,** 356
318. Ryan, R. R., Mastin, S. H. and Reisfeld, M. J. (1971). *Acta Crystallogr.*, **B27,** 1270
319. Gorbunova, Yu. E., Pakhomov, V. I., Kuznetsov, V. G. and Kovaleva, E. S. (1972). *Zh. Strukt. Khim.*, **13,** 165
320. Haase, W. and Hoppe, H. (1968). *Acta Crystallogr.*, **B24,** 282
321. Andersson, S. and Galy, J. (1969). *Acta Crystallogr.*, **B25,** 847
322. Petillon, F., Yoninou, M. T. and Guerchais, J. E. (1969). *C. R. Acad. Sci. Ser. C*, **268,** 615
323. Mathern, G. and Weiss, R. (1971). *Acta Crystallogr.*, **B27,** 1610
324. Mathern, G., Weiss, R. and Rohmer, R. (1969). *Chem. Commun.*, 70
325. Markin, V. N. (1968). *Vestn. Leningrad Univ. Fiz. Khim.*, **23,** 156
326. Hazell, A. C. (1968). *J. Inorg. Nucl. Chem.*, **30,** 1981
327. Form, G. E., Raper, E. S., Oughtred, R. and Shearer, H. M. M. (1972). *Chem. Commun.*, 945
328. Florian, L. R. and Corey, E. R. (1968). *Inorg. Chem.*, **7,** 722
329. Day, V. W. and Hoard, J. L. (1968). *J. Amer. Chem. Soc.*, **90,** 3374
330. Atovmyan, L. O., Tkachev, V. V. and Shishova, T. G. (1972). *Dokl. Akad. Nauk. SSSR*, **205,** 609
331. Atovmyan, L. O., D'yachenko, O. A. and Lobkovskii, E. B. (1970). *Zh. Strukt. Khim.*, **11,** 469
332. Knox, J. R. and Prout, C. K. (1968). *Chem. Commun.*, 1227
333. Knox, J. R. and Prout, C. K. (1969). *Acta Crystallogr.*, **B25,** 1857
334. König, E. (1969). *Inorg. Chem.*, **8,** 1278
335. Otovmyan, L. A., Krasochka, O. N. and Rahlin, M. Ya. (1971). *Chem. Commun.*, 610
336. Park, J. J., Glick, M. D. and Hoard, J. L. (1969). *J. Amer. Chem. Soc.*, **91,** 301
337. Sergienko, V. S., Porai-Koshits, M. A. and Khodashova, T. S. (1972). *Zh. Strukt. Khim.*, **13,** 461
338. Schlupp, R., LeCarpentier, J. M. and Weiss, R. (1970). *Rev. Chim. Miner.*, **7,** 63
339. Shandles, R., Schlemper, E. O. and Murmann, R. K. (1971). *Inorg. Chem.*, **10,** 2785
340. Fenn, R. H. and Graham, A. J. (1971). *J. Chem. Soc. A*, 2880
341. Murmann, R. K. and Schlemper, E. O. (1971). *Inorg. Chem.*, **10,** 2352
342. Garif'yanov, N. S., Kamenev, S. E. and Ovchinnikov, I. V. (1969). *Zh. Fiz. Khim.*, **43,** 1091
343. Müller, A., Glemser, O., Diemann, E. and Hofmeister, H. (1967). *Z. Anorg. Allg. Chem.*, **371,** 74
344. Calvo, C., Jayadevan, N. C. and Lock, C. J. L. (1969). *Can. J. Chem.*, **47,** 4213
345. Glowniak, T., Lis, T. and Jezowska-Trzebiatowska, B. (1972). *Bull. Acad. Pol. Sci., Ser. Sci. Chim.*, **20,** 199
346. Harel, M., Knobler, C. and McCullough, J. D. (1969). *Inorg. Chem.*, **8,** 11
347. Plaksin, P. M., Stoufer, R. C., Mathew, M. and Palenik, G. (1972). *J. Amer. Chem. Soc.*, **94,** 2121
348. Atovmyan, L. O. and Porai-Koshits, M. A. (1969). *Zh. Strukt. Khim.*, **10,** 853
349. Müller, A., Krebs, B. and Höltje, W. (1967). *Spectrochim. Acta*, **A23,** 2753
350. Clark, R. J. H. (1968). *The Chemistry of Titanium and Vanadium*, (Amsterdam: Elsevier)
351. Muto, F., Oonishi, I. and Komiyama, Y. (1968). *Asahi Garasu Kogyo Gijutsu Shorei-Kai Kenkyu Hokoku*, **14,** 487
352. Lastochkina, A. A., Sheka, I. A. and Malinko, L. A. (1968). *Izv. Sib. Otd. Akad. Nauk. SSSR, Ser. Khim. Nauk*, 37
353. Chamberland, B. L. (1971). *Mater. Res. Bull.*, **6,** 311
354. Davies, J. E. D. and Long, D. A. (1968). *J. Chem. Soc. A*, 2560
355. Ravez, J. and Hagenmuller, P. (1971). *Bull. Soc. Chim. Fr.*, 800
356. Reznichenko, V. A. (1970). *Izv. Akad. Nauk. SSSR, Metal.*, 62
357. Toptygina, G. M. and Dergacheva, N. P. (1972). *Zh. Neorg. Khim.*, **17,** 270
358. Sheka, I. A., Lastochkina, A. A. and Malinko, L. A. (1968). *Zh. Neorg. Khim.*, **13,** 2974

359. Kanaeva, I. G., Mel'nik, L. A. and Serebrennikov, V. V. (1968). *Zh. Neorg. Khim.*, **13,** 1974
360. Feltz, A. (1968). *Z. Anorg. Allg. Chem.*, **358,** 21
361. Savenko, N. F., Sheka, I. A., Matyash, I. V. and Kalinichenko, A. M. (1972). *Ukr. Khim. Zh.*, **38,** 146
362. Gorbunova, Yu. E., Kuznetsov, V. G. and Kovaleva, E. S. (1968). *Zh. Neorg. Khim.*, **13,** 102
363. Khan, M. M. and Malik, A. U. (1972). *Ind. J. Chem.*, **10,** 234
364. Kolenkova, A. A., Lainer, A. I., Sazhina, V. A., Zakirova, A. V., Popov, A. I. and Eroshin, P. K. (1969). *Tsvet. Metal.*, **42,** 66
365. Macarovici, C. G. and Strajescu, M. (1968). *Rev. Roum. Chim.*, **13,** 1477
366. Nekhamkin, L. G., Kozlova, V. K., Laube, I. G. and Novoselova, G. P. (1968). *Nauch. Tr. Gos. Nauch.-Issled. Proekt. Inst. Redkomental. Prom.*, **20,** 119
367. Paris, G., Szabo, G. and Paris, R. A. (1968). *C. R. Acad. Sci. Ser. C*, **266,** 554
368. Petit, G. and Bourlange, C. (1969). *C. R. Acad. Sci. C*, **269,** 657
369. Broadbent, D., Dollimore, D. and Dollimore, J. (1969). *Analyst (London)*, **94,** 543
370. Vinarov, I. V., Kovaleva, E. I. and Gertsenshtein, E. Sh. (1968). *Zh. Neorg. Khim.*, **13,** 294
371. Dmitrevskii, G. E., Belitskaya, A. A., Savchenko, M. I. and Kharchenko, L. P. (1968). *Zh. Neorg. Khim.*, **13,** 2663
372. Vasil'ev, V. P. and Vorob'ev, P. N. (1970). *Zh. Fiz. Khim.*, **44,** 1181
373. Gutmann, V. and Michlmayr, M. (1968). *Monatsh. Chem.*, **99,** 316
374. Barker, M. G. and Wood, D. J. (1972). *J. C. S. Dalton Trans.*, 9
375. Ranade, A. C., Müller, A. and Diemann, E. (1970). *Z. Anorg. Allg. Chem.*, **373,** 258
376. Müller, A., Ranade, A. C. and Rao, V. V. K. (1971). *Spectrochim. Acta*, **A27,** 1973
377. Ahlborn, E., Diemann, E. and Müller, A. (1972). *Chem. Comm.*, 378
378. Griffiths, I. M. and Nicholls, D. (1970). *Chem. Commun.*, 713
379. Nicholls, D., Griffith, I. M. and Seddon, K. R. (1971). *J. Chem. Soc. A*, 2513
380. Drake, J. E., Vekris, J. E. and Wood, J. S. (1968). *J. Inorg. Nucl. Chem.*, **30,** 3380
381. Davidovich, R. L., Sergienko, V. I. and Murzakhanova, L. M. (1968). *Izv. Sib. Otd. Akad. Nauk. SSSR, Ser. Khim. Nauk*, 58
382. Davidovich, R. L., Sergienko, V. I. and Murzakhanova, L. M. (1968). *Zh. Neorg. Khim.*, **13,** 3186
383. Khalilova, N. K. and Morozov, I. S. (1968). *Zh. Neorg. Khim.*, **13,** 1337
384. Khalilova, N. K. and Morozov, I. S. (1968). *Zh. Neorg. Khim.*, **13,** 988
385. Kobets, L. V., Vorob'ev, N. I., Pechkovskii, V. V. and Komyak, A. I. (1971). *Zh. Prkl. Spektrosk.*, **15,** 682
386. Morozov, A. I. and Karlova, E. V. (1972). *Zh. Neorg. Khim.*, **17,** 669
387. Nicholls, D. and Wilkinson, D. (1970). *J. Chem. Soc. A*, 1103
388. Pausewang, G. and Dehnicke, K. (1969). *Z. Anorg. Allg. Chem.*, **369,** 265. Pausewang, G. (1971). *Z. Anorg. Allg. Chem.*, **381,** 189
389. Sengupta, A. K. and Bhaumik, B. B. (1970). *Sci. Cult.*, **36,** 123
390. Sengupta, A. K. and Bhaumik, B. B. (1972). *Z. Anorg. Allg. Chem.*, **390,** 61
391. Sengupta, A. K. and Bhaumik, B. B. (1971). *Z. Anorg. Allg. Chem.*, **384,** 251
392. Sengupta, A. K. and Bhaumik, B. B. (1971). *Z. Anorg. Allg. Chem.*, **384,** 255
393. Sengupta, A. K. and Bhaumik, B. B. (1970). *J. Ind. Chem. Soc.*, **47,** 89
394. Piovesana, O. and Selbin, J. (1969). *J. Inorg. Nucl. Chem.*, **31,** 433
395. Galy, J., Anderson, S. and Portier, J. (1969). *Acta Chem. Scand.*, **23,** 2949
396. Djordjevic, C. and Sevdic, D. (1968). *J. Less Common Metals*, **16,** 233
397. Laitinen, H. A. and Lieto, L. R. (1972). *Croat. Chem. Acta*, **44,** 275
398. Lastochkina, A. A., Sheka, I. A. and Malinko, L. A. (1969). *Izv. Akad. Nauk. SSSR, Neorg. Mater.*, **5,** 757
399. Buslaev, Yu. A., Ii'in, E. G., Kopanev, V. D. and Gavrish, O. G. (1971). *Izv. Akad. Nauk SSSR, Ser. Khim.*, 1139
400. Nicholls, D. and Bennett, B. G. (1971). *J. Chem. Soc. A*, 1204
401. Boehland, H. and Schneider, F. M. (1972). *Z. Anorg. Allg. Chem.*, **390,** 53
402. Jahr, K. F., Preuss, F. and Rosenhahn, L. (1969). *J. Inorg. Nucl. Chem.*, **31,** 297
403. Karayannis, M. N., Strocko, M. J., Mikulski, C. M., Bradshaw, E. E., Pytlewski, L. L., Labes, M. M. (1970). *J. Inorg. Nucl. Chem.*, **32,** 3962
404. Nikolov, G. S. and Nikolova, B. M. (1968). *J. Inorg. Nucl. Chem.*, **38,** 331
405. Preuss, F., Ussat, W. and Wegener, K. (1969). *Chem. Ber.*, **102,** 3046
406. Preuss, F. and Ussat, W. (1969). *Chem. Ber.*, **102,** 3057
407. Schmauss, G. and Specker, H. (1969). *Z. Anorg. Allg. Chem.*, **364,** 1

408. Babko, A. K. Mazurenko, E. A. and Nabivanets, B. I. (1968). *Zh. Neorg. Khim.*, **13,** 718
409. Brnicevic, N. and Djordjevic, C. (1968). *Inorg. Chem.*, **7,** 1936
410. Brnicevic, N. and Djordjevic, C. (1971). *J. Less Common Metals*, **23,** 10
411. Morozov, I. S. and Lipatova, N. P. (1968). *Zh. Neorg. Khim.*, **13,** 2128
412. Ackermann, G. and Koch, S. (1969). *Talanta*, **16,** 284
413. Buslaev, Yu. A. and Kokunov, Yu. V. (1968). *Izv. Akad. Nauk. SSSR, Neorg. Mater.*, **4,** 537
414. Lutz, B. and Wendt, H. (1970). *Ber. Bunsenges. Phys. Chem.*, **74,** 372
415. Müller, A. and Krebs, B. (1967). *Spectrochim. Acta*, **A23,** 1591
416. Dietsch, J. F., Muller, M. and Dehand, J. (1971). *C. R. Acad. Sci. Ser. C*, **272,** 541
417. Edge, R. A. (1969). *J. Less Common Metals*, **18,** 325
418. Fotiev, A. A. and Golovkin, B. G. (1970). *Tr. Inst. Khim. Akad. Nauk. SSSR, Ural. Filial* No. 20, 3
419. Ivakin, A. A. and Voronova, E. M. (1970). *Tr. Inst. Khim. Akad. Nauk. SSSR, Ural. Filial.*, No. 17, 139
420. Ivakin, A. A. and Voronova, E. M. (1969). *Zh. Neorg. Khim.*, **14,** 1557
421. Ivakin, A. A. and Yatsenko, A. P. (1971). *Zh. Neorg. Khim.*, **16,** 1689
422. Mazurenko, E. A. and Nabivanets, B. I. (1969). *Zh. Neorg. Khim.*, **14,** 3286
423. Morozov, A. I. (1969). *Zh. Prikl. Khim.* (*Leningrad*), **42,** 757
424. Pyatnitskii, I. V. and Sereda, E. S. (1968). *Ukr. Khim. Zh.*, **34,** 1162
425. Smirnova, E. K., Vasil'kova, I. V. and Prokhorova, L. I. (1968). *Zh. Neorg. Khim.*, **13,** 1791
426. Smirnova, E. K. and Tsintsius, W. M. (1971). *Zh. Neorg. Khim.*, **16,** 1454
427. Surat, L. L. and Golovkin, B. G. (1970). *Tr. Inst. Khim. Akad. Nauk SSSR, Ural. Filial*, No. **20,** 8
428. Vasil'kova, I. V., Smirnova, E. K. and Smirnova, E. D. (1969). *Vestn. Leningrad Univ. Fiz. Khim.*, 107
429. Voronova, E. M. and Ivakin, A. A. (1969). *Zh. Neorg. Khim.*, **14,** 1564
430. Zhurenkov, E. M. and Pobezhimovskaya, D. N. (1970). *Radio-Khimiya*, **12,** 105
431. Dartiguenave, Y., Dartiguenave, M. and Walter, J. P. (1969). *Bull. Soc. Chim. Fr.*, 2287
432. Gutmann, V. and Laussegger, H. (1968). *Monatsh. Chem.*, **99,** 947
433. Gutmann, V. and Laussegger, H. (1968). *Monatsh. Chem.*, **99,** 963
434. McCleverty, J. A., Locke, J., Ratcliff, B. and Wharton, E. J. (1969). *Inorg. Chim. Acta*, **3,** 283
435. Astakhina, N. S. and Gurevich, V. G. (1970). *Ukr. Khim. Zh.*, **36,** 886
436. Scholder, R., Schwochow, E. F. and Schwarz, H. (1968). *Z. Anorg. Allg. Chem.*, **363,** 10
437. Reau, J. M. and Fouassier, C. (1971). *Bull. Soc. Chim. Fr.*, 398
438. Kessler, H., Hatterer, A. and Ringenbach, C. (1970). *C. R. Acad. Sci. Ser. C*, **270,** 815
439. Kessler, H. and Hatterer, A. (1972). *C. R. Acad. Sci. Ser. C*, **274,** 623
440. Jain, D. V. S. (1970). *Ind. J. Chem.*, **8,** 945
441. Müller, A., Diemann, E., Krebs, B. and Leroy, M. J. F. (1968). *Angew. Chem. Int. Ed.*, **7,** 817
442. Müller, A., Diemann, E. and Heidborn, U. (1969). *Z. Anorg. Allg. Chem.*, **371,** 136
443. Müller, A. and Diemann, E. (1969). *Chem. Ber.*, **102,** 3277
444. Müller, A. and Diemann, E. (1969). *Chem. Ber.*, **102,** 2044
445. Müller, A. and Diemann, E. (1969). *Chem. Ber.*, **102,** 945
446. Müller, A. and Diemann, E. (1969). *Chem. Ber.*, **102,** 2603
447. Müller, A., Diemann, E. and Schulze, H. (1970). *Z. Anorg. Allg. Chem.*, **376,** 120
448. Müller, A. and Diemann, E. (1970). *Z. Anorg. Allg. Chem.*, **373,** 57
449. Müller, A. and Heinsen, H. H. (1972). *Chem. Ber.*, **105,** 1730
450. Müller, A., Heinsen, H. H. and Vandrish, G. (1974). *Inorg. Chem.*, in the press
451. Müller, A., Ahlborn, E. and Heinsen, H. H. (1971). *Z. Anorg. Allg. Chem.*, **386,** 102
452. Müller, A. and Diemann, E. (1971). *Chem. Commun.*, 65
453. Müller, A., Diemann, E. and Heinsen, H. H. (1971). *Chem. Ber.*, **104,** 975
454. Aymonino, P. J., Ranade, A. C. and Müller, A. (1969). *Z. Anorg. Allg. Chem.*, **371,** 295
455. Aymonino, P. J., Ranade, A. C., Diemann, E. and Müller, A. (1969). *Z. Anorg. Allg. Chem.*, **371,** 300
456. Müller, A., Sievert, W. and Schulze, H. (1972). *Z. Naturforsch.*, **B27,** 720
457. Ahlborn, E., Diemann, E. and Müller, A. (1972). *Z. Naturforsch.*, **B27,** 1108
458. Diemann, E., Ahlborn, E. and Müller, A. (1972). *Z. Anorg. Allg. Chem.*, **390,** 217
459. Karyakin, Yu. V. and Kryachko, E. N. (1968). *Tr. Voronezh. Tekhnol. Inst.*, **17,** 110
460. Opalovskii, A. A. and Batsanov, S. S. (1968). *Zh. Neorg. Khim.*, **13,** 2135
461. Youinou, M. T., Petillon, F. and Guerchais, J. E. (1968). *Bull. Soc. Chim. Fr.*, 503, 2375
462. Saha, H. K. and Banerjee, A. K. (1968). *J. Ind. Chem. Soc.*, **45,** 660

463. Colton, R. and Rose, G. G. (1968). *Aust. J. Chem.*, **21,** 883
464. Youinou, M. T. and Guerchais, J. E. (1968). *Bull. Soc. Chim. Fr.*, 40
465. Glinkina, M. I. (1969). *Geokhimiya*, 797
466. Buslaev, Yu. A., Petrosyants, S. P. and Chagin, V. I. (1972). *Zh. Neorg. Khim.*, **17,** 704
467. Marinina, L. K., Rakov, E. G., Bratishko, V. D., Gromov, B. V. and Kokanov, S. A. (1970). *Zh. Neorg. Khim.*, **15,** 3279
468. Pausewang, G. and Rüdorff, W. (1969). *Z. Anorg. Allg. Chem.*, **364,** 69
469. Pausewang, G. (1971). *Z. Naturforsch.*, **B26,** 1218
470. Saha, H. K. and Banerjee, A. K. (1969). *J. Ind. Chem. Soc.*, **46,** 680
471. Saha, H. K. and Banerjee, A. K. (1971). *J. Inorg. Nucl. Chem.*, **33,** 2989
472. Akena, A. M., Brown, D. S. and Tuck, D. G. (1971). *Can. J. Chem.*, **49,** 1505
473. Armstrong, R. and Gibson, N. A. (1968). *Aust. J. Chem.*, **21,** 897
474. Camelot, M. (1969). *Rev. Chim. Miner.*, **6,** 853
475. Feltz, A. and Sennewald, E. (1968). *Z. Anorg. Allg. Chem.*, **358,** 29
476. Kepert, D. L. and Mandyczewsky, R. (1968). *J. Chem. Soc. A*, 530
477. Marov, I. N., Belyaeva, V. K., Dubrov, Yu. N. and Ermarkov, A. N. (1970). *Zh. Neorg. Khim.*, **15,** 3031
478. Saha, H. K. and Halder, M. C. (1971). *J. Inorg. Nucl. Chem.*, **33,** 3719
479. Saha, H. K. and Banerjee, A. K. (1972). *J. Inorg. Nucl. Chem.*, **34,** 1861
480. Saha, H. K., Roy, S. and Chakraverty, S. (1972). *J. Ind. Chem. Soc.*, **49,** 299
481. Saha, H. K. and Halder, M. C. (1972). *J. Inorg. Nucl. Chem.*, **34,** 3097
482. Winfield, J. M., McFarlane, W. and Noble, A. M. (1971). *J. Chem. Soc. A*, 948
483. Youinou, M. T., Petillon, F. and Guerchais, J. E. (1969). *Bull. Soc. Chim. Fr.*, 1589
484. Gard, G. L. and Gerlach, J. N. (1971). *Inorg. Chem.*, **10,** 1541
485. Gopalakrishnan, J., Viswanathan, B. and Srinivasan, V. (1970). *J. Inorg. Nucl. Chem.*, **32,** 2565
486. Marov, I. N., Dubrov, Yu. N., Ermarkov, A. N. and Martynova, G. N. (1968). *Zh. Neorg. Khim.*, **13,** 3247
487. Mitchell, P. C. H. (1969). *J. Chem. Soc. A*, 146
488. Saha, H. K. and Banerjee, A. K. (1972). *J. Inorg. Nucl. Chem.*, **34,** 697
489. Spivack, B. and Dori, Z. (1970). *J. Chem. Soc. A*, 1716
490. Willis, C. J. (1972). *J. C. S. Chem. Commun.*, 944
491. Lamache-Duhameaux, M. (1968). *Rev. Chim. Miner.*, **5,** 1001
492. Baldea, I. (1969). *Stud. Univ. Babes-Bolyai, Ser. Chem.*, **14,** 77
493. Baldea, I. and Niac, G. (1968). *Inorg. Chem.*, **7,** 1232
494. Kordes, E. and Nolte, G. (1969). *Z. Anorg. Allg. Chem.*, **371,** 156
495. Lin, C. T. and Beatti, J. K. (1972). *J. Amer. Chem. Soc.*, **94,** 3011
496. Marinina, L. K., Rakov, E. G., Bratishko, V. D., Gromov, B. V. and Kokanov, S. A. (1970). *Tr. Mosk. Khim.-Tekhnol. Inst., Nr.*, **67,** 83, 86
497. Marov, I. N., Dubrov, Yu. N., Ermarkov, A. N. and Martynova, G. N. (1969). *Zh. Neorg. Khim.*, **14,** 438
498. Marov, I. N., Dubrov, Yu. N., Ermarkov, A. N. and Martynova, G. N. (1968). *Dokl. Akad. Nauk. SSSR*, **181,** 111
499. Marov, I. N., Dubrov, Yu. N., Evtikova, G. A., Belyaeva, V. K., Ermarkov, A. N. and Korovaikov, P. A. (1970). *Zh. Neorg. Khim.*, **15,** 2227
500. Nazarenko, V. A. and Shelikhina, E. I. (1971). *Zh. Neorg. Khim.*, **16,** 166
501. Opalovskii, A. A. and Batsanov, S. S. (1968). *Zh. Neorg. Khim.*, **13,** 533
502. Opalovskii, A. A., Batsanov, S. S., Kuznetsova, Z. M. and Nesterenko, M. N. (1968). *Izv. Sib. Otd. Akad. Nauk. SSSR, Ser. Khim. Nauk.*, 15
503. Park, I. H. (1972). *Bull. Chem. Soc. Jap.*, **45,** 2749, 2753
504. Pyatnitskii, I. V. and Kravtsova, L. F. (1968). *Ukr. Khim. Zh.*, **34,** 231
505. Pyatnitskii, I. V. and Kravtsova, L. F. (1968). *Ukr. Khim. Zh.*, **34,** 86
506. Van den Akker, A. W. M., Koster, A. S. and Rieck, G. D. (1970). *J. Appl. Crystallogr.*, **3,** 389
507. Wendling, E. and Lavillandre, de J. (1968). *Bull. Soc. Chim. Fr.*, 866
508. Hepler, K. G. (1969). *Can. J. Chem.*, **47,** 3469
509. Karov, Z. G. and Bagov, I. Kh. (1969). *Uch. Zap. Kabardino-Balkar. Gos. Univ.*, No. 41, 435
510. Linge, H. G. and Jones, A. L. (1968). *Aust. J. Chem.*, **21,** 2189
511. Shidlovskii, A. A., Balakireva, T. N. and Voskresenskii, A. A. (1971). *Zh. Fiz. Khim.*, **45,** 1857
512. Smith, J. A. and Metz, R. C. (1970). *Proc. Indiana Acad. Sci.*, **80,** 159
513. Prasad, T. P., Diemann, E. and Müller, A. (1973). *J. Inorg. Nucl. Chem.*, **35,** 1895

514. Jones, A. L. and Linge, H. G. (1969). *Aust. J. Chem.*, **22,** 663
515. Krishnaiah, K. S. R. (1968). *Proc. Ind. Acad. Sci. A*, **67,** 222
516. Lamache-Duhameaux, M. (1968). *Rev. Chim. Miner.*, **5,** 459
517. Piovesana, O. (1969). *Gazz. Chim. Ital.*, **99,** 86
518. Rakowska, E. (1968). *Rocz. Chem.*, **42,** 1567
519. Sladky, J. (1969). *Czech. J. Phys.*, **19,** 123
520. Chretien, A. and Duquenoy, G. (1969). *C. R. Acad. Sci. Ser. C*, **268,** 509
521. Duquenoy, G. (1971). *Rev. Chim. Miner.*, **8,** 683
522. Duquenoy, G. (1969). *C. R. Acad. Sci. Ser. C*, **268,** 828
523. Le Flem, G., Olazcuaga, R., Parant, J. P., Reau, J. M. and Fouassier, C. (1971). *C. R. Acad. Sci. Ser. C*, **273,** 1358
524. Olazcuaga, R., Reau, J. M. and LeFlem, G. (1972). *C. R. Acad. Sci. C*, **275,** 135
525. Lee, M. R. and Freundlich, W. (1969). *C. R. Acad. Sci. Ser. C*, **268,** 2302
526. Baud, G. and Capestan, M. (1968). *Bull. Soc. Chim. Fr.*, 3999
527. Baran, E. J. and Müller, A. (1969). *Z. Anorg. Allg. Chem.*, **368,** 186
528. Semenov, G. A., Nikolaev, E. N. and Opendak, I. G. (1972). *Zh. Neorg. Khim.*, **17,** 1819
529. Petrov, K. I., Bardin, V. A. and Kalyazhnaya, V. G. (1968). *Dokl. Akad. Nauk. SSSR*, **178,** 1097
530. Bol'shakov, K. A., Sinitsyn, N. M., Petrov, K. I., Travkin, V. F. and Rubtsov, M. V. (1968). *Zh. Neorg. Khim.*, **13,** 3082
531. Bol'shakov, K. A., Sinitsyn, N. M., Travkin, V. F. and Morozova, L. M. (1972). *Izv. Vyssh. Ucheb., Khim. Khim. Tekhnol.*, **15,** 334
532. Paul, R. C., Puri, J. K., Kapila, V. P. and Malhotra, K. C. (1971). *Ind. J. Chem.*, **9,** 1387
533. Majumdar, S. K., Pacer, R. A. and Rulfs, C. L. (1969). *J. Inorg. Nucl. Chem.*, **31,** 33
534. Frigerio, A. N. (1969). *J. Amer. Chem. Soc.*, **91,** 6200
535. Covington, A. K., Freeman, J. G. and Lilley, T. H. (1969). *Trans. Faraday Soc.*, **65,** 3136
536. Lenz, E. and Murmann, R. K. (1968). *Inorg. Chem.*, **7,** 1880
537. Bol'shakov, K. A., Sinitsyn, N. M. and Travkin, V. F. (1969). *Dokl. Akad. Nauk. SSSR*, **185,** 338
538. Borisova, L. V., Marov, I. N., Dubrov, Yu. N. and Ermakov, A. N. (1971). *Zh. Neorg. Khim.*, **16,** 3026
539. Edwards, D. A. and Ward, R. T. (1970). *J. Chem. Soc. A*, 1617
540. Edwards, D. A. and Ward, R. T. (1972). *J. C. S. Dalton Trans.*, 89
541. Spitsyn, V. I., Glinkina, M. I. and Kuzina, A. F. (1971). *Dokl. Akad. Nauk. SSSR*, **200,** 1372
542. Swarnakar, R. D. and Chakrabarty, D. K. (1972). *Ind. J. Chem.*, **10,** 528
543. Chakravorti, M. C. and Chaudhuri, M. K. (1972). *J. Inorg. Nucl. Chem.*, **34,** 3479
544. Selig, H. and Karpas, Z. (1971). *Israel J. Chem.*, **9,** 53
545. Mitschler, A., LeCarpentier, J. M. and Weiss, R. (1968). *Chem. Commun.*, 1260
546. Morozov, A. I. and Leonova, I. I. (1972). *Zh. Neorg. Khim.*, **17,** 2128
547. Rohwer, E. F. C. H. and Cruywagen, J. J. (1969). *J. S. Afr. Chem. Inst.*, **22,** 198
548. Zeltmann, A. H. and Morgan, L. O. (1971). *Inorg. Chem.*, **10,** 2739
549. Krebs, B. and Müller, A. (1967). *Z. Chem.*, **7,** 243
550. Müller, A., Böschen, I. and Baran, E. J. (1973). *Monatsh. Chem.*, **104,** 821
551. Lastochkina, A. A., Sheka, I. A. and Malinko, L. A. (1970). *Izv. Akad. Nauk. SSSR, Neorg. Mater.*, **6,** 897
552. Mathieu, B., Ladriere, J., Cambier, J., Apers, D. and Capron, (1972). *Bull. Soc. Chim. Belg.*, **81,** 343
553. Petit, R. H., Briat, B., Müller, A. and Diemann, E. (1973). *Chem. Phys. Lett.*, **20,** 540
554. Petit, R. H., Briat, B., Müller, A. and Diemann, E. (1974). *Mol. Phys.*, in the press
555. Toppen, D. L. and Murmann, R. K. (1970). *Inorg. Nucl. Chem. Lett.*, **6,** 139
556. Tamaki, M., Masuda, I. and Shinra, K. (1972). *Chem. Lett.*, 165
557. Münze, R. (1968). *Z. Phys. Chem. (Leipzig)*, **238,** 364
558. Boldyrev, V. V., London, G. B. and Zhuravlev, V. K. (1968). *Phys. Status Solidi*, **30,** K13
559. Erenburg, B. G., Senchenko, L. N., Boldyrev, V. V. and Malysh, A. V. (1972). *Zh. Neorg. Khim.*, **17,** 2154
560. Gross, M., Lemoine, P. and Brenet, J. (1968). *C. R. Acad. Sci. Ser. C*, **267,** 1384
561. Gross, M., Lemoine, P. and Brenet, J. (1970). *Electrochim. Acta*, **15,** 251
562. Henze, G. and Geyer, R. (1968). *Z. Chem.*, **8,** 437
563. Lektorskaya, N. A. and Nguyen Van Ngok (1968). *Izv. Vyssh. Ucheb. Zaved. Khim. Tekhnol.*, **11,** 383

564. Mathur, P. K. and Venkateswarlu, K. S. (1970). *Report Bhabha Atom Res. Cent.*, I–82
565. Popov, G. R. (1968). *Khim. Ind.* (*Sofia*), 166
566. Rubinskaya, T. Ya. and Mairanovskii, S. G. (1971). *Elektrokhimiya*, **7**, 1403
567. Shropshire, J. A. (1968). *J. Electroanal. Chem. Interfac. Electrochem.*, **16**, 275
568. Giurgiu, M. (1971). *Stud. Univ. Babes-Bolyai, Ser. Chem.*, **16**, 127
569. Beard, J. H., Calhoun, C., Casey, J. and Murmann, R. K. (1968). *J. Amer. Chem. Soc.*, **90**, 3389
570. Baran, E. J. and Aymonino, P. J. (1968). *Monatsh. Chem.*, **99**, 606
571. Brough, B. J., Habboush, D. A. and Kerridge, D. H. (1972). *Inorg. Chim. Acta*, **6**, 366
572. Busey, R. H., Bevan, R. B. and Gilbert, R. A. (1972). *J. Chem. Thermodyn.*, **4**, 77
573. Haugen, G. R. and Friedman, H. L. (1968). *J. Phys. Chem.*, **72**, 4549
574. Herbstein, F. H., Ron, G. and Weissman, A. (1971). *J. Chem. Soc. A*, 1821
575. Ianovic, E. and Zaitseva, N. G. (1969). *J. Inorg. Nucl. Chem.*, **31**, 2669
576. Kornilova, N. V., Dymarchuk, N. P. and Mishchenko, K. P. (1971). *Zh. Prikl. Khim.* (*Leningrad*), **44**, 2121
577. Lux, H., Renauer, E. and Findeiss, W. (1972). *Z. Anorg. Allg. Chem.*, **390**, 303
578. Oblivantsev, A. N. and Boldyrev, V. V. (1968). *Kinet. Katal.*, **9**, 930
579. Petriashvili, L. D. and Agladze, R. I. (1969). *Elektrokhim. Morgantsa*, **4**, 284
580. Ratner, Y. E., Tsvetkov, Yu. V. and Berezkina, L. G. (1968). *Zh. Neorg. Khim.*, **13**, 1516
581. Semenov, G. A. and Shalkova, E. K. (1969). *Vestn. Leningrad Univ., Fiz. Khim.*, 111
582. Skudlarski, K. and Lukas, W. (1972). *Nukleonika*, **17**, 189
583. Veprek-Siska, J. and Ettel, V. (1969). *J. Inorg. Nucl. Chem.*, **31**, 789
584. Ettel, V. and Veprek-Siska, J. (1969). *Collect. Czech. Chem. Commun.*, **34**, 2182
585. Haissinsky, M. and Dran, J. C. (1968). *J. Chim. Phys. Physichochim. Biol.*, **65**, 321
586. Ishiyama, T. and Matsumara, T. (1970). *Annu. Rep. Radiat., Cent. Osaka Prefect.*, **11**, 50
587. Pawson, D. and Griffith, W. P. (1972). *Chem. Ind.* (*London*), 609
588. Bol'shakov, K. A., Sinitsyn, N. M., Borisov, V. V. and Borbat, V. F. (1969). *Dokl. Akad. Nauk. SSSR*, **188**, 815
589. Broomhead, J. A., Dwyer, F. P., Goodwin, H. A., Kane-Maguire, L. and Reid, I. (1968). *Inorg. Synth.*, **11**, 70
590. Spencer, A. and Wilkinson, G. (1972). *J. C. S. Dalton Trans.*, 1570
591. Bardin, M. B. and Goncharenko, V. P. (1970). *Zh. Neorg. Khim.*, **15**, 490
592. Connery, J. G. and Cover, R. E. (1968). *Anal. Chem.*, **40**, 87
593. Müller, A. and Bollmann, F. (1968). *Z. Naturforsch.*, **B23**, 1539
594. Müller, A., Diemann, E. and Bollmann, F. (1968). *Naturwissenschaften*, **55**, 443
595. Baran, E. J., Müller, A., Kebabcioglu, R., Bollmann, F. and Aymonino, P. J. (1970). *An. Asoc. Quim. Argent.*, **58**, 247
596. Poddar, S. N. and Podder, N. G. (1968). *J. Ind. Chem. Soc.*, **45**, 562
597. Venkatadri, A. S., Wagner, W. F. and Bauer, H. H. (1971). *Anal. Chem.*, **43**, 1115
598. Mercer, E. E. and Farrar, T. D. (1969). *Can. J. Chem.*, **47**, 581
599. Gonzales-Vilchez, F. and Griffith, W. P. (1972). *J. C. S. Dalton Trans.*, 1416
600. Königer, F. and Müller, A., unpublished results
601. McGinnety, J. A. (1972). *Acta Crystallogr.*, **B28**, 2845
602. Hackert, M. L. and Jacobson, R. A. (1971). *J. Solid State Chem.*, **3**, 364
603. Carter, R. L. and Margulis, T. N. (1972). *J. Solid State Chem.*, **5**, 75
604. Koz'min, P. A. and Surazhskaya, M. D. (1968). *Zh. Strukt. Khim.*, **9**, 917
605. Gatehouse, B. M. and Leverett, P. (1969). *J. Chem. Soc. A*, 1857
606. Stephens, J. S. and Cruickshank, D. W. J. (1970). *Acta Crystallogr.*, **B26**, 437
607. Thrierr-Sorel, A. and Lantemant, M. (1969). *C. R. Acad. Sci. Ser. C*, **268**, 1748
608. Mattausch, Hj. and Müller-Buschbaum, H. K. (1972). *Z. Naturforsch.*, **B27**, 739
609. Panagiotopolos, N. Ch. and Brown, I. D. (1972). *Acta Crystallogr.*, **B28**, 1352
610. Brandon, J. K. and Brown, I. D. (1968). *Can. J. Chem.*, **46**, 933
611. Gatehouse, B. M. and Leverett, P. (1969). *J. Chem. Soc. A*, 849
612. Koster, A. S., Kools, F. X. N. M. and Rieck, G. D. (1969). *Acta Crystallogr.*, **B25**, 1704
613. Gonschorek, W., Hahn, Th. and Müller, A. (1973). *Z. Kristallogr.*, **138**, 380
614. Krebs, B., Buss, B. and Ferwanah, A. (1972). *Z. Anorg. Allg. Chem.*, **387**, 142
615. Müller, A., Weinstock, N., Krebs, B., Buss, B. and Ferwanah, A. (1971). *Z. Naturforsch.*, **B26**, 268
616. Krebs, B., Müller, A. and Kindler, E. (1970). *Z. Naturforsch.*, **B25**, 222
617. Atovmyan, L. O. and D'yachenko, O. A. (1969). *Zh. Strukt. Khim.*, **10**, 504

618. Palenik, G. J. (1967). *Inorg. Chem.*, **6,** 507
619. Hardy, A. and Fourre, B. (1971). *C. R. Acad. Sci. Ser. C*, **273,** 1508
620. Matausch, M. and Müller-Buschbaum, H. (1971). *Z. Anorg. Allg. Chem.*, **386,** 1
621. Kalman, A. (1971). *J. Chem. Soc. A*, 1857

4
Complexes of Anionic Species with the Transition Metals

R. E. HESTER
University of York

4.1 INTRODUCTION

This review deals with a selection of the many papers treating aspects of the topic defined by the title and published during the period late 1970 to early 1973. The topic may further be defined as the chemistry of transition metal complexes with polyatomic anionic species wherein the specific metal–ligand interaction occurs through oxygen or nitrogen atoms. The range of ligand species involved represents an extension of the set covered by Chapter 1 of Vol. 5 of Series One[1] and includes some species, such as alkoxides, which were specifically excluded from the earlier work due to their being treated separately in other chapters. Again a more clear-cut arrangement of the material has been possible on the basis of the periodic classification of the metals than the alternative of ligand types. These latter range from simple oxoanions, such as NO_3^- and CO_3^{2-}, through to large polydentate species, such as nitrilotriacetate (NTA) and porphyrins. Although the literature coverage is far from exhaustive, it is intended to be representative, covering synthetic, thermodynamic and kinetic studies as well as the predominant structural work.

A number of general reviews containing material relating to this chapter have appeared during the period covered though, naturally, these relate only to original literature of earlier date. Particularly thorough among these are the excellent *Specialist Periodical Reports* of the Chemical Society[2,3]. Other reviews of a more particular nature relating to the material of this chapter treat criteria for distinguishing between inner- and outer-sphere complexes[4], water-exchange kinetics in labile aquo and substituted aquo transition metal ions by means of ^{17}O n.m.r. studies[5], kinetics and mechanisms of isomerisation and racemisation processes of 6-coordinate chelate complexes[6], the role of the carboxylate group in octahedral substitution[7], the geometry of five coordinate complexes[8], optical and geometrical isomerisation of β-diketone complexes[9], metal complexes of thio-β-diketones[10,11], structural aspects of metal complexes with some tetradentate Schiff bases[12], *N,N'*-ethylene-bis(salicylideneiminato) transition metal ion chelates[13], the coordination chemistry of metalloporphyrins[14], stereochemistry of bis-chelate metal(II) complexes[15], bond character of β-diketonate metal chelates[16], structural aspects of coordinated nitrate groups[17] and the hydroxyl ion as a ligand[18].

4.2 THE SCANDIUM GROUP

Exclusion of the lanthanide and actinide elements reduces this group to the elements scandium and yttrium. A thorough review of the coordination chemistry of scandium has been published[19], including β-diketonates, carboxylates, alcoholates (alkoxides) and hydroxoscandates, as well as complexes with nitrogen donor ligands and both nitrogen and oxygen donor ligands. The first mixed-ligand chelates of Sc^{III} have been investigated[20]. These complexes involved the 1:1 scandium(III)ethylenediaminetetra-acetic acid (EDTA) chelates with various secondary ligands, a potentiometric method being used to determine the nature and stability of the mixed-ligand species formed. It was found that 1:1:1 Sc^{III}–EDTA–secondary ligand chelates were more stable than the corresponding Y^{III} chelates, but they were more susceptible to hydrolysis as a result of the smaller ionic radius of scandium. Secondary ligands used in this study were 2,3-dihydroxynaphthalene-6-sulphonate (DHN), 1,2-dihydroxybenzene-3,5-disulphonate (Tiron) and pyrocatechol (PY), all of these being bidentate.

Both Sc^{III} and Y^{III} have been found[21] to form monoethylester complexes by reaction of MCl_3 with diethylpyridine-2-phosphate, diethyl-4-methylpyridine-2-phosphate and diethyl-6-methylpyridine-2-phosphate. Evidence from i.r. spectroscopy, solubility and thermal stability characteristics of these new complexes suggested that they were monomeric hexa-coordinated metal chelates, with the monoacid phosphonate esters acting as bidentate O,N-

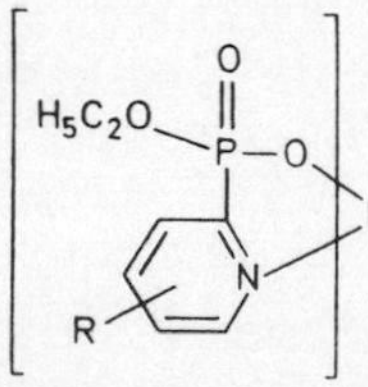

Figure 4.1 Proposed structure for the ethoxypyridine- or methylpyridine-2-phosphonato complexes of Sc^{III} and Y^{III} [21]

ligands as shown in Figure 4.1. The carbonate ion also has been found[22] to act as a bidentate ligand in coordination with Sc^{III} in the complex anion $[Sc(CO_3)_4]^{5-}$. Raman polarisation data have been used to distinguish this arrangement from monodentate carbonate. Other simple complexes studied include $Sc(ClO_3)_3$ and $Sc(BrO_3)_3$, for which stoichiometric stability constants have been determined at several temperatures[23]. The reaction

$$Sc^{3+} + H_2PO_4^- \rightleftharpoons ScPO_4 + 2H^+$$

has been reported[24] to have an equilibrium constant of 9.4, as determined from spectrophotometric measurements, and the related complexes $Sc(H_2PO_4)^{2+}$ and $Sc(H_2PO_4)_2^+$ have also been investigated[25]. The structures of phosphite and hypophosphite complexes of scandium, $Sc(H_2PO_3)_3$, $Sc_2(HPO_3)_3$, $Sc_2(HPO_3)_3 \cdot 4H_2O$, and $Sc(H_2PO_2)_3$, have been studied by i.r. and n.m.r. methods[26]. A trifluoroacetate of Sc^{III} has been synthesised[27] by reaction between freshly-prepared $Sc(OH)_3$ and a cooled aqueous solution of CF_3CO_2H. Yttrium has been studied along with the lanthanide elements[28] and a number of its mixed-ligand complexes with fluorinated β-diketones and neutral donors such as Bu_3PO_4, Bu_3PO and Bu_2SO have been determined by gas chromatography.

4.3 THE TITANIUM GROUP

4.3.1 Titanium

The structures of two crystals of the binuclear complex di-μ-oxo-bis[diacetylacetonatotitanium(IV)], $[TiO(C_5H_7O_2)_2]_2$, have been determined by x-ray single-crystal methods[29]. The two crystals differed from one another in that one contained two molecules of dioxane, but both showed the complex to be based on a planar cyclic di-μ-oxo-dititanium ring slightly distorted from a square, with the acetylacetonate rings being very nearly planar. The structure displayed in Figure 4.2 shows that the octahedral

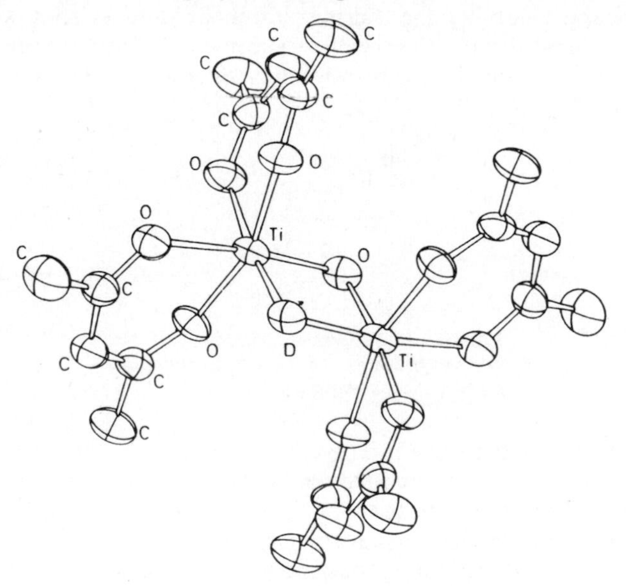

Figure 4.2 ORTEP drawing of $\{TiO(acac)_2\}_2$. (From Smith[29], by courtesy of the American Chemical Society.)

coordination around titanium is somewhat distorted; angles expected to be 90 degrees vary from 83 degrees to 100 degrees, while those expected to be 180 degrees differ from this by as much as 20 degrees. The tris(hexafluoroacetyl-acetonato)titanium(III) complex, $Ti(hfa)_3$, has been subjected to photo-electron spectroscopic (PE) study[30], yielding the spectrum shown in Figure

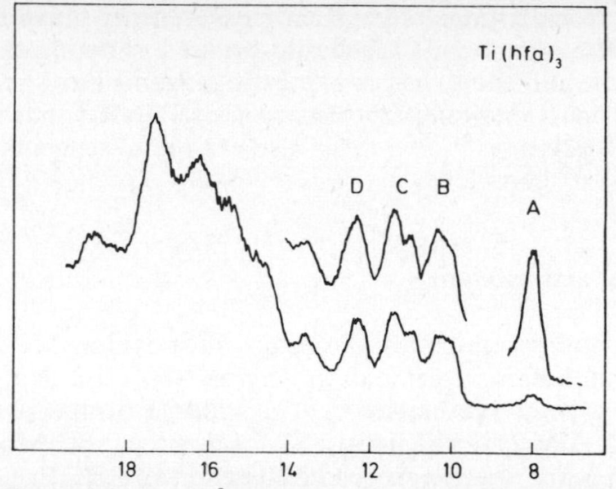

Figure 4.3 The He^I photo-electron spectrum of Ti(hfa)$_3$ (the electron count rate is plotted against ionisation energy in eV). (From Evans[30], by courtesy of Gordon and Breach Science Publishers, Ltd.)

4.3. This shows a high ionisation energy structure above about 14 eV characteristic of the enol form of the free protonated ligand and associated with ionisation of the ligand σ-electron framework and the fluorine lone-pair electrons. Bands between 10 and 13 eV are believed to arise from the leading metal–ligand bonding orbitals. Comparison with the PE spectrum of Sc(hfa)$_3$ reveals that the additional band (A) observed at 7.98 eV in the Ti(hfa)$_3$ spectrum must be due to ionisation of the single unpaired electrons localised mainly on the metal.

The heat of combustion and enthalpy of formation of tetraethoxytitanium(IV) have been determined[31], and a number of trisalkoxytitanium(IV) complexes investigated[32, 33]. These latter compounds, of general form (RO)$_3$TiL, where L is an oxygen-chelated β-diketonate ligand, were earlier thought to involve 5-coordinate titanium(IV), though the new evidence suggests mixtures of Ti(OR)$_4$ and Ti(OR)$_2$L$_2$ for most cases. Other alkoxides studied include the orange-red compound Ti(OPh)$_4 \cdot$PhOH, which has been shown[34] to have a dimeric structure with octahedrally-coordinated alkoxytitanium units, and the series of compounds[35] Ti(OR)$_{4-x}$L$_x$, with $x = 1$–4, R = Pr^i when RH = oxime, and R = Et, Pr^i, or Bu^i when LH = Et$_2$NOH.

A number of double salts of titanium involving simple oxoanions, including the species BaTi(BO$_3$)$_2$, prepared[36] by cooling molten solutions of BaO$\cdot x$B$_2$O$_3$ and TiO$_2$, the compounds MTi$_2$(PO$_4$)$_3$ (M = Li, Na, Ag, Rb, or Tl), prepared[37] by heating together the salts M$_2$CO$_3$, (NH$_4$)$_2$HPO$_4$ and TiO$_2$, and BaTi(SO$_4$)$_3$[38] have been studied. The dissociation constant of the Ti(C$_2$O$_4$)$_2^-$ ion in aqueous HCl has been determined (p$K = 9$ at 298 K and infinite dilution)[39].

A number of amido compounds of titanium have been studied. The compound Ti[N(SiMe$_3$)$_2$]$_3$ has been prepared[40] from TiCl$_3$(NMe$_3$)$_2$ and LiN(SiMe$_3$)$_2$ and shown to involve trigonal-planar (D_{3h}) coordination about Ti^{III}. Dialkylamides of the form [(π-Cp)$_{3-n}$Ti(NR$_2$)$_n$], with Cp = cyclopentadienyl, $n = 1$, 2, or 3, have been prepared and characterised[41]. A wide range of physical measurements have been used to establish that these compounds are amido-bridged, showing strong metal–metal interactions, and their uses as synthetic reagents have been explored. Controlled disproportionation of the compounds $Ti(NR_2)_3$ has been used to produce Ti^{II} dialkylamides[41], and these also are useful reagents for the synthesis of other Ti d^2 complexes[42].

4.3.2 Zirconium and hafnium

Several volatile double ethoxides and isoproproxides, $MHf_2(OR)_9$ and $M_2Hf_3(OR)_{14}$, of hafnium with alkali metals (M = Li, Na, or K) and $KHf(OBu^t)_5$ have been synthesised[43]. The volatile double isopropoxides $HfM(OPr^i)_7$ and $HfM_2(OPr^i)_{10}$ have also been prepared (M = Al or Ga), and their i.r. and n.m.r. spectra recorded. Reaction of $Zr(OPr^i)_4$ with acetylacetone yielded a variety of mixed alkoxide-β-diketonate complexes[44]. Confirmation of the 7-coordinate nature of $Zr(acac)_3Cl$ has been obtained[45], and rearrangement reactions of the pentagonal-bipyramidal (π-Cp)$Zr(hfac)_3$ (hfac = hexafluoroacetylacetonate) have been studied[46]. Similar kinetic studies of the zirconium and hafnium (π-Cp)$M(acac)_2X$ (X = halogen) have been made[47], characterising the stereochemical rearrangements which interchange non-equivalent acetylacetonate ligands.

Eight-coordinate Zr^{IV} has been established in the complex sulphates $K_2[Zr(SO_4)_3]\cdot 2H_2O$[48] and $Na_2[Zr(SO_4)_3]\cdot 3H_2O$[49]. A complex oxalate of stoichiometry $[Hf(C_2O_4)_5]^{6-}$ has been proposed on the basis of evidence from ion-exchange studies[50], and stereospecific effects in the 8-coordinate complexes of Zr^{IV} and Hf^{IV} with α-hydroxycarboxylates, lactates, mandelates and isopropylmandelates have been investigated[51]. Raman intensity data and depolarisation ratios have been obtained for the totally symmetric M–O vibrations of 8-coordinate β-diketonate complexes of Zr^{IV} and related to the M—O bond covalency[52]. A general review of the d^1 complexes of zirconium and hafnium(III) has been published[53].

The use of zirconium salts as active components in ion-exchange materials has been investigated. Ion-exchange properties and separation coefficients for Cs^+, Sr^{2+} and Ca^{2+} of two structurally-different phosphates prepared from $ZrOCl_2\cdot 8H_2O$ and $Zr(SO_4)_2\cdot 4H_2O$ solutions by precipitation with H_3PO_4 have been compared[54], and the optimum conditions for using zirconium phosphate[55] and zirconium silicate[56] have been determined.

Nitrogenous anionic species coordinated with zirconium and hafnium include dialkyamides in complexes of the type $M(NR_2)_4$. Insertion reactions of these with a wide range of simple AB dipolar molecules have yielded several new compounds[57]. A variety of new complexes with octa-ethylporphines have been reported[58, 59].

4.4 THE VANADIUM GROUP

4.4.1 Vanadium

Oxoanion complexes of vanadium in all the oxidation states II–V have been studied. An example of a V^{II} complex is that formed[60] with the tridentate (SOO) thiodiacetate ligand, $V[S(CH_2CO_2)_2]\cdot(H_2O)_3$. The magnetic properties of basic vanadium(III) acetate have been investigated[61] and interpreted in terms of a trinuclear cluster complex ion of the type shown in Figure 4.4.

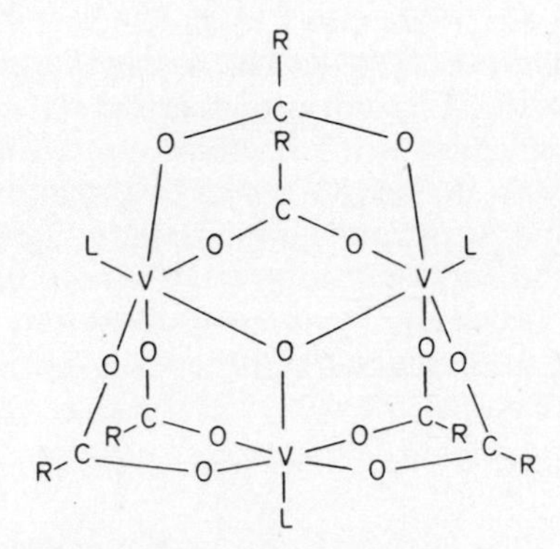

Figure 4.4 Structure of the trinuclear cation $\{V_3O(O_2CR)_6L_3\}^+$

Related spectroscopic studies of vanadium(III) carboxylates of general formula $V_3(OH)(O_2CR)_8$, and the preparation and characterisation of oxovanadium(IV) carboxylates have been reported[62]. A series of new coordination complexes of oxovanadium(IV) and oxovanadium(V) with tridentate *ONO* donor ligands have been examined[63]. The oxovanadium(IV) complexes were of the type: (i) $VOL(H_2O)$, (ii) $VOL(py)_2$, (iii) VOL(phen), (iv) $VOL'\cdot xH_2O$ and (v) VOL′(phen) (where LH_2 = 2-hydroxynaphthylideneamino acid, a tridentate dibasic ligand with *ONO* donor atoms; py = pyridine; phen = *ortho*-phenanthroline; $L'H_2$ = 2-hydroxynaphthylidene-*ortho*-aminophenol). The $VOL(H_2O)$, $VOL(py)_2$, VOL(phen) and VOL′(phen) complexes have normal magnetic moments at room temperature, but the $VOL'\cdot xH_2O$ complexes exhibit sub-normal magnetic moments at room temperature and at lower temperatures. As expected, the oxovanadium(V) complexes were found to be diamagnetic, and conductance measurements showed them to be non-electrolytes. Similar studies of the magnetochemistry of vanadyl complexes with carboxylic acid derivatives having increased coordination capacity have been reported[64]. Evidence for vanadium(IV) VOL_4 species coordinating an additional ligand *trans* to the V=O bond has been obtained[65] from dipole moment and molar Kerr constant measurements on $VO(acac)_2$ in solution with 1,4-dioxan. The 1:1 addition compound formed has been shown by x-ray analysis[66] to contain dioxan molecules bridging two $VO(acac)_2$ molecules by coordination in the sixth position.

An unusual coordination arrangement has been discovered[67] in the oxovanadium(V) compound $VO(NO_3)_2 \cdot CH_3CN$. Structural studies have shown a pentagonal-bipyramidal configuration about the vanadium atom, with five oxygen atoms from two bidentate and one monodentate nitrato groups forming the equatorial plane, and the vanadyl oxygen and CH_3CN nitrogen forming the apices. Raman data were used in these studies to distinguish between the two types of coordinated nitrato groups. Irregular octahedral coordination has been identified from x-ray crystallographic studies[68–70] of the compounds $(NH_4)_3[O_2V(C_2O_4)_2] \cdot 2H_2O$, $(NH_4)[O_2VAH_2] \cdot 3H_2O$ and $Na_3(O_2VA) \cdot 4H_2O$, with A = EDTA. The VO_2^+ group in each of these complexes has an OVO interbond angle in the range 104–107 degrees. Significant lengthening of V—O bonds *trans* to V=O was determined. Infrared spectra of the complexes VO_2Y, where Y represents ethylenediamine-*N*,*N*,*N'*,*N'*-tetra-acetate (EDTA), ethylenediamine-*N*,*N'*-diacetate (EDDA), or *N*,*N'*-dimethylethylenediamine-*N*,*N'*-diacetate(DMEDDA), have also shown the existence of VO_2 units having a *cis* conformation[71].

The compound $V[N(SiMe_3)_2]_3$ has been prepared[40] by a parallel reaction to that reported for the corresponding titanium compound, i.e. between $VCl_3(NMe_3)_2$ and $LiN(SiMe_3)_2$. The brown crystalline compound is paramagnetic, but gives no e.s.r. signal in toluene solution. It is, however, believed to be a d^2-trigonal system.

4.4.2 Niobium and tantalum

A number of compounds of the form $Nb(AA)_4$, where AA represents bidentate chelating ligands such as the β-diketones, benzoyltrifluoroactone, tropolone, or 8-quinolinol, have been prepared and established as 8-coordinate complexes[72]. These preparations were effected by direct reaction with $NbCl_4$, but the corresponding reactions with $TaCl_4$ were complicated by oxygen abstraction, and the only analogous tantalum complex formed was Ta(dibenzoylmethane)$_4$. A coordination number of 8 has also been proposed for Nb^V and Ta^V complexes of composition $M(NCS)_2(OR)_3(bipy)$ M = Nb or Ta; R = Me or Et; bipy = bipyridyl) on the basis of molecular weight determinations and spectral evidence[73]. The i.r. spectra from these compounds were interpreted as indicating the presence of two bridging alkoxo groups, nitrogen-bonded isothiocyanates in the terminal positions and normal bidentate coordination of bipyridyl, making up the asymmetrical ligand spheres shown in Figure 4.5.

The solubility of freshly precipitated tantalum hydroxide in citric acid at different pH values, from 0.9 to 2.3, has been investigated[74]. It was concluded from this work that the $H_2C_6O_7H_5^-$ ion plays the chief role in forming a tantalum complex. A study of the kinetics of some substitution reactions of a 7-coordinate niobium(V) complex, $Nb(BPHA)_3$, where HBPHA is benzoylphenylhydroxylamine, has been reported[75]. The reactions involved replacement of $BPHA^-$ by tropolone, 8-quinolinol and several derivatives of

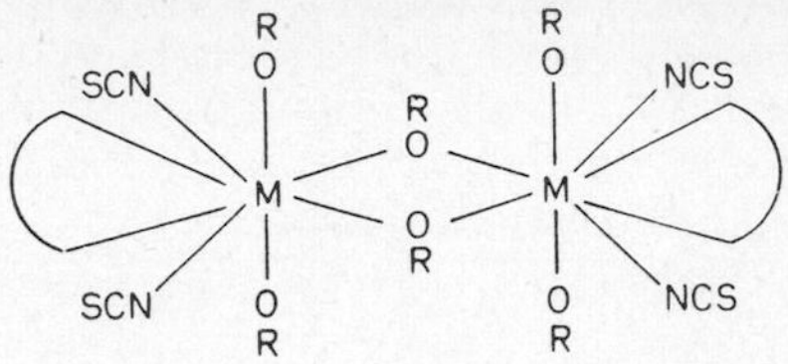

Figure 4.5 Structures of the thiocyanato-alkoxo-bipyridyl complexes of niobium(V) and tantalum(V)[73]

8-quinolinol, and were performed in chloroform and dichloroethane. The mechanistic studies were unable to produce a definitive assignment, the reactions being too complex for this, but evidence was obtained for dissociative rate-determining steps, with the entering ligand becoming associated with the niobium complex prior to the rate-determining step. Another kinetic study was concerned with the intramolecular exchange of ligands in niobium and tantalum penta-alkoxides[76]. Variable-temperature n.m.r. measurements on dimeric methoxides, ethoxides and isobutoxides showed the activation energy for scrambling of the terminal and bridging alkoxy group to be fairly constant for all the compounds, though for the tantalum derivatives the variation in entropy of activation was found to be quite dependent on ligand bulk, and it has been suggested that steric effects play an important role in the activation process. Terminal-site scrambling was found to occur at a faster rate than terminal-bridging scrambling, as reflected in a lower activation energy.

4.5 THE CHROMIUM GROUP

4.5.1 Chromium

Much of the new work on chromium complexes has been with carboxylate ligands. The basic trinuclear chromium(III) acetate, $[Cr_3O(CH_3CO_2)_6(H_2O)_3]Cl \cdot 6H_2O$, has been the subject of a polarised electronic spectroscopy study[77]. The exchange coupling between the three chromium ions in this complex (see Figure 4.6) was directly observed in the 690–750 nm region where transitions originating from different ground-state spin levels were identified. Kinetic studies of aquation reactions of four $[Cr(NH_3)_5(O_2CR)]^{2+}$ complex ions, where $R=CCl_3$, $CHCl_2$, CH_2Cl, or CH_3, have shown[78] the rates depend on the base strength of the acid anion, RCO_2^-. For $R=CCl_3$, the reaction observed was aquation of the acido group and was acid-catalysed, a comparison with the $R=CF_3$ analogue showing that the rate varied inversely with the RCO_2^- base strength. On the other hand, for the

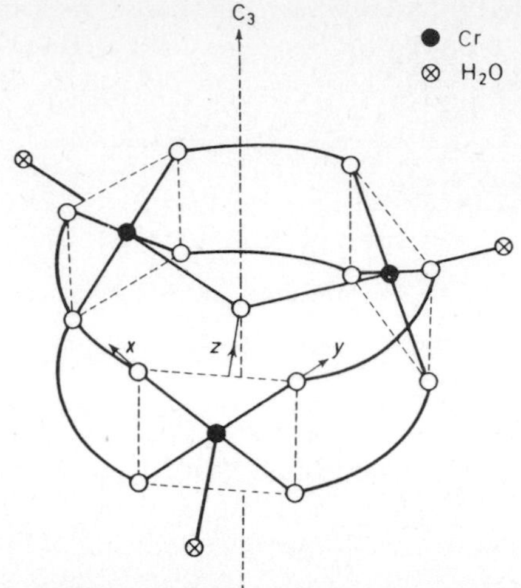

Figure 4.6 Schematic drawing of the structure of $\{Cr_3O(CH_3CO_2)_6(H_2O)_3\}^+$. (From Dubicki[77], by courtesy of the American Chemical Society.)

three other species, NH_3 aquation was the primary step and rates increased with RCO_2^- basicity. Both photochemical and thermal reactions were characterised. The kinetics and mechanism of electron transfer between the chromium-(II) aquo cation and some aquo-oxalatochromium(III) complex ions have been investigated[79] by use of ^{51}Cr labelling. For the system $Cr(C_2O_4)(H_2O)_4^+–Cr_{aq}^{2+}$, the electron-transfer reaction was found to be first order in each species and independent of acid concentration up to 2.0 $HClO_4$, but for the system *cis*-$Cr(C_2O_4)_2(H_2O)_2^-–Cr_{aq}^{2+}$ the reaction was first order in acid concentration as well as in each of the other two reactants. Electron transfer was found to be accompanied by transfer of a single oxalate ligand, so that in effect catalysed aquation of the Cr^{III} complex occurred. Mechanisms involving an electron-exchange intermediate with oxalate doubly bridged between the two chromium ions, such as is shown in Figure 4.7, were found to provide

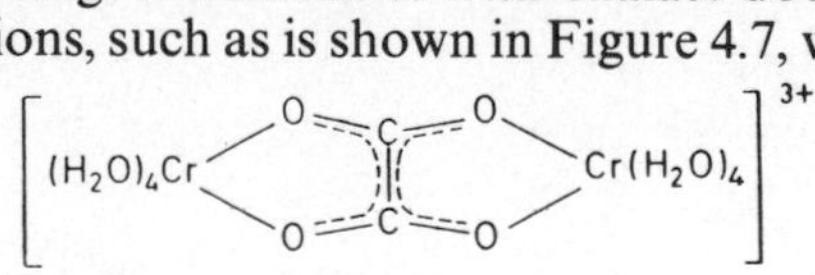

Figure 4.7 Proposed form of the intermediate for electron exchange in the $Cr(C_2O_4)(H_2O)_4^+–Cr_{aq}^{2+}$ system[79]

a satisfactory explanation of the observations. Other kinetic studies of chromium oxalate species[80, 81] have been concerned with the racemisation of the trisoxalatochromium(III) anion, $Cr(ox)_3^{3-}$, and the effect of pressure on the kinetics of formation of the oxalatochromium(III) ions in solution[82].

Closely related to the work on oxalate complexes is a study of the kinetics of the acid-catalysed aquation of *cis*-bis(malonato)diaquochromate(III) and the isomerisation of the corresponding *trans* compound[83]. The mechanism of this aquation reaction has been shown to involve a rapid protonation preequilibrium followed by rate-determining attack by water on the protonated intermediate. The possible mechanisms for the *trans–cis* isomerisation of *trans*-$Cr(C_3H_2O_4)_2(H_2O)_2^-$ have been postulated as shown in Figure 4.8.

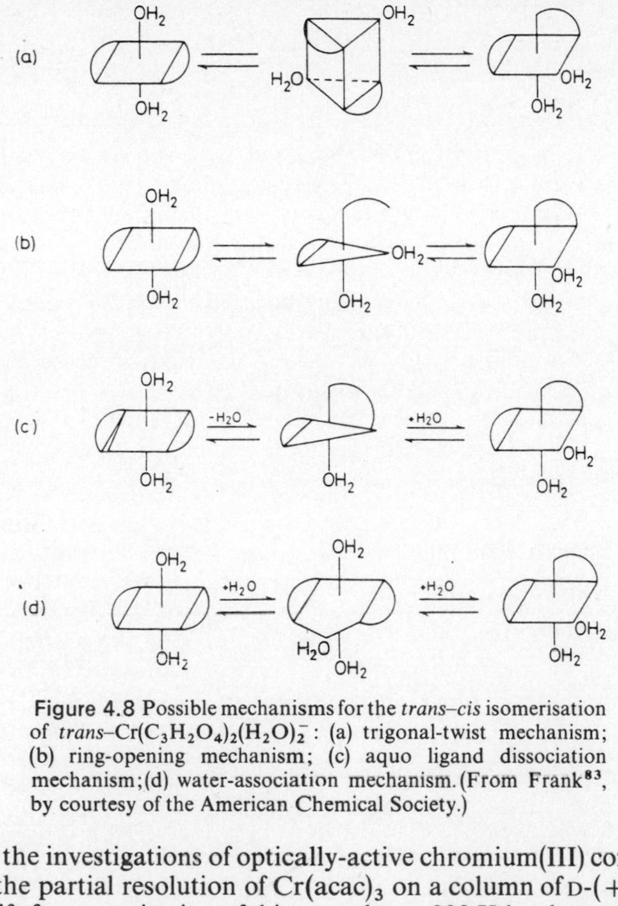

Figure 4.8 Possible mechanisms for the *trans–cis* isomerisation of *trans*-$Cr(C_3H_2O_4)_2(H_2O)_2^-$: (a) trigonal-twist mechanism; (b) ring-opening mechanism; (c) aquo ligand dissociation mechanism; (d) water-association mechanism. (From Frank[83], by courtesy of the American Chemical Society.)

Among the investigations of optically-active chromium(III) complexes is a report of the partial resolution of $Cr(acac)_3$ on a column of D-(+)-lactose[84]. The half-life for racemisation of this complex at 298 K has been calculated as 700 years, in contrast with the comparatively short value of 14 years for the $Co(acac)_3$ compound, the activation entropy difference evidently being responsible. Racemisation and ^{18}O exchange with the complexes $\{Cr(C_2O_4)L_2\}^+$, with L = bipyridyl, *o*-phenanthroline, or ethylenediamine, have been shown to follow different mechanistic routes[85]. Absolute configurations have been assigned for a number of Cr^{III} species, including the (+)-$[Cr(C_2O_4)_3]^{3-}$ and (+)-$[Cr(malonate)_3]^{3-}$ complex ions[86].

An inner-sphere perchlorate complex believed to involve 5-coordinate Cr^{III} has been prepared[87]. This brown-black compound, of formula $[Cr_3(TPPO)_8(OClO_3)_3](ClO_4)_6$, with TPPO = triphenylphosphine oxide, has an electronic spectrum closely resembling that of $CrCl_3 \cdot 2NH_3$. Studies of the anion $CF_3SO_3^-$ in coordination with Cr^{III} have suggested that it has advantages over ClO_4^- as a non-complexing anion, in that it is a very weak nucleophile and offers a high resistance to hydrolysis and reduction[88].

The tris(glycinato)chromium(III) complex has been prepared as a monohydrate and its crystal structure determined[89], showing the Cr^{III} to be octahedrally-coordinated with the three nitrogen atoms mutually *cis*. A structural and magnetic characterisation of the binuclear complex di-μ-hydroxotetraglycinato-dichromium(III), $[Cr(gly)_2(OH)]_2$, has been reported[90], again showing approximately octahedral coordination of Cr^{III} and a planar Cr—O—Cr—O unit.

A new crystal structure determination of $Cr_2(CH_3CO_2)_4 \cdot 2H_2O$ has yielded a short Cr—Cr distance consistent with the diamagnetism of the compound[91]. A dimeric unit also has been proposed for the carbonate salts $M_2Cr(CO_3)_2 \cdot nH_2O$, with M = Li, Na, K, NH_4, Rb, Cs, or $\frac{1}{2}$Mg, with bridging carbonato groups being indicated by magnetic and i.r. spectroscopic data[92].

Short Cr—N bonds (*ca.* 1.87 Å) have been interpreted in terms of ligand–metal π-bonding in the compound tris(di-isopropylaminato)chromium(III), $Cr(NPr^i_2)_3$. X-Ray data for this compound have shown it to have trigonal geometry, the $Cr(NC_2)_3$ group having almost D_3 symmetry[93]. Alcoholysis of $Cr(NEt_2)_4$ has produced several new Cr^{IV} tetra-alkoxides[94]. The compound $Cr[N(SiMe_3)_2]_3$ has been shown to give a low-temperature e.s.r. spectrum consistent with a d^3 system of axial symmetry[95].

4.5.2 Molybdenum and tungsten

The geometrical configuration of the octahedral dioxo-bis(2,4-pentanedionato)molybdenum(VI) complex in the crystalline state has been established as *cis* by an x-ray diffraction study[96]. This shows the Mo—O bonds to the two acetylacetonato oxygen atoms which are *trans* to the oxo ligands to be longer than the other Mo—O bonds. The variable-temperature n.m.r. spectra were found to show methyl line broadening and coalescence effects indicative of considerable intramolecular lability of the complex in solution, however, suggesting as one possible explanation a rapid equilibration with the *trans* isomer in solution. Another series of β-diketonate complexes reported involve molybdenum and tungsten in the +4 oxidation state[97]. These are compounds of the general formula $M(\beta\text{-dik})_2Cl_2$, with M = Mo or W. The β-diketones investigated include acetylacetone, 3,5-heptanedione, 4,6-nonanedione, benzoylacetone and 2-acetylcyclohexanone. The compounds were formed by direct reaction between a metal halide and the appropriate solid or liquid β-diketone, the structures of the products then being investigated by i.r. spectroscopy. Other tungsten(IV) compounds reported include a series of 8-quinolinol complexes of the type WL_4 (L = 8-quinolinol or one of its derivatives)[98]. Syntheses of many of these complexes were achieved by sealed-tube melt reactions between either $K_3W_2Cl_9$ or $W(CO)_6$ and the appropriate quinolinol derivative, though salts of $W_2Cl_9^{3-}$ were required for the parent complex. The magnetic and spectroscopic properties of the compounds have been interpreted as evidence for the first unequivocal series of inert, completely chelated 8-coordinate complexes of a d^2 spin-paired electronic configuration. Electrochemical investigations of redox processes involving molybdenum(V, VI) complexes with 8-quinolinol in dimethylsulphoxide solution have also been described[99].

Further molybdenum complexes of biochemical interest have been studied. Complexes of Mo^V with a series of carbohydrates in aqueous solution have been prepared and their ORD spectra obtained[100]. Similar structures have been established[101] from x-ray diffraction data for the amino-acid complexes Mo_2O_4(L-histidine) and Mo_2O_4(L-cysteine), and a new series of complexes of formula $[Mo(\pi\text{-}Cp)_2(A)]X$ have been prepared[102] with HA = glycine, alanine, phenylalanine, valine, leucine, isoleucine, proline, methionine, sarcosine, *N*-methylalamine, or cysteine, and X = Cl or PF_3.

A number of molybdenum complexes with chelating ligands have been reported. Evidence for an oxygen-bridged dimeric structure has been obtained[103] for the anion $[Mo_2O_4(EDTA)]^{2-}$ in a variety of salts and in the crystalline acid. Complexes of diethylenetriaminepenta-acetic acid and triethylenetetraminehexa-acetic acid with molybdenum(VI) have been prepared and their stability constants determined[104], and a series of $\{MoO_2L_2\}^{2-}$ complexes with L deriving from σ-diphenols have been characterised[105].

4.6 THE MANGANESE GROUP

4.6.1 Manganese

A recent addition to the small class of simple anhydrous compounds of manganese(III) is the nitrate[106], $Mn(NO_3)_3$. The syntheses of this salt and its dinitrogen pentoxide 'adduct' have been achieved by reaction of N_2O_5 with MnF_3. High-spin Mn^{III} is indicated by the magnetic susceptibility of the compounds, and infrared spectra have shown the 'adduct' to be the nitronium salt, $[NO_2]^+[Mn(NO_3)_4]^-$.

An unusual structure has been formed[107] for the acetylacetonate complex $Mn(acac)_2Br_2$. This comprises a polymeric chain of $\{MnBr_4O_2\}$ units of octahedral geometry linked together by double Br bridges as shown in Figure

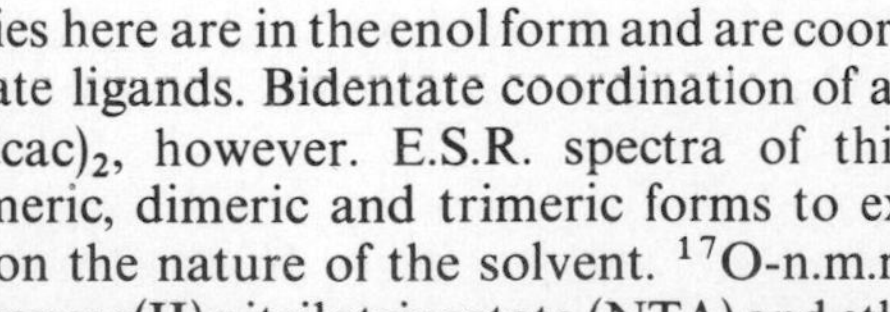

Figure 4.9 Structure[107] of the complex $Mn(acac)_2Br_2$

4.9. The acac species here are in the enol form and are coordinated to manganese as monodentate ligands. Bidentate coordination of acac clearly exists in anhydrous $Mn(acac)_2$, however. E.S.R. spectra of this compound have shown[108] monomeric, dimeric and trimeric forms to exist in solution, the form depending on the nature of the solvent. ^{17}O-n.m.r. studies have been made[109] on manganese(II) nitrilotriacetate (NTA) and ethylenediaminetetra-acetate (EDTA) complexes where the ligands are multidentate. Complex species $Mn(NTA)(H_2O)_2^-$ and $Mn(EDTA)(H_2O)^{2-}$ (containing 7-coordinate Mn^{II}) have been shown to exist in solution, and various kinetic parameters (*k*,

ΔH^*, ΔS^*) characterising the exchange of water molecules have been determined.

The complex anion *trans*-1,2-diamino-cyclohexanetetra-acetatomanganate(III)(CyDTAMn)$^-$ has been shown[110] to exist in aqueous methanol (2.5% water v/v) in three forms: Mn(CyDTA)(H_2O)$^-$ in neutral solution, Mn(CyDTA)(OH)$^{2-}$ in base, and HMn(CyDTA)(H_2O) in acid. In basic methanol the complex decomposes by disproportionation, but in acidic solution the methanol is oxidised to formaldehyde. Reaction of the complex with hindered phenols results in the generation of aryloxy radicals, some of which are stable and others of which undergo further reaction.

The thermal decomposition of the Mn^{II} salt of squaric acid has been found to occur by a two-step process[111]. An initial, reversible dehydration step is followed by break-up of the anion to yield CO_2, CO and carbon:

$$Mn(C_4O_4)\cdot 2H_2O \rightleftharpoons Mn(C_4O_4) + H_2O$$

$$Mn(C_4O_4) \rightarrow Mn + (4-2x)CO + xCO_2 + xC$$

Stability constants for Mn^{II} chelates with *N*-phenylbutyrohydroxamic acid[112] and with 1-hydroxy-2-nitrosonaphthalene-4-sulphonic acid[113] have been determined on a comparable basis with the corresponding complexes of other transition metal ions. Of biochemical interest are complexes with the glycinate ion[114] and with various porphyrins[115]. A review of the latter includes manganese porphyrin–protein complexes, manganese chlorophylls and phthalocyanines[115].

4.6.2 Technetium and rhenium

No interesting new work with complexes of anionic species with technetium has been located, though several such complexes of rhenium have been studied. The crystal and molecular structure of the β-diketone complex of Re^I, bis(μ-*O*-1,3-diphenylpropane-1,3-dionatotricarbonylrhenium(I)), has been determined[116]. The complex is dimeric, the two units being held together by two bridging oxygen atoms, one from each of the β-ketoenolate groups, as shown in Figure 4.10. The distance of the bridging oxygen from the rhenium

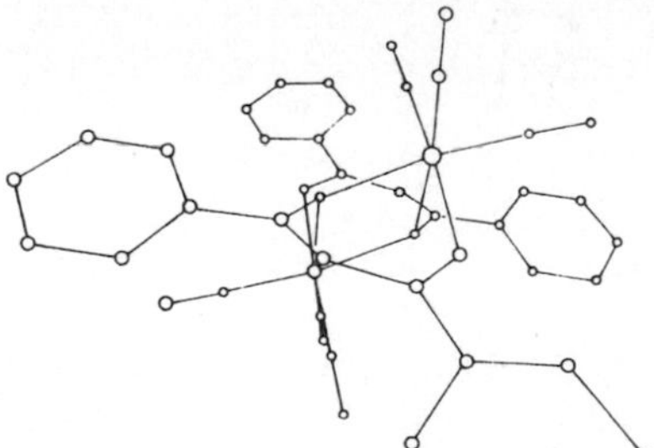

Figure 4.10 The molecular structure of $Re_2(CO)_6L_2$, where L = 1,3-diphenylpropane-1,3-dione. (From Barrick[116], by courtesy of the National Research Council of Canada.)

atom to which it is chelated is shorter by 0.042 Å than the distance to the other rhenium atom. The more conventional form of bonding occurs in the complex *trans*-$Re(acac)_2Cl_2$, which has been studied[117] by spectroscopic and classical physical methods and shown to be monomeric, in spite of earlier reports that it was dimeric with bridging acac groups. Other monomeric β-diketonate complexes of Re^I have also been prepared for the first time[117]. The physical properties of pure samples of the difficult-to-prepare compounds $Re(acac)_3$ and $Re(hfac)_3$ (hfacH = 1,1,1,5,5,5-hexafluoropentane-2,4-dione) have been studied[118]. The influence of the metal atom on the breakdown pattern of the compounds in a mass spectrometer has been analysed.

A binuclear mixed-ligand fluoro-oxalato complex anion of rhenium(VII) has been prepared[119] in the form of an acid and a series of salts. On the basis of the conductance and molecular weight of the potassium salt, the analytical data and i.r. spectra, the constitution $H_2[(C_2O_4)(OH)_3Re\cdot O\cdot Re(OH)_3F_2]$ has been proposed for the acid.

4.7 THE IRON GROUP

4.7.1 Iron

The complex nitrato anion $[Fe(NO_3)_4]^-$, prepared as the tetraphenylarsonium salt, has been shown to have all four nitrate groups bidentate, thus giving the first example of an 8-coordinate iron(III) complex[120]. The arrangement of oxygen atoms about the metal is dodecahedral. Aqueous $Fe(NO_3)_3$ mixed with aqueous K_2WO_4 at *ca.* pH 6 has been shown[121] to yield an amorphous precipitate of $KFe(WO_4)_2\cdot 4H_2O$. Spectroscopic (e.s.r. and i.r.) data from the complexes $FeL_4(ClO_4)_3$ and $FeL_2(NO_3)_3$ (L = Ph_3PO) suggest outer-sphere perchlorate interaction with the planar FeL_4^{2+} ion, but inner-sphere nitrate complexation in 5-coordinate structures[122].

The magnetic properties of iron complexes are of central interest in a number of studies reported. A Mössbauer spectroscopic investigation of the complex $[Fe_3O(CH_3CO_2)_6(H_2O)_3]Cl\cdot 6H_2O$ has shown antiferromagnetic interactions in the triangular Fe^{3+} clusters which are different for each Fe^{3+} ion[123]. Several weakly antiferromagnetic compounds of the type $[L_2Fe(OR)]_2$, where L is the enolate of acetylacetonate or dipivaloylmethane (HDPM) and R is CH_3, C_2H_5, and i-C_3H_7, have been prepared and characterised[124]. The i.r. and optical spectroscopic properties of these compounds are consistent with their having the dimeric double-alkoxy-bridged structure shown in Figure 4.11. Magnetic measurements, together with Mössbauer and e.s.r. data, from the compounds FeCl(RCOCHCOR′) (R = R′ = Ph, and R = Me, R′ = Ph) have shown them to contain high-spin iron(III)[125]. An earlier report[126] of the crystal structure determination of $Fe(acac)_2Cl$ established the coordination geometry of this 5-coordinate compound as essentially square-pyramidal, the Fe atom being located slightly above the plane of the four acetylacetonate oxygen atoms. The mixed O,N-donor complex $\{Fe(salen)\}_2O\cdot CH_2Cl_2$, which has the structure shown in

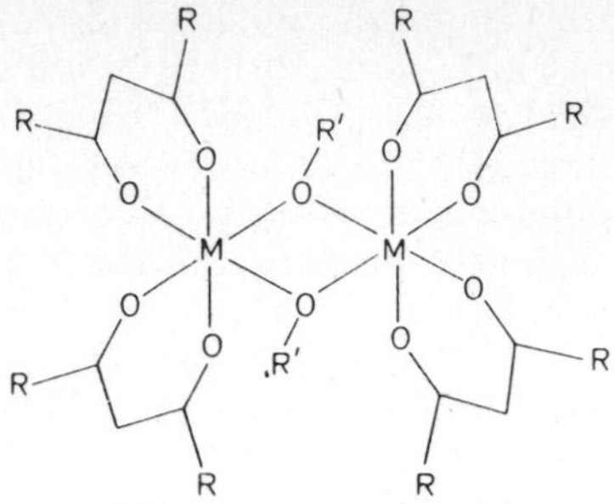

Figure 4.11 Proposed structure of $[(\beta\text{-dik})_2M(OR)]_2$. (From Wu[124], by courtesy of the American Chemical Society.)

Figure 4.12,has magnetic properties indicative of spin–spin exchange involv-

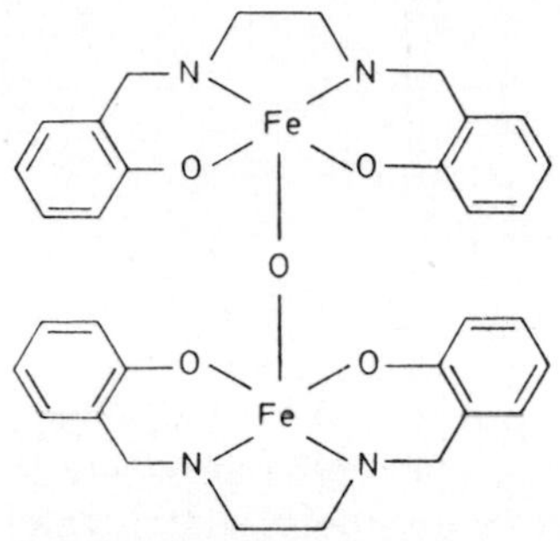

Figure 4.12 Structure of the $\{Fe(salen)_2\}O{\cdot}CH_2Cl_2$ complex[127]

ing a 'dipolar coupling' mechanism[127].

In weakly acidic solution, several amino acids have been found to form monodentate complexes with Fe^{2+} by coordination through the carboxylate group only. Similar findings have been reported for dipeptide complexes with iron(II)[128]. Formation constants for iron(II) and iron(III) complexes with phenylalanine have been determined on a comparative basis with those for cobalt(II), nickel(II) and copper(II)[129].

Reviews have been published on some aspects of the coordination chemistry of iron(III)[130], and on high-spin behaviour and chemical properties of organometallic derivatives of iron(III)[131].

4.7.1 Ruthenium and osmium

The product of the reaction between $RuCl_3 \cdot 3H_2O$ and acetic acid/sodium acetate, originally but incorrectly formulated as a dimer, has been shown[132] to be a complex of the basic acetate oxo-centred triangular type, with the formula $[Ru_3O(CH_3CO_2)_6(H_2O)_3](CH_3CO_2)$. The complex shows amphoteric behaviour in solution, due to the ionisation possibilities of the coordinated water, and the water molecules may be replaced by pyridine to give $[Ru_3O(CH_3CO_2)_6(py)_3]^+$. In the presence of triphenylphosphine, the aquo complex gives the reduced species $Ru_3O(CH_3CO_2)_6(PPh_3)_3$. Both one- and two-electron reductions of the aquo and pyridine complexes have been observed, giving complexes in which the metal atoms are in formal oxidation states $+2\frac{2}{3}$ or (III, III, II) and +2 (II, II, II). In the +2 state the central oxygen atom is lost from the triangle of Ru^{II} atoms, but this may be re-inserted by using molecular oxygen or a more specific reagent such as pyridine-*N*-oxide. Analogous complexes of other carboxylic acids have also been described[132].

The reaction of oxalic acid with the complex $\{Ru(py)_6\}^{2+}$ has been studied[133]. X-Ray analysis has established an oxalate-bridged structure for the product $[(py)_4Ru(C_2O_4)Ru(py)_4](BF_4)_3$. Monomeric carboxylate complexes of Ru^{II} have been obtained[134] by reaction of carboxylic acids with $Ru(CO)_2(PPh_3)_2$. I.r. and n.m.r. studies have been used to confirm that the carboxylate ligands are monodentate in the complexes $Ru(CO)_2\ (PPh_3)_2$—$(CO_2R)_2$, where R=H, CH_3, or C_2H_5.

The optical isomers of $Ru(acac)_3$ have been partially resolved by column chromatography on D(+)-lactose, a further tenfold increase in resolution being achieved by crystallisation of the partially resolved racemate from mixed benzene/hexane solution[135]. The effects of chelate ring substituents on the polarographic redox potentials of tris(β-diketonato)ruthenium(II, III) complexes have been examined[136]. This study was initiated as a consequence of the somewhat unexpected finding that $Ru(tfac)_3$ (tfac = trifluoroacetylacetonate) was readily reduced to $Ru(tfac)_3^-$, in contrast with $Ru(acac)_3$. Mono-anions $M(sacsac)_3^-$ (sacsac = dithioacetylacetonate) have also been produced electrochemically for Fe^{III}, Ru^{III} and Os^{III} [137]. Two separate isomers have been separated by thin-layer chromatography from a preparation of $Ru(tfac)_3$[138].

A potentiometric titration of the complex Ruthenium Brown, $[(NH_3)_5RuORu(NH_3)_4ORu(NH_3)_5]^{7+}$, with OH^- and Ce^{4+} has confirmed that the change to Ruthenium Red, $[(NH_3)_5RuORu(NH_3)_4ORu(NH_3)_5]^{6+}$, is a one-electron reduction which proceeds according to the equation[139]:

$$\text{Ruthenium Brown} + OH^- \rightarrow \text{Ruthenium Red} + \tfrac{1}{4}O_2 + \tfrac{1}{2}H_2O$$

Oxidative addition of tetra-*X*-1,2-benzoquinone to $Ru^{II}Cl_2(PPh_3)_2$ has been shown[140] to yield the green ruthenium(IV) complex

$$Ru(\sigma\text{-}C_6X_4O_2)Cl_2(PPh_3)_2.$$

4.8 THE COBALT GROUP

4.8.1 Cobalt

A number of complexes of simple oxyanions have been studied. Electronic absorption spectra of cobalt(II) in Li_2SO_4–DMSO solution and in a $NaHSO_4$–$KHSO_4$ melt have been interpreted in terms of a blue dodecahedral $[Co(O_2SO_2)_4]^{6-}$ species and a pink octahedral $[Co(OSO_3)_6]^{10-}$ species[141]. Similar studies of various carbonato complexes with cobalt(III) have been made, and covalency factors determined for the Co—O binding[142]. The nature of the triscarbonatocobaltate(III) species in aqueous solution and in some salts has been studied by i.r. spectroscopy[143], and the tris-chelated species $[Co(CO_3)_3]^{3-}$ has been resolved using $(+)[Co(en)_3]^{3+}$. A kinetic study of ^{18}O exchange between water and carbon-atopenta-amminecobalt(III) in solution has been made[144]. The rate-controlling step in this exchange reaction evidently involves hydration of CO_2.

Some unusually interesting structures have been determined for cobalt chelates. A cubane-type complex containing both Co^{II} and Co^{III}, for example, di-μ-acetato-tetrakis-μ_3-methoxo-2,4-pentanedionatocobalt(II, III), $Co_4(OMe)_4(OAc)_2(acac)_4$, has been prepared[145]. The structure, which is shown in Figure 4.13, has 6-coordinate cobalt ions and methoxide oxygens at alternate corners of the central distorted cube; each cobalt also is chelated by an acac group, and the bidentate acetate groups link pairs of cobalt atoms across the top and bottom faces of the cube. The magnetic properties of this complex suggest superexchange through the bridging acetate group. A cubane-type skeleton also has been established for $Co_4(OMe)_4(acac)_4(MeOH)_4$ and its nickel analogue[146].

The absolute configurations of a number of β-diketone complexes of cobalt (III) have been determined[147]. The absorption, ORD and CD spectra of $(+)\{Cr(en)_2(dik)\}^{2+}$ were recorded in the u.v. and visible region and the electronic transitions assigned. Several azido complexes of bis(acetylacetonato)cobalt(III) have been synthesised and shown to undergo a facile *cis–trans* isomerisation reaction in solution[148]. The absolute configurations of the complexes $[Co(malonate)_2(en)]$ and $[Co(ox)_3]^{3-}$ have been assigned on the basis of x-ray results[149], and some ^{18}O-exchange studies on the trisoxalatocobalt(III) anion reported[150].

Amino-acid complexes of cobalt have been extensively investigated. Separations of the diastereoisomers of L-phenylalanineato-bis(acetylacetonato)-cobalt(III) and L-valineato-bis(acetylacetonato)cobalt(III) have been achieved by fractional crystallisation and column chromatography on D-lactose[151]. Further studies in a long series devoted to optically-active coordination compounds deal with a variety of mono-dipeptide complexes of cobalt(III)[152, 153], and demonstrate the partial resolution of 1,2,4-triglycinatocobalt(III) by a bacterial method[154]. The thermodynamics and kinetics of complex formation between CO^{II}, Ni^{II} and Cu^{II} and glycyl-L-leucine and L-leucylglycine have been studied potentiometrically[155]. The results of this work are consistent with metal coordination via the amino end-group and the *O* atom of the amide group. The pH dependence of the rate of loss of chloride ion from *cis*-chloro-bis(ethylenediamine)glycinato-*N*-cobalt(III) has been shown to indicate weak S_N2 character involving participation of the free carboxylate group of the monodentate glycinato ligand[156].

Studies of cobalt carboxylate complexes include the characterisation[157] of salts of μ-amido-μ-acetato-bis[tetra-amminecobalt(III)] and μ-amido-μ-formato-bis[tetra-amminecobalt(III)]. The doubly-bridged binuclear structure of these complex ions is shown in Figure 4.14. Electron transfer between acetato complexes of cobalt(III) and iron(II) has been studied[158], and the kinetics of Cr^{2+} and V^{2+} reductions of some cobalt(III) acetato and formato complexes investigated[159]. These reactions evidently proceed via outer-

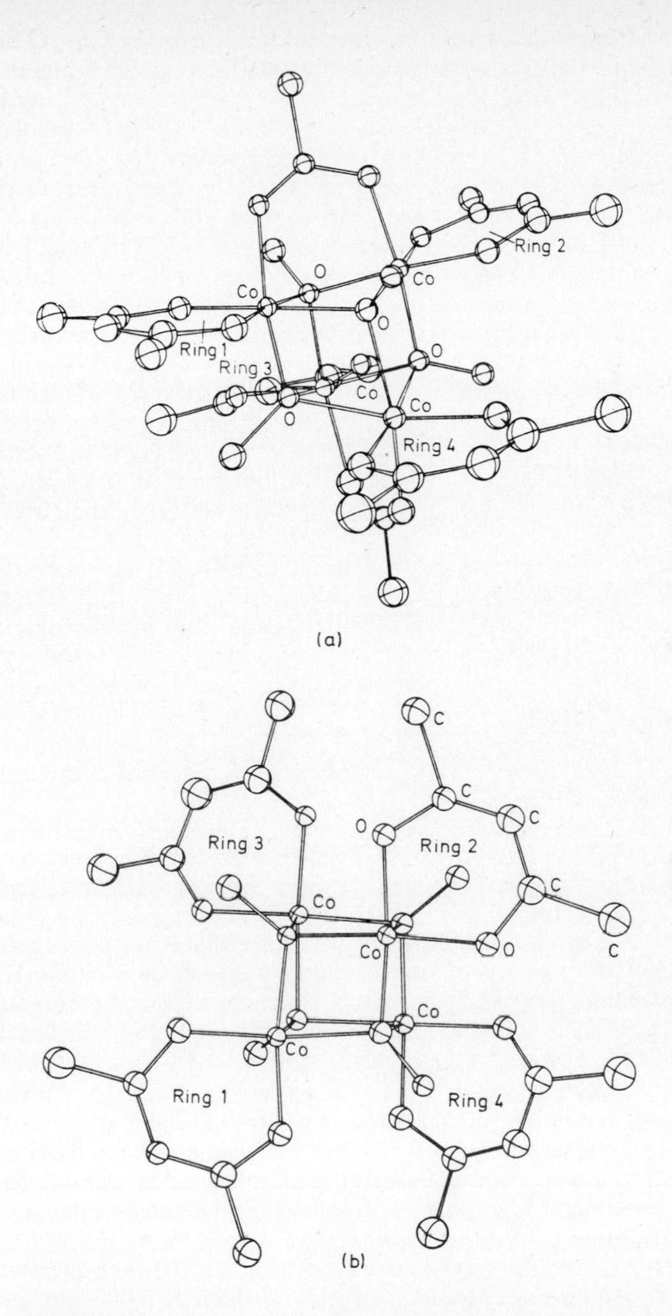

Figure 4.13 A perspective drawing of the molecular structure of $Co_4(OMe)_4(OAc)_2(acac)_4$. (b) The structure of $Co_4(OMe)_4(OAc)_2(acac)_4$ viewed down the pseudo-twofold axis. The acetate groups have been omitted for clarity. (From Bertrand[145], by courtesy of The American Chemical Society.)

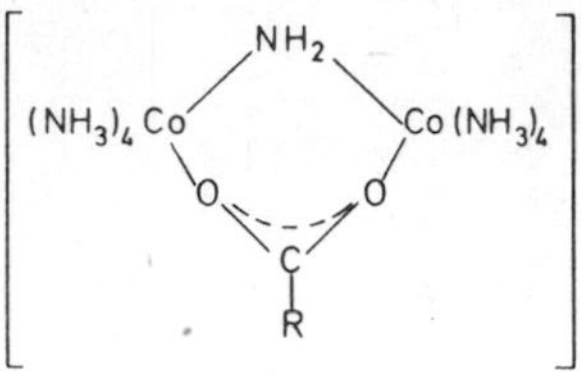

Figure 4.14

sphere mechanisms. Kinetic studies of a number of hydrolysis reactions of cobalt complexes have been reported, including catalysed reactions of the *trans*-diacetato-bis(ethylenediamine)cobalt(III)[160], and *trans*-chloroacetato- and *trans*-chlorobenzoato-bis(ethylenediamine)cobalt(III) complex ions[161], and a review of the hydrolysis reactions of some other carboxylato-bis(ethylenediamine)cobalt(III) complexes has been prepared[162]. Another review dealing with bonding effects in circularly dichroic cobalt(III) complexes includes a section on oxalato systems[163], and the kinetics of electron-transfer reactions between oxalatocobalt(III) complexes and iron(II) in acidic aqueous solutions have been investigated[164]. Stereoselective complex formation has been studied[165] through the reactions of optically-active lactic acid with *rac*-$[Co(en)_2CO_3]^+$ and optically-active pantoyl lactone with *trans*-$[Co(en)_2Cl_2]ClO_4$. A variety of spectroscopic data suggest that the hydroxy acid anions coordinate in a preferred conformation.

Recent developments in the field of organometallic derivatives of cobalt chelates have been reviewed[166], including those aspects relevant to systems of biological interest such as vitamin B_{12} and its analogues.

4.8.2 Rhodium and iridium

The complexes $M(OClO_3)(CO)(PPh_3)_2$, with M = Rh or Ir, have been prepared[167] and represent the only perchlorato complexes known of a d^8 metal in the +1 oxidation state. The perchlorato ligand, which was inserted by the action of $AgClO_4$ on a benzene solution of the chloro complexes $MCl(CO)(PPh_3)_2$, is readily displaced from the complexes by polar solvents or by a further molecule of PPh_3. Syntheses of some new rhodium(III) and iridium(III) nitrito and nitrato complexes of the form $K_3[Ir(NO_2)_3(NO_3)_3]$, $K_3[Rh(NO_3)_6]$ and $K_3[Ir(NO_3)_6]$ have been reported[168], and a new oxide nitrate of iridium, $Ir_3O(NO_3)_{10}$, has been prepared[169]. The i.r. spectrum of this latter compound suggests its formulation as $[Ir_3O(NO_3)_9]^+$ NO_3^-, and the oxygen-centred trinuclear structure shown in Figure 4.15 has been suggested for the cation.

The reaction between SO_2 and Rh—O_2 and Ir—O_2 complexes has been shown to yield sulphato complexes of the type $MX(SO_4)(PPh_3)_3$ and

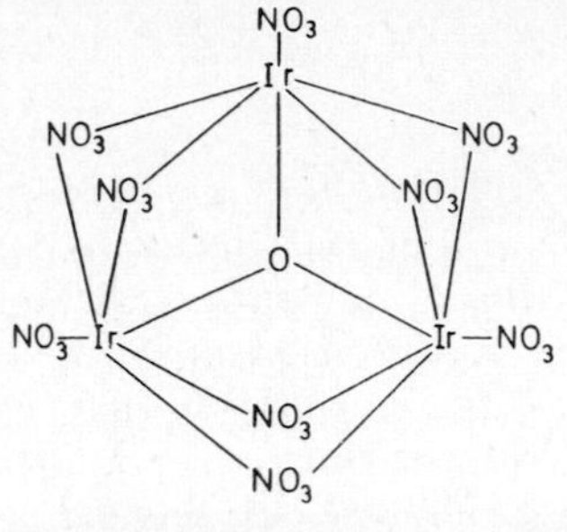

Figure 4.15 Proposed structure for $Ir_3O(NO_3)_9^+$. (From Harrison, by courtesy of The Chemical Society.)

$MX(SO_4)(CO)(PPh_3)_2$, where X is a halogen and M is Rh or Ir[170, 171]. Analogous reactions with NO_2 have been shown to yield nitrato complexes[172]. Coordination of monodentate SO_3 groups through the S atom has been proposed on the basis of the i.r. spectra of a series of sulphito dioximes of rhodium(III) and cobalt(III)[173]. A wide range of physical measurements has confirmed the sulphato complex known as Delipine's salt, $K_4NIr_3(SO_4)_6 \cdot 3H_2O$, as containing two iridium(IV) and one iridium(III) atoms linked by bridging bidentate sulphate groups[174].

A bidentate acetato complex of rhodium(II), which also contains bidentate azobenzene coordinated through a metal–carbon bond, has been characterised and its structure determined by an x-ray study[175]. Reactions between carboxylic acids and the complex *trans*-$IrCl(N_2)(PPh_3)_2$ have been found to result in formation of hydridocarboxylato complexes according to[176]:

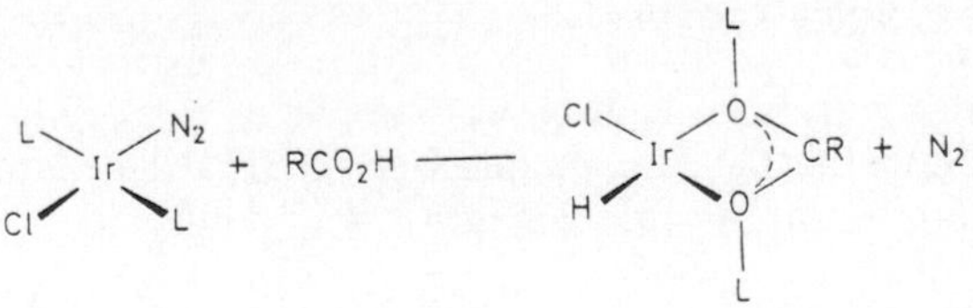

Confirmation of this has been obtained by running the reaction with deuteroacetic acid, when an Ir—D bond resulted. The low *trans*-labilising effect of the carboxylato ligand is shown in the unusually high Ir—H stretching frequencies observed for these compounds.

Dimeric rhodium(II)[177] and monomeric rhodium(III)[178] complexes of dimethylglyoxime have been shown to contain the chelating ligands in an essentially planar arrangement around the metal atoms. The autocatalytic reduction of the complex $[Rh^{III}(DH)(DH_2)Cl_2]$ (DH_2 = dimethylglyoxime) in alkaline aqueous ethanol to a rhodium(I) complex has been studied[179].

4.9 THE NICKEL GROUP

4.9.1 Nickel

The acetato groups in nickel(II) acetate tetrahydrate have been shown to be monodentate, four other coordination sites around the Ni^{II} being occupied by water molecules[180, 181]. Phenoxyacetato and methoxyacetato groups similarly are monodentate in their hydrated nickel(II) salts. The pseudo-octahedral configuration of the nickel(II) in these complexes is also believed to be present in several pyridinecarboxylate complexes[182]. On the basis of i.r. spectroscopic studies, which showed two sets of C=O stretching vibrations, two distinct types of carboxylate ligand have been proposed for the complex Ni(benzoate)$_2 \cdot 3H_2O$, one covalently bonded and one ionic[183]. From kinetic data on the *cis*-dichlorodiammineplatinum(II)–oxalate reaction, it has been determined that the oxalate group is a very weak nucleophile compared with neutral amine molecules[184].

A variety of chelating ketonate complexes of nickel have been studied. Particular interest centres around the magnetic properties of polynuclear species, and in this context the synthesis of a new set of 1,3,5,7-tetraketonates[185] is worthy of note. These have the general form shown in Figure 4.16, where M is a divalent metal ion and B is a neutral ligand

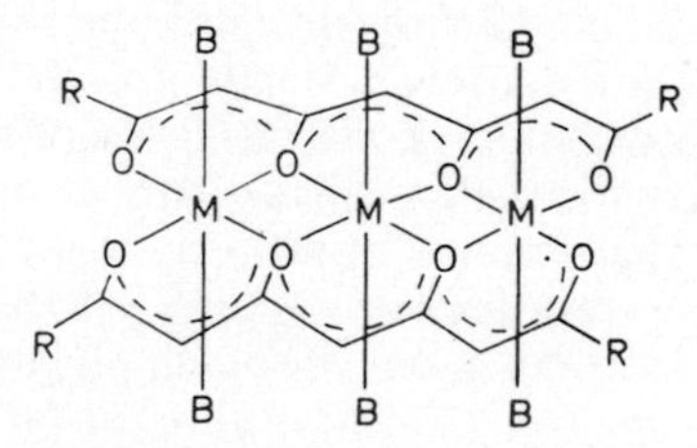

Figure 4.16 Trinuclear metal-1,3,5,7-tetraketonates[185]

such as H_2O or pyridine. Analogous complexes with only two metal ions coordinated appear to have the metals separated rather than adjacent to one another, structures (a) or (b) of Figure 4.17 being likely ones for these

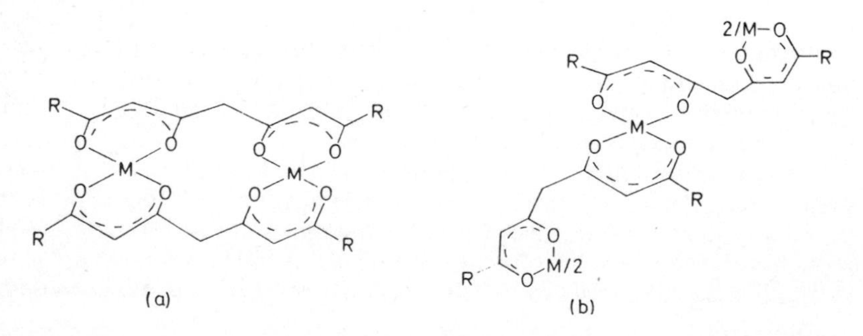

Figure 4.17 Possible structural forms of binuclear metal-1,3,5,7-tetra-ketonates[185]

complexes[185]. Exchange coupling in magnetic cluster complexes has been reviewed[186], including the linear trimer bis(acetylacetonato)nickel(II), which has the structure shown in Figure 4.18 and was the first cluster complex

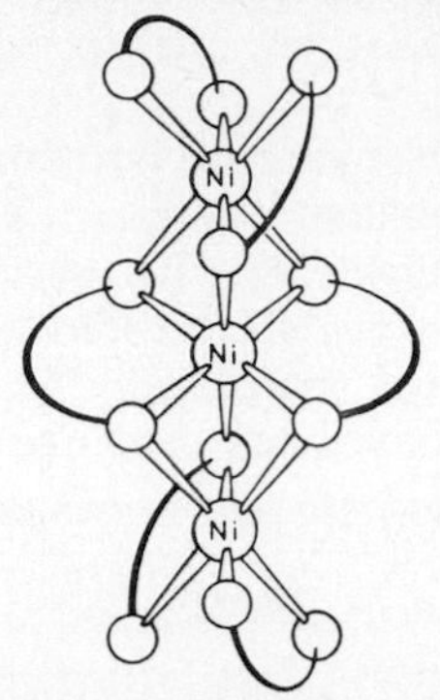

Figure 4.18 Molecular structure of $[Ni(acac)_2]_3$[186]

demonstrated to have parallel spin alignment in its ground state, and nickel tetramers of the 'cubane' type, resembling the cobalt complexes shown in Figure 4.13. The bis(β-keto-enolate)nickel(II) adducts of piperidine, piperazine, methylpiperazine and morpholine have been isolated and their magnetic and spectroscopic properties analysed in terms of the *trans*-octahedral configuration[187]. The standard enthalpy of formation of the parent bis(acetylacetonato)nickel(II) and the Ni—O bond energy in this complex have been determined[188].

Among the complexes with nitrogen donors, the macrocyclic quadridentate systems have received much attention. Several metalloporphyrins, including nickel, have been subjected to an ambitious normal-coordinate analysis treating the 18 in-plane vibrations[189]. Bands due to CC stretching, CN stretching, CH bending and CCN bending or coupled vibrations between these modes have been assigned. The cationic species $[Ni^{III}(TPP)]^+$ and the radical cation $[Ni^{IV}(TPP)\cdot]^+$, where TPP = tetraphenylporphyrin, have been identified by e.s.r. spectroscopy as oxidation products of $Ni^{II}(TPP)$[190]. A variety of reactions of nickel corrinoids have been reported[191].

Many mixed-donor complexes of nickel have been studied. ^{17}O n.m.r. studies of nickel(II)–EDTA complexes in aqueous solution have shown the structures to contain 6-coordinate Ni^{II}, but one coordination site is evidently occupied by an H_2O molecule at certain pH values[192]. Distorted trigonal-bipyramidal configurations have been established for the nickel(II) complexes of the pentadentate ligands bis(salicylidene-γ-iminisopropyl)methylamine (Figure 4.19)[193] and the corresponding n-propyl derivative[194]. Geometrical

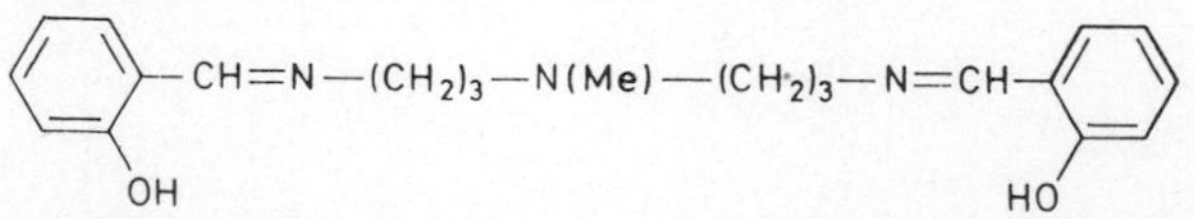

$$\text{CH}{=}\text{N}{-}(\text{CH}_2)_3{-}\text{N(Me)}{-}(\text{CH}_2)_3{-}\text{N}{=}\text{CH}$$

OH HO

Figure 4.19 The pentadentate bis(salicylidene-γ-iminisopropyl)methylamine ligand[193]

constraints within the ligands evidently are responsible for this unusual coordination geometry.

4.9.2 Palladium and platinum

The molecular oxygen adducts $M(O_2)(PPh_3)_2$ have been found to react with SO_2 to give sulphato complexes, with NO to give N-bonded nitrito complexes, and with N_2O_4 to give nitrato complexes of palladium and platinum[171].

A series of bridging *cis*-di-μ-carboxylato complexes of palladium(II) and platinum(II) of the general types $[X(Me_2PhE)M(O_2CR)]_2$ and $[X_2(Me_2PhE)_2M_2\{O_2C(CH_2)_nCO_2\}]$ (X = halogen; E = P, As; M = Pd, Pt, R = Me, CH_2Cl, CH_2Br, CF_3, CCl_3, CMe_3, CPh_3; $n = 3$–8) have been synthesised and characterised structurally[195]. These complexes, which are represented in Figure 4.20, have been shown to undergo rapid intramolecular

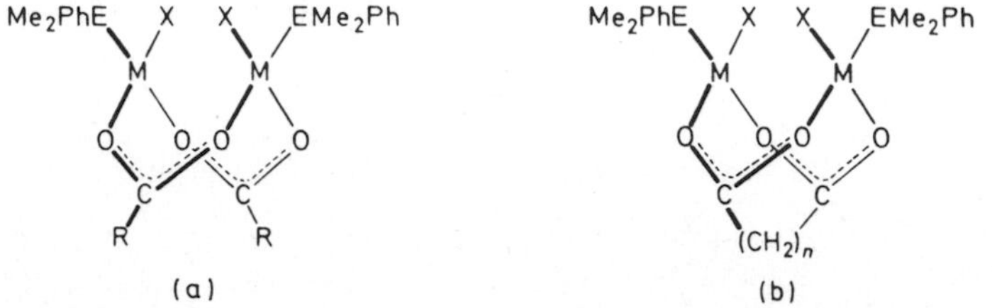

Figure 4.20 Cis-di-μ-carboxylato complexes of palladium(II) and platinum(II)[195]

ligand exchange by a mechanism which involves solvolysis of the metal–carboxylate bond, resulting in a mono-μ-carboxylate intermediate. Reactions of the carboxylate complexes $Pt(O_2CMe)_2(PPh_3)_2$ and $Pt(O_2CCF_3)_2(PPh_3)_2$ with CO and some olefins and acetylenes in alcoholic solution have been shown to follow a mechanism involving formation of an intermediate alkoxide complex, $Pt(O_2CR)(OR')(PPh_3)_2$ (R = Me or CF_3, R' = Me or Et)[196]. Reactions of SO_2 with the complex $Pt(O_2CMe)_2(PPh_3)_2$ have also been investigated.

A study of the kinetics of the reaction of the bis(oxalato)platinate(II) ion with SCN^- has been made and the results interpreted in terms of a ring-opening mechanism involving intermediates containing monodentate oxalato ligands[197]. The isolated intermediate species, *trans*-$Pt(C_2O_4)_2$—$(SCN)_2 \cdot 3H_2O$, is evidently produced by the two-step process shown in Figure 4.21. The very weak basic character of this intermediate (c) is ac-

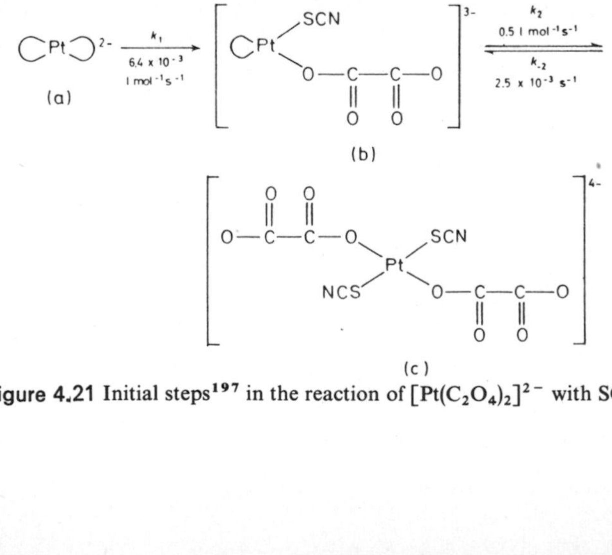

Figure 4.21 Initial steps[197] in the reaction of $[Pt(C_2O_4)_2]^{2-}$ with SCN^-

counted for in terms of the hydrogen-bonded aquo complex structure shown in Figure 4.22. Oxalatoplatinate(II) complexes also have received attention in

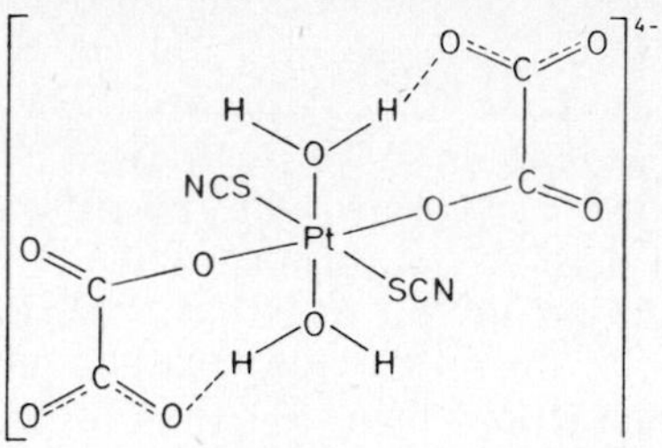

Figure 4.22 Proposed structure[197] for an intermediate in the reaction between $[Pt(C_2O_4)_2]^{2-}$ and SCN^-

the solid state, where they can exist as partly-oxidised species such as $K_{1.6}Pt(C_2O_4){\cdot}xH_2O$ and $Mg_{0.82}Pt(C_2O_4){\cdot}5.3H_2O$. Single crystals of these materials show electrical conduction properties consistent with a phonon-assisted hopping model, and their chain-like structures are of interest as possible one-dimensional 'metals'[198].

A number of interesting peroxo complexes of platinum(II) have been prepared. Reaction of $Pt(O_2)(PPh_3)_2$ with excess $(CF_3)_2CO$ has yielded the complex shown in Figure 4.23(a). This is seen to contain a 7-membered ring. A

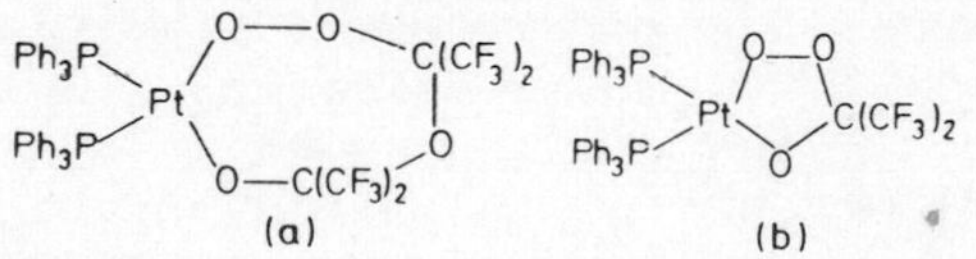

Figure 4.23 Reaction products from $Pt(O_2)(PPh_3)_2$ and $(CF_3)_2CO$[199]

quite different product, shown in Figure 4.23(b), resulted from reaction between equimolar amounts of the above reactants and contained a 5-membered ring[199]. Analogous reactions have been reported[200] with α-diketones, the products of these providing a route to the synthesis of dicarboxylates of the type $Pt(O_2CR)_2(PPh_3)_2$.

I.R. spectra and normal-coordinate analyses have been reported for some glyoximato complexes of nickel(II), palladium(II) and platinum(II)[201]. The derived force constants for the metal–nitrogen bonds decrease in the order Pt > Pd > Ni. A number of neutral bis(acetylacetone-ethylenedi-imine) complexes of palladium(II) and platinum(II), of the general form shown in

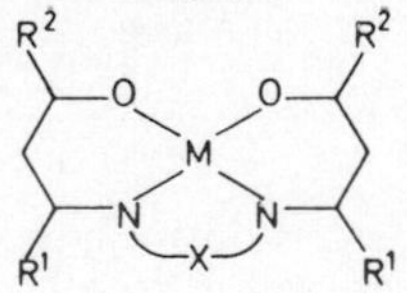

Figure 4.24 Bis(acetylacetone-ethylenedi-imine) complexes of palladium(II) and platinum(II)[202]. {$R^1 = R^2 = Me$; or $R^1 = Me$, $R^2 = CF_3$; $X = (CH_2)_2$ or CH_2CHMe}

Figure 4.2.4, have been prepared and separated in pure form by gas chromatography[202].

A number of amino-acid complexes of Pd^{II} and Pt^{II} have been investigated. The *cis-trans*-isomerisation reaction of $Pd(glycinato)_2$ in aqueous solution has been studied[203], and the irreversible *cis-trans* photo-isomerisation of *cis*-$[Pt(glycinato)_2]$ has been analysed theoretically[204]. A series of β-phenyl-β-alanine complexes have been prepared by reacting alkaline solutions of this amino acid with $K_2[PtCl_4]$[205]. Absorption and CD spectra of the complexes PdL_2, with L = *N*-methyl-L-amino acids, have been interpreted together with spectra from a series of dipeptide complexes of palladium(II) in terms of possible π-electron interactions being active in promoting optical activity in the ligand field bands of these complexes[206].

4.10 THE COPPER GROUP

4.10.1 Copper

Since the tentative proposals for the structure of the binuclear anion $[Cu_2(dl\text{-tartrate})_2]^{4-}$ were presented in the earlier review of this series[1], an x-ray crystallographic study of this species has been completed[207]. The rough estimate of the Cu—Cu distance in the complex made on the basis of e.s.r. spectral data was *ca.* 3.5 Å, and this has now been shown to be too large, the true figure appropriate to the structure shown in Figure 4.25 is 2.9869(7) Å.

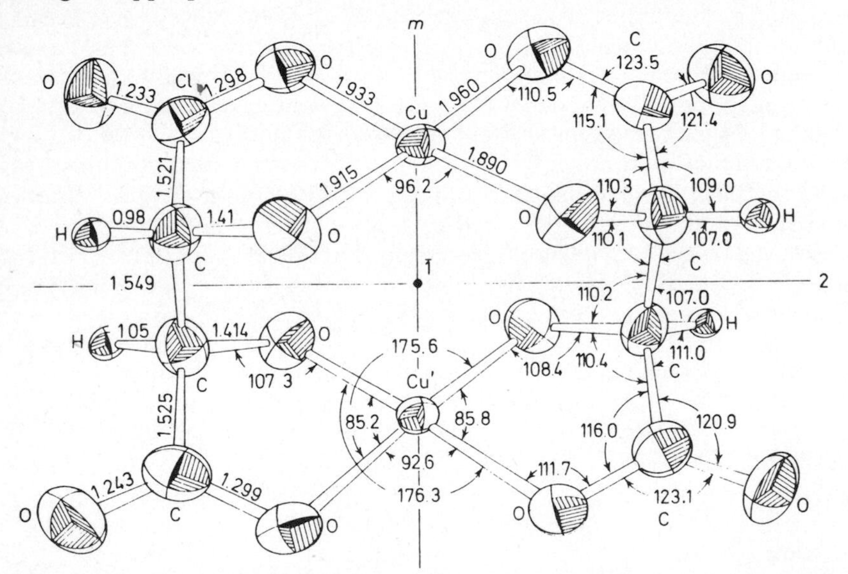

Figure 4.25 Bond lengths and angles in the anion $[Cu_2(dl\text{-tartrate})_2]^{4-}$. The ellipsoids show the thermal motion of the atoms with a probability of 50%. (From Missavage[207], by courtesy of Gordon and Breach Science Publishers, Ltd.)

Moreover, the Cu^{II} atoms are only 4-coordinate in this structure, being slightly displaced from planar coordination *towards* the other half of the dimer. In the crystal of $Na_2[Cu(dl\text{-}C_4O_6H_2)] \cdot 5H_2O$, all the water molecules form hydrogen bonds in a regular manner to the tetranegative anion so as to leave small holes above the Cu^{II} atoms, but these are not large enough to accommodate a fifth coordination to Cu. Similar structural units have been proposed for some 3-ketoglutarates of Cu^{II} and other divalent metals, as shown in Figure 4.26, on the basis of i.r. evidence[208].

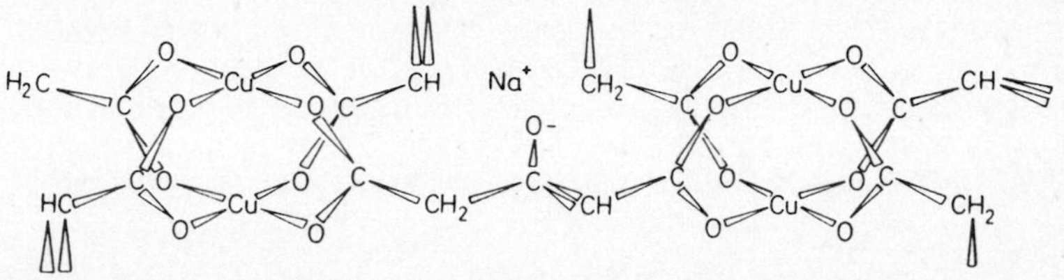

Figure 4.26 Metal-bridged structure for NaCu(3-ketoglutarate) polymers. (From Yerhoff[208], by courtesy of The National Research Council of Canada.)

A polymeric acetatocopper(II) complex containing asymmetric bridging acetate groups has been subjected to x-ray analysis[209]. Bridging trifluoroacetate groups have been similarly proposed in a number of phenylphosphine and olefin complexes of copper(I) trifluoroacetate[210]. An extensive set of data from some 140 compounds has been examined in an investigation of antiferromagnetism in binuclear copper(II) carboxylate complexes[211]. Most of the data have been fitted by a singlet–triplet–singlet model applied to the low-lying excited states of these complexes. Weak spin–spin coupling between Cu^{II} atoms in polymeric dioxalatocuprate(II) salts has been characterised through e.s.r. and magnetic susceptibility measurements[212], and a mechanism has been proposed involving the weak interaction being transmitted by the oxalate bridges. Strong antiferromagnetism in crystals of the oxalatocopper-(II) complex of formula $Cu(NH_3)(C_2O_4)$ similarly has been attributed to a superexchange mechanism operating through bridging oxygen atoms of oxalate groups[213].

The reaction between copper(II) acetate and thio-bis(β-diketones) (Figure

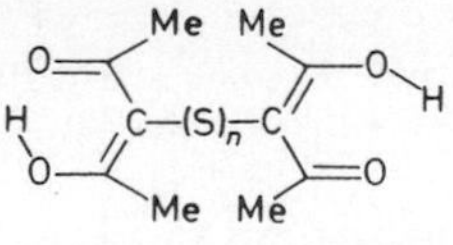

Figure 4.27 The thio-bis(β-diketone) ligand[214] ($n = 1$, 2 or 3)

4.27) in acetone/ethanol as a solvent has been shown to yield highly polymeric complexes[214], while the simple monomeric bis-acetylacetone complex is produced by reaction of $Cu(acetate)_2$ with phenyl- and tolyl-acetylacetones[215]. Structures with 6-coordinate copper have been reported for a number of hexafluoroacetylacetonatocopper(II) complexes of the type $Cu(hfac)_2L$ and $Cu(hfac)_2$, where L is a bidentate ligand, the former containing bidentate hfac and the latter monodentate hfac ligands[216]. When L is a monodentate ligand such as a tertiary phosphine, however, the 1:1

adducts $Cu(hfac)_2L$ contain 5-coordinate Cu^{II} [217]. Mixed acetylacetonate–carboxylate complexes of copper(II) have been prepared[218].

Mixed *O*, *N*-donor complexes of copper(II) have been formed by reaction of pyrazine carboxylates with $Cu(OAc)_2 \cdot H_2O$[219]. The pyrazine-2,3,5-tricarboxylate is thought to form the trinuclear complex shown in Figure 4.28,

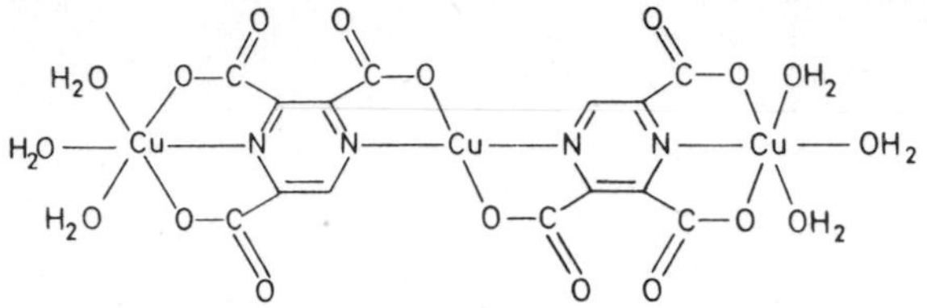

Figure 4.28 Probable structure of bis(pyrazine-2,3,5-tricarboxylato)-tricopper(II) hexahydrate[219]

while the polymeric product depicted in Figure 4.29 results from the pyrazine-

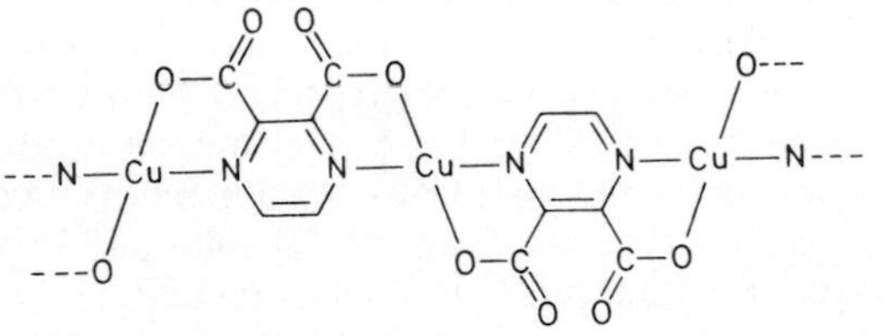

Figure 4.29 Probable structure of the pyrazine-2,3-dicarboxylato-copper(II) coordination polymer[219]

2,3-dicarboxylate reaction. Interest in the use of nitrilotriacetate as a possible ingredient of commercial detergents has led to a sharp rebuke being administered[220] for an alleged misinterpretation of data used for the characterisation of the nitrilotriacetate complex of copper. The prevailing view that the complex is of the form CuL rather than the alternative protonated CuHL form has been upheld.

A study of geometrical isomerism in a wide range of bis(aminoacidato)copper(II) complexes has been reported in two parts[221, 222]. Adenine ligands have been shown to act as bridging groups in the trinuclear complex octachloro-bis(adeninium)tricopper(II) tetrahydrate, which has a structure based on a 6-coordinate central copper atom and 5-coordinate terminal atoms[223]. Mixed complexes of copper(II) with pairs of amino-acid ligands have been established as more stable than the parent binary complexes, and this stabilisation has been attributed primarily to statistical effects[224]. The formation of a tetrameric copper(II) complex of 3,4-dihydroxyphenylglycine with the unusual macrocyclic structure shown in Figure 4.30 has been reported[225].

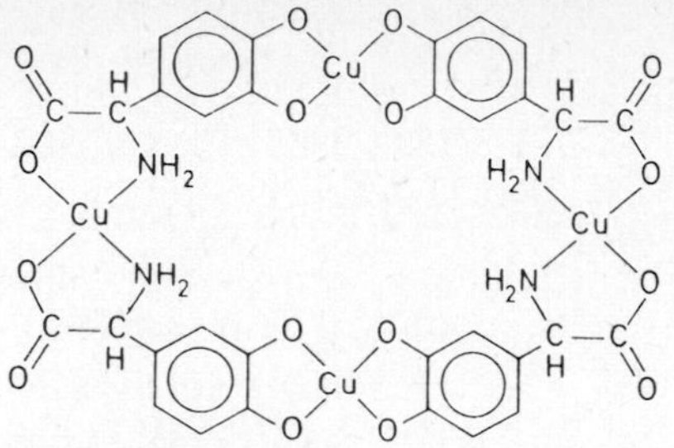

Figure 4.30 Probable structure of the tetrameric 3,4-dihydroxy-phenylglycinato-copper(II) complex[225]

Several new investigations of simple oxoanion complexes with copper(II) have been reported. X-Ray studies have shown the complex $Cu(py)_3(NO_3)_2$ (py = pyridine) to be monomeric in the crystalline state, with strongly asymmetric metal–nitrate coordination[226]. A thorough vibrational spectroscopic study of the nature of the metal–ligand environment in complexes $Cu(py)_n(NO_3)_2$, where $n=2$, 3 and 4, has been made[227]. The Cu–O stretching vibration in these complexes has been shown to shift to higher frequency as the covalency of the nitrate bonding increases, while the Cu–N stretching vibration moves to lower frequency. The carbonate ion, which is isoelectronic with nitrate, has been shown to exist in the bidentate coordinated form in anhydrous sodium copper carbonate[228]. This structure contains copper atoms coordinated by four carbonate oxygens in a square-planar arrangement and linked by bridging carbonates in an infinite two-dimensional polymeric puckered array.

4.10.2 Silver and gold

A review of recent advances in the coordination chemistry of gold has appeared[229], and some new tetranitrato-aurates(III) have been prepared and characterised[230]. $KAu(NO_3)_4$ evidently has all four nitrate groups in the anion bonded to gold in the same monodentate manner, providing the first authenticated example of a square-planar complex containing only monodentate nitrate ligands. Nitrate coordination to silver(I) has been indicated in the complexes $[AgL_2(NO_3)]$ (L = 3-chloropyridine, 4-cyanopyridine or 3,5-lutidine) by their i.r. spectra and the fact that they are non-electrolytes[231]. The corresponding $[AgL_2](ClO_4)$ complexes, however, evidently do not involve silver–perchlorate coordination.

A new series of silver(I) complexes with a wide range of O-donor ligands, such as dimethylsulphoxide, dimethylformamide, or triethylphosphate, has been prepared, many of these being photosensitive[232]. Some non-ionic silver(I) olefin complexes of the type (olefin)(β-diketonato)silver(I) which are soluble in relatively non-polar solvents such as dichloromethane or cyclohexane have been described[233]. The thio-β-diketone of formula PhCOCHMe-CSPh has been shown to complex silver(I), but i.r. spectroscopic evidence suggests that the ligand is not coordinated to silver through the carbonyl group[234]. Oxygen coordination does appear to exist, however, in the silver(I) complex [Ag(pyz-2-CO_2)] formed by reacting $AgNO_3$ with neutralised pyrazine-2-carboxylic acid, and in the silver(II) complex [Ag(pyz-2-CO_2)$_2$] which results when the free acid is used[235]. A mixed Ag^I–Ag^{II} complex of formula $Ag^I_2Ag^{II}[C_4H_2N_2(CO_2)_2]_2$ is formed when pyrazine-2,3-dicarboxylic acid is used[235].

Among the pure nitrogen-donor complexes of gold, the bis(trimethylsilyl) amine compounds $(R_3E)AuN(SiMe_3)_2$ ($ER_3 = PMe_3$, PPh_3 or $AsPh_3$) are particularly noteworthy. These have been formed by addition of $HN(SiMe_3)_2$ to the compounds $AuCl(ER_3)$ and are the first compounds made with the Au—N—Si linkage[236].

A variety of coordination configurations have been reported for amino-acid complexes of silver(I)[237]. In [Ag(gly)] [Figure 4.31(a)], each silver atom

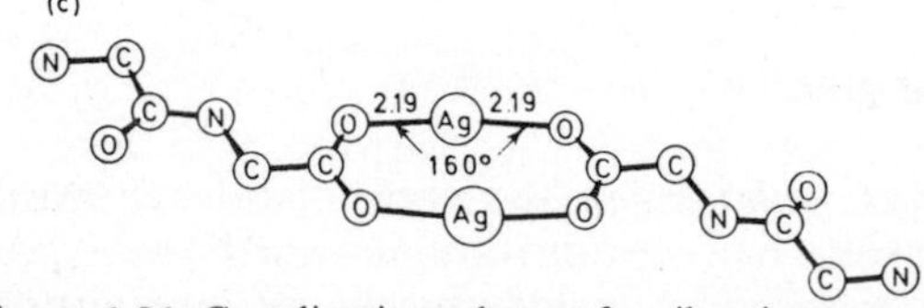

Figure 4.31 Coordination schemes for silver in some amino-acid complexes. (From Acland[237], by courtesy of The Chemical Society.)

is bound to a carboxyl oxygen of one glycinate and to the amino nitrogen of another; in [Ag(gly)]·0.5H_2O [Figure 4.31(b)], alternate Ag ions are bound to two O and two N atoms, respectively; in [Ag(H gly–gly)](NO_3), double oxygen coordination of the type shown in Figure 4.31(c) has been assigned. Silver(I) complexes with L-cysteine and *S*-methyl-L-cysteine and their methyl esters, and with DL-methionine all appear to involve Ag—S bonding[238].

4.11 THE ZINC GROUP

The kinetic behaviour of aminopolycarboxylate ligands in substitution reactions of zinc(II) complexes has been investigated[239]. Kinetic differences for reaction of maleonitriledithiolate (mnt^{2-}) with $Zn(NTA)^-$ and for similar reactions with $Cu(NTA)^-$ and $Ni(NTA)^-$ suggest differences in structure between these NTA complexes (NTA^{3-} = nitrilotriacetate). Proton transfer from $NTAH^{2-}$ is believed to control the rates of reaction with $Zn(H_2O)_6^{2+}$ and $Zn(dto)_2^{2-}$ =dithio-oxalate). The EDTA reaction with zinc complexes has been compared with those of two analogues that form 6-membered chelate rings.

Stability constants of some zinc and cadmium polyamine and aminocarboxylate mixed-ligand complexes have been determined[240]. The complexation reaction considered was $ML+X \rightleftharpoons MLX$, where $M = Zn^{II}$ or Cd^{II}, $L = H_3NTA$ or β,β',β''-triaminotriethylamine (tren), and X = ethylenediamine(en), glycine (Hgly), malonic acid (H_2ma) or iminodiacetic acid (H_2ida). Stability constants of the monohydroxy complexes of $Zn(NTA)^-$ and $Zn(tren)^{2+}$ have also been determined[240]. Further stability constants for the complexes Zn(NTA)L, where L is mercaptoacetic, thiolactic or thiomalic acid, have been reported[241]. Polarographic measurements have demonstrated the formation of *trans*-1,2-cyclohexanediaminetetra-acetic acid complexes of cadmium(II) in dimethylformamide solution[242]. The kinetics of the partial dissociation of zine and cadmium complexes of 1,3-propylenediaminetetra-acetic acid have been studied in solution using dynamic n.m.r. methods[243]. The method depends on the fact that the acetate methylenic protons are non-equivalent and give rise to an AB triplet pattern in the n.m.r. spectrum which collapses to an extent dependent upon the average lifetime before interchange.

The much simpler cadmium diacetate complex has been the subject of a crystal and molecular structure determination[244]. In the solid state the cadmium is coordinated to seven oxygen atoms, with both acetate groups being bidentate and one of them having an oxygen atom in a bridging position between neighbouring cadmiums. Acetato complexes of mercury(I) and mercury(II) have been studied[245] by i.r. and Raman spectroscopy in the frequency region below 400 cm^{-1}. Both Hg—Hg and Hg—O stretching frequencies have been assigned for $Hg_2(OAc)_2$. Mixed oxalato–maleato complexes of cadmium(II) have been characterised by polarographic measurements in solution[246]. Notwithstanding earlier evidence to the contrary, the complex dinitrato-trispyridinecadmium(II) has been shown by x-ray analysis to be monomeric in the solid state, containing cadmium atoms made 7-coordinate by the bidentate nature of the two nitrato groups[247].

Acknowledgement

I should like to thank Reuben B. Girling for his help with the literature search.

References

1. Hester, R. E. (1972). *MTP International Review of Science, Series One, Inorganic Chemistry*, Vol. 5, Chapter 1 (London: Butterworths)
2. Johnson, B. F. G. (senior reporter) (1972). *Specialist Periodical Reports, Inorganic Chemistry of the Transition Elements*, Vol. 1, (London: The Chemical Society)

3. Greenwood, N. N. (senior reporter) (1972). *Specialist Periodical Reports, Spectroscopic Properties of Inorganic and Organometallic Compounds*, Vol. 5 (London: The Chemical Society)
4. Ahrland, S. (1972). *Coord. Chem. Rev.*, **8,** 21
5. Hunt, J. P. (1971). *Coord. Chem. Rev.*, **7,** 1
6. Serpone, N. and Bickley, D. G. (1972). *Progr. Inorg. Chem.*, **17,** 391
7. Dasgupta, T. P. and Tobe, M. L. (1972). *Coord. Chem. Rev.*, **8,** 103
8. Hoskins, B. F. and Whillans, F. D. (1972–1973). *Coord. Chem. Rev.*, **9,** 365
9. Fortman, J. J. and Sievers, R. E. (1971). *Coord. Chem. Rev.*, **6,** 331
10. Cox, M. and Darken, J. (1971). *Coord. Chem. Rev.*, **7,** 29
11. Livingstone, S. E. (1971). *Coord. Chem. Rev.*, **7,** 59
12. Calligaris, M., Nardin, G. and Randaccio, L. (1972). *Coord. Chem. Rev.*, **7,** 385
13. Hobday, M. D. and Smith, T. D. (1972–1973). *Coord. Chem. Rev.*, **9,** 311
14. Hambright, P. (1971). *Coord. Chem. Rev.*, **6,** 247
15. Holm, R. H. and O'Connor, M. J. (1971). *Progr. Inorg. Chem.*, **14,** 241
16. Book, B., Flatan, K., Junge, H., Kuhr, M. and Musso, H. (1971). *Angew. Chem. Int. Ed. Engl.*, **10,** 225
17. Addison, C. C., Logan, N., Wallwork, S. C. and Garner, C. D. (1971). *Quart. Rev. Chem. Soc.*, **25,** 289
18. Baran, V. (1971). *Coord. Chem. Rev.*, **6,** 65
19. Melson, G. A. and Stotz, R. W. (1971). *Coord. Chem. Rev.*, **7,** 133
20. Söylemez, Z. and Özer, U. Y. (1973). *J. Inorg. Nucl. Chem.*, **35,** 545
21. Speca, N. N., Karayannis, N. M. and Pytlewski (1972). *Inorg. Chim. Acta*, **6,** 639
22. Taravel, B., Fromage, F., Delorme, P. and Lorenzelli, V. (1972). *C. R. Acad. Sci. Paris, Ser. B*, **275,** 589
23. Morris, D. F. C., Haynes, F. B., Lewis, P. A. and Short, E. L. (1972). *Electrochim. Acta*, **17,** 2017
24. Men'kov, A. A. (1971). *Fiz.-Khim. Metody Issled. Anal. Biol. Ob'ektov Nekot. Tekh. Mater.*, 31 (from (1972). *Chem. Abstr.*, **77,** 172113v)
25. Filatova, L. N. and Novichkova, S. L. (1971). *Tr. Vses. Nauch. Issled. Inst. Khim. Reaktiv. Osobo Chist. Khim. Veshchestv*, **33,** 47 (from (1972). *Chem. Abstr.*, **78,** 23369b)
26. Komissarova, L. N., Teterin, E. G., Mel'nikov, P. P. and Chuvaev, V. F. (1972). *Zh. Strukt. Khim.*, **13,** 837 (from (1972). *Chem. Abstr.*, **78,** 22056k)
27. Pokorny, J. and Petra, F. (1972). *Sb. Vys. Sk. Chem. Technol. Praze, Anorg. Chem. Technol.*, **B15,** 45 (from (1972). *Chem. Abstr.*, **78,** 15444g)
28. Siek, R. F. and Banks, C. V. (1972). *Analyt. Chem.*, **44,** 2307
29. Smith, G. D., Caughlan, C. N. and Campbell, J. A. (1972). *Inorg. Chem.*, **11,** 2989
30. Evans, S., Hamnett, A. and Orchard, A. F. (1972). *J. Coord. Chem.*, **2,** 57
31. Shaulov, Yu. Kh., Genchel, V. G., Aizatallova, R. M. and Petrova, N. V. (1972). *Zh. Fiz. Khim.*, **46,** 2382 (from (1972). *Chem. Abstr.*, **77,** 169520v)
32. Alyea, E. C. and Merrell, P. H. (1973). *Inorg. Nucl. Chem. Lett.*, **9,** 69
33. Holloway, C. E. and Sentek, A. E. (1971). *Can. J. Chem.*, **49,** 519
34. Svetich, G. W. and Voge, A. A. (1971). *Chem. Commun.*, 676
35. Gupta, V. D. and Mehrotra, R. C. (1971). *J. Chem. Soc.* (*A*), 2440
36. Schultze, D., Wilke, K. T. and Waligora, C. H. (1971). *Z. Anorg. Allg. Chem.*, **380,** 37
37. Masse, R. (1970). *Bull. Soc. Fr. Mineral. Crystallogr.*, **93,** 500
38. Soliev, L. and Goroshchenko, Ya. G. (1970). *Zh. Neorg. Khim.*, **15,** 2515
39. Pan, K., Lai, C. C. and Huang, T.-S. (1971). *Bull. Chem. Soc. Jap.*, **44,** 93
40. Bradley, D. C. and Copperthwaite, R. G. (1971). *Chem. Commun.*, 766
41. Lappert, M. F. and Sanger, A. R. (1971). *J. Chem. Soc.* (*A*), 1314
42. Keppie, S. A., Lappert, M. F., McMeeking, J., Sanger, A. R. and Srivastava, R. C. (1971). *U.S. Govt. Res. Develop. Rep.*, **71,** 54
43. Mehrotra, R. C. and Mehrotra, A. (1972). *J. Chem. Soc.* (*Dalton*), 1203
44. Puri, D. M. (1970). *J. Indian Chem. Soc.*, **47,** 525
45. Fay, R. C., Von Dreele, R. B. and Stezowski, J. J. (1971). *J. Amer. Chem. Soc.*, **93,** 2887
46. Howe, J. J. and Pinnavaia, T. J. (1970). *J. Amer. Chem. Soc.*, **92,** 7343
47. Pinnavaia, T. J. and Lott, A. L. (1971). *Inorg. Chem.*, **10,** 1388
48. Mumme, W. G. (1971). *Acta Cryst.*, **B27,** 1373
49. Bear, I. J. and Mumme, W. G. (1971). *Acta Cryst.*, **B27,** 494
50. Navratil, O. and Smola, J. (1971). *Collect. Czech. Chem. Commun.*, **36,** 1659

51. Larsen, E. M. and Homeier, E. H. (1972). *Inorg. Chem.*, **11**, 2687
52. Wiedenheft, C. J. (1971). *Inorg. Nucl. Chem. Lett.*, **7**, 439
53. Miller, D. A. and Bereman, R. D. (1972/73). *Coord. Chem. Rev.*, **9**, 107
54. Dolmatov, Yu. D., Bulavina, Z. N. and Dolmatova, M. Yu. (1972). *Radiokhimiya*, **14**, 530 (from (1972). *Chem. Abstr.*, **78**, 8241k)
55. Pai, K. R. and Krishnaswamy, N. (1972). *Indian J. Technol.*, **10**, 229
56. Lad, K. V. and Baxi, D. R. (1972). *Indian J. Technol.*, **10**, 224
57. Chandra, G., Jenkins, A. D., Lappert, M. F. and Srivastava, R. C. (1970). *J. Chem. Soc.* (*A*), 2550
58. Buchler, J. W., Eikelmann, G., Puppe, L., Rohbock, K., Schneehage, H. H. and Week, D. (1971). *Justus Liebigs Ann. Chem.*, **745**, 135
59. Buchler, J. W. and Rohbock, K. (1972). *Inorg. Nucl. Chem. Lett.*, **8**, 1073
60. Podlaka, J. and Podlahova, J. (1970). *Inorg. Chim. Acta*, **4**, 521
61. Allin, B. J. and Thornton, P. (1973). *Inorg. Nucl. Chem. Lett.*, **9**, 449
62. Bennett, B. G. and Nicholls, D. (1972). *J. Inorg. Nucl. Chem.*, **34**, 673
63. Syamal, A. and Theriot, L. J. (1973). *J. Coord. Chem.*, **2**, 193
64. Kabinnikov, V. T., Zelentsov, V. V., Ubozhenko, O. D. and Aminov, T. G. (1972). *Dokl. Akad. Nauk SSSR*, **206**, 627 (from (1972). *Chem. Abstr.*, **78**, 21723p)
65. Aroney, M. J., Chio, H., James, J. M., Le Fevre, R. J. W., Pierens, R. K. and Skamp, K. R. (1972). *J. Chem. Soc.* (*Dalton*), 712
66. Dichmann, K., Hamer, G., Nyburg, S. C. and Reynolds, W. F. (1970). *Chem. Commun.*, 1295
67. Einstein, F. W. B., Enwall, E., Morris, D. M. and Sutton, D. (1971). *Inorg. Chem.*, **10**, 678
68. Scheidt, W. R., Tsui, C. and Hoard, J. L. (1971). *J. Amer. Chem. Soc.*, **93**, 3867
69. Scheidt, W. R., Collins, D. M. and Hoard, J. L. (1971). *J. Amer. Chem. Soc.*, **93**, 3873
70. Scheidt, W. R., Countryman, R. and Hoard, J. L. (1971). *J. Amer. Chem. Soc.*, **93**, 3878
71. Amos, L. W. and Sawyer, D. T. (1972). *Inorg. Chem.*, **11**, 2692
72. Deutscher, R. L. and Kepert, D. L. (1970). *Inorg. Chim. Acta*, **4**, 645
73. Vuletic, N. and Djordjevic, C. (1972). *J. Chem. Soc.* (*Dalton*), 2322
74. Pets, L. I. (1972). *Tr. Tallin. Politekh. Inst.*, **No. 319**, 73 (from (1972). *Chem. Abstr.*, **78**, 20753m)
75. Johnson, R. C. and Yamal, A. S. (1971). *J. Inorg. Nucl. Chem.*, **33**, 2547
76. Holloway, C. E. (1971). *J. Coord. Chem.*, **1**, 253
77. Dubicki, L. and Day, P. (1972). *Inorg. Chem.*, **11**, 1868
78. Zinato, E., Furlani, C., Lanna, G. and Riccieri, P. (1972). *Inorg. Chem.*, **11**, 1746
79. Spinner, T. and Harris, G. M. (1972). *Inorg. Chem.*, **11**, 1067
80. Odell, A. L. and Shooter, D. (1972). *J. Chem. Soc.* (*Dalton*), 135
81. Odell, A. L., Olliff, R. W. and Rands, D. B. (1972). *J. Chem. Soc.* (*Dalton*), 752
82. Schenk, C. and Kelm, H. (1972). *J. Coord. Chem.*, **2**, 71
83. Frank, M. J. and Huchital, D. H. (1972). *Inorg. Chem.*, **11**, 776
84. Fay, R. C., Girgis, A. Y. and Klabunde, U. (1970). *J. Amer. Chem. Soc.*, **92**, 7056
85. Broomhead, J. A., Kane-Maguire, N. and Lauder, J. (1971). *Inorg. Chem.*, **10**, 955
86. Snow, M. R. and Butler, K. R. (1971). *Chem. Commun.*, 550
87. Karayannis, N. M., Mikulski, C. M., Strocko, M. J., Pytlewski, L. L. and Labes, M. M. (1970). *J. Inorg. Nucl. Chem.*, **32**, 2629
88. Scott, A. and Taube, H. (1971). *Inorg. Chem.*, **10**, 62
89. Bryan, R. F., Greene, P. T., Stokely, P. F. and Wilson, Jr., E. W. (1971). *Inorg. Chem.*, **10**, 1468
90. Hodgson, D. J., Veal, J. T., Hatfield, W. E., Jeter, D. Y. and Hemel, J. C. (1972). *J. Coord. Chem.*, **2**, 1
91. Cotton, F. A., De Boer, B. G., La Prade, M. D., Pipal, J. R. and Ucko, D. A. (1971). *Acta Cryst.*, **B27**, 1664
92. Ouahes, R., Amiel, J. and Suquet, H. (1970). *Rev. Chim. minerale*, **7**, 789; Ouahes, R., Perezat, H. and Gayoso, J., *ibid.*, 849; Ouahes, R., Devallez, B. and Amiel, J., *ibid.*, 855
93. Bradley, D. C., Hursthouse, M. B. and Neuring, C. W. (1971). *Chem. Commun.*, 411
94. Alyea, E. C., Basi, J. S., Bradley, D. C. and Chisholm, M. H. (1971). *J. Chem. Soc.* (*A*), 772
95. Bradley, D. C., Copperthwaite, R. G., Cotton, S. A., Sales, K. D. and Gibson, J. F. (1973). *J. Chem. Soc.* (*Dalton*), 409
96. Craven, B. M., Ramey, K. C. and Wise, W. B. (1971). *Inorg. Chem.*, **11**, 2626
97. Doyle, G. (1971). *Inorg. Chem.*, **10**, 2348
98. Bonds, Jr., W. D. and Archer, R. D. (1971). *Inorg. Chem.*, **9**, 2057
99. Isbell, Jr., A. F. and Sawyer, D. T. (1971). *Inorg. Chem.*, **10**, 2449

100. Brown, D. H. and MacPherson, J. (1970). *J. Inorg. Nucl. Chem.*, **32,** 3309
101. Delbaere, L. T. J. and Prout, C. K. (1971). *Chem. Commun.*, 162
102. Gore, E. S. and Green, M. L. H. (1970). *J. Chem. Soc.* (*A*), 2315
103. Kloubek, J. and Podlaha, J. (1971). *Inorg. Nucl. Chem. Lett.*, **7,** 67
104. Lund, W. (1971). *Analyt. Chim. Acta*, **53,** 295
105. Soni, R. N. and Bartusek, M. (1971). *J. Inorg. Nucl. Chem.*, **10,** 2557
106. Johnson, D. W. and Sutton, D. (1972). *Can. J. Chem.*, **50,** 3326
107. Koda, S., Ooi, S., Kuroya, H., Nakamura, Y. and Kawaguchi, S. (1971). *Chem. Commun.*, 280
108. Hudson, A. and Kennedy, J. de G. (1971). *Inorg. Nucl. Chem. Lett.*, **7,** 333
109. Zetter, M. S., Grant, M. W., Wood, E. J., Dodgen, H. W. and Hunt, J. P. (1972). *Inorg. Chem.*, **11,** 2701
110. Poh, B. L. and Stewart, R. (1972). *Can. J. Chem.*, **50,** 3432, 3437
111. Bailey, R. A., Willis, W. N. and Tangredi, W. J. (1971). *J. Inorg. Nucl. Chem.*, **33,** 2387
112. Shukla, J. P. and Taudon, S. G. (1972). *Bull. Chem. Soc. Jap.*, **45,** 3073
113. Lingaiah, P. and Sundaram, E. V. (1972). *Indian J. Chem.*, **10,** 670
114. Izatt, R. M., Johnson, H. D. and Christensen, J. J. (1972). *J. Chem. Soc.* (*Dalton*), 1152
115. Boucher, L. J. (1972). *Coord. Chem. Rev.*, **7,** 289
116. Barrick, J. C., Fredette, M. and Locke, C. J. L. (1973). *Can. J. Chem.*, **51,** 317
117. Courrier, W. D., Lock, C. J. L. and Turner, G. (1972). *Can. J. Chem.*, **50,** 1797
118. Courrier, W. D., Forster, W., Lock, C. J. L. and Turner, G. (1972). *Can. J. Chem.*, **50,** 8
119. Chakravorti, M. C. and Chaudhuri, M. K. (1973). *J. Inorg. Nucl. Chem.*, **35,** 949
120. King, T. J., Logan, N., Morris, A. and Wallwork, S. C. (1971). *Chem. Commun.*, 554
121. Zayats, M. N., Mokhosoev, M. V. and Get'man, E. I. (1971). *Russian J. Inorg. Chem.*, **16,** 84
122. Cotton, S. A. and Gibson, J. F. (1971). *J. Chem. Soc.* (*A*), 859
123. Takano, M. (1972). *J. Phys. Soc. Jap.*, **33,** 1312
124. Wu, C.-H. S., Rossman, G. R., Gray, H. B., Hammond, G. S. and Shagar, H. J. (1972). *Inorg. Chem.*, **11,** 990
125. Cox, M., Fitzsimmons, B. W., Smith, A. W., Larkworthy, L. F. and Rogers, K. A. (1971). *J. Chem. Soc.* (*A*), 2158
126. Lindley, P. F. and Smith, A. W. (1970). *Chem. Commun.*, 1355
127. Coggon, P., McPhail, A. T., Mabbs, F. E. and McLachlan, V. N. (1971). *J. Chem. Soc.* (*A*), 1014
128. Bowles, A. M., Szarck, W. A. and Baird, M. C. (1971). *Inorg. Nucl. Chem. Lett.*, **7,** 25
129. Williams, D. R. and Yeo, P. A. (1972). *J. Chem. Soc.* (*Dalton*), 1988
130. Cotton, S. A. (1972). *Coord. Chem. Rev.*, **8,** 185
131. Floriani, C. and Calderazzo, F. (1972). *Coord. Chem. Rev.*, **8,** 57
132. Spencer, A. and Wilkinson, G. (1972). *J. Chem. Soc.* (*Dalton*), 1570
133. Cheng, P. T., Loescher, B. R. and Nyburg, S. C. (1971). *Inorg. Chem.*, **10,** 1275
134. Johnson, B. F. G., Johnston, R. D., Lewis, J. and Williams, I. G. (1971). *J. Chem. Soc.* (*A*), 689
135. Fay, R. C., Girgis, A. Y. and Klabunde, U. (1970). *J. Amer. Chem. Soc.*, **92,** 7056
136. Patterson, G. S. and Holm, R. S. (1972). *Inorg. Chem.*, **11,** 2285
137. Gordon, J. G., O'Connor, M. J. and Holm, R. H. (1971). *Inorg. Chim. Acta*, **5,** 381
138. Potvin, C., Manoli, J. M., Dereigne, A. and Pannetier, G. (1972). *Bull. Soc. Chim. Fr.*, 3078
139. Earley, J. E. and Fealey, T. (1971). *Chem. Commun.*, 331
140. Balch, A. L. and Sohn, Y. S. (1971). *J. Organometal. Chem.*, **30,** C31
141. Dickinson, J. R. and Stone, M. E. (1972). *Can. J. Chem.*, **50,** 2946
142. Sastri, V. S. (1972). *Inorg. Chim. Acta*, **6,** 264
143. Gillard, R. D., Mitchell, P. R. and Price, M. G. (1972). *J. Chem. Soc.* (*Dalton*), 1211
144. Francis, D. J. and Jordan, R. B. (1972). *Inorg. Chem.*, **11,** 1170
145. Bertrand, J. A. and Hightower, T. C. (1973). *Inorg. Chem.*, **12,** 206
146. Bertrand, J. A., Ginsberg, A. P., Kaplan, R. I., Kirkwood, C. E., Martin, R. L. and Sherwood, R. C. (1971). *Inorg. Chem.*, **10,** 240
147. Boucher, L. J. (1972). *Inorg. Chem.*, **11,** 29
148. Boucher, L. J. and Herrington, D. R. (1972). *Inorg. Chem.*, **11,** 1772
149. Butler, K. R. and Snow, M. R. (1971). *Chem. Commun.*, 550; *J. Chem. Soc.* (*A*), 565
150. Odell, A. L. and Rands, D. B. (1972). *J. Chem. Soc.* (*Dalton*), 749
151. Laurie, S. H. (1972). *J. Chem. Soc.* (*Dalton*), 573
152. Gillard, R. D. and Spencer, A. (1972). *J. Chem. Soc.* (*Dalton*), 902
153. Browning, I. G., Gillard, R. D., Lyons, J. R., Mitchell, P. R. and Phipps, D. A. (1972). *J. Chem. Soc.* (*Dalton*), 1815

154. Gillard, R. D., Lyons, J. R. and Thorpe, C. (1972). *J. Chem. Soc. (Dalton)*, 1584
155. Pasternack, R. F., Gipp, L. and Sigel, H. (1972). *J. Amer. Chem. Soc.*, **94**, 8031
156. Comley, H. M. and Higginson, W. C. E. (1972). *J. Chem. Soc. (Dalton)*, 2522
157. Scott, K. L. and Sykes, A. G. (1972). *J. Chem. Soc. (Dalton)*, 2364
158. Cannon, R. D. and Gardiner, J. (1972). *J. Chem. Soc. (Dalton)*, 887
159. Scott, K. L. and Sykes, A. G. (1972). *J. Chem. Soc. (Dalton)*, 1832
160. Dasgupta, T. P. and Tobe, M. L. (1972). *Inorg. Chem.*, **11**, 1011
161. Dasgupta, T. P., Fitzgerald, W. and Tobe, M. L. (1972). *Inorg. Chem.*, **11**, 2046
162. Farago, M. E., Smith, M. A. R. and Keefe, I. M. (1972). *Coord. Chem. Rev.*, **8**, 95
163. Katzin, L. I. (1972). *Coord. Chem. Rev.*, **7**, 331
164. Ohashi, K. (1972). *Bull. Chem. Soc. Jap.*, **45**, 3093
165. Kipp, E. B. and Haines, R. A. (1972). *Inorg. Chem.*, **11**, 271
166. Costa, G. (1972). *Coord. Chem. Rev.*, **8**, 63
167. Peone, J. and Vaska, L. (1971). *Angew. Chem. Int. Ed. Engl.*, **10**, 511
168. Shubochkin, L. K., Nefodov, V. I., Shubochkina, E. F., Golubnichaya, M. A. and Sorokina, L. D. (1972). *Zh. Neorg. Khim.*, **17**, 2852 (from (1973). *Chem. Abstr.*, **78**, 23416q)
169. Harrison, B. and Logan, N. (1972). *J. Chem. Soc. (Dalton)*, 1587
170. Valentine, J., Valentine, D. and Collman, J. P. (1971). *Inorg. Chem.*, **10**, 219
171. Levison, J. J. and Robinson, S. D. (1971). *J. Chem. Soc. (A)*, 762
172. Clark, G. R., Reed, C. A., Roper, W. R., Skelton, B. W. and Waters, T. N. (1971). *Chem. Commun.*, 758
173. Syrtsova, G. P., Kharitonov, Yu. Ya. and Bolgar, T. S. (1972). *Zh. Neorg. Khim.*, **17**, 2719 (from (1973). *Chem. Abstr.*, **78**, 21917e)
174. Brown, D. B., Robin, M. B., McIntyre, J. D. E. and Peck, W. F. (1970). *Inorg. Chem.*, **9**, 2315
175. Hoare, R. J. and Mills, O. S. (1972). *J. Chem. Soc. (Dalton)*, 2138
176. Smith, S. A., Blake, D. M. and Kubota, M. (1972). *Inorg. Chem.*, **11**, 660
177. Caulton, K. G. and Cotton, F. A. (1971). *J. Amer. Chem. Soc.*, **93**, 1914
178. Cotton, F. A. and Norman, J. G. (1971). *J. Amer. Chem. Soc.*, **93**, 80
179. Miller, J. D. and Oliver, F. D. (1972). *J. Chem. Soc. (Dalton)*, 2469
180. Prout, C. K., Walker, C. and Rossotti, F. J. C. (1971). *J. Chem. Soc. (A)*, 556
181. Downie, T. C., Harrison, W., Rasper, E. S. and Hepworth, M. A. (1971). *Acta Cryst.*, **B27**, 706
182. Anagnostopoulos, A., Matthews, R. W. and Walton, R. A. (1972). *Can. J. Chem.*, **50**, 1307
183. Pavkovic, S. F. (1971). *J. Inorg. Nucl. Chem.*, **33**, 1475
184. Teggins, J. E., Lee, K. E., Baker, J. M. and Smith, E. D. (1971). *J. Coord. Chem.*, **1**, 215
185. Andrelczyk, B. and Lintvedt, R. L. (1972). *J. Amer. Chem. Soc.*, **94**, 8633
186. Ginsberg, A. P. (1971). *Inorg. Chim. Acta Rev.*, **5**, 45
187. Marcotrigiano, G., Battistuzzi, R. and Pellacani, G. C. (1972). *Can. J. Chem.*, **50**, 2557
188. Iqumenov, I. K. and Popov, A. N. (1972). *Izv. Sib. Otd. Akad. Nauk. SSSR, Ser. Khim. Nauk*, 38 (from (1972). *Chem. Abstr.*, **77**, 169564n)
189. Ogoshi, H., Saito, Y. and Nakamoto, K. (1972). *J. Chem. Phys.*, **57**, 4194
190. Wolberg, A. and Manassen, J. (1970). *Inorg. Chem.*, **9**, 2365
191. Hamilton, A. and Johnson, A. W. (1971). *Chem. Commun.*, 523
192. Grant, M. W., Dodgen, H. W. and Hunt, J. P. (1970). *Chem. Commun.*, 1446
193. DiVaira, M., Orioli, P. L. and Sacconi, L. (1971). *Inorg. Chem.*, **10**, 553
194. Selborg, M., Holt, S. L. and Post, B. (1971). *Inorg. Chem.*, **10**, 1501
195. Powell, J. and Jack, T. (1972). *Inorg. Chem.*, **11**, 1039
196. Barlex, D. M. and Kemmitt, R. D. W. (1972). *J. Chem. Soc. (Dalton)*, 1436
197. Giacomelli, A. and Indelli, A. (1972). *Inorg. Chem.*, **11**, 1033
198. Gomm, P. S., Underhill, A. E. and Watkins, D. M. (1972). *J. Chem. Soc. (Dalton)*, 2309
199. Hayward, P. J. and Nyman, C. J. (1971). *J. Amer. Chem. Soc.*, **93**, 617
200. Hayward, P. J. and Nyman, C. J. (1971). *Inorg. Chem.*, **10**, 1311
201. Bigotto, A., Galasso, V. and Atti, G. D. (1971). *Spectrochim. Acta*, **27A**, 1659
202. Belcher, R., Pravica, M., Stephen, W. I. and Uden, P. C. (1971). *Chem. Commun.*, 41
203. Coe, J. S. and Lyons, J. R. (1971). *J. Chem. Soc. (A)*, 829
204. Richardson, F. S., Shillady, D. D. and Waldrop, A. (1971). *Inorg. Chim. Acta*, **5**, 279
205. Volshtein, L. M. and Dikanskaya, L. D. (1971). *Russian J. Inorg. Chem.*, **16**, 223
206. Wilson, E. W. and Martin, R. B. (1971). *Inorg. Chem.*, **10**, 1197
207. Missavage, R. J., Belford, R. L. and Paul, I. C. (1972). *J. Coord. Chem.*, **2**, 145
208. Yerhoff, F. W. and Larson, D. W. (1972). *Can. J. Chem.*, **50**, 826
209. Skelton, B. W., Waters, T. N. and Curtis, N. F. (1972). *J. Chem. Soc. (Dalton)*, 2133

210. Dines, M. B. (1972). *Inorg. Chem.*, **11,** 2949
211. Jotham, R. W., Kettle, S. F. A. and Marks, J. A. (1972). *J. Chem. Soc.* (*Dalton*), 428
212. Jeter, D. Y. and Hatfield, W. E. (1972). *J. Amer. Chem. Soc.*, **94,** 523
213. Calvalca, L., Villa, A. C., Manfredotti, A. G., Mangia, A. and Tomlinson, A. A. G. (1972). *J. Chem. Soc.* (*Dalton*), 391
214. Jones, R. D. G. and Power, L. F. (1971). *Aust. J. Chem.*, **24,** 735
215. Graddon, D. P. and Heng, K. B. (1971). *Aust. J. Chem.*, **24,** 1059
216. Fenton, D. E., Nyholm, R. S. and Truter, M. R. (1971). *J. Chem. Soc.* (*A*), 1577
217. Zelonka, R. A. and Baird, M. C. (1972). *Can. J. Chem.*, **50,** 1269
218. Aly, M. M. (1973). *Inorg. Nucl. Chem. Lett.*, **9,** 253
219. Walters, R. W. and Walton, R. A. (1971). *Inorg. Chem.*, **10,** 1433
220. McBryde, W. A. E. and Cheam, V. (1973). *Inorg. Nucl. Chem. Lett.*, **9,** 95
221. Herlinger, A. W., Wenhold, S. L. and Long, T. V. (1970). *J. Amer. Chem. Soc.*, **92,** 6474
222. Herlinger, A. W. and Long, T. V. (1970). *J. Amer. Chem. Soc.*, **92,** 6481
223. De Meester, P. and Skapski, A. C. (1972). *J. Chem. Soc.* (*Dalton*), 2400
224. Ting Po I and Nancollas, G. H. (1972). *Inorg. Chem.*, **11,** 2414
225. Gorton, J. E. and Jameson, R. F. (1972). *J. Chem. Soc.* (*Dalton*), 307
226. Cameron, A. F., Taylor, D. W. and Nuttall, R. H. (1972). *J. Chem. Soc.* (*Dalton*), 1603
227. Choca, M., Ferraro, J. R. and Nakamoto, K. (1972). *J. Chem. Soc.* (*Dalton*), 2297
228. Healy, P. C. and White, A. H. (1972). *J. Chem. Soc.* (*Dalton*), 1913
229. Johnson, B. F. G. (1971). *Gold Bull.*, **4,** 9
230. Addison, C. C., Brownlee, G. S. and Logan, N. (1972). *J. Chem. Soc.* (*Dalton*), 1440
231. Misra, M. K. and Rao, D. V. R. (1970). *J. Indian Chem. Soc.*, **47,** 1162
232. Paul, R. C. and Chadha, S. L. (1971). *Indian J. Chem.*, **9,** 175
233. Partenheimer, W. and Johnson, E. H. (1972). *Inorg. Chem.*, **11,** 2840
234. Uhlemann, E. and Eckelmann, U. (1971). *Z. Anorg. Allg. Chem.*, **383,** 321
235. Matthews, R. W. and Walton, R. A. (1971). *Inorg. Chem.*, **10,** 1433
236. Shiotani, A. and Shmidbaur, H. (1970). *J. Amer. Chem. Soc.*, **92,** 7003
237. Acland, C. B. and Freeman, H. C. (1971). *Chem. Commun.*, 1016
238. Natursch, D. F. S. and Porter, L. J. (1971). *J. Chem. Soc.* (*A*), 2527
239. Pearson, R. G. and DeWit, D. (1973). *J. Coord. Chem.*, **2,** 175
240. Rabenstein, D. L. and Blakney, G. (1973). *Inorg. Chem.*, **12,** 128
241. Panchal, B. R. and Bhattacharya, P. K. (1972). *J. Inorg. Nucl. Chem.*, **34,** 3932
242. Dhar, S. K. and Stroz, D. A. (1972). *Inorg. Nucl. Chem. Lett.*, **8,** 1019
243. Rabenstein, D. L. and Fuhr, B. J. (1972). *Inorg. Chem.*, **11,** 2430
244. Harrison, W. and Trotter, J. (1972). *J. Chem. Soc.* (*Dalton*), 956
245. Cooney, R. P. J. and Hall, J. R. (1972). *J. Inorg. Nucl. Chem.*, **34,** 1519
246. Khurana, S. C. and Gupta, C. M. (1973). *J. Inorg. Nucl. Chem.*, **35,** 209
247. Cameron, A. F., Taylor, D. W. and Nuttall, R. H. (1972). *J. Chem. Soc.* (*Dalton*), 1608

5

Phosphine, Arsine, Stibine, Bismuthine, Thioether, Selenide and Telluride Complexes

R. J. CROSS
Glasgow University

5.1 INTRODUCTION

The ligands discussed in this section are all soft bases and thus form their most stable complexes with soft acids or class B acceptors. Consequently, many of their coordination compounds involve metals from the second and third transition series, low oxidation state metals from the end of these series being most common. Nevertheless, complexes of nearly every transition element are known, and their oxidation numbers vary from -1 to $+6$, so as a ligand class, phosphines, thioethers and their heavier analogues are extremely versatile.

This review is concerned with simple, formally electroneutral ligand species only. Thus while R_3P and R_2S complexes fall within its scope, derivatives of R_2P^- and RCS_2^-, for example, do not. This means effectively that of the group VI elements, only the secondary derivatives R_2S, R_2Se and R_2Te are considered, as the hydrides generally yield thiolate, selenate or tellurate (RM^-) complexes. Similarly, the group V ligands are mostly tertiary (though the presence of hydrogen on the ligand atom is occasionally encountered here).

The number of known complexes of tertiary phosphines far outweighs those of the other ligands reported here. Accordingly, the largest section of this review is devoted to them. It is worth keeping in mind, however, that so far as ligand–metal bonding and electronic properties are concerned, all the ligands in this group are quite similar, and conclusions reached for one ligand type may well apply to the others.

5.2 PHOSPHINE COMPLEXES

5.2.1 Reviews

Coordination complexes of tertiary phosphines were reviewed in 1964[1]. Since that time, growth in the field has been so rapid that comprehensive coverage is not possible, and recent reviews tend to concentrate on specific areas of interest. Examples include spectroscopic studies of metal–phosphorus bonding[2] (which contains many transition–element examples), and complexes of bidentate[3] and multidentate[4] phosphines.

5.2.2 The nature of metal–phosphorus bonds

Perhaps stimulated by their role in the early development of transition–metal organometallic compounds, a great deal of emphasis has been placed on the relative magnitude of π–bonding from metal to ligand, compared to the conventional σ–donor part of the bond. Figure 5.1. depicts the classical

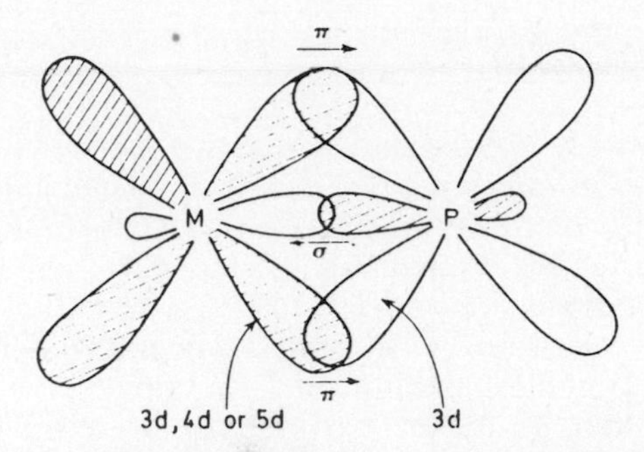

Figure 5.1 Classical representation of metal– phosphorus bonding

picture of metal–phosphorus bonding. The situation has been recently reappraised, and a variety of arguments, based on structural, spectroscopic or theoretical studies have been advanced. These have been discussed in some detail for metal–phosphine complexes in general[3], phosphine–platinum complexes in particular[5], and for related transition–metal complexes of silicon, germanium, tin and lead[6], and the reader is referred to those works for the development of these arguments. No unambiguous picture or general agreement has yet emerged, but most authors are of the opinion that whilst a π-component of the bond is present, it is greatly inferior in its strength and effects to the σ-component. The controversy continues, and references concerning the π-bonding ability of the ligands will be encountered from time to time in this review. Meanwhile, it has recently become apparent that metal→ ligand π-bonding is irrelevant (except in its contribution to overall bond strength) to the stabilisation of metal–carbon bonds[7].

Accurate crystal and molecular structure determinations on the complexes *trans*-$(PhMe_2P)_2MCl_4$ (M = W, Re, Os and Ir), *trans*-$(Et_3P)_2PtCl_4$, and *mer*-$(PhMe_2P)_3MCl_3$ (M = Re, Os and Ir) have led to some interesting conclusions on the nature of the bonds[8]. Unit change in the number of d-electrons results in a change in the M—P bond lengths of *ca.* 5 pm, compared to *ca.* 1 pm in M—Cl. On the other hand change in oxidation number causes a difference of *ca.* 3 pm in M—Cl but leaves M—P virtually unchanged. At these oxidation levels a significant dπ-contribution can be discounted, and the variations are accounted for on the basis of the metal-phosphorus bond being more covalent in character than the metal-chlorine link. The bond lengths to phosphorus are very sensitive to the degree of s-orbital character, and this increases from W—P to Pt—P (with a corresponding increase in p-orbital character in the M—Cl bonds, a result confirmed by chlorine n.q.r. spectrometry). Interestingly the *trans*influence of the phosphines on the metal–chlorine bonds is comparable in these octahedral complexes (*ca.* 10 pm) to that found in square-planar d^8 compounds, reinforcing the idea that the primary bond effect is sigma.

5.2.3 Spectroscopic studies

5.2.3.1 I.R. and Raman spectra

While reports on the vibrational spectra of phosphine complexes are common, difficulty in interpretation is frequently encountered, particularly with arylphosphine complexes, arising from skeletal vibrations of the ligands themselves. Nevertheless, assignments of the metal–phosphorus stretching modes in many complexes have been made (see Ref. **9** and refs. therein for assignments on a variety of phosphine, arsine and thioether complexes of palladium(II), platinum(II), platinum(IV) and gold(I)).

Metal-isotope shifts in the i.r. spectra of some complexes have led to a reappraisal of some assignments, the metal–phosphorus stretching modes being observed at a lower frequency than expected. Table 5.1 lists some of these new assignments[10]. Skeletal bending modes are much less sensitive to metal isotope substitution, but assignments have been made in some cases.

Table 5.1 I.R. frequencies of some metal–phosphine complexes obtained from isotope substitutions[10]

	ν(M—P)/cm^{-1}	δ(M—P)/cm^{-1}
trans-$(Et_3P)_2\,^{58}NiCl_2$	273.4	161.5
trans-$(Et_3P)_2\,^{58}NiBr_2$	265.0	—
cis-$(Ph_2PC_2H_4PPh_2)^{58}NiBr_2$	365.0, 308.4	—
cis-$(Ph_2PC_2H_4PPh_2)^{58}NiI_2$	352.9, 277.6	—
C_{2v}-$(Ph_3P)_2\,^{58}NiCl_2$	189.6, 164.0	—
C_{2v}-$(Ph_3P)_2\,^{58}NiBr_2$	196.8, 189.5	—
trans-$(Ph_2EtP)_2\,^{58}NiBr_2$	243.0	—
C_{2v}-$(Ph_2EtP)_2\,^{58}NiBr_2$	194.5, 182.0	—
trans-$(Et_3P)_2\,^{104}PdCl_2$	234.5	152.0
trans-$(Ph_3P)_2\,^{104}PdCl_2$	191.2	143.0
trans-$(Ph_3P)_2PdBr_2$	185	142

5.2.3.2 Electron spectroscopy

The photoelectron spectra of triphenylphosphine and a number of complexes $(Ph_3P)_2MCl_2$ (M = Ni, Cd or Pd) reveal an almost constant electron density at P[11]. This is indicative of a considerable degree of $d\pi$ back bonding (Figure 5.1) in these complexes. A similar result is suggested for $(Ph_3P)_2PdX_2$ (X = Cl, Br, I or CN) from a photoelectron study of a series of palladium complexes[12], and further confirmation of significant back-donation from low valency metals to Ph_3P is obtained from an E.S.C.A. study of the core level energies (Pt $^4f_{7/2}$) in a number of Pt^0 and Pt^{II} compounds[13].

5.2.3.3 N.M.R. spectra

^1H N.M.R. has been used now for many years to elucidate the structure of tertiary phosphine complexes. Undoubtedly one of the most useful simplifying concepts is the principle of virtual coupling between chemically equivalent *trans*-oriented phosphine ligands. This is now well established[14]. A more recent structural application is the detection of diamagnetic anisotropy induced in the phosphine ligands by multiple metal–metal bonds[15]. This has been observed in the complexes $Mo_2Cl_4L_4$ and $Re_2Cl_6L_2$ (L = PEt_3, PPr_3, PBu_3, $PPhMe_2$, or $P(OMe)_3$), where quadruple bonds are believed to be present. The ^1H n.m.r. of paramagnetic $(Ph_2MeP)_{4-n}MBr_n$ (M = Co, $n = 0, 1, 2$; M = Ni, n = 1,2) are interpreted in terms of shifts arising mainly from π spin density in the phenyl rings[16]. These results suggest that dπ bonding is an important spin-transfer mechanism in the lower valency compounds, but rapidly becomes unimportant at higher oxidation levels.

Increasing sophistication of instrumentation is leading to more reports of ^{31}P n.m.r. spectra of transition-metal complexes. The spectra of *mer*-$(R_3P)_3RhCl_3$ and *mer*-$(R_3P)_3IrCl_3$ give $^2J(P—P)(cis)$ as *ca.* 20 Hz, compared to *ca.* 430 Hz for $^2J(P—P)$ (*trans*)[17]. The ^{31}P n.m.r. spectra of $(R_3{}^1P)(R_3{}^2P)PtX_2$, with non-equivalent phosphines, give evidence for a ligand *cis*-effect[18]. The spectrum of *trans*-$(Bu^n{}_3P)(Et_3P)PtCl_2$ is not first order. The spectra of $(R_3P)_2CdI_2$ have allowed measurements of $^1J(^{111}Cd—^{31}P)$ and $^1J(^{113}Cd—^{31}P)$[19]. The Cd—P bonds are weaker than Hg—P in analogous mercury compounds, and cooling is necessary to slow the rate of phosphine exchange.

5.2.4 Structural variety of phosphine ligands

The variety of phosphorus compounds which have been used as ligands serves to illustrate the richness of this field of study. Hydrogen, the halogens and many alkyl and aryl radicals are some of the more common substituents attached to the ligating P atom. Literally hundreds of variations have been used and are hardly worthy of comment. A recent exotic example is tris-trimethylsilylmethylphosphine, $(Me_3SiCH_2)_3P$[20]. Reference 3 lists over 60 bidentate phosphines commonly encountered and reviews their preparation and known complexes. A number of polydentate phosphines have also recently been synthesised and used as ligands[4]. Examples of tri-, tetra- and hexa-dentate ligands are shown in Figure 5.2. Five-coordinate structures are often produced from use of such ligands and these will be discussed later.

The ultimate limit in polydentate phosphines, a resin-bound polymeric polytertiary phosphine, has recently been reported[21]. Made by the sequential bromination, lithiation and Ph_2PCl treatment of cross-linked polystyrene, this infinite polymer with triarylphosphine side-chains complexes to a

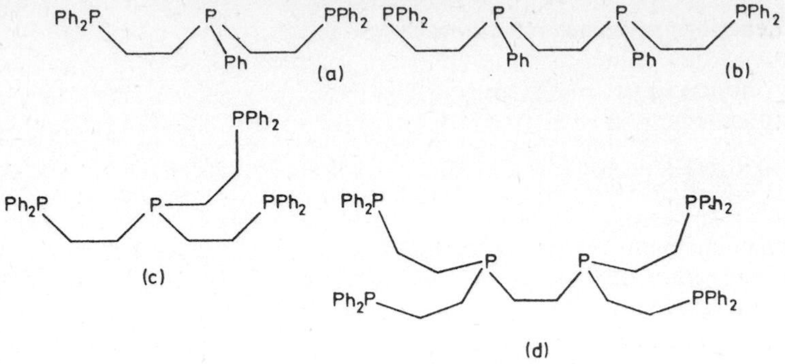

(a) (b)

(c)

(d)

Figure 5.2 Some polydentate phosphine ligands[4]

number of metals. Some of the complexes may have considerable value as heterogeneous versions of powerful homogeneous catalysts.

Complexes prepared from optically active phosphines are well known, and some of these have been used in recent years as catalysts in asymmetric syntheses. Chirality at phosphorus can be achieved by employing three different organic substituents, and these ligands have been made via the reductions of the corresponding phosphine oxide. A number of such phosphines (*o*-anisylmethylcyclohexylphosphine being the most outstanding) have been used as rhodium complexes to catalyse the hydrogenation of asymmetric olefins. Optical purity of over 90% was achieved for the products[22]. The halogen-bridged platinum complexes $Cl(R_3P^*)Pt(\mu\text{-}Cl)_2Pt(P^*R_3)Cl$, ($^*PR_3$ = (R)-(+)-benzylmethylphenylphosphine or (R)-(−)-methylphenyl-n-propylphosphine) catalyse the addition of $MeCl_2SiH$ to alkylphenylketones to give optically-active silylethers[23].

The source of the chirality does not need to be at phosphorus. Chiral substituents can have a similar effect. Thus the use of (−)-2,3-*o*-isopropylidene-2,3-dihydroxy-1,4-bisdiphenylphosphinobutane, again as its rhodium complex, has produced optical purity of 80% as an asymmetric hydrogenation catalyst[24]. The commercial availability of chiral phosphines is likely to give a substantial boost to research of this type.

An incipient chirality of complexes of triphenylphosphine is indicated by a recent report[25]. The phenyl groups of individual Ph_3P groups usually adopt a propeller-like configuration, and as rotation about the aryl-P bonds is hindered, each individual group is chiral (Figure 5.3). Effects of this chirality have been observed in low-temperature n.m.r. spectra.

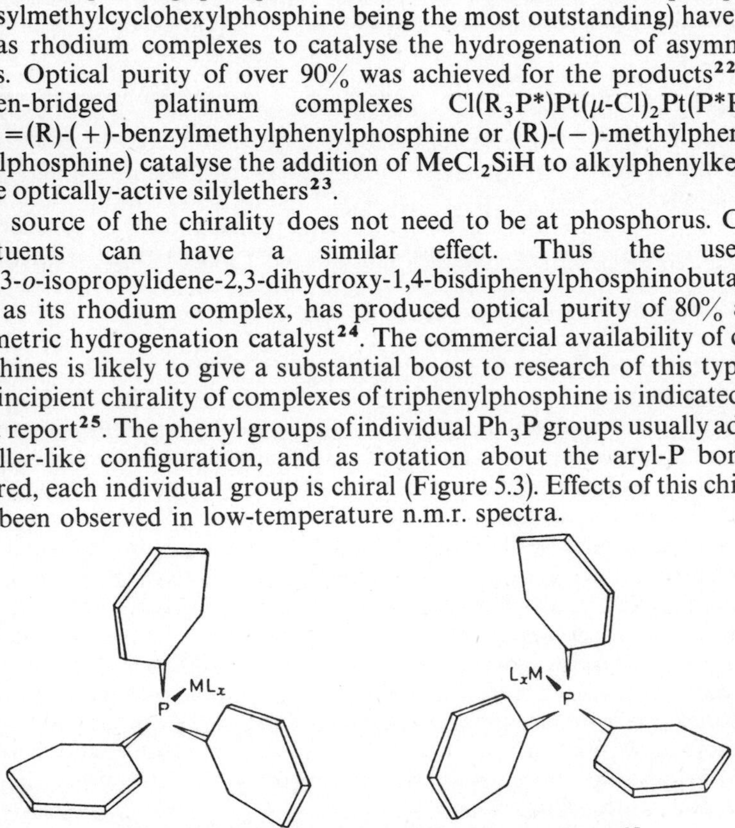

Figure 5.3 Chirality in triphenylphosphine complexes[25]

5.2.5 Reactions of coordinated phosphines

F_2PNR_2 bonds to the platinum of $PtCl_2$ by its phosphorus atom, but when the complex is heated with excess ligand, exchange of substituents occurs[26].

$$cis\text{-}[R_2NPF_2]_2PtCl_2 + 2R_2NPF_2 \rightarrow cis\text{-}[(R_2N)_2PF]PtCl_2 + 2PF_3$$

Probably the most common reaction of coordinated phosphines is the intramolecular metallation of a substituent carbon atom. These reactions are not confined to phosphine ligands, and the products are usually organometallic derivatives, so they fall under the scope of Volume 6 of this series. They will therefore be mentioned only briefly here. Most (but not all) reactions are found with square-planar d^8 complexes of metals at the end of the transition series. They are best represented as the internal oxidative addition of a suitably-placed ligand C—H bond (either aromatic or aliphatic) to the metal atom[27]. Frequently this step is followed by the elimination of HX (X = halogen) from the metal.

It is not yet possible to predict just which phosphines will or will not undergo these metallations, but very bulky ligands seem particularly prone, so relief of steric crowding may be an important consideration. Thus, for example, $PBu^t(o\text{-tolyl})_2$ and $PBu^t_2(o\text{-tolyl})$ form complexes with platinum(II) or palladium(II)[28], and undergo metallation at the tolyl methyl group.

$$2PBu^t_2(o\text{-tolyl}) + PtCl_2(NCPh)_2 \rightarrow (Bu^t_2(o\text{-tolyl})P)Cl\overline{Pt{-}P(Bu^t_2)C_6H_4 \cdot CH_2}$$

Dimethyl(1-naphthyl)phosphine is readily metallated in the 8-(*peri*-)position by Ir^{III} or Rh^{III} [29]. The reversible *ortho*-metallations of triphenylphosphine and triphenylphosphite can be used to introduce deuterium in the 2,6-positions[30].

5.2.6 Zeropositive complexes

Most of these and related low-valency derivatives also contain CO in the coordination sphere and again fall under the jurisdiction of Volume 6. The phosphine complexes of nickel(0), palladium(0) and platinum(0) are worthy of mention here, however, as many derivatives are known which contain only phosphine ligands. They make valuable starting materials for the synthesis of other derivatives.

The complexes are of the formula ML_4, where L can be any of a variety of phosphines, but Ph_3P is most frequently encountered. The structures are tetrahedral. $M(PF_3)_4$ (M = Ni or Pt) have been examined by electron diffraction and each PF_3 is twisted by about 40 deg. from the eclipsed position with MP_3[31]. Photoelectron spectra suggest that in these complexes PF_3 is a better π-acceptor than CO[32]. Even strained chelating biphosphines such as 1,2-biscyclohexylphosphinomethane produce structures approaching T_d symmetry[33].

The zeropositive compounds most often undergo reactions after an initial loss of one coordinated ligand (several ML_3 are well characterised) to produce a 16-electron molecule (see Ref. 27 and Refs. therein). ML_3 is then prone to oxidative addition processes, usually involving final loss of another L.

$$ML_4 \rightleftharpoons ML_3 + L \xrightarrow{X_2} ML_3X_2 \rightarrow ML_2X_2 + L$$

The ease of ligand removal has led to $Ni(PF_3)_4$ being used as a PF_3 generator[34].

Low oxidation state transition metals often bind oxygen molecules, and these complexes can act as oxidation catalysts. The species oxidised may be the phosphine ligands themselves[35], or other coordinated groups such as SO_2[36]. $(Ph_3P)_3Pt$ forms adducts with BCl_3, Al_2Me_6 and SiF_4[37]. Their structure is not yet understood.

$Ni(PHPh_2)_4$ and $Pd(PHPh_2)_4$ are rare examples of complexes of secondary phosphines[38].

5.2.7 Solution dynamics

Towards the end of the transition series 16-electron and 18-electron complexes have similar energies for metals in low oxidation levels[27]. Facile interchange between these configurations leads to solution equilibria between 3- and 4- and 5-coordinate species, and related ligand-exchange and isomerisation phenomena. Many known examples involve phosphine ligands, and it may well be that the particular bonding properties of these ligands are in some way responsible.

5.2.7.1 Ligand exchange, substitution and isomerisation

The conversion between $(R_3P)_4M$ and $(R_3P)_3M + R_3P$ has been mentioned already (Section 5.2.6) (M = Ni, Pd, Pt). The relative stabilities of the 3- and 4-coordinate molecules depends on R_3P. When the phosphine is Ph_2MeP, a rapid exchange in solution has been detected by v.t.n.m.r.[39]. ^1H and ^{19}F n.m.r. studies have detected rapid phosphine exchange in $RhCl(PPh_3)(PF_2NMe_2)_2$ and $RhCl(PF_2NMe_2)_3$[40]. ^{31}P N.M.R. data show ligand lability in AgL_4^+ as low as 200 K[41]. The ligand lability is higher when L = $(o\text{-tolyl})_3P$ than for L = $(EtO)_3P$. Studies on related copper(I) systems suggest this difference may be primarily steric in origin[42].

The *cis–trans* isomerisation of L_2PtX_2 catalysed by free L has long been believed to involve ligand exchange and the intermediacy of such species as $[L_3PtX]^+X^-$. A recent n.m.r. investigation of this reaction, and the thermodynamic stabilities of L_2PtX_2, LL^1PtX_2 and $L^1{}_2PtX_2$ when different phosphines are involved, substantiate this mechanism[43].

5.2.7.2 Five-coordination

The low-spin 5-coordinate $[Co[(Me_2N)_2PF]_3I_2]$ can be isolated from CoI_2 and $(Me_2N)_2PF$, but in solution it exists in equilibrium with the high-spin pseudotetrahedral $[Co[(Me_2N)_2PF]_2I_2]$[44]. Similar solution equilibria have been determined between the diamagnetic species $[Ni(PHEt_2)_3X]^+$, $[Ni(PHEt_2)_4X]^+$ and $[Ni(PHEt_2)_3X_2]$[45], and $(biL)_2NiX^+$ and $(biL)NiX_2$ ($biL = Et_2PC_2H_4PEt_2$ or $Ph_2PC_2H_4PPh_2$)[46]. Palladium shows no tendency to form 5-coordinate species under the same conditions[46a]. The structures of the $[(biL)_2NiX]^+$ complexes appear to be square pyramidal, but trigonal bipyramidal geometries for a number of monodentate phosphine complexes L_3NiX_2 have been substantiated[47], with 'intermediate' geometries ascribed to $(PhMe_2P)_3Ni(CN)_2$ and $(Ph(OEt)_2P)_3Ni(CN)_2$[48].

Probably by reducing the effects of ligand dissociation, chelating ligands have been employed in the isolation of many five-coordinate species. Particular success has been achieved using 'tripod-like' tetradentate ligands of the type shown in Figure 5.2c. A variety of ligating atoms have been employed, but most examples include at least one P atom. Examples include $(Ph_2PC_2H_4)_3N$[49] and $(Ph_2PC_2H_4)_2(Et_2NC_2H_4)N$[50]. Square-pyramidal geometry has been determined for the nickel complexes formed from $(Me_2AsC_3H_6)_2AsMe$ and $(Me_2AsC_3H_6)_2PPh$[51]. Fluorinated ligand substituents sometimes reduce the tendency towards 5-coordination, even with chelating systems[52].

Trigonal bipyramidal complexes $[(Me_2AsCH_2)_3Sb]NiX^+$ show marked (and often linear) changes in their visible–u.v. spectra with pressure, and they have been suggested as internal pressure calibrants[53].

5.2.7.3 Isomerism between square-planar and tetrahedral geometries

The stereochemistry at nickel of $(R_3P)_2NiX_2$ may be either square-planar (diamagnetic complexes) or tetrahedral (paramagnetic complexes), depending mainly on the steric requirements of the ligands. When R_3P is intermediate in size, both forms can coexist (e.g. $(Ph_2(PhCH_2)P)_2NiBr_2$ crystallises with both planar and tetrahedral molecules in its unit cell)[54] and interconversion in solution can be observed by v.t.n.m.r. techniques. Earlier reports[55] noted only one signal, corresponding to a weighted average between the diamagnetic and isotropically-shifted paramagnetic environments. Resonances from both forms of $(MePh(p\text{-anisyl})P)_2NiBr_2$ were seen at low temperatures, however, and a conversion rate of $2.4 \times 10^6\ s^{-1}$ determined in $CDCl_3$ at 298 K[56]. An examination of 26 such complexes shows that the tetrahedral form is favoured by larger halogens, as expected, but the equilibrium position is sensitive to *para*-substitution when R (of R_3P) is aryl, so that electronic factors are also involved[57]. The series of complexes $(RPh_2P)_2NiX_2$ reveals that the interconversion rate decreases in the orders $X = I > Cl > Br$ and $R = Et > Bu > Pr > Me$[58]. Free ligand also increases the rate.

Bisbenzyldiphenylphosphine-dibromonickel(II) in its pseudotetrahedral form has been converted to the square planar complex by high pressure[59].

5.2.8 Molecular structure determinations

These are now very numerous. Often the main source of interest lies elsewhere than the phosphine ligand, and the structures are not reported here in detail.

Table 5.2 lists a selection of recently elucidated compounds (almost all by x-ray diffraction), and notes some salient features. It must be kept in mind that wide variations in M—P distances can be found, even within a given molecule, due to ligand *trans*-influences. Sizeable effects are expected in 4-coordinate square planar compounds, and such a *trans*-influence was recently confirmed for $(Ph_3P)AuCl_3$[83]. Discrepancies are noticeable in 5-, 6- and 7-coordinate molecules also, however. Bond weakening effects have even been documented in mercury(II) derivatives, and assigned to σ-bond influences[84].

The small but significant lengthening of M—P bonds at higher oxidation numbers might be ascribed to diminishing covalent character and π-bonding.

One of the most remarkable series of compounds to be identified recently is the low-oxidation state gold clusters. They have no obvious counterparts

Table 5.2 Structural details of some phosphine complexes

Compound	*Metal atom (oxidation number)*	*Coordination number (idealised geometry)*	**M – P** *distance* **(***mean, in* **pm)**	*Re*
cis-mer-$(PhMe_2P)_3MoCl_2O$	Mo^{IV}	$6(O_h)$	253	6
$(PhMe_2P)_3MoCl_4$	Mo^{IV}	7(dist. C_{3v})	258	6
$(PhMe_2P)_3MoBr_4$	Mo^{IV}	$7(C_{3v})$	255	6
$(Ph_3P)_2Mo(NO)_2Cl_2$	Mo	$6(O_h)$	261	6
trans-$(PhMe_2P)_2WCl_4$	W^{IV}	$6(O_h)$	255	
$(PhMe_2P)_3ReCl_2(N_2Ph)$	Re^{III}	$6(O_h)$	243	6
mer-$(PhMe_2P)_3ReCl_3$	Re^{III}	$6(O_h)$	244	
trans-$(PhMe_2P)_2ReCl_4$	Re^{IV}	$6(O_h)$	250	
$(Ph_3P)_2Ru(NO)_2Cl$	Ru	5(sq. pyr.)	242	6
mer-$(PhMe_2P)_3OsCl_3$	Os^{III}	$6(O_h)$	238	
trans-$(PhMe_2P)_2OsCl_4$	Os^{IV}	$6(O_h)$	245	
$(Ph_3P)Co(CO)_2NO$	Co^0	$4(T_d)$	222	6
$[(cis\text{-}Ph_2PC_2H_2PPh_2)_2CoO_2]BF_4$	Co^I	5(dist. trig. bipy.)	224	6
$(Ph_3P)Co[N(Me_3Si)_2]_2$	Co^{II}	3(trigonal)	248	6
$[Ph_2PC_2H_4N(C_2H_4NEt_2)_2CoCl](ClO_4)$	Co^{II}	5(dist. trig. bipy.)	242	[illegible]
$(PF_3)_3Rh(NO)$*	Rh	$4(T_d)$	224	[illegible]
$[(Ph_2PC_2H_4PPh_2)_2RhO_2]PF_6$	Rh^I	5(dist. trig. bipy.)	235	[illegible]
$[(Ph_2PC_2H_4PPh_2)_2IrO_2]PF_6$	Ir^I	5(dist. trig. bipy.)	236	[illegible]
$[(Ph_2PC_2H_4PPh_2)_2IrS_2]Cl$	Ir^I	5(dist. trig. bipy.)	235	[illegible]
mer-$(PhMe_2P)_3IrCl_3$	Ir^{III}	$6(O_h)$	232	
trans-$(PhMe_2P)_2IrCl_4$	Ir^{IV}	$6(O_h)$	239	
$[(C_6H_{11})_2PCH_2P(C_6H_{11})_2]_2Ni$	Ni^0	$4(T_d)$	221	[illegible]
$(Ph_3P)Ni[N(Me_3Si)_2]_2$	Ni^I	3(trigonal)	222	[illegible]
$[(Ph_2PCH_2)_3CCH_3]NiI$	Ni^I	$4(T_d)$	221	[illegible]
trans-$(Bu^t_2FP)_2NiBr_2$	Ni^{II}	4(sq. planar)	223	
trans-$(PhPC_8H_{14})_2NiCl_2$	Ni^{II}	4(sq. planar)	223	
$\{[(Ph_2PC_2H_4)_2NC_2H_4NEt_2]NiI\}I$	Ni^{II}	5(dist. trig. bipy.)	223	
trans-$(PhMe_2P)_2NiBr_3$	Ni^{III}	5(trig. bipy.)	226	
cis-$(PhMe_2P)_2Pd(C_2N_4H_3)_2$	Pd^{II}	4(sq. planar)	226	
trans-$(Et_3P)_2PdHCl$	Pd^{II}	4(sq. planar)	231	
$(Ph_3P)_2PtC_2H_4$	Pt^0	4(planar)	227	
trans-$(Et_3P)_2Pt(NO_2)_2$	Pt^{II}	4(sq. planar)	231	
trans-$(Et_3P)_2PtCl_4$	Pt^{IV}	$6(O_h)$	239	
$(Ph_2PC_2H_4PPh_2)_2Cu_2Cl_2$	Cu^I	$4(T_d)$	229	
$(Ph_3P)_3Cu_2Cl_2$	Cu^I	{4 3	224 218	}
$(Ph_3P)AuCl_3$	Au^{III}	4(sq. planar)	234	
$[(EtO)_2PO]HgCl$	Hg^{II}	—	240	

* Electron diffraction used

elsewhere in the periodic table, though some transition-metal carbonyl clusters have points of similarity. Prepared via the $NaBH_4$ reduction of R_3P-Au^I-X (X is usually a poor ligand such as NO_3^-), several basic formula groups have been recognised[85]. These include $[Au_6(PR_3)_6]^{2+}$, $[Au_9(PR_3)_8]^{3+}$, and $Au_{11}(PR_3)_7L_3$ (R is aryl; L is a strongly coordinating ligand such as I^- or SCN^-). Figure 5.4 shows the skeletons of the $Au_6P_6^{2+}$, $Au_{11}P_7I_3$ and $Au_9P_8^{3+}$ clusters. Gold–gold bonds vary from 303 pm in the Au_6 derivatives to 268 pm, the centre-to-periphery distance in the Au_{11} clusters. $Au-P$ is 230 pm in Au_6P_6 and Au_9P_3, but varies from 221 to 229 pm in $Au_{11}P_7I_3$.

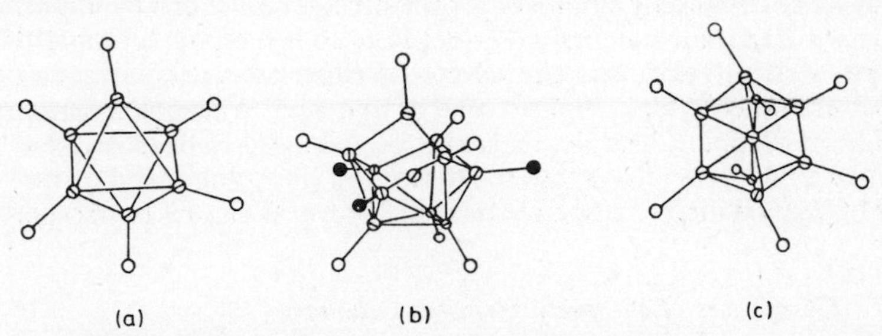

Figure 5.4 Phosphine–gold cluster compounds[85] (a) Skeleton of $[Au_6(P[p\text{-tolyl}]_3)_6](BPh_4)_2$. (b) Skeleton of $Au_{11}[P(p\text{-}FC_6H_4)_3]_7I_3$. (c) Skeleton of $[Au_9[P(p\text{-tolyl})_3]_8](PF_6)_3$. (○ = P; ⊘ = Au; ● = I).

5.2.9 Miscellaneous

The cobalt(II) complex $(Ph_2PC_2H_4PPh_2)_2Co(CN)_2$ resembles $Co(CN)_5^{3-}$, and is probably 5-coordinate with one of the biphosphine ligands monodentate[86]. In methanol solution it absorbs molecular oxygen to form a bridged peroxo-complex, of surprising activity. It reacts with the solvent to produce formaldehyde and Co^{III}. 1,2-Dichloroethane is oxidised by O_2 to CO_2 in the presence of this compound.

The stability of Cu^I-P bonds, compared to $Cu^{II}-P$, is illustrated by the use of R_3P to reduce copper(II) to copper(I). Triphenylphosphine reacts with copper(II) carboxylates in ethanol to produce $(Ph_3P)_3CuOCOR$ or $(Ph_3P)_2CuO_2CR$, where the carboxylates are mono- or bi-dentate, respectively[87]. Similarly, copper(II) nitrate has been reduced by a variety of tertiary phosphines[88]. In contrast, careful reaction of bishexafluoroacetylacetonatocopper(II) with *one equivalent* of PR_3 afforded the first copper(II) phosphine derivatives[89]. Formulated $Cu(biX)_2PR_3$, the dark green compounds are monomeric in chloroform solution and have magnetic moments in the range 1.65–1.76 BM.

Zwitterion complexes formed from positively charged phosphine (or other group V donor) species have been examined from time to time (see Ref. 90 and Refs. therein). Commonly a bidentate phosphine is employed, with one donor atom quarternised. Examples[90] are $Et_2HN^+C_2H_4PPh_2Co^-X_3$, in which normal $Co-P$ bonding is found and the compounds are non-electrolytes. Recently a positively charged phosphine–metal complex has been employed as L^+ in the formation of such species[91]. $[(Ph_2PC_2H_4PPh_2)_2Co(CN)_2]^+$, obtained from the oxidation of $(Ph_2PC_2H_4PPh_2)_2Co(CN)_2$ (see earlier), bonds through a CN group nitrogen atom to $CoCl_3^-$ to produce $(Ph_2PC_2H_4PPh_2)_2Co^{III}(CN)_2Co^{II}Cl_3$.

5.2.9.1 *Synthetic methods*

Many papers on phosphine compounds, particularly those concerned with spectroscopic or structural determinations, do not report full preparative details of the materials involved. While simple ligand addition or replacement processes are commonly employed, experimental details can be important in some cases. Experimental method is emphasised in a report on cobalt(II) and copper(I) derivatives[92], and the effects of experimental conditions on the product are noted[93] for complexes of Ru^{II} and Ru^{III}. A complete series of Rh^{I} and Rh^{III} phosphine complexes, many of catalytic importance, has been described[94]. A convenient synthesis of tertiary phosphine (and arsine) complexes of W^{IV} via direct zinc amalgam reduction of WCl_6 has been reported[95].

5.2.9.2 *Complexes of the early transition elements*

Phosphine derivatives of groups III, IV and V are quite uncommon and their chemistry relatively obscure. The first phosphine complexes of scandium(III) were isolated quite recently[96]. Prepared from $ScCl_3$ or $ScBr_3$ and $Ph_2PC_2H_4PPh_2$, they appear from their vibrational spectra to be polymeric. The formulations $[(Ph_2PC_2H_4PPh_2)_{1/2}ScCl_2(Ph_2PC_2H_4PPh_2)_{1/2}]_n^{n+}nCl^-$, and $[(Ph_2PC_2H_4PPh_2)_{1/2}ScBr(Ph_2PC_2H_4PPh_2)]_n^{2n+}2nBr^-$, with bridging diphosphine and 4-coordinate scandium, have been proposed.

More has been reported on titanium, though few complexes of Ti^{III} are known. $TiCl_3(PR_3)$ ($R_3 = H_2Me$, HMe_2, Me_3 and Et_3) were prepared by direct combination of the components in hot toluene[97]. They are monomeric in benzene. $TiCl_3(Ph_2PC_2H_4PPh_2)$ has been reported earlier[98]. Of Ti^{IV}, complexes $TiCl_4L$ ($L = PH_2Me$, $PHMe_2$ or Me_3P) have been characterised. They are trigonal bipyramidal with Me_3P occupying an axial site, whereas PH_2Me and $PHMe_2$ adopt equatorial positions[99]. ^1H and ^{31}P n.m.r. studies on similar complexes ($L = PMe_2Ph$, PMe_3 or PBu^n_3) indicate a fast solution exchange reaction between $TiCl_4L$, $TiCl_4L_2$ and free L[100]. There is some evidence for binuclear complexes, formulated as $[Cl_3Ti(\mu\text{-}Cl)_2TiCl_3L]$, at low temperatures.

Triethylphosphine reacts with $VOCl_2$ in toluene to yield square pyramidal $(Et_3P)_2VOCl_2$[101]. The complexes $(R_3P)MCl_5$ (R = Bu or Ph; M = Nb, or Ta) have been described[102]. Bu_3PNbCl_5 decomposes at 748 K to produce polymeric $NbPCl_2$.

5.3 ARSINE, STIBINE AND BISMUTHINE COMPLEXES

Most work on tertiary arsine complexes tends to confirm the expected similarity between these and analogous phosphine derivatives. Indeed arsenic analogues of many phosphine complexes are often reported in the same publications, emphasising this relationship. Antimony and bismuth analogues are quite uncommon, and their chemistry relatively obscure, so predictions of similarities down the entire series P, As, Sb, Bi are made only at considerable risk of error.

5.3.1 Diversity of ligand species

A number of derivatives R_3M have been employed for As, Sb and Bi, but only for arsenic is the series extensive. Here, a variety of polydentate species have also been used. Phosphorus atoms in some of the polyphosphines depicted in Figure 5.2 have been replaced by As, and zeropositive platinum derivatives prepared from them[103]. The best documented bidentate arsine, commonly abbreviated to 'diars', is *ortho*-phenylene-bis-dimethylarsine. Recent work suggests a completely delocalised electronic structure for this ligand[104]. 1,8-Bisdimethylarsino-naphthalene has also been used as a chelating ligand with Pd^{II} and Pt^{II} [105].

Optically active arsines are also known. The isomers of MeEtPhAs were separated by fractional crystallisation of the internal diasteriomeric complex [{(±)-MeEtPhAs}PtCl(biL)]Cl, where biL was (−)-stilbenediamine[106]. The $PdCl_2$ complexes of both *meso* and *racemic-dl*-$MePhAsC_2H_4AsMePh$ were separated by silica-gel chromatography[107].

The preparation of some arsine, phosphine, and mixed arsine-phosphine ligands is presented in Refs. **4** and **108**.

5.3.2 Bonding properties

The π-bond controversy applies to these ligands as well as phosphines. An examination of the equilibria

$$LMX + L \rightleftharpoons L_2M^+ + X^-$$

in nitrobenzene (M = Ag or Au; L = Ph_3P, Ph_3As or Ph_3Sb) led to the conclusion that π-bonding was stronger between L—Au, than L—Ag, a result that might be expected from the 'soft' character of the ligands[109]. Evidence of dπ-bonding is also obtained from the electronic spectra of low-spin square-pyramidal $[Ni(diars)_2X]^{n+}$ (X = Cl, Br, I, SCN, thiourea or NO_2)[110a]. Assignments of the ligand-field bands recorded between 77 K and 300 K, have been made and the nickel d-orbital energy sequence $xy < xz,yz < z^2 \ll x^2-y^2$ deduced. The order of d_{xy} and d_{xz} is the reverse to that expected from simple crystal-field considerations, but can be accounted for by Ni—As back-bonding.

E.S.R. studies on the nickel(III) complexes $[Ni(diars)_2Cl_2]Cl$ and $[Ni(diars)_2Cl_2]ClO_4$, measured in samples diluted by the diamagnetic host material $[Co(diars)_2Cl_2]ClO_4$, also indicate that the unpaired electron is strongly delocalised over metal and all six donor atoms[110b].

Mössbauer studies on the Fe^{IV} complex $[Fe(diars)_2X_2](BF_4)_2$ are in agreement with a D_{4h} structure and a low-spin (d_{xy}^2, d_{xz}^1, d_{yz}^1) ground-state[111].

Here there is little evidence for electron delocalisation via metal–arsenic π-bonding, as might be expected of a metal in such a high oxidation level. On the other hand, a series of complexes of uranium(V), UCl_5L (L includes Ph_3P and Ph_3As) have been isolated, and e.s.r. studies indicate that the unpaired electron is extensively delocalised through the ligand π-bond structure[112].

Electron spin resonance studies on $VOCl_2$ and Bu_3Sb in toluene show that only one antimony atom is coordinated to vanadium[113] (in THF, no interaction is observed). This contrasts with the situation for R_3As[114] and R_3P[101], where two ligand species are attached to vanadium. The observation that no complex of $VOCl_2$ and tri-n-butylbismuthine could be isolated fits the sequence $P > As > Sb > Bi$ for coordinating ability to this centre.

The x-ray photoelectron spectra for a number of arsenic derivatives have been reported and 3p and 3d orbital chemical shifts noted. The complexes studied include $Ph_3AsHgCl_2$ and $(Ph_3As)_2HgI_2$[115].

5.3.3 Five-coordinate complexes

Bearing in mind the ability of poly-tertiary phosphines to stabilise 5-coordinate derivatives, particularly of nickel, it is not surprising that many analogous arsenic compounds have also been reported.

A series of low-spin trigonal-bipyramidal complexes of nickel, palladium and platinum, [M(tetraL)X]Y has been reported[116] (tetraL = (o-$Me_2As \cdot C_6H_4)_3As$; X = Cl, Br, I, NCS, NO_2, N_3, NO_3 or $O\bar{A}c$; Y = X, BF_4 or BPh_4), and square pyramidal $[Ni(diars)_2X]^{n+}$ has been mentioned[110]. The ligands bis-3-dimethylarsinopropylmethylarsine and dimethyl(o-dimethylarsinophenyl)stibine produce the 5-coordinate nickel complex [Ni(biL)-(triL)]Br_2 with a unique As_4Sb coordination set[117]. The geometry is probably square pyramidal.

The coordinating ability of two tridentate arsines, 1,1,1-trisdimethylarsinomethylethane and 1,1,1-trisdiphenylarsinomethylarsine have been compared at nickel(I) and nickel(II). Depending on conditions, the former ligand produces complexes NiL_2X_2 (X = Cl, Br, I, ClO_4, BF_4 or BPh_4) or $NiLX_2$ (X = CNS) which may be high-spin 6-coordinate or low-spin 5- or 4-coordinate[118]. In the presence of sodium tetrahydroborate, the phenyl analogue reacts with Ni^{II} to produce NiLBr or NiLI. These are tetrahedral nickel(I) complexes.

A series of complexes of the triarsine bis-(o-dimethylarsinophenyl)methylarsine with Ni^{II}, Pd^{II} and Pt^{II} have been examined[119]. The complexes $(triL)MX_2$ (X = Cl, Br or I) are 5-coordinate monomers for M = nickel in non-polar solvents. In hydroxylic solvents, however, they disproportionate to $[(triL)_2Ni]^{2+}$ and NiX_4^{2-}. The palladium and platinum compounds are 1:1 electrolytes in solution, suggesting 4-coordination, though they do tend towards association under some conditions. A nickel(III) derivative of the same ligand, $(triL)NiBr_3$, was also isolated. Tris-(o-dimethylarsinophenyl)-stibine forms a series of trigonal bipyramidal complexes [(tetraL)PdX]Y (X = Cl, Br, I or SCN; Y = Cl, SCN, or BPh_4)[120]. The Pd—Sb bond is compressed as a result of chelation.

The copper(II) derivatives of P and As donors, $LCu(CF_3COCHCOCF_3)_2$ (where L = PPh_3, $PMePh_2$, PMe_2Ph, PEt_3, PBu_3 or Ph_3As) appear to be 5-coordinated[121].

5.3.4 Molecular structures

Some recent determinations are listed in Table 5.3.

Table 5.3 Molecular structures of some arsine complexes

Compound	*Metal (oxidation level)*	*Coordination No. (idealised stereochemistry)*	M–As (pm)	*Ref.*
$[(Ph_2As)_2CH_2]_2Mo(CO)_2Br_2$	Mo^{II}	7 (capped octahedron)	262	122
$(diars)WOCl_4$	W^{VI}	7 (pentagonal bipyramid)	267	123
$[(diars)_2CoCl_2]Cl$	Co^{III}	6 (O_h)	233	124
$[(diars)_2NiCl_2]Cl$	Ni^{III}	6 (O_h)	234	124
$[o\text{-}(Me_2As)(Me_2N)C_6H_4]_2Cu_2I_2$	Cu^{I}	4 (dist. T_d)	233	125
$[o\text{-}(Me_2As)(MeS)C_6H_4]_2PdI_2$	Pd^{II}	4 (sq. planar)	239	129
$(Ph_3As)Hg(SCN)_2$	Hg^{II}	3 (trigonal)	260	126

5.3.5 Miscellaneous

Although arsine complexes of uranium(V) have been characterised, *diars* reacts with UCl_5L (L = trichloroacryloyl chloride) to produce a uranium(IV) derivative, UCl_4 (diars)[127]. This unstable and air-sensitive material is the first U^{IV} arsenic derivative reported. Six- and eight-coordinate complexes have been formed between $TiCl_4$ and various substituted diarsine ligands[128]. The choice of stereochemistry seems to depend wholly on the bulk of the groups attached to As, and not at all on electronic effects, crystal forces, or the solvent used.

5.4 THIOETHER, SELENIDE AND TELLURIDE COMPLEXES

5.4.1 Survey of compound types

The distribution of complexes of these group VI donors is generally similar to those of group V, though perhaps the number of derivatives reported for the earlier transition metals (particularly in high oxidation levels) is proportionately higher.

Diethylsulphide reacts with VCl_3 or VBr_3 to give *trans* trigonal bipyramidal complexes $VX_3(SEt_2)_2$[130]. In Et_2S solution, these complexes are in equilibrium with the 6-coordinate $VX_3(SEt_2)_3$. In the analogous Cr^{III} systems, only 6-coordinate species are observed. A series of niobium(IV) compounds NbX_4L and NbX_4L_2 has been reported (X = Cl, Br or I; L = Me_2S, Et_2S or $(CH_2)_4S$)[131]. For the Me_2S adducts an equilibrium between the 5- and

6-coordinate species in benzene was detected, and the stability of the bis-adduct with respect to the mono- decreased in the order I > Br > Cl. *Cis*-configurations for NbX_4L_2 are indicated by their i.r. spectra. Solid NbX_4L is halogen-bridged (possibly with Nb—Nb bonding also), but the compound is monomeric in solution. The complexes $MX_4(CH_3SC_2H_4SCH_3)_2$ (M = Nb; X = Cl, Br or I; M = Zr; X = Cl) are 8-coordinated with a triangular dodecahedral ligand configuration[132].

The stability constants of MCl_5L (M = Nb or Ta) have been measured by n.m.r. methods with the interesting result that stronger complexes are formed when L = R_2S than with L = R_2O. This assigns a 'soft acid' behaviour to $NbCl_5$, an assignment suspected for some time from preparative work. The Raman spectra of crystalline MX_5L (where L = Me_2S) have been assigned[134] in terms of a symmetry lower than C_{4v}.

Sulphide complexes of W^{VI} and W^{V} have been isolated. Compounds of stoichiometry WCl_6L, WCl_6L_2 and WCl_5L_2 were obtained[135]. The sulphide ligands show little tendency to reduce tungsten under mild conditions, unlike nitrogen and phosphorus donors, and an explanation has been advanced in terms of the spatial requirements of the ligands.

Where a variety of sulphide complexes of Rh^{III} and Ir^{III} have been isolated[136], compounds of these metals in their unipositive oxidation state are difficult to isolate and there seems to be little similarity between sulphides and phosphines as far as these metals are concerned. The complexes MCl_3L_3 (L = Me_2S, Et_2S, $(CH_2)_4S$ or $(CH_2)_5S$) have all been assigned octahedral meridional configurations on the basis of spectroscopic studies[136b] but agreement over the assignments of some compounds is not complete.

A comparison of the bonding of R_3P, R_3As and R_2S has been made for the D_{4h} complexes L_2IrCl_4 by c.t. absorption and magnetic c.d. spectra[137]. Charge transfer bands from phosphines and arsines to Ir^{IV} occur at very low energy (9200 cm^{-1}), but those from R_2S are at significantly higher energies (15 100 cm^{-1}). This difference reflects the reducing abilities of the ligands, and may explain why R_3P and R_3As are less often found in complexes of high oxidation levels. π-Bonding appears to be insignificant with all these ligands at iridium(IV).

The vibrational spectra below 800 cm^{-1} have been assigned[138] for the Me_2S complex species LAuX, MX_3L^-, *trans*-$[L_2MX_2]$, PtL_3X^+, *cis*-$[PtL_2X_2]$ and $Pd_2X_4L_2$ (X is halogen, M is Pd or Pt). Many different solid forms of *cis*-L_2PtX_2 were detected, possibly arising from the pyramidal conformation at sulphur (see Section 5.4.2). Vibrational assignments[136c] of various sulphide and selenide complexes, *trans*-(L_2MX_2) (M = Pt or Pd) and *fac*-$[L_3MX_3]$ (M = Ru, Os, Rh or Ir), place (M—S) at *ca.* 300 cm^{-1} and (M—Se) at *ca.* 220 cm^{-1}. Both sulphur and selenium appear to be less effective π-acceptors than phosphorus. Metal-isotope shift studies assign ν(Ni—S) to 260–210 cm^{-1} in a variety of compounds[139].

5.4.2 Effects of the pyramidal geometry of the ligand atoms

The pyramidal nature of sulphur has been well established in the solid by crystallographic studies and in solution by n.m.r. studies. The ^1H n.m.r.

spectrum of $[(MeSC_2H_4SMe)_2RhCl_2]Cl$ shows a complicated methyl resonance pattern (all doublets, due to coupling with ^{103}Rh) which can be simplified by irradiating at seven different ^{103}Rh frequencies[140]. The seven isomers are accounted for by different conformations about sulphur. The presence of *meso*- and racemic *dl*-isomers of $(MeSC_2H_4SMe)MCl_2$ (M = Pd or Pt) can be detected in solution[141] (see Figure 5.5a) and the mixed-donor-

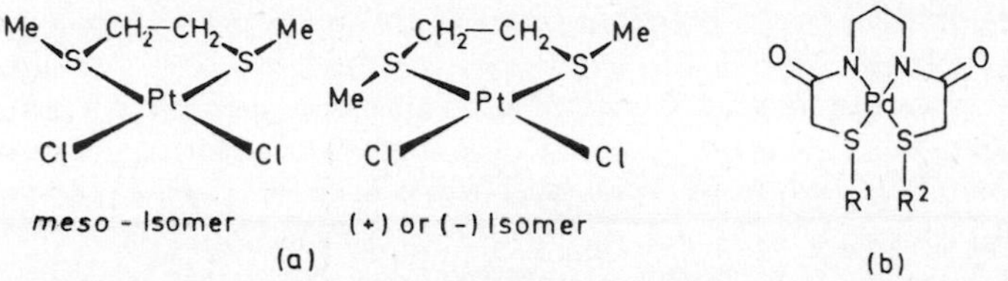

Figure 5.5 (a) *Meso*- and *dl*-configurations for $(MeSC_2H_4SMe)PtCl_2$[141] (b) Similar conformations exist for 1-R^1-11-R^2-1,11-dithia-4,8-diazaundecane-3,9-dionatopalladium ($R^1R^2 = Et_2$, Ph_2, or Et, Ph)[142]

atom complexes depicted in Figure 5.5b also show this phenomenon in their ^1H n.m.r. spectra[142].

At higher temperatures inversion at sulphur causes interconversion of the isomers, and a simplification of the n.m.r. spectra[141, 143]. The low activation energy for this process, and the retention of $^{195}Pt—S—C—^1H$ coupling at higher temperatures has led to the postulation of an inversion process involving simultaneous bond making with the S atom lone pair as the original coordinate bond breaks[143] (Figure 5.6).

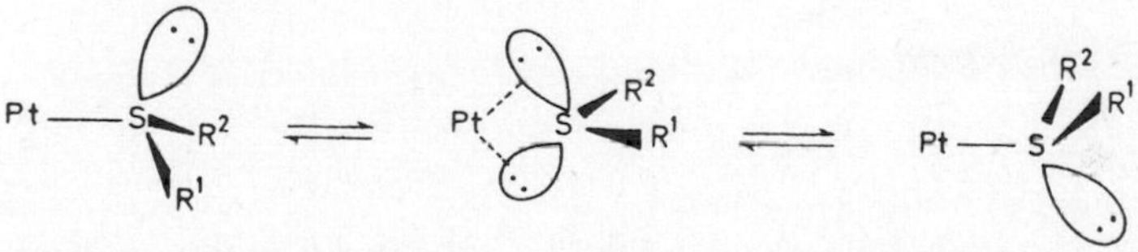

Figure 5.6 Proposal inversion mechanism for coordinated sulphides[143]

Sulphur inversion in a series of chelate complexes $(RSC_2H_4SR)MX_2$ (R = Me, Et, Pr^i, Pr^n or Bu^n; M = Pt or Pd; X = Cl, Br or I) has been compared[144]. The interconversions of *meso*- and *dl*-forms takes place via inversion at single sulphur atoms. Inversion at Pd is more rapid than at Pt, suggesting weaker Pd—S bonds (a dissociation-recombination mechanism cannot be discounted here). Also, the inversion rate increases as the *trans*-influence of X increases. This is paralleled by a drop in J(Pt—H), indicating a reduction in the s-character of the Pt—S bonds.

A similar inversion at sulphur has been found in $(PhCH_2)_2SAuCl_3$ and $(PhCH_2)_2SAuCl$[145]. The inversion rate at square-planar gold(III) is comparable to that at Pt^{II}, but the rate at linear Au^I is much faster. This confirms a weaker Au^I—S bond. Selenides and tellurides[146] appear to behave in a similar manner to sulphides when bonded to Pt^{II} or Pd^{II}.

5.4.3 Formation and displacement reactions

Formation constant measurements on $(RSC_2H_4SR)PdI_2$ and $(RSC_3H_6SR)PdI_2$ show that 5-member chelate rings are more stable than six in these systems[147]. The nature of the R group has a greater influence on the 5-atom ring compounds however. The displacement of chloride from $(bipy)PtCl_2$ by Et_2S and a number of bidentate ligands follows the expected second-order kinetics[148]. The results for the bidentate ligands are consistent with the formation of a transition state involving interaction between one end of the bidentate ligand and a Cl atom of the substrate. The second-order substitution reactions of py_2PtCl_2 with 4,4′-disubstituted diphenylsulphides are markedly dependent on bond making between Pt and S as the driving force[149]. The results are in agreement with the apparent inability of 4,4′-dinitrodiphenylsulphide to coordinate to platinum(II).

The kinetics of ring opening and displacement of the chelating ligands in $(PhSC_2H_4SPh)PdCl_2$[150] and $(PhSeC_2H_4SePh)PdCl_2$[151] by various amines have been reported. The sequence followed is

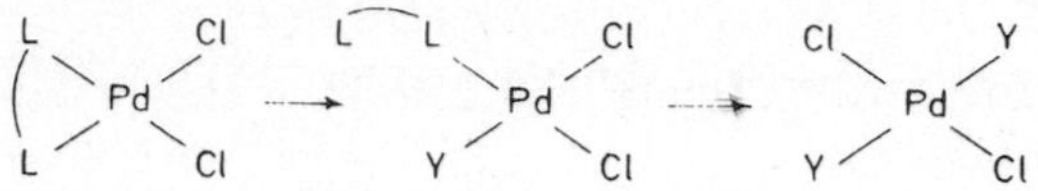

The selenide is much less sensitive to steric effects, as expected for the bigger ligand atom. The results also suggest that π-bonding from Pd^{II} to Se is stronger than from Pd^{II} to S. This has been claimed before, along with the reverse order for platinum(II)[152].

5.4.4 Miscellaneous

5.4.4.1 Polydentate sulphides

Though not as numerous as those of P or As, some polydentate sulphur ligands have been described. They include some macrocyclic derivatives. 1,4,8,11-tetrathiacyclotetradecane forms complexes of nickel(II), $Ni(tetraL)X_2$[153]. These are diamagnetic square-planar 1:2 electrolytes when X^- is BF_4^- or ClO_4^-, but when X^- has a stronger coordinating power (Cl^-, Br^-, I^-, NCS^-) paramagnetic octahedral non-electrolytes result. The octadentate ligand 1,4,8,11,15,18,22,25-octathiacyclo-octacosane has been isolated[154]. This can coordinate simultaneously to two Ni^{II} ions (Figure 5.7).

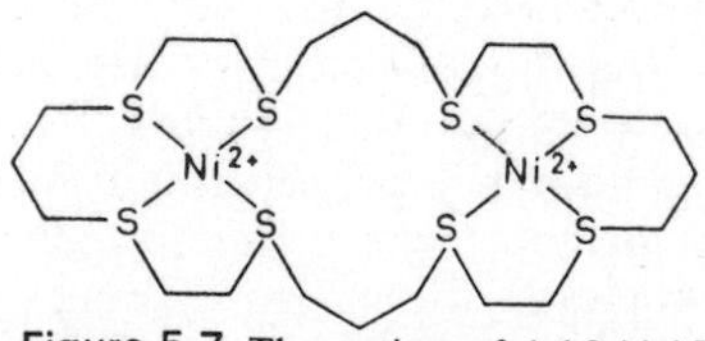

Figure 5.7 The cation of 1,4,8,11,15,18,22,25-octathiacyclo-octacosanodinickel(II) tetrakis(tetrafluoroborate)[154]

5.4.4.2 Molecular structures

There are few compared to the previous sections. Chloro(bis-{2-[2-pyridylmethylamino]ethyl}disulphide)nickel(II) perchlorate is octahedrally coordinated at Ni with one Ni—S at 247 pm[155]. The palladium(II) selenides $(Et_2Se)_2PdCl_2$[156] and $Pr^iSeC_2H_4SePr^i)PdCl_2$[157] have Pd—Se distances of 242 pm and 238 pm respectively. The longer bond distance of the *trans*-complex probably reflects the higher *trans*-influence of Se.

5.4.4.3 Sulphide complexes as catalysts

There have been a number of reports and behaviour generally is similar to the more extensively studied phosphine derivatives. $(Et_2S)_3RhCl_3$ catalyses the hydrogenation of olefins in *N,N'*-dimethylacetamide under mild conditions[158]. The first step is reduction to a rhodium(I) species, and the weak coordinating ability of sulphur may be advantageous here in allowing olefin and H_2 to coordinate. Despite this, an inhibiting effect of the sulphide ligands has been noted. The use of a basic solvent is important for the first reduction sequence, displacing sulphide from Rh^{III} and enhancing the reaction.

5.4.4.4 Thermolysis reactions

The thermal decomposition of Fe^{III}, Co^{II}, Ni^{II}, Cu^{I} and Zn^{II} complexes of 1,5-bis-(2-mercaptoethylthio)pentane have been investigated in the absence of air[159]. After dehydration, ligand decomposition to give metal sulphides was noted and a correlation exists between the decomposition temperatures and the stability of the sulphide complexes. In the presence of oxygen, decomposition is initiated at lower temperatures, and ligand oxidation is observed. Oxidation of the metal is also seen with cobalt(II).

The thermolysis of $(Bu^t_2S)HgCl_2$ produces Hg_2SCl_2, Bu^tSHgCl, 2-methylpropene, HCl and Bu^tCl. The reaction is first order and an intra-molecular decomposition route involving an interaction between ligand side-chain and Cl is proposed[160].

5.5 MIXED-LIGAND COMPLEXES

Many reports fall into this category; space permits the inclusion here of only a few.

Two main features emerge from these studies. The first is that in a 'ligand-deficient environment', ligands of greatly differing character will coordinate to a given metal ion. Thus even ligands which bond only weakly will form coordinate bonds rather than leave the metal ion coordinately unsaturated. Naturally, chelating ligands with mixed donor atoms greatly add to the stability of such coordination systems. Examples include a $N_2O_2S_2$ octahedral coordination-shell[161] at Co^{III}, and a square planar $N_2S_2Ni^{II}$

macrocycle, prepared *in situ* by a 'template synthesis' method[162]. The tetradentate N_2As_2 donor N,N'-bisdiphenylarsinoethyl-*N*,*N'*-dimethylethylenediamine forms 4-coordinate complexes with Ni^{II} in the absence of strongly coordinating anions, but octahedral (tetraL)NiX_2 if the anions have donor ability[163]. Similar behaviour is reported for NAs bidentate ligands at Ni^{II}, Pd^{II}, Cu^{II} [164] and Cu^{I} [165].

The use of polyphosphine and mixed arsine–phosphine ligands to isolate 5-coordinate nickel and cobalt complexes has been referred to. Similar mixed-donor ligands (often of 'tripod' configuration) involving O and N donor atoms as well as P or As also give rise to these structures[166].

The second main feature is, as expected, when a sufficiency of ligand atoms can compete for coordination sites, then those that win are generally of a similar character and most suited to the metal ion in question according to the principle of hard and soft acids and bases. The application of this principle to metal complexes has recently been reviewed, with many examples[167]. An example is the complex di-iodo-bis(dimethyl-*o*-methylthiophenylarsine)-palladium(II)), which is *trans*-square planar with a $PdAs_2I_2$ skeleton and no Pd—S bonding[129].

The hard and soft acid-base theory also operates when available donor atoms are restricted by ligand geometry. Dimethyl sulphoxide complexes fall into this bracket. Donation from O or S is possible, but not from both for steric reasons. Recent interest has centred on the hard-soft borderline, where both O- and S-bonded ligands can be encountered, often in the same molecules[168]. Rh^{III} and Pd^{II} appear to be on this borderline.

References

1. Booth, G. (1964). *Advan. Inorg. Chem. Radiochem.*, **6,** 1
2. Verkade, J. G. (1972). *Coord. Chem. Rev.*, **9,** 1
3. Levason, W. and McAuliffe, C. A. (1972). *Advan. Inorg. Chem. Radiochem.*, **14,** 173
4. King, R. B. (1972). *Accounts Chem. Res.*, **5,** 177
5. Hartley, F. R. (1973). *The Chemistry of Platinum and Palladium*, (London: Applied Science Publishers)
6. Brooks, E. H. and Cross, R. J. (1970). *Organometal. Chem. Rev. A*, **6,** 227
7. Braterman, P. S., and Cross, R. J. (1972). *J. Chem. Soc. Dalton Trans.*, 657
8. (a) Aslanov, L., Mason, R., Wheeler, A. G. and Whimp, P. O. (1970). *Chem. Commun.*, 30. (b) Mason, R. and Randaccio, L. (1971). *J. Chem. Soc. A*, 1150
9. (a) Duddel, D. A., Goggin, P. L., Goodfellow, R. J., Norton, M. G., and Smith, J. G. (1970). *J. Chem. Soc. A*, 545 (b) Adams, D. M. and Chandler, P. J. (1967). *J. Chem. Soc. A*, 1009
10. Nakamoto, K. (1972). *Angew. Chem. Internat. Ed.*, **11,** 666 and Refs. therein
11. Blackburn, J. R., Nordberg, R., Stevie, F., Albridge, R. G. and Jones, M. M. (1970). *Inorg. Chem.*, **9,** 2374
12. Kumar, G., Blackburn, J. R., Albridge, R. G., Moddeman, W. E. and Jones, M. M. (1972). *Inorg. Chem.*, **11,** 296
13. Cook, C. D., Wan, K. Y., Gelius, U., Hamrin, K., Johansson, G., Olsson, E., Siegbahn, H., Nordling, C. and Siegbahn, K. (1971). *J. Amer. Chem. Soc.*, **93,** 1904
14. Siddall, T. H., and Stewart, W. E. (1969). *Prog. Nucl. Mag. Res. Spectrosc.*, **5,** 33 and Refs. therein.
15. San Filippo, Jr., J. (1972). *Inorg. Chem.*, **11,** 3140
16. La Mar, G. L., Sherman, E. O. and Fuchs, G. A. (1971). *J. Coord. Chem.*, **1,** 289
17. Mann, B. E., Masters, C. and Shaw, B. L. (1972). *J. Chem. Soc. Dalton Trans.*, 704
18. Allan, F. H. and Sze, S. N. (1971). *J. Chem. Soc. A*, 2054

19. Mann, B. E. (1971). *Inorg. Nucl. Chem. Lett.*, **7,** 595
20. Hsieh, A. T. T., Ruddick, J. D. and Wilkinson, G. (1972). *J. Chem. Soc. Dalton Trans.*, 1966
21. Collman, J. P., Hegedus, L. S., Cooke, M. P., Norton, J. R., Dolcetti, G. and Marquardt, D. N. (1972). *J. Amer. Chem. Soc.*, **94,** 1789
22. Knowles, W. S., Sabacky, M. J. and Vineyard, B. D. (1972). *Chem. Commun.*, 10
23. Yamamoto, K., Hayashi, T. and Kumada, M. (1972). *J. Organometal. Chem.*, **46,** C65
24. Kagan, H. B. and Dang, T. P. (1972). *J. Amer. Chem. Soc.*, **94,** 6429
25. Brown, J. M. and Mertis, K. (1973). *J. Organometal. Chem.*, **47,** C5
26. Nixon, J. F. and Sexton, M. D. (1969). *Chem. Commun.*, 827
27. Braterman, P. S. and Cross, R. J. (1973). *Chem. Soc. Reviews*, **2,** 271
28. (a) Cheney, A. J. and Shaw, B. L. (1972). *J. Chem. Soc. Dalton Trans.*, 754. (b) (1972). 860
29. Duff, J. M., and Shaw, B. L. (1972). *J. Chem. Soc. Dalton Trans.*, 2219
30. Parshall, G. W., Knoth, W. H. and Schunn, R. A. (1969). *J. Amer. Chem. Soc.*, **91,** 4990; Barefield, E. K. and Parshall, G. W. (1972). *Inorg. Chem.*, **11,** 964
31. Marriott, J. C., Salthouse, J. A., Ware, M. J. and Freeman, J. M. (1970). *Chem. Commun.*, 595
32. Green, J. C., King, D. I. and Eland, J. H. D. (1970). *Chem. Commun.*, 1121
33. Krüger, C. and Tsay, Y-H. (1972). *Acta Crystallogr.*, **B28,** 1941
34. King, R. B. and Efraty, A. (1972). *J. Amer. Chem. Soc.*, **94,** 3768
35. See, for example, Graham, B. W., Laing, K. R., O'Connor, C. J. and Roper, W. R. (1972). *J. Chem. Soc. Dalton Trans.*, 1237 and Refs. therein
36. Robinson, S. D. and Levison, J. J. (1972). *J. Chem. Soc. Dalton Trans.*, 2013
37. Durkin, T. R. and Schram, E. P. (1972). *Inorg. Chem.*, **11,** 1048, 1054
38. Weston, C. W., Bailey, G. W., Nelson, J. H. and Johanssen, H. B. (1972). *J. Inorg. Nucl. Chem.*, **34,** 1752
39. Clark, H. C. and Itoh, K. (1971). *Inorg. Chem.*, **10,** 1707
40. Clement, D. A. and Nixon, J. F. (1973). *J. Chem. Soc. Dalton Trans.*, 195
41. Muetterties, E. L. and Alegranti, C. W. (1972). *J. Amer. Chem. Soc.*, **94,** 6386
42. Lippard, S. J. and Mayerle, J. J. (1972). *Inorg. Chem.*, **11,** 753
43. Cooper, D. G. and Powell, J. (1973). *J. Amer. Chem. Soc.*, **95,** 1102
44. Nowlin, T. and Cohn, K., (1972). *Inorg. Chem.*, **11,** 560
45. Rigo, P. and Bressau, M. (1972). *Inorg. Chem.*, **11,** 1314
46. (a) Alyea, E. C. and Meek, D. W. (1972). *Inorg. Chem.*, **11,** 1029. (b) Morassi, R. and Dei, A. (1972). *Inorg. Chim. Acta*, **6,** 314
47. (a) Meek, D. W., Alyea, E. C., Stalick, J. K. and Ibers, J. A. (1969). *J. Amer. Chem. Soc.*, **91,** 4920; (b) Haugen, L. P., and Eisenberg, R. (1969). *Inorg. Chem.*, **8,** 1072
48. Stalick, J. K. and Ibers, J. A. (1969). *Inorg. Chem.*, **8,** 1084, 1090
49. Orioli, P. L. and Sacconi, L. (1969). *Chem. Commun.*, 1012
50. Bianchi, A., Dapporto, P., Fallani, G., Ghilardi, C. A. and Sacconi, L. (1973). *J. Chem. Soc. Dalton Trans.*, 641
51. McAuliffe, C. A., Workman, M. O. and Meek, D. W. (1972). *J. Coord. Chem.*, **2,** 137
52. Eller, P. G. and Meek, D. W. (1972). *Inorg. Chem.*, **11,** 2518
53. Ferraro, J. R. (1970). *Inorg. Nucl. Chem. Lett.*, **6,** 823
54. Kilbourn, B. T. and Powell, H. M. (1970). *J. Chem. Soc. A*, 1688
55. Shaw, S. Y. and Dudek, E. P. (1969). *Inorg. Chem.*, **8,** 1360; Gerlach, D. H. and Holm, R. H. (1969). *J. Amer. Chem. Soc.*, **91,** 3457: Powers, C. R. and Everett, G. W., jun. (1969). *J. Amer. Chem. Soc.*, **91,** 3468
56. Pignolet, L. H. and Horrocks, W. DeW., jun. (1969). *J. Amer. Chem. Soc.*, **91,** 3976
57. Pignolet, L. H., Horrocks, W. DeW., jun. and Holm, R. H. (1970). *J. Amer. Chem. Soc.*, **92,** 1855
58. La Mar, G. N. and Sherman, E. O. (1970). *J. Amer. Chem. Soc.*, **92,** 2691
59. Ferraro, J. R., Nakamoto, K., Wang, J. T. and Lauer, L. (1973). *Chem. Commun.*, 266
60. Manojlović-Muir, L. (1971). *J. Chem. Soc. A*, 2796
61. Manojlović-Muir, L. (1973). *Inorg. Nuclear Chem. Lett.*, **9,** 59
62. Drew, M. G. B., Wilkins, J. D. and Wolters, A. P. (1972). *Chem. Commun.*, 1278
63. Visscher, M. O. and Caulton, K. G. (1972). *J. Amer. Chem. Soc.*, **94,** 5923
64. Duckworth, V. F., Douglas, P. G., Mason, R. and Shaw, B. L. (1970). *Chem. Commun.*, 1083
65. Pierpont, C. G. and Eisenberg, G. (1972). *Inorg. Chem.*, **11,** 1088
66. Ward, D. L., Caughlan, C. N., Voecks, G. E. and Jennings, P. W. (1972). *Acta Crystallogr*, **B28,** 1949
67. Terry, N. W., tert., Amma, E. L. and Vaska, L. (1972). *J. Amer. Chem. Soc.*, **94,** 653

68. Bradley, D. C., Hursthouse, M. B., Smallwood, R. J. and Welch, A. J. (1972). *Chem. Commun.*, 872
69. Dapporto, P. and Fallani, G. (1972). *J. Chem. Soc. Dalton Trans.*, 1498
70. Bridges, D. M., Rankin, D. W. H., Clement, D. A. and Nixon, J. F. (1972). *Acta Crystallogr.*, **B28,** 1130
71. McGinnety, J. A., Payne, N. C. and Ibers, J. A. (1969). *J. Amer. Chem. Soc.*, **91,** 6301
72. Bonds, W. D., jun. and Ibers, J. A. (1972). *J. Amer. Chem. Soc.*, **93,** 3413
73. Dapporto, P., Fallani, G., Midollini, S. and Sacconi, L. (1972). *Chem. Commun.*, 1161
74. Sheldrick, W. S. and Stelzer, O. (1972). *J. Chem. Soc. Dalton Trans.*, 926
75. Smith, A. E. (1972). *Inorg. Chem.*, **11,** 3017
76. Bianchi, A., Dapporto, P. Fallani, G., Ghilardi, C. A. and Sacconi, L. (1973). *J. Chem. Soc. Dalton Trans.*, 641
77. Ansell, G. B. (1973). *J. Chem. Soc. Dalton Trans.*, 371
78. Schneider, M. L. and Shearer, H. M. M. (1972). *J. Chem. Soc. Dalton Trans.*, 354
79. Cheng, P. -T. and Nyburg, S. C. (1972). *Can. J. Chem.*, **50,** 912
80. Graziani, R., Bombieri, G. and Forsellini, E. (1972). *Inorg. Nucl. Chem. Lett.*, **8,** 701
81. Albano, V. G., Bellon, P. L. and Ciani, G. (1972). *J. Chem. Soc. Dalton Trans.*, 1938
82. Albano, V. G., Bellon, P. L., Ciani, G. and Manassero, M. (1972). *J. Chem. Soc. Dalton Trans.*, 171
83. Bandoli, G., Clemente, D. A., Marangoni, G. and Cattalini, L. (1972). *J. Chem. Soc. Dalton Trans.*, 886
84. Bennet, J., Pidcock, A., Waterhouse, C. R., Coggon, P. and McPhail, A. T., (1970). *J. Chem. Soc. A*, 2094
85. (a) Bellon, P. L., Manassero, M., Naldini, L. and Sansoni, M. (1972). *Chem. Commun.*, 1035; (b) Cariati, F., and Naldini, L. (1972). *J. Chem. Soc. Dalton Trans.*, 2286. (c) Bellon, P. L., Manassero, M., and Sansoni, M. (1972). *J. Chem. Soc. Dalton Trans.*, 1481. (d) Bellon, P. L., Cariati, F., Manassero, M., Naldini, L. and Sansoni, M. (1971). *Chem. Commun.*, 1423
86. Rigo, P., Bressan, M., Corain, B. and Turco, A. (1970). *Chem. Commun.*, 598
87. Hammond, B., Jardine, F. H. and Vohra, A. G. (1971). *J. Inorg. Nucl. Chem.*, **33,** 1017
88. Anderson, W. A., Carty, A. J., Palenik, G. J. and Schreiber, G. (1971). *Can. J. Chem.*, **49,** 761
89. Zelonka, R. A. and Baird, M. C. (1971). *Chem. Commun.*, 780
90. Taylor, R. C. and Kolodny, R. A. (1970). *Chem. Commun.*, 813
91. Rigo, P., Longato, B. and Favero, G. (1972). *Inorg. Chem.*, **11,** 300
92. Tayim, H. A., Thabet, S. K. and Karkanawi, M. U. (1972). *Inorg. Nucl. Chem. Lett.*, **8,** 235
93. Switkes, E. S., Ruis-Ramires, L., Stephenson, T. A. and Sinclair, J. (1972). *Inorg. Nucl. Chem. Lett.*, **8,** 593
94. Intille, G. M. (1972). *Inorg. Chem.*, **11,** 695
95. Butcher, A. V., Chatt, J., Leigh, G. J. and Richards, P. L. (1972). *J. Chem. Soc. Dalton Trans.*, 1064
96. Greenwood, N. N. and Tranter, R. L. (1969). *J. Chem. Soc. A*, 2878
97. Schmulbach, C. D., Kolich, C. H. and Hinckley, C. C. (1972). *Inorg. Chem.*, **11,** 2841
98. Chatt, J. and Hayter, R. G. (1963). *J. Chem. Soc.*, 1343
99. Beattie, I. R. and Collis, R. (1969). *J. Chem. Soc. A*, 2960
100. Calderazzo, F., Losi, S. A. and Susz, B. P. (1971). *Helv. Chim. Acta*, **54,** 1156
101. Henrici-Olivé, G. and Olivé, S. (1970). *Angew. Chem. Int. Ed.*, **9,** 957
102. Glushkova, M. A., Ershova, M. M., Ovichinikova, N. A. and Buslaev, Yu., A. (1972). *Russ. J. Inorg. Chem.*, **17,** 77
103. King, R. B. and Kapoor, P. N. (1972). *Inorg. Chem.*, **11,** 1524
104. Preer, J. R., Tsay, F. -D. and Gray, H. B. (1972). *J. Amer. Chem. Soc.*, **94,** 1875
105. Ros, R. and Tondello, E. (1971). *J. Inorg. Nucl. Chem.*, **33,** 245
106. Bosnich, B. and Wild, S. B. (1970). *J. Amer. Chem. Soc.*, **92,** 459
107. Gordon, J. G., jun. and Holm, R. H. (1970). *J. Amer. Chem. Soc.*, **92,** 5319
108. Chia, L. S. and Cullen, W. R. (1972). *Can. J. Chem.*, **50,** 1421
109. Westland, A. D. (1969). *Can. J. Chem.*, **47,** 4135
110. (a) Preer, J. R. and Gray, H. B. (1970). *J. Amer. Chem. Soc.*, **92,** 7306; (b) Bernstein, P. K. and Gray, H. B. (1972). *Inorg. Chem.*, **11,** 3035
111. Páez, E. A., Oosterhuis, W. T. and Weaver, D. L. (1970). *Chem. Commun.*, 506
112. Selbin, J., Ahmad, N. and Pribble, M. J. (1969). *Chem. Commun.*, 759
113. Henrici-Olivé, G. and Olivé, S. (1972). *J. Organometal. Chem.*, **46,** 101
114. Henrici-Olivé, G. and Olivé, S. (1969). *Chem. Commun.*, 596

115. Stec, W. J., Morgan, W. E., Albridge, R. G. and van Wazer, J. R. (1972). *Inorg. Chem.*, **11**, 219
116. Headley, O. St. C., Nyholm, R. S., McAuliffe, C. A., Sindellari, L., Tobe, M. L. and Venanzi, L. M. (1970). *Inorg. Chim. Acta*, **4**, 93
117. Dalton, J., Levason, W. and McAuliffe, C. A. (1972). *Inorg. Nucl. Chem. Lett.*, **8**, 797
118. Midollini, S. and Cecconi, F. (1973). *J. Chem. Soc. Dalton Trans.*, 681
119. Cunninghame, R. G., Nyholm, R. S. and Tobe, M. L. (1972). *J. Chem. Soc. Dalton Trans.*, 229
120. Baraceo, L. and McAuliffe, C. A. (1972). *J. Chem. Soc. Dalton Trans.*, 948
121. Zelonka, R. A. and Baird, M. C. (1972). *Canad. J. Chem.*, **50**, 1269
122. Drew, M. G. B. (1972). *J. Chem. Soc. Dalton Trans.*, 626
123. Drew, M. G. B. and Mandyczewsky, R. (1970). *Chem. Commun.*, 292
124. Bernstein, P. K., Rodley, G. A., Marsh, R. and Gray, H. B. (1972). *Inorg. Chem.*, **11**, 3040
125. Graziani, R., Bombieri, G. and Forsellini, E. (1971). *J. Chem. Soc. A*, 2331
126. Makhija, R. C., Beauchamp, A. L. and Rivest, R. (1972). *Chem. Commun.*, 1043
127. Selbin, J., Ahmad, N. and Pribble, M. J. (1970). *J. Inorg. Nucl. Chem.*, **32**, 3249
128. Crisp, W. P., Deutscher, R. L. and Kepert, D. L. (1970). *J. Chem. Soc. A*, 2199
129. Beale, J. P. and Stephenson, N. C. (1970). *Acta Crystallogr.*, **B26**, 1655
130. Clark, R. J. H. and Natile, G. (1970). *Inorg. Chim. Acta*, **4**, 533
131. Hamilton, J. B. and McCarley, R. E. (1970). *Inorg. Chem.*, **9**, 1333
132. Hamilton, J. B. and McCarley, R. E. (1970). *Inorg. Chem.*, **9**, 1339
133. Merbach, A. and Bünzli, J. -C. (1971). *Chimia (Switz.)*, **25**, 222
134. Fowles, G. W. A., Gadd, K. F., Rice, D. A., Tomkins, I. B. and Walton, R. A. (1970). *J. Mol. Struct.*, **6**, 412
135. Boorman, P. M., Islip, M., Remier, M. M. and Reimer, K. J. (1972). *J. Chem. Soc. Dalton Trans.*, 890
136. (a) Chatt, J., Leigh, G. T., Storace, A. P., Squire, D. A. and Starkey, B. J. (1971). *J. Chem. Soc. A*, 899. (b) Allen, E. A., and Wilkinson, W. (1972). *J. Chem. Soc. Dalton Trans.*, 613. (c) Aires, B. E., Fergusson, J. E., Howarth, D. T. and Miller, J. M. (1971). *J. Chem. Soc. A*, 1144
137. Rowe, M. D., McCafferty, A. J. Gale, R. and Copsey, D. N. (1972). *Inorg. Chem.*, **11**, 3090
138. Goggin, P. L., Goodfellow, R. J., Haddock, S. R., Reed, F. J. S., Smith, J. G. and Thomas, K. M. (1972). *J. Chem. Soc. Dalton Trans.*, 1904
139. Schläpfer, C. W., Saito, Y. and Nakamoto, K. (1972). *Inorg. Chim. Acta*, **6**, 284
140. McFarlane, W. (1969). *Chem. Commun.*, 700
141. Abel, E. W., Bush, R. P., Hopton, F. J., and Jenkins, C. R. (1966). *Chem. Commun.*, 58
142. Hill, H. A. O. and Simpson, K. A. (1970). *J. Chem. Soc. A*, 3266
143. Haake, P. and Turley, P. C. (1967). *J. Amer. Chem. Soc.*, **89**, 4611, 4617
144. Cross, R. J., Dalgleish, I. G., Smith, G. J. and Wardle, R. (1972). *J. Chem. Soc. Dalton Trans.*, 992
145. Coletta, F., Ettorre, R. and Gambaro, A. (1972). *Inorg. Nucl. Chem. Lett.*, **8**, 667
146. Cross, R. J., Green, T. and Keat, R. Unpublished observations
147. Cattalini, L., Cassal, A., Marangoni, G., Rizzardi, G. and Rotondo, E. (1969). *Inorg. Chim. Acta*, **3**, 681
148. Rotondo, E., Marsala, V., Cattalini, L. and Coe, J. S. (1972). *J. Chem. Soc. Dalton Trans.*, 2546
149. Gaylor, J. R. and Senoff, C. V. (1971). *Can. J. Chem.*, **49**, 2390
150. Cattalini, L., Marangoni, G., Coe, J. S., Vidali, M. and Martelli, M. (1971). *J. Chem. Soc. A*, 593
151. Cattalini, L., Coe, J. S., Faraone, F., Marala, V. and Rotondo, E. (1972). *Inorg. Chim. Acta*, **6**, 303
152. Pluśćec, J. and Westland, A. D. (1965). *J. Chem. Soc.*, 5371
153. Rosen, W. and Busch, D. H. (1969). *Chem. Commun.*, 148
154. Travis, K. and Busch, D. H. (1970). *Chem. Commun.*, 1041
155. Riley, P. E. and Seff, K. (1972). *Inorg. Chem.*, **11**, 2993
156. Skakke, P. E. and Rasmussen, S. E. (1970). *Acta Chem. Scand.*, **24**, 2634
157. Whitfield, H. J. (1970). *J. Chem. Soc. A*, 113
158. James, B. R. and Ng, F. T. T. (1972). *J. Chem. Soc. Dalton Trans.*, **335**, 1321
159. Steger, H. F. (1972). *J. Inorg. Nucl. Chem.*, **34**, 175
160. Biscarini, P., Fusina, L. and Nivellini, G. D. (1972). *J. Chem. Soc. Dalton Trans.*, 1921
161. Worrell, J. H., Goddard, R. A., Gupton, E. M., jun. and Jackman, T. A. (1972). *Inorg. Chem.*, **11**, 2734
162. Alcock, N. W. and Tasker, P. A. (1972). *Chem. Commun.*, 1239

163. Sacconi, L. and Gatteschi, D. (1972). *J. Coord. Chem.*, **2,** 107
164. Chiswell, B., Plowman, R. A. and Verral, K. (1972). *Inorg. Chim. Acta*, **6,** 113, 275
165. Volponi, L., Zarli, B. and De Paoli, G. G. (1972). *Inorg. Nucl. Chem. Lett.*, **8,** 309
166. (a) Dapporto, P., Fallani, G. and Sacconi, L. (1971). *J. Coord. Chem.*, **1,** 269; (b) Sacconi, L. and Dei, A. (1971). *J. Coord. Chem.*, **1,** 229; (c) Bianchi, A., Ghilardi, C. A., Mealli, C. and Sacconi, L. (1972). *Chem. Commun.*, 651
167. Garnovskii, A. D., Osipov, O. A. and Bulgarevich, S. B. (1972). *Russ. Chem. Rev.*, **41,** 341
168. (a) Kukushkin, Yu. N. and Khokhryakov, K. A. (1972). *Russ. J. Inorg. Chem.*, **17,** 136; (b) Price, J. H., Williamson, A. N., Schramm, R. F. and Wayland, B. B. (1972). *Inorg. Chem.*, **11,** 1280

6
Complexes of Neutral and Positively Charged Oxygen- and Nitrogen-containing Ligands

S. M. NELSON
Queen's University, Belfast

Abbreviations used in the text

L	ligand
L-H	deprotonated ligand
L^+	positively charged ligand
M	metal atom or ion
X	univalent anionic ligand (usually halide)
Y	univalent anion
Me	methyl
Et	ethyl
Pr	propyl
Bu	butyl
Ph	phenyl
thf	tetrahydrofuran
diox	1,4-dioxan
thiox	1,4-thioxan
dme	1,2-dimethoxyethane
cp	cyclopentadienyl
Hacac	acetylacetone
acac	acetylacetone anion
py	pyridine
iq	isoquinoline
pyNO	pyridine -N-oxide
Ph_3PO	triphenylphosphine oxide
dmso	dimethyl sulphoxide
en	ethylenediamine
pn	1,2-diaminopropane
tn	1,3-diaminopropane
dpt	3,3′-diaminodipropylamine
dien	diethylenetriamine
dpt	dipropylenetriamine
trien	triethylenetetramine(1,8-diamino-3,6-diazaoctane)
2,2,2-tet	triethylenetetramine(1,8-diamino-3,6-diazaoctane)
2,3,2-tet	1,9-diamino-3,7-diazanonane
3,2,3-tet	1,10-diamino-4,7-diazadecane
3,3,3-tet	1,11-diamino-4,8-diazaundecane
tetren	tetraethylenepentamine
tren	tris(2-aminoethyl)amine
trpn	tris(3-aminopropyl)amine
pic	picoline
dip	2,2′dipyridyl
phen	1,10-phenanthroline
naphth	1,8-naphthyridine
pz	pyrazine
dipyam	bis-(2-pyridyl)amine
tripyam	tris-(2-pyridyl)amine

6.1 INTRODUCTION

The coverage of this review is similar to that of Chapter 5 of Volume 5 of Series 1[1] except that complexes of anionic ligands are excluded. Once again, certain important classes of neutral oxygen- and nitrogen-containing ligands are not considered because a comprehensive treatment of this vast topic in the limited space available is impossible. Thus, for example, the important and growing chemistry of transition metal macrocyclic compounds is omitted. Also omitted are complexes of polydentate ligands which contain, in addition to oxygen or nitrogen, other donor atoms such as phosphorus, arsenic, antimony or sulphur, except in a few cases where it seems appropriate to draw comparisons. The Review covers work published during 1971 and 1972. Emphasis is on structure although conformational aspects of coordinated ligands are not discussed in the detail they deserve[2]. Optical activity is not treated. In selecting material an attempt has been made to reflect the more novel features of the subject; however, the selection is necessarily subjective.

Several reviews on certain specialised aspects of coordination chemistry which deal in part with oxygen- and nitrogen-containing ligands have appeared. Specific mention may be made of the reviews by Black and Hartshorn[3] (ligand design and synthesis), Lindoy[4] (metal ion control in ligand synthesis), Calligaris *et al.*[5] (tetradentate Schiff bases), Holm and O'Connor[6] (bis-chelates), Boucher[7] (manganese porphyrin complexes), Sacconi[8], Wood[9] and Hoskins and Whillans[10] (pentacoordinate complexes), Barbucci *et al.*[11] (thermodynamics of Cu^{II}–polyamine complexes), Waltz and Sutherland[12] and Simmons and Wendlandt[13] (photochemistry), and Addison *et al.*[14] (coordinated nitrate).

6.2 OXYGEN-CONTAINING LIGANDS

6.2.1 Alcohols

The low thermodynamic and kinetic stability of the ROH–metal interaction is responsible for the small number of simple alcohol complexes isolable in the solid state though complexes containing a molecule of coordinated alcohol in company with other stronger ligands are common; in such cases the alcohol usually has its origin in the solvent of preparation. Coordination enhances the acidity of alcohol and alkoxides are common products of reaction particularly with 'hard' metal ions of high charge, even in non-basic media. Where chelation is possible, stability is usually increased because of the favourable entropy term as in ethyleneglycol complexes. Rapid interconversion between bidentately and monodentately coordinated ethyleneglycol has been shown to occur in acidic solutions of Cr^{III} [15]. The greater size of alcohols compared to H_2O usually renders them poorer complexing agents but the situation is apparently reversed in adduct formation with certain tris-(β-diketonates) of Nd^{3+} and Er^{3+}, the relative coordinating strengths being $EtOH > H_2O > Et_2O$ [16]. It may be that a balance of both polar and non-polar forces is required for the most effective penetration of the donor molecule to the metal ion in certain cases. Where the second donor atom of a chelate ring is nitrogen, stability is further enhanced and many complexes of amino-alcohols have been characterised[17–19]. Again, deprotonation may occur. The results of a study[17] of Co^{II}, Ni^{II} and Cu^{II} with mono-, di- and tri-ethanolamines indicate a progressive decrease in coordinating ability in the order $N > O^- > OH$. Both coordinated and uncoordinated hydroxyl groups occur in these complexes as judged by i.r. data. The related ligand tris(hydroxymethyl)aminomethane behaves similarly in that it can form one or two, but not three, 5-membered '*NO*' chelate rings[19]. Both free and bonded OH groups have been observed for complexes of *N*-hydroxyethylethylenediamine, i.e. depending on whether it acts as a bidentate or as a tridentate ligand. Several of the Ni^{II} complexes react with acetone through condensation with a coordinated NH_2 group to give complexes of the linear hexadentate ligand, 7,9,9-trimethyl-3,6,10,13-tetra-azapentadeca-6-en-1,15-diol[20].

6.2.2 Ethers

Cyclic ethers appear to coordinate better than open-chain ethers, probably partly for steric reasons and partly as a result of entropy factors. Tetrahydrofuran gives both 1:1 and 1:2 adducts with $MeTiCl_3$ whereas only 1:1 adducts are given by 1,4-dioxan or 1,2-di-methoxyethane. The insolubility of the dioxan complex suggests that both oxygen atoms are bonded in a polymeric 6-coordinate structure while dme is presumed to be chelated[21]. The mode of coordination of thioxan in its 1:2 adduct is uncertain though the positions of the methylene protons in the ^1H n.m.r. spectrum favour *S*-coordination. Reduction of cyclopentadienyltitanium(IV) trihalides in thf yields the Ti^{III} complexes [$cpTiX_2$(thf)] believed to have monomeric pseudotetrahedral structures[22]. The antisymmetric COC stretch occurs *ca.* 35 cm^{-1} lower than in the free ether. Low-temperature ^{19}F n.m.r. spectra of $TiCl_4$ and TiF_4 in dme has shown that halogen redistribution occurs to yield all possible octahedral mixed complexes having a single chelated ether[23]. The i.r. spectral changes shown by diox, thiox and dme on coordination have been summarised by Karayannis *et al.*[24] and correlated with the mode of bonding. In complexes with first-row transition metal perchlorates a variety of bonding modes are exhibited—monodentate *O*-bonded (diox and thiox), monodentate *S*-bonded (thiox), monodentate *O*-bridging (diox), monodentate *S*-bridging (thiox), chelating *O* and *O* (dme) and chelating *O* and *S* (thiox). A 4-coordinate high-spin [$FeCl_3$(thf)] complex has been prepared by direct reaction under anhydrous conditions. Molecular weight and other measurements indicate some concentration-dependent association in benzene and 1,2-dichloroethane[25]. Tetrahydrofuran and certain other ligands with little or no ability to stabilise low valence states cause a disproportionation of Mo^{II} in $Mo(CO)_4X_2$ into MoO and MoL_3X_2. The same ether competes effectively with dinitrogen for the metal atom in the complex fragment π-cyclopentadienyl(dicarbonyl)Mn^I when $cpMn(CO)_2N_2$ is dissolved in thf[27].

6.2.3 Carbonyl compounds

N-methyl-γ-butyrolactam forms both tetrakis and hexakis complex ions with Co^{II}, isolable as the perchlorate salts. Tetrahedral and octahedral geometries, respectively, were demonstrated by the magnetic and spectral properties[28]. The CO stretching vibration falls by 50 cm^{-1} on coordination indicating that the carbonyl oxygen is the donor atom. The ligand field splitting parameters (10 *Dq*) were evaluated as 5130 cm^{-1} (CoL_4^{2+}) and 8510 cm^{-1} (CoL_6^{2+}). The complexing properties of a series of α-nitroketones with several transition metal ions have been examined; these normally coordinate as chelating *O*-donors in octahedral structures[29].

Complexes of the mono-anions of β-diketones are well known. Recently, a number of complexes containing uncharged acetylacetone have been prepared by direct reaction of the diketone with the anhydrous metal halide or by reaction of the metal diketonate with dry HBr, or, in one case, by reaction of Hacac with $CrCl_3(thf)_3$. Tetrahedral $Co(Hacac)Cl_2$ and $Zn(Hacac)_2Cl_2$ and octahedral $Mn(Hacac)_2Br_2$, $Ni(Hacac)_2Br_2$, $Cr(acac)(Hacac)Br_2$ and $Cr(acac)(Hacac)Cl_2$ have been prepared. In the Mn^{II} complex which contains bridging Br^- ions, the ligand is in the enol form and is coordinated monodentately, while in the others the ligand is ketonic and bidentate[30].

Tridentate di-2-pyridylketone may exist as the ketone hydrate ($L \cdot H_2O$) or as the derived anion ($L \cdot OH^-$) in its Co^{III} complexes, all of which have pseudo-octahedral 'N_4O_2' ligand fields. The iron (III) complex $[Fe(L \cdot OH)_2](NO_3)_3$ is low-spin[31]. Pyridine-2-carbaldehyde similarly co-ordinates (to Cu^{II}) as the hydrate anion and also as the neutral ketone[32]. In the Cu^{II} bis-complexes of *N*-benzoylhydrazine the carbonyl oxygen and the terminal hydrazinic nitrogen are the donor atoms in the 5-membered chelate rings[33]. The i.r. spectrum of $Co_3[Fe(CN)_5(CO)]_2$ is consistent with the presence of a CO group forming a linear bridge between Fe(*C*-bonded) and Co(*O*-bonded)[34].

6.2.4 Amides

Amides possess two potential donor atoms, the carbonyl oxygen and the amide nitrogen, and much interest has been shown in the mode of bonding in transition metal complexes. The accumulated evidence is that the oxygen atom is the preferred coordination site in most of the cases studied. I.R. spectroscopy has been widely used as a means of distinguishing between the two possibilities. A shift in the carbonyl stretching frequency to lower values is usually taken to indicate a metal–oxygen linkage while a shift in ν(CO) to higher frequencies indicates a metal–nitrogen bond. The universal validity of this spectral criterion has recently been questioned. In an x-ray study Trefonas and co-workers[35] have shown that in hexakis(ethyleneurea)Cd^{II} perchlorate the bonding is to the oxygen atom yet there is no shift in ν(CO) relative to the free ligand, and that in bis(ethyleneurea) Hg^{II} chloride, which shows a positive $\Delta\nu(CO)$ of 49 cm^{-1}, the ligand is bonded to two metal atoms, one via the oxygen and the other (less strongly) via the nitrogen. Clearly, special care must be taken in the application of i.r. spectra to structure assignment in such cases; geometrical distortions and hydrogen bonding, as well as coordination, can affect $\nu(CO)$.

Acrylamide, which on the evidence of mass, i.r., n.m.r. and photo-electron spectra has been claimed[36] to be *N*-bonded in $SnCl_4 \cdot 2L$, is considered, from i.r., e.s.r. and electron spectra to be *O*-bonded in several transition metal complexes[37]. The presence of bidentate (bridged) as well as monodentate urea in $[Co(urea)_4](NO_3)_2$ has been suggested[38]. There is evidence for an enhancement of the basicity of the amide nitrogen on coordination of the carbonyl group. Bubbling H_2 through a dimethylformamide solution of $RuCl_3 \cdot 3H_2O$ yielded the complex $RuCl_3 \cdot HL$ where HL is the protonated amide[39]. Deprotonation of the amide nitrogens renders these atoms the preferred coordination sites in the bis-biureto and bis-oxamidato complexes of Ni^{II} and Cu^{II} and, further, facilitates metal oxidation with formation of $K[M^{III}L_2]$[40]. The amido-hydrogen of ligand (1) readily dissociates in the presence of transition metal ions to yield *N*-bonded complexes. Alkaline conditions are required for deprotonation of (2) which uses the oxygen atom in coordination when neutral and the nitrogen atom when anionic[41].

N C(=O) NH–$(CH_2)_n$– N

(1) $n = 1$
(2) $n = 2$

The preparation of a variety of amide complexes with the lanthanide metals[42], U^{IV}, Th^{IV}[43] and Nb^{IV}[44] tetrahalides has been described; in all cases the physical properties indicate that the donor atom is oxygen. This is true also for acetamide in $NiL_4Cl_2 \cdot 2H_2O$ whose structure has been determined by x-ray analysis[45].

Ethyleneurea is *O*-bonded in its coordination compounds with Sn^{IV} and Ti^{IV} halides while ethylenethiourea is apparently *N*-bonded. Taking the magnitude of the $\nu(CO)$ and $\nu(NH)$ shifts as a measure of relative donor power the following orders were obtained (a) for amides, ethyleneurea > dimethylformamide > *N,N'*-dimethylurea > oxamide, and (b) for thioamides, ethylenethiourea > dithio-oxamide > diphenylthiourea > thiourea[46]. Peyronel and co-workers[47,48] have studied the coordinating properties of dithio-oxamides. In neutral or alkaline solution dithio-oxamide and its *N,N'*-substituted derivatives form insoluble polymeric inner complexes in which a proton has been lost from each nitrogen. In strongly acid media mononuclear Ni^{II} complexes of the neutral ligand are obtained in which the ligands are chelating using one nitrogen and one sulphur[47]. With tetramethyl- and tetraethyl-dithio-oxamide cationic octahedral complexes containing bidendate *S*-bonded ligands are obtained with Mn^{II}, Fe^{II}, Co^{II}, Ni^{II} and Cu^{II} and neutral or cationic planar complexes with Pd^{II}[48].

Bonding through the nitrogen atom in thioureas is a little unusual, *S*-bonding being more often encountered, at least in compounds of transition metals having several d electrons. Presumably, this is because covalency effects take over from the electronegativity factor (O > N > S) which might be expected to predominate with 'hard' acceptors[49]. Prediction of the donor atom is often difficult. *N,N'*-alkyl- or aryl-substituted selenoureas are *Se*-bonded in the tetrahedral NiL_2X_2 (X = halide) and planar $[NiL_4](ClO_4)_2$ complexes[50]. Thiomorpholin-3-one (3) uses the oxygen atom when coordinated to Zn^{II} and Cd^{II} and the sulphur atom in the Hg^{II} complex while thiazolidine-2-thione (4) is *N*-bonded with all three metals[51].

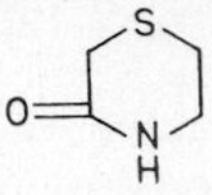

(3) 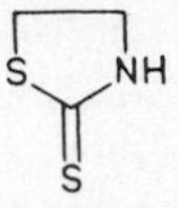 (4)

The varied coordinating potentialities of this class of ligand are well illustrated by the behaviour of 2-acetamidothiazole and 2-acetamidobenzothiazole in combination with Ni^{II} and Cu^{II}. I.R. spectra distinguished three types of structure in which the donor atoms are: (a) the carbonyl oxygen and the thiazole nitrogen; (b) the thiazole nitrogen and the amide nitrogen; and (c) the carbonyl oxygen and both nitrogen atoms. In (a) and (c) the thiazole nitrogen and oxygen atoms are members of 6-membered chelate rings whereas in (b) and (c) the two nitrogens are probably bonded to adjacent metal ions. *N,N,N',N'*-tetrasubstituted thiodiacetamides act as tridentate '*OSO*' ligands giving 5-coordinate Cu^{II} complexes, $CuLX_2$ (X = Cl, Br, NO_3)[53]. When the sulphide group is replaced by an ether group the ligand is bidentate in its complex with $CuCl_2$.

The origin of the small paramagnetism ($\mu_{eff} \simeq 1.0$ B.M. at 20 °C) of $[Co(urea)_6](ClO_4)_3$ has been resolved[54]. It is not due to population of a low lying high-spin quintet state as suggested earlier[55] but to a thermal decomposition to a high-spin 6-coordinate Co^{II} species.

6.2.5 Amine oxides

Among the new compounds of trialkylamine *N*-oxides are some Co^{II} complexes of triethylamine *N*-oxide and tripropylamine *N*-oxide. Two types of complexes were obtained, CoL_2X_2 (X = halide) and $[CoL_4](ClO_4)_2$, both having essentially tetrahedral geometry in the solid and in MeCN solution. As found for previously studied trimethylamine *N*-oxide systems, and in contrast to complexes of the pyridine *N*-oxides, there is little change in the NO stretching frequency on coordination though a splitting of the band is usually observed. Addition of excess ligand to solutions of the complexes produced a new 5-coordinate CoL_5^{2+} species in the case of the ethyl compound[56].

Most di- and tri-positive first-row transition metal ions can accommodate up to six pyridine *N*-oxide molecules in the first coordination shell. This is true also of 2-substituted derivatives in many cases showing that the steric effect of the 2-substituent is small. Until recently Cu^{2+} appeared exceptional in that only unsubstituted pyNO formed a hexakis-complex. It has now been shown[57] that 4-Me-pyNO also gives $[CuL_6](ClO_4)_2$ as well as $[CuL_4]$-

$(ClO_4)_2$; similarly, $[Cu(4\text{-}CN\text{-}pyNO)_6](BF_4)_2$ can be prepared. $[ML_6](ClO_4)_n$ are also given by Fe^{II} with the 2-, 3- and 4-cyanopyridine *N*-oxides and by Fe^{III} with 3- and 4-cyanopyridine *N*-oxide[58]. The i.r. spectra of a number of 2-, 4- and 2,6-substituted pyridine *N*-oxides have been discussed in relation to coordination number, stereochemistry and bonding[59]. Some new 5-coordinate and distorted octahedral complexes containing 2,6-lutidine have been described[60]. The distortion parameters D_s and D_t in the hexakis-pyNO and hexakis-antipyrine complexes of Mn^{III} have been evaluated[60]. The stereochemistry of the 8-coordinate $[La(pyNO)_8]^{3+}$ ion shows a distortion from square prismatic towards cubic geometry[61].

Cu^{II} complexes of pyridine *N*-oxides exhibit varied magnetic properties. These often have their origin in super-exchange through bridging oxygen atoms. The 4.2–37.5 K magnetic susceptibilities of the $CuLCl_2$ complex, L = 4-nitroquinoline *N*-oxide, assumed to be *O*-bridged, fit the van Vleck equation for exchange-coupled Cu^{2+} ions in a dimeric unit but the mechanism is uncertain[62]. Kettle and co-workers[63] have examined the susceptibility *v.* temperature data for a large number of dinuclear Cu^{II} complexes having bridging *N*-oxide ligands. A model is proposed which allows for thermal population of excited states other than those predicted by the vector coupling model. The mechanism of super-exchange is considered to be dominated by two paths which promote an effective metal–metal overlap in opposite senses yet co-operate in the antiferromagnetic exchange process. In the yellow modification of $Cu(4\text{-}Me\text{-}pyNO)_2Cl_2$ the *O*-bridged dimeric molecule contains Cu ions in distorted square-based pyramids. The *O*-bridge system contains unequal Cu—O bonds of 1.937 and 2.153 Å. It is predicted that the magnetic moment should be larger than found for the 1:1 dimeric complexes[64]. N.Q.R. spectra of chlorine and bromine in several pyNO complexes of $CuCl_2$ and $CuBr_2$ have been reported and related to structure[65].

6.2.6 Phosphoryl compounds as ligands

Triphenylphosphine oxide is among the most extensively studied monodentate oxygen-donors. New complexes reported during 1971–1972 are $ML_n(BF_4)_2$, where M = Mg, Ca, Mn, Fe, Co, Ni, Cu, Zn, Cd ($n = 4$) and Co, Ni ($n = 5$). Tetrahedral (Co, Zn), planar (Cu), square pyramidal (Mn, Ni, Cd) and distorted octahedral (Mg, Ca, Fe, Co, Ni) structures were proposed on the evidence of i.r. and electronic spectra. The BF_4^- ion is believed to be coordinated in some cases[66]. The cation in $[Mn(Ph_3PO)_4I][Mn(CO)_4I_2]$ is square pyramidal with an unusually long Mn—I (apical) bond (2.811 Å)[67]. The Mn^{III} compounds MnL_3Cl_3 (L = pyNO, Ph_3PO, Ph_3AsO) are stable in air at room temperature[68]. The coordination geometry about the metal atom in $Cu(Ph_3PO)_2Cl_2$ is that of a slightly elongated tetrahedron (Cu—Cl = 2.170 Å, Cu—O = 1.958 Å, CuOP = 150.9 °)[69] and not the flattened tetrahedron suggested earlier[70]. In $Ce(Ph_3PO)_2(NO_3)_4$ the metal is 10-coordinate, the nitrate groups being bidentate; the Th^{IV} compound is isostructural. From the e.s.r. and vibrational spectra of $FeL_4(ClO_4)_3$ and FeL_2X_3 (L = Ph_3PO, Ph_3AsO; X = Cl, Br, NCS, NO_3) distortion parameters have been deduced[72]. The perchlorate complexes contain planar FeL_4^{3+} ions, the arsine oxide exercising the larger field. The halides have the salt structure *trans*-$[FeL_4X_2][FeX_4]$ while the nitrate and thiocyanate are neutral, monomeric and 5-coordinate. The complex $CuL_4(BF_4)_2$ (L=trimorpholinophosphine oxide) is tetrahedral (strong band in the electronic spectrum at 7640 cm^{-1}) with ionic BF_4^- ions[73].

A novel approach to the study of complex formation by phosphoryl compounds has been the study of i.r. spectra of the ligands chemisorbed onto $CuCl_2$ and $FeCl_2$ surfaces[74]. The ligands included alkyl- and chlorophosphates, phosphonates, phosphinates and phosphine oxides. Complex formation was indicated mainly by large reductions in the frequency of the P=O stretch. The major factor governing the strength of the metal–ligand interaction was considered to be the electronegativity of the substituents on the phosphorus atom, this controlling the availability of oxygen electrons for back-bonding to phosphorus. When halogen is attached directly to phosphorus, bonding no longer occurs through the phosphoryl oxygen; instead, an ion-pair complex results, e.g. $[POCl_2]^+[FeCl_4]^-$.

Considerable attention has been directed recently to diphosphoryl compounds as ligands, the phosphorus atoms being linked by carbon, nitrogen or oxygen

$$R^1P(=O)(R^2)\cdot X \cdot P(=O)(R^2)R^1$$

(5) $X = CH_2$ $R^1 = R^2 = Ph$
(6) $X = (CH_2)_2$ $R^1 = R^2 = Ph$
(7) $X = (CH_2)_4$ $R^1 = R^2 = Ph$
(8) $X = CH_2$ $R^1 = R^2 = O^iPr$
(9) $X = (CH_2)_2$ $R^1 = R^2 = O^iPr$
(10) $X = (CH_2)_4$ $R^1 = R^2 = O^iPr$
(11) $X = O$ $R^1 = R^2 =$ Me, Et, Ph
(12) $X = CH_2$ $R^1 = R^2 = NMe_2$
(13) $X = O$ $R^1 = R^2 = NMe_2$
(14) $X = O$ $R^1 = R^2 = OCHMe_2$
(15) $X = NMe$ $R^1 = R^2 = NMe_2$
(16) $X = O$ $R^1 = NMe_2$; $R^2 =$ Me, Et, Pr^i

One of these, bis-(1,1′-diphenylphosphinyl)methane (5) forms octahedral 1:3 complexes, CoL_3^{2+}, with Co^{II}[75]. When the carbon chain linking the phosphorus (or arsenic) atoms is lengthened as in (6) and (7), there is some doubt as to whether chelation occurs in some octahedral complexes of Co^{II} perchlorate and nitrate[76, 77]. With Co^{II} and Cu^{II} halides (6) gives polymeric pseudotetrahedral 1:1 complexes in which the ligand is bridging, as deduced from vibrational and electronic spectra, and other physical properties[76], and in one case by x-ray structure determination[78]. The relative efficiencies for the extraction of Ce^{III} from aqueous HNO_3 of the diphosphonates (8)–(10) is 100:10:1. This is attributed to differences in the chelating abilities of the three ligands[79]. More recent structural studies of the adducts of the lanthanide nitrates have confirmed this general conclusion, showing that (8) and (9) form chelates but that (10) does not[80]. The electronic spectra of dialkyl- and diphenyl-phosphinic anhydride (11) complexes of Co^{II} and Ni^{II} show them to be 6-coordinate of the type $[ML_3](ClO_4)_2$. They appear rather less stable than corresponding complexes of octamethylpyrophosphoramide though the ligand field splitting parameter *Dq* is similar[81].

Phosphoramides also coordinate via the phosphoryl oxygen. The coordination compounds of hexamethylphosphoric triamide have been reviewed[82]. Many of the investigations of these ligands have concerned their complexes with the lanthanide metal oxy-acid salts. The interest here turns mainly on the coordination number of the metal[83]. The Cu^{II} chelates of the *tris*-complexes of several bidentate phosphoramides (12), (13), (16) have been studied by Joesten *et al.*[84]. The 297 K e.s.r. spectra all indicate an isotropic *g* value of 2.25 whereas at 88 K $g_{\parallel} = 2.42–2.52$ (among the highest values known for Cu^{2+}) and $g_{\perp} = 2.08–2.13$. An x-ray structure determination of the complex of ligand (12) shows it to be a slightly distorted octahedron with each chelate ring in the boat conformation[85]. From measurements of the polarised crystal spectra down to 5 K of the *tris*-complexes of Co^{II} and Ni^{II} with ligand (13) the ligand field parameters *Dq*, B and C were evaluated, respectively, as 765, 855 and 3680 cm^{-1} (Co^{II}), and 731, 950 and 3740 cm^{-1} (Ni^{II})[86]. Very similar values have been obtained for the *tris*-complexes of the chelating ligand (16) with Ni^{II}. This ligand contains asymmetric phosphorus atoms and the ^{31}P n.m.r. spectra show two peaks of equal intensity which are attributed to *racemic* and *meso* forms[87].

Diethylpyridine 2-phosphonate (17) reacts with certain $lanthanide^{III}$ chlorides in the absence of solvent at 120–150 °C to give *tris*-chelates of the monomethylester (18) and the by-products HCl, EtCl and ethylene[88].

$$(17) \xrightarrow{MCl_3} (18)$$

(17) (18)

6.2.7 Sulphoxides

As outlined in Series 1[1] sulphoxides have ambidentate character showing oxygen-coordination in most of the known metal complexes but also using the sulphur atom with some of the heavier metals. The two types of complex are usually distinguished on the basis of the frequency of the SO stretching vibration in the i.r. (reduced for M—O and increased for M—S) and also, less frequently, of the assignment of the metal–donor atom (O or S) stretching frequencies. Recent studies support earlier conclusions that the 'harder' metal ions prefer to bond to oxygen. Thus, tetramethylene sulphoxide is *O*-bonded in its hexa-coordinate complexes with Al^{3+}, Cr^{3+}, Fe^{3+}, Mn^{2+}, Co^{2+}, Ni^{2+} and Zn^{2+} [89], and in various complexes of high coordination number with lanthanide(III) perchlorates[90] and nitrates[91], and in some vanadyl and uranyl complexes of diphenylsulphoxide[92]. Reductions in ν(SO) of up to *ca.* 80 cm^{-1} have been observed. A particularly large increase (145 cm^{-1}) was found[93] for $[AuCl_3(dmso)]$. This and the occurrence of a band at 430 cm^{-1}

(assigned to the Au—S stretch) indicates coordination by the sulphur atom. Wayland and co-workers[94] have continued their studies on Pd^{II} and Pt^{II} complexes containing both *O*-bonded and *S*-bonded sulphoxide ligands. X-Ray confirmation is now available[95]. New examples have been reported and when L = isoamylsulphoxide in $[PdL_4](BF_4)_2$ there is exclusive *O*-bonding. Supporting the view that increased steric restriction leads to a change from *S*- to *O*-coordination are the results of an ^1H n.m.r. study of $MeNO_2$ solutions of $[Pt(dmso)_4](ClO_4)_2$. This complex contains both kinds of sulphoxide coordination and retains its structure in solution. *O*-Bonded sulphoxide exchanges rapidly with free ligand but *S*-bonded sulphoxide is kinetically stable. Line broadening techniques have shown that $[Co(dmso)_6]^{2+}$ and $[Ni(dmso)_6]^{2+}$ exchange too fast while $[Fe(dmso)_6]^{3+}$ and $[Co(dmso)_6]^{3+}$ exchange too slowly for the exchange parameters to be measured by this method[96]. A dissociative mechanism is involved, at least in the case of the former complexes[97]. For other aspects of dimethyl sulphoxide chemistry see the collection of articles edited by Jacob *et al.*[98].

6.3 COMPLEXES OF NITROGEN-CONTAINING LIGANDS

6.3.1 Ammines

The single-crystal x-ray structure of $Co(NH_3)_3(NO_2)_3$ has been re-determined. The complex is the *meridial* isomer having one pair of *trans*-NO_2 and one pair of *trans*-NH_3 groups[99]. The coordination geometry of $[Cu(dip)_2(NH_3)](BF_4)_2$ is approximately trigonal-bipyramidal[100]. The thermal decomposition product of $[Co(NH_3)_6](NO_3)_2$ and of $[(NH_3)_5Co—O—O—Co(NH_3)_5](NO_3)_4$ *in vacuo* is $Co(NH_3)_2(NO_3)_2$, shown by spectroscopic and magnetic measurements to be 6-coordinate with bidentate nitrate[101]. The halides of Pu^{III} and Pu^{IV} react with gaseous or liquid NH_3 to give, initially, products containing *ca.* 9 and 8, respectively, NH_3 ligands. These are unstable and lose NH_3 over a period of weeks to give quasi-stable species such as PuI_3 (5.5–6.0) NH_3[102]. Reaction of metallic Pd with HNO_3 followed by treatment with NH_3 has yielded the novel complex $[Pd(NH_3)_3(NO_2)]_2[Pd(NH_3)_4](NO_3)_4$. The metal is square planar in both cations and there is no metal–metal interaction[103] $Cu(NH_3)_2CO_3$ shows antiferromagnetism but the origin is uncertain. Either superexchange through bridging CO_3 groups or direct metal–metal interaction between pairs of Cu^{2+} ions in different chains, or both, may be involved[104]. Drago and Chiang[105] have given a detailed analysis of the ligand field spectra of the tetragonal complexes, NiL_2X_2 (L = NH_3, aniline, py, H_2O; X = Cl, Br).

6.3.2 Monodentate aliphatic amines

The rates of base hydrolysis of $[Co(RNH_2)_5Cl]^{2+}$ (R = alkyl) complexes are $>10^5$ greater than for the corresponding ammine[106]. This is attributed to the greater relief of steric strain as the reactant proceeds to the 5-coordinate intermediate via a dissociative mechanism. A single crystal x-ray structure determination of $[Co(MeNH_2)_5Cl](NO_3)_2$ confirms the occurrence of steric interactions, notably deviations of 11–15° in the Co—N—C angles and of up to 7° in the N—Co—N angles[107]. Steric effects of alkyl groups in alkylamine complexes are reflected also in the stability constants (determined by potentiometric titration) and formation and dissociation rate constants (temperature-jump method)[108]. The tenfold decrease in the formation constants for the 1:1 Ni^{II} complexes in aqueous solution at 25 °C parallels the increase in the bulk of the substituent, $NH_3 > MeNH_2 > EtNH_2 > Pr^iNH_2 > Me_2NH$; dissociation rate constants are less affected. In the case of Ag^+ complexes in aqueous solution neither ΔG° nor ΔH° is much affected by change in chain length of the amine; the small variations observed closely parallel those found for the corresponding protonation reactions[109]. Some bis-adducts of long chain primary alkyl- and alkenyl-amines with Co^{II} and Ni^{II} acetylacetonates have been prepared[110]. UCl_4 and UBr_4 form 1:2, 1:3 or 1:4 complexes with a variety of amines, both aliphatic and unsaturated heterocyclic[111], stoichiometries are rationalised in terms of the electroneutrality principle. Allylamine is known to coordinate in a bidentate manner to Cu^I and Pt^{II}, using the nitrogen and the π-electrons of the double bond[112]. In Rh^{III} complexes such as RhL_3Cl_3 only the nitrogen is coordinated, no shift in $\nu(C{=}C)$ being observed[113]. However, i.r. evidence was found for the participation of the double bond during reaction with ethanol (in the presence of HCl in some cases) which gave propionaldehyde, acrolein and propylene among the products. Chelate complexes of 2-aminoethanol, as well as of its anion, have been known since 1932. A single-crystal structure determination on $CoNi(OC_2H_4NH_2)_3(HOC_2H_4NH_2)_3I_2$ has now been reported[114]. This novel binuclear species contains two octahedra linked face-to-face by H-bonds, the three equivalent O—O distances being 2.51 Å. The i.r. spectra of methylamines chemisorbed on transition metal salt surfaces have been recorded[115]. Trimethylamine splits the chloro- or bromo-bridges in various dimeric and polymeric Pd^{II} and Pt^{II} complexes to yield *trans*-$M(NMe_3)X_2$ and $[M(NMe_3)X_3]^-$ species; ^1H n.m.r., i.r., Raman and electronic spectra are reported[116]. The molecular structures of 5-coordinate $Ti(NMe_3)_2Br_3$ and $Cr(NMe_3)_2Cl_3$ have been determined and compared with the previously determined structure of $V(NMe_3)_2Cl_3$ and those of corresponding trimethylphosphine complexes[117]. Both molecules are basically trigonal bipyramidal but with some distortion from the equatorial XMX angle of 120 deg, particularly where M = Cr. This is thought to have an electronic, not a steric, origin.

McWhinnie *et al.*[118] have prepared the methylamine complex $[Ru(MeNH_2)_6]^{2+}$ isolated as the I, Br or BF_4 salt. The coordinated amine is much more susceptible to oxidation than is the free base and produces coordinated cyanide on exposure to oxygen under ambient conditions.

The single crystal electronic and e.s.r. spectra of tetra-(6-amino-hexanoic acid)Cu^{II} perchlorate have been examined by two groups of workers[119, 120]. This interesting molecule is 8-coordinate, the Cu ion being coordinated to the oxygens of four carboxylate groups forming four short and four long bonds[121]. The nitrogen atoms lie 3.76 Å from the nearest metal atom and are clearly not coordinated as was thought earlier[122, 123]. The spectra give the relative orbital energies, $d_{xy} > d_{x^2-y^2} > d_{z^2} > d_{xz}, d_{yz}$.

6.3.3 Arylamines

Ni^{II} complexes of the type NiL_nX_2 (L = aniline or substituted aniline; X = halide; $n = 2$, 4 or 6) have 6-coordinate structures, polymeric in the case of $n = 2$[124]. With *o*-substituted anilines only NiL_4X_2 complexes are formed while with p-substituted anilines and X = NCS, only $NiL_6(NCS)_2$. In the latter compounds it is considered that the thiocyanate ions are not coordinated. Spectral measurements on RuL_6^{2+} (L = various substituted anilines) place the ligands between en and H_2O in the spectrochemical series[125]. *Trans*-structures have been assigned to a number of Ru^{III} complexes of formula $[RuL_2X_4]^-$ (L = aniline or derivative; X = Cl, Br, I, $\frac{1}{2}$oxalate). Aniline forms 1:1 (5-coordinate) and 1:2 (6-coordinate) complexes with $TiCl_4$, and 1:2 complexes with $TiBr_4$. *O*-Allylaniline forms a 1:1 adduct in which the allyl group is not coordinated[126]. Electrochemical studies on the Pd—N_4 chelate (19) derived from 1,3-di-(2-aminophenylamino)propanol provide evidence for the series of complexes where z = +2, +1, 0, −1 and −2[121].

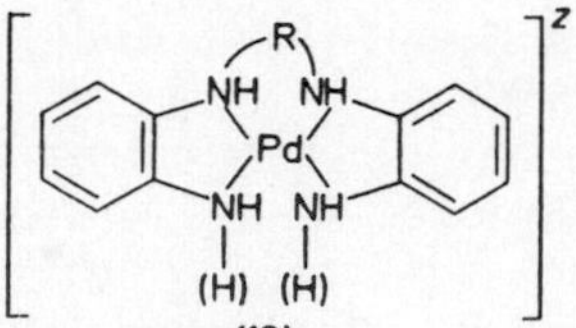

(19)

6.3.4 Saturated heterocyclic amines

Some piperidine complexes of Co^{II}, Ni^{II} [127] and Ir^{III} [128] have been synthesised and their i.r., and in the case of Ir^{III}, the n.m.r. spectra reported. Quinuclidine (and pyridine) forms 1:1 adducts with Cu^{II} hexafluoroacetylacetonates. The ^{14}N hyperfine splitting in the e.s.r. spectrum is inconsistent with axial ligation and suggests a structure (basically trigonal bipyramidal or square pyramidal) in which the nitrogen occupies an equatorial or basal plane position[129]. Morpholine coordinates via the nitrogen atom in its bis-complex with $PdCl_2$[130] and its 1:1 adducts, presumably 5-coordinate, with bis-(*o,o'*-dialkyldithiophosphato) Ni^{II} [131]. The ring nitrogen and oxygen atoms of *N*-2-aminoethyl-substituted piperazine, pyrrolidine and morpholine are probably uncoordinated in the octahedral NiL_2Cl_2 complexes[132].

Quagliano and co-workers[133] have extended their studies on the donor properties of positively charged amines. The mono-protonated ditertiary amine 1,4-diazabicyclo[2.2.2]octane, (20), forms a series of 5-coordinate $[M(LH^+)_2X_3]X$ (M = Mn^{II}, Fe^{II}, Ni^{II} and Cu^{II}; X = Cl, Br) complexes.

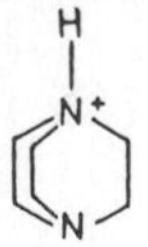

(20) 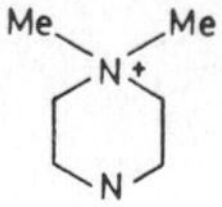 (21)

Electronic spectra of the Ni^{II} complexes indicate a trigonal bipyramidal structure with little distortion from D_{3h} symmetry. It seems likely that the C_3 axis of each protonated ligand is alligned with the C_3 axis of the coordination sphere. The Mn^{II} and Fe^{II} compounds are isostructural. A characteristic of the complexes is their remarkably high thermal stability, a property probably related to strong electrostatic and H-bonded interactions involving the NH^+ group. The related 1,1,4-trimethylpiperazinium cation (21) behaves rather differently, yielding tetrahedral $[M(L^+)X_3]$ complexes ($M = Co^{II}$, Ni^{II}, Cu^{II} and Zn^{II}; X = Cl, Br, I). Here the donor properties of the tertiary nitrogen appear relatively weak since under certain conditions of preparation the salts $(L^+)_2[MX_4]$ may be obtained. Also, in the presence of traces of water protonation may occur to give $(LH^{2+})[MX_4]$[134].

6.3.5 Aliphatic diamines

While the usual mode of coordination of ethylenediamine is chelation, monodentately coordinated (usually in the monoprotonated form) ethylenediamine is now well known[1]. A recent new example[135] is found in $[Cr(en)(enH)H_2OF_2]^{2+}$, an intermediate isolated by ion-exchange chromatography, in the aquation of $[Cr(en)_2F_2]^+$ to *cis*-$[Cr(en)_2(H_2O)F]^{2+}$ and $[Cr(en)(H_2O)_2F_2]^+$. A solid phase displacement of coordinated enH^+ by lattice Cl^- in $[Pd(en)(enH)Cl]Cl_2$ has been inferred from i.r. measurements[136].

A second alternative mode of coordination of ethylenediamine is as a bridging ligand. Nakamoto[137] has summarised much of the information on en complexes of this kind. Both Raman and i.r. spectra have been used to detect the presence of bridging en in $MenX_2$ complexes (M = Zn, Cd, or Hg; X = Cl, Br, or NCS)[138].

A large volume of new information on complexes containing chelated ethylenediamine has appeared; not all can be included here. Much of the work concerns spectroscopic properties. The polarised electronic spectra of tetragonal *trans*-$[Cr(en)_2X_2]ClO_4$ (X = Cl, Br) and *trans*-$[Cr(en)_2(H_2O)_2]Cl \cdot H_2O$ have been measured and the quartet ligand field bands assigned[139]. The i.r. and electronic spectra of the tetragonal CuL_2X_2 series (L = en, pn or an *N*-alkyl-substituted derivative) indicate a correlation between the square of the highest M—N stretching vibration and the energy of the main d–d band. The position of a complex on the line is a measure of its tetragonal distortion and of the strength of the in-plane field; 1,3-tn forms weaker complexes than en[140]. In the case of *N,N*-diethylethylenediamine and X = ClO_4, BF_4 or NO_3, the tetragonal distortion is temperature dependent leading to thermochroism[141]. The mixed ligand complexes of Ni^{II} and Cu^{II} with 3,3′-diaminodipropylamine and en or tn, $[M(dpt)(en)]X_2$ and $[M(dpt)(tn)]X_2$ are 5-coordinate[142]. The α-form of $[Co(en)(dpt)Cl]^{2+}$ has an approximately octahedral structure with the three nitrogen atoms of the triamine in one plane. The orientation about the secondary amine nitrogen is such that the NH proton is adjacent to the chlorine in contrast to the β-isomer in which this proton is remote from the chlorine. In both isomers the fused 6-membered chelate rings adopt a chair–boat conformation[143]. $Cu(tn)_2(SCN)_2$ is an elongated tetragonal octahedron with *S*-bonded thiocyanate. Replacement of one NCS by ClO_4 gives the trigonal bipyramidal $[Cu(tn)_2(NCS)]^+$ ion in which the remaining coordinated NCS is now *N*-bonded[144]. The preparation and some properties of new tn complexes of Cr^{II} [145] and Cr^{III} [146] have been described.

N- and *C*-substitution in diamines may be expected to introduce a steric inhibition to chelation, although this may often be small. From free energy and enthalpy data for equilibria (6.1) and (6.2)

$$Cu^{2+} + nL \rightleftharpoons CuL_n^{2+} \tag{6.1}$$

$$CuL_2^{2+} + X^- \rightleftharpoons CuL_2X^+ \tag{6.2}$$

(n=1 or 2; L=en, *NN'*-Me_2en, *N,N*-Me_2en or *N,N,N'*,-$e_4$en;

X=Cl, Br, I or NCS)

in aqueous and non-aqueous solution it has been concluded that increasing methyl substitution reduces the strength of the M—N bond[147]. *NN*-Me_2en apparently forms weaker complexes than *NN'*-Me_2en probably because in the symmetrical isomer both methyl groups occupy the position with the lower steric hindrance (probably the equatorial position) whereas in the unsymmetrical isomer one methyl group is forced into the more hindered position. Similar conclusions regarding the relative steric effects in *N*-alkylsubstituted diamines have been reached on the basis of differences in the rates of formation of Ni^{II} complexes[148]. However, formation constants for mono- and bis-Cu^{II} and Ni^{II} chelates with *gem*-dialkyl-1,3-diamines are larger than those for the complexes with unsubstituted tn; this is attributed to a more favourable entropy term[149]. Increasing *C*-methyl substitution in en causes the bis-complexes of Ni^{II} (diamagnetic) to be relatively the more stable *vis-a-vis* the tris-complexes (paramagnetic)[150]. *N,N'*-Dibenzylethylenediamine forms both 5- and 6-coordinate CuL_2X_2 or CuL_2XY complexes depending on the nature of X and Y; $[CuL_2Br]^+$ exists in both trigonal-bipyramidal and square-pyramidal forms[151]. Mixed Ni^{II} chelates of *N N N'N'*-Me_4en and some β-diketones may be square planar or 6-coordinate (paramagnetic), the latter showing configurational equilibria in solution[152]. 2,3-Diaminobutane exists in both *meso* and optically active forms[153]. The complexed *meso*-isomer will have one methyl group in an axial environment and one in an equatorial environment while the optically-active form has both groups in similar environments. Mono-, bis- and tris-complexes of both forms with Ni^{II} have been prepared and the rate of racemisation of $[Ni(meso\text{-}L)_3]^{2+}$ measured. On heating the solids or ethanolic solutions of $CuLCl_2$ or $CuLBr_2$ (L = *NNN'N'*-Et_4en) the complexes Cu(*NNN'*-Et_3en)X_2 are produced with evolution of acetaldehyde[154].

It is well known that coordination may enhance the acidity of a ligand which bears a proton on the donor atom. $[Pten_2]^{2+}$ is not deprotonated by NH_3 but when one en molecule is replaced by a π-bonding ligand such as 2,2'-dipyridyl or 1,10-phenanthroline the acidity of the remaining molecule of coordinated en is considerably increased. This is attributed to a *trans*-effect in which the NH bond is labilised. The effect has been studied by chemical reactions and by ^1H n.m.r. spectroscopy of solutions and by thermogravimetric analysis of solids[155]. Low-spin $Na_2[Fe(CN)_4L]$ complexes, where L = en, NN'-Me_2en, or *o*-phenylenediamine) have been synthesised by the addition of four equivalents of CN^- to high-spin tris-(diamine)Fe^{II} complexes[156]. They are reactive to oxidising agents (O_2, H_2O_2, Cl_2) yielding α-diimine ligands. The formation of a Fe^{III} intermediate in the oxidative dehydrogenation was demonstrated by the isolation of $[Fe^{III}(CN)_4en]^-$ salts under acidic conditions. The i.r. and electronic spectra of octahedral Ni^{II} halide complexes of NN'-Me_2en have been compared with those of corresponding complexes of 2,5-dithiahexane and 2-(ethylthio)ethylamine; Ni—S, Ni—N and Ni—Cl band assignments were made[157].

$$Et_2N \cdot CH_2 \cdot CH_2 \cdot NH \cdot CH_2 \cdot CH_2 \cdot \underset{\overset{\|}{O}}{P}Ph_2$$

(22)

The ligand (22) acts as a tridentate chelating agent in the distorted trigonal bipyramidal $CoL(NCS)_2$ complexes[158].

6.3.6 Linear aliphatic polyamines

Tridentate diethylenetriamine (dien) has sufficient flexibility to coordinate in either a planar or vicinal (bent) manner. In $[Ni(dien)_2]Cl_2 \cdot H_2O$ the two ligand molecules coordinate equatorially with the secondary nitrogen atoms slightly closer (Ni—N = 2.06 Å) to the metal than the primary nitrogen atoms (Ni—N = 2.13 Å and 2.18 Å)[159]. The polarised single-crystal spectra can be interpreted in terms of D_{2h} symmetry though 'octahedral' splittings are very small[160]. *Dq* is given as 1176 cm^{-1} and *B* as 784 cm^{-1}. Both planar and vicinal arrangements are suggested for some tetragonally-distorted mono-dien Cu^{II} complexes[160]. The potentially tetradentate *N*,*N*-bis(2-diethylaminoethyl)-2-methylthioethylamine (23) does not use the sulphur atom in its distorted trigonal bipyramidal complexes with $Co(NCS)_2$ and $Ni(NCS)_2$.

$$(Et_2N \cdot CH_2 \cdot CH_2)_2N(CH_2 \cdot CH_2 \cdot SMe) \cdot CH_2 \cdot CH_2 \cdot NEt_2$$

(23)

The ligand occupies two equatorial and one (the tertiary nitrogen) apical position. When the thiomethyl group is replaced by a diphenylarsino group the ligand remains tridentate[162]. Rates of hydrogen exchange and racemisation at the secondary NH in *trans*-$[Co(dien)_2]^{3+}$ are similar to those found for $[Co(NH_3)_4(N\text{-Meen})]^{3+}$ indicating that coupling of chelate rings across the secondary NH centre in the diene molecule has little effect on either process[163]. Dien chelates of Ln^{3+} ions are of two types, $[M(dien)_3](NO_3)_3$ and $[M(dien)_2(NO_3)_2](NO_3)$. In the former the nitrate groups are ionic and a coordination number of 9 is inferred for the metal. In the latter complexes both ionic and coordinated nitrate groups occur. Since the denticity of the coordinated nitrate could not be established the coordination number of the metal in these cases is uncertain[164].

3,3′-Diamino-dipropylamine (dpt) having two 6-membered chelate rings in its tridentate complexes has somewhat more flexibility than dien. The planar arrangement seems the preferred one although the conformation of the chelate rings may vary, both chair–chair and chair–(distorted)boat being known. The latter has been found recently in catena-μ-acetato-di(3,3′di-aminodipropylamine)Cu^{II} perchlorate[165]. The complex $Cu(Et_4dien)N_3Br$ is a distorted trigonal-bipyramid with bromine in the equatorial plane and the azide group in an apical position[166].

The linear polyamine 1,8-diamino-3,6-diazaoctane (trien or 2,2,2-tet) has three consecutively linked 5-membered chelate rings when complexed as a tetradentate ligand. Like dien, linear tetradentate ligands of this class are facultative in that they can assume a planar arrangement (all four donor atoms in the same plane, or nearly so) or a non-planar arrangement. The relative stabilities of the geometrical isomers of octahedral Co^{III} complexes of linear tetramines is a subject of much interest. The α-*cis* and β-*cis* forms apparently involve less ring strain than the *trans*-form in the case of 2,2,2-tet, whereas with 2,3,2-tet the *trans*-isomer is the more stable[167]. In *cis*-complexes of 2,3,2-tet the β-configuration is preferred over the α-configuration[167, 168]. In the case of 3,2,3-tet experiment and molecular models both suggest that there is less steric strain in the α-*cis* than in the β-*cis* form, though the *trans*-configuration is preferred in $[Co(3,2,3\text{-tet})X_2]^{n+}$, where X = unidentate ligand[167], as in $[Co(3,2,3\text{-tet})(NO_2)_2]Br$[169]. The solution electronic spectrum of *trans*-$[Co(3,2,3\text{-tet})Cl_2]^+$ has been resolved into the four bands expected for pseudo-D_{4h} symmetry[170]. Analysis of the spectrum indicates that 10 Dq for the N_4 equatorial field is *ca.* 1500 cm^{-1} less than in *trans*-$[Coen_2Cl_2]^+$, due, presumably, to a reduced Co-amine interaction.

The heats of formation in solution of Ni^{II}, Cu^{II} and Zn^{II} complexes of 2,3,2-tet and 3,3,3-tet have been determined and compared with previously determined functions for other polyamines[171]. Evidence for two different kinds of steric constraint is adduced. Firstly, there is the strain present in an individual chelate ring and, secondly, there is the strain arising from the linking of consecutive rings. The former is greater in 6-membered rings while cumulative ring strain is greater in 5-membered ring systems. Standard thermodynamic functions for equilibrium (6.3) in methanol at 25 °C have been measured. $\Delta H°$ values are all positive

$$CuL^{2+} + X^- \rightleftharpoons CuLX^+ \tag{6.3}$$

(L = en_2, tn_2, 2,2,2-tet, 2,3,2-tet, 3,3,3-tet) (X = I, NCS, N_3)

or only slightly negative. For a given anion X, $\Delta H°$ and $\Delta S°$ decrease according to the order

$$(en)_2 > 2,2,2\text{-tet} \simeq 2,3,2\text{-tet} > (tn)_2 > 3,3,3\text{-tet}$$

The series of complexes $[M(2,2,2\text{-tet})(en)]Y_2$ (M = Ni^{II} or Cu^{II} and Y = various univalent anions), are octahedral in both solid and solution[172].

The fully *N*-methyl-substituted ligand Me_6-2,2,2-tet, despite its greater bulk, appears to display coordination behaviour essentially the same as 2,2,2-tet itself in the complexes (M(Me_6-2,2,2-tet)X_2 where M = Ni^{II} or Cu^{II}. I.R. and electronic spectra, magnetic moments and electrical conductances indicate octahedral structures for the Ni^{II} compounds with X = Cl, Br, NO_3 and NCS, and 5-coordinate structures for the Cu^{II} compounds with X = Br, NCS and $\frac{1}{2}SO_4$. Where X = ClO_4 or BPh_4 the complexes of both ligands are square planar[172].

^{14}N N.M.R. and electronic spectra have shown that the predominant species in MeCN solutions of Co^{II} salts and 2,2,2-tet are $[Co(MeCN)_6]^{2+}$, $[Co(2,2,2\text{-tet})(MeCN)_n]^{2+}$ ($n = 1$ or 2), $[Co(2,2,2\text{-tet})_3]^{2+}$ and $[Co(2,2,2\text{-tet})_2]^{2+}$. The lability of coordinated MeCN is increased by a factor greater than 1000 when 2,2,2-tet is also coordinated[173]. The new tetradentate ligand (24) having four optically active centres of different relative configuration is strongly stereoselective for the *trans*-isomer in $[CoLCl_2]^+$. The other isomer observed, *β-cis*, readily isomerises to the *trans*-geometry[174].

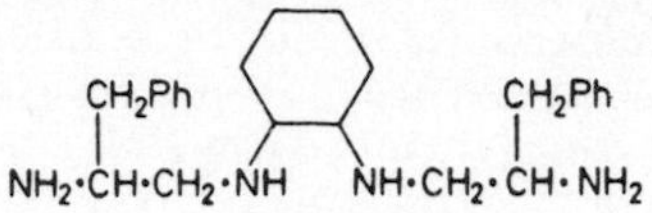

CH₂Ph CH₂Ph
NH₂·CH·CH₂·NH NH·CH₂·CH·NH₂

(24)

Square pyramidal cations occur in the complexes MLXY where L is the ligand (25) and M = Co^{II} (X = Br or I), Ni^{II} (X = Cl or Br and probably NCS) and Cu^{II} (X = Cl, Br, I and NCS), and Y = ClO_4 or PF_6. In addition a 5-coordinate species $CoLCl_2$ in which the ligand is tridentate (free NH_2 group) has also been characterised.

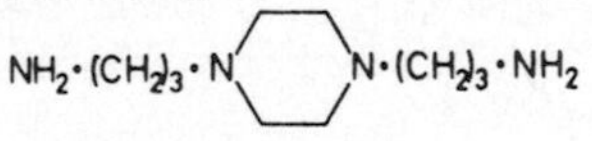

$NH_2\cdot(CH_2)_3\cdot N$ $N\cdot(CH_2)_3\cdot NH_2$

(25)

Ligand (25) is non-facultative when tetradentate and must coordinate in an essentially planar manner. Indeed planar $[MN_4]^{2+}$ species have been identified for M = Ni^{II} and Cu^{II} and probably Co^{II}[175]. Cation exchange chromatography of partially aged solutions in the aquation of *α-cis* $[Cr(2,2,2\text{-tet})Cl_2]^+$ to $[Cr(H_2O)_6]^{3+}$ has permitted the isolation of intermediates in which the ligand has a denticity of three, two or one. The partially 'unwrapped' tetramines are stabilised by proton uptake[176].

The crystal structures of a pair of *αβ*-chloro(tetraethylenepentamine)-Co^{III} cations have been determined and the conformational behaviour of this linear polyamine discussed[177]. The mixed nitrogen–sulphur ligand (26) has been synthesised[178].

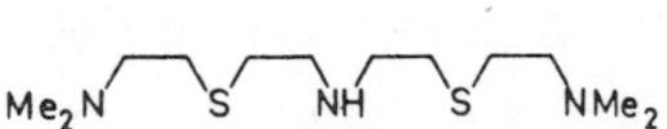

Me_2N S NH S NMe_2

(26)

6.3.7 Branched polyamines

Rates of the ligand exchange of dien, trien and tetren with Cu(tren) complexes have been measured as a function of pH using the stopped flow technique[179] (tren = tris(2-aminoethyl)amine). The results suggest that two unprotonated nitrogen atoms on the incoming ligand are sufficient for displacement of tren and that one nitrogen on the incoming ligand is bonded prior to the rate determining step. Two geometrical isomers (α and β) of $[Co(tren)(NH_3)Cl]Cl_2$ have been isolated after careful decomposition of the dinuclear complex $[(NH_3)(tren)Co-O-O-Co(tren)(NH_3)]Cl_4$ in concentrated aqueous NH_4Cl[180]. Both mono- and bis-chelates of tren with M^{III} lanthanides have been prepared in MeCN solution[181]. I.R. spectra indicate coordination of all NO_3 groups (probably bidentately) in the non-electrolyte complexes $M(tren)(NO_3)_3$. It is inferred from calorimetric measurements that all four nitrogen atoms of tren are engaged in bonding; thus, a coordination number of 10 is indicated. Some of the bis-complexes (those with the larger ions, La^{3+}–Nd^{3+}) have one of the NO_3 groups coordinated while in the complexes of the smaller ions, Sm^{3+}–Yb^{3+}, none are coordinated. Conductance values for the bis-chelates are low and suggest that the thermodynamically favoured species in solution is the mono-chelated complex.

Both $[Zn(tren)Cl]BPh_4$[182] and $Ni(Me_6tren)(NCS)_2H_2O$[183] have a basically trigonal-bipyramidal geometry. There is considerable distortion in the case of the complex of the substituted ligand. The polarised single-crystal electronic spectrum of this complex has been reported[184].

Introduction of one more methylene group into each arm of tren to give tris-(3-aminopropyl)amine (trpn) increases the facultative character allowing it to achieve more nearly tetrahedral coordination. Thus, the trigonal bipyramidal cation is distorted towards tetrahedral geometry (27), the axial Co—Br distance being unusually long (2.658 Å) as compared with the corresponding bond length in $[Co(Me_6tren)Br]Br$ (2.431 Å)[184]. Further, the N_{ax}—Co—N_{eq} angle is 91.3 deg compared with 81.1 deg in the Me_6tren compound. The same conclusions have been reached

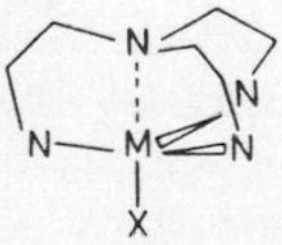

(27)

also by Dei and Morassi[185] on the basis of electronic spectra for the series of complexes $[M(Me_6trpn)X]BPh_4$ ($M = Co^{II}$ or Ni^{II}, X = Cl, Br or I). The complex $[Co(Me_6trpn)](BPh_4)_2$ has a typically pseudotetrahedral spectrum.

The branched, potentially pentadentate, ligand 4-(2-aminoethyl)-1,4,7,10-tetrazadecane (trenen) uses all five nitrogen atoms in octahedral $[Co(trenen)N_3](NO_3)_2 \cdot H_2O$; the conformations of the chelate rings have been discussed[185].

6.3.8 Heterocyclic aromatic amines

6.3.8.1 Pyridine

Although pyridine and related unsaturated heterocyclic amines are poorer bases (towards H^+) than most aliphatic amines they are excellent coordinating ligands. Having unoccupied $p_{\pi*}$ orbitals of suitable symmetry for overlap with non-bonding transition metal d orbitals the possibility of $d_{\pi} \rightarrow p_{\pi*}$ back coordination exists and no doubt this contributes to the stability of the metal–ligand bond in most cases. However, direct evidence for π-bonding is often difficult to obtain. From a consideration of the M—N bond lengths in $Tipy_3Cl_3$, Collins and Drew[186] found no support for any significant π-contribution, a not surprising result for a +3 oxidation state d^1 system. In pyridine complexes of the later transition metals π-bonding may be expected to be more important and its consequences more readily recognised. Electronic, i.r. and ^1H n.m.r. spectra of the pentammine Ru^{III} complex of 4-formylpyridine show it to be >90% in the hydrated form (28) while the corresponding Ru^{II} complex exists to >90% in the carbonyl form (29).

Ru^{III}—N(C₅H₄)—C(OH)₂H (28) Ru^{II}—N(C₅H₄)—C(=O)H (29)

This is explained[187] in terms of a greater degree of π-bonding in the Ru^{II} complex and also when the ligand is in the carbonyl form. A molecular orbital treatment[188] of the charge transfer spectra of $[Ru(NH_3)_5L]^{2+}$ and *cis*- and *trans*-$[Ru(NH_3)_4L_2]^{2+}$ complexes (L = py, isonicotinamide, methylisonicotinamide, or pyrazine) similarly suggests significant M→L back bonding, increasing in the ligand order given. On the other hand, Tong and Brewer[189] noted no significant change in the position of the Cu—Br charge-transfer band in a series of CuL_2Br_2 complexes as the electron-withdrawing substituent in the 4-position of the pyridine ligand L was varied.

A substantial number of new pyridine complexes have been synthesised and the structures of several new as well as previously known complexes have been elucidated by single-crystal x-ray studies. Among these are several pyridine and α-picoline complexes of dipositive metal nitrates[190]. Here the interest lies mainly in the coordinating characteristics of the nitrate ion and in the occurrence of formally 7-coordinate species in some cases. The bis-chelate of Cu^{II} with ω-acetophenone forms a pseudo-octahedral bis-adduct with γ-picoline and a tetragonal pyramidal mono-adduct with α-picoline. In the former complex the pyridines occupy *trans*-equatorial positions and in the latter the molecule of base occupies one corner of the square[191]. The adduct of 4-aminopyridine with $Cu(acac)_2$ has a similar structure; the amino-group is not coordinated but is H-bonded to an oxygen atom of an adjacent molecule[192]. Cu^{II} chloroacetate forms a purple 1:2 complex with α-picoline which has a centrosymmetric tetragonally-distorted octahedral geometry. The equatorial plane consists of two *trans*-Cu—N bonds and two *trans*-Cu—O bonds (1.975 Å). Two long Cu—O bonds (2.707 Å) are formed above and below this plane[193]. Cu^{II} dichloroacetate forms two 1:2 addition compounds with α-picoline, one purple and the other blue. The former has similar electronic properties to the complex described above and presumably has a similar structure. The blue isomer is not centrosymmetric and in contrast to the monochloroacetate complex the α-methyl substituents lie *cis* to one another with respect to the Cu atom. Two non-equivalent 'long' axial Cu—O distances (2.711 Å and 2.493 Å) occur and the complex could reasonably be considered 5-coordinate[193].

Spectrophotometrically-determined formation constants for the association (6.4) in benzene, acetone or MeCN solution at 22 °C increase

$$Cu(acac) + py \rightleftharpoons Cu(acac)_2py \tag{6.4}$$

with the electron-withdrawing capacity of the substituent at the 3-position of acac[194]. In the case of the 3-NO_2- and 3-CN-compounds bis-pyridine adducts form at high pyridine concentrations. The pyridine adducts of $VO(acac)_2$ may have *cis* or *trans* structures depending on the nature of the pyridine substituent[195]. The different isomers may be recognised by the magnitude of the reduction in $\nu(V{=}O)$, this being 42 ± 4 cm^{-1} for *cis*-structures and 29 ± 4 cm^{-1} for *trans*-structures. No obvious correlation between the geometry of the adduct and the size, position or electronic nature of the pyridine substituent is apparent. The correctness of the i.r.-based structure assignment has been verified by a single-crystal x-ray structure determination for the adduct with 4-phenylpyridine. The electronic spectra of the complexes $[VL_4](NCS)_3$ (L = py, β-pic, γ-pic, 3,4-lutidine or 3,5-lutidine) are consistent with a tetrahedral geometry. The three predicted spin-allowed bands were observed giving values of Dq and β of 950–970 cm^{-1} and 0.94–0.98, respectively[196]. Reduction of $Cu(ClO_4)_2 \cdot 6H_2O$ in pyridine (or γ-picoline or quinoline) with copper produces $[Cu(py)_4]ClO_4$. The coordination about the metal is almost exactly tetrahedral[197]. No Cu—N stretching vibration could be located in the far i.r. region. Tetra-coordination of Cu^I by several heterocyclic and other monofunctional amines in propylene carbonate has also been recognised using polarographic techniques[198]. With α-picoline the trigonally hybridised Cu^I complex $[Cu(\alpha\text{-pic})_3]ClO_4$ has been prepared[199]. The crystal structure of CuL_2Cl_2 (L = 4-ethylpyridine) has been compared with that of the corresponding 4-vinylpyridine complex which undergoes solid-state polymerisation; the packing of the ethyl and vinyl groups in the two structures is quite different[200]. *Cis*-$Ru(py)_4X_2$ (X = Cl or Br) complexes have recently been synthesised for the first time by displacement of oxalate by X^- in aqueous solution[201]. The binuclear oxalate-bridged $[(py)_4Ru{-}C_2O_4{-}Ru(py)_4](BF_4)_2$ has also been prepared and its crystal and molecular structure determined[202].

On the basis of the temperature dependence of the quadrupole splitting of the Mössbauer spectra of certain tetrakispyridine complexes of Fe^{II} approximate splittings of the d_{xy}, d_{xz} and d_{yz} orbitals have been calculated. From this information it was proposed[203] that $Fe(py)_4Cl_2$ has a *cis*-configuration whereas $Fe(py)_4I_2$ and $Fe(py)_4(NCS)_2$ are *trans*. However, a limited crystallographic study[204] has shown that $Fe(py)_4Cl_2$ has tetragonal symmetry. A comparison of the lattice constants with those of $Ni(py)_4Cl_2$ and $Co(py)_4Cl_2$, known to have *trans*-configurations, strongly indicates a *trans*-structure for $Fe(py)_4Cl_2$ also. This assignment is supported by the observed temperature dependences of the Mössbauer spectra of $Fe(py)_4X_2$ and $Fe(iq)_4X_2$ complexes which agree with the theoretically predicted dependence for D_{4h} symmetry[205]. Two different preparations of $Fe(py)_2Cl_2$ show different magnetic properties and electronic spectra at low temperatures while having identical x-ray powder patterns at room temperature, these also being identical with that of the modification of $Co(py)_2Cl_2$ known to have symmetrical Co—Cl—Co bridges. Mössbauer spectra show that the crystalline form of $Fe(py)_2Cl_2$ (prepared by a rapid precipitation method) undergoes a transformation at low temperature to a different form thought to have asymmetrical Fe—Cl—Fe bridges[206].

The polymeric distorted octahedral structures of $Cu(py)_2Cl_2$ and $Cu(py)_2Br_2$ are well known. The corresponding bromo-complex of the sterically hindered α-picoline has a tetragonal pyramidal dimeric structure with the metal atoms linked by bromide bridges[207, 208]. The Cu—Cu separation is 4.926 Å. The magnetic properties (4.2–300 K) of both the bromide and the chloride obey the van Vleck equation for exchange coupled Cu^{2+} ions and yield singlet–triplet separations of 7.4 cm^{-1} (Br) and 5 cm^{-1} (Cl). The tetrahedral$\rightleftharpoons$octahedral configuration equilibria in solutions of $Co(py)_nX_2$ ($n = 2$ or 4) complexes in organic solvents have been re-investigated by ^1H n.m.r. techniques[210]; good agreement between equilibrium constants determined by the n.m.r. and spectrophotometric methods was observed. In a kinetic study of the same system evidence was found for a 5-coordinate $Co(py)_3Br_2$ intermediate[211]. Electronic, i.r. and e.s.r. spectra of some complexes of formula $ML_4X_2 \cdot 2H_2O$ ($M = Mn^{II}$ or Co^{II}; X = β-pic or 3-ethylpyridine; X = Br or I) are consistent with the salt structure $[ML_4(H_2O)_2]X_2$. Pyrolysis of PtL_4I_2 (L = py, β-pic, γ-pic, 4-ethylpyridine or 3,4-lutidine) is a convenient route to the otherwise difficulty obtainable PtL_2I_2 complexes[213]. The low-spin *trans*-bis(pyridine)diphenylglyoximato – Fe^{II} complex readily undergoes reversible substitution of one or both axial pyridines by CO. The mechanism is dissociative. Similar behaviour has been observed for L = imidazole, nicotinamide and PPh_3[214].

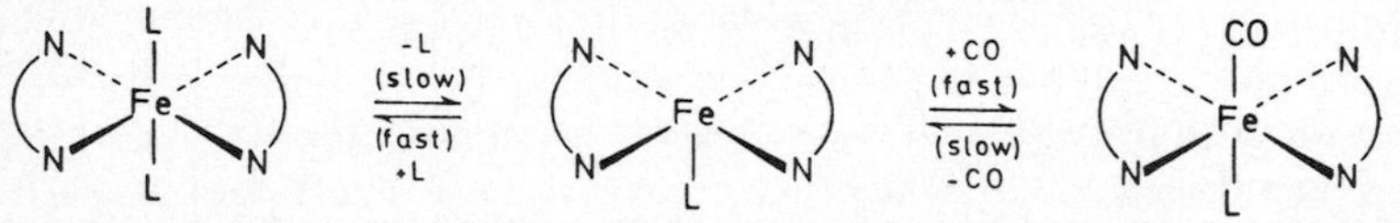

Complexes of 2-pyridyldiphenylphosphine may use either the phosphorus or the nitrogen atom in coordination. When the metal is Pd^{II} or Pt^{II}, phosphorus is the preferred donor atom though in the case of Co^{II} complexes there is some evidence for a Co—N bond[215]. From i.r. spectra it has been deduced that the donor atoms in 6-methyl-2-thioamidopyridine are the sulphur atom and the pyridine nitrogen in $[CoL_2(H_2O)_2]ClO_4$, NiL_2Br_2, $[CuL_2]I$, $[Ru_2L_4Cl_2O]Cl_2$, OsL_2Cl_2, $[AuL][AuCl_2]$ and $HgLI_2$ whereas in $[FeLCl_3]_2$ and $[ZnL_2](ClO_4)_2$ chelation is through both nitrogen atoms[216]. It has been reported[217] that 2-cyanopyridine complexes of Co^{II}, Ni^{II} and Cu^{II} chlorides are subject to nucleophilic attack by, e.g. OR^- in alcohol solution as a result of activation of the CN group by chelation producing chelates of *O*-alkylpyridine-2-carboximidate (30)[217].

6.3.8.2 2,2′-Dipyridyl and related ligands

The *cis*-configuration of the complex cation $[Cu(dip)_2(H_2O)_2]^{2+}$ in aqueous solution is usually attributed to a steric interaction between the 6,6′ hydrogen atoms which would exist in the *trans*-arrangement. Sigel[218], however, believes the reason may be electronic, namely, that the *cis*-configuration may allow a stronger charge transfer between metal and ligand. Noack and Gordon[219] who demonstrated that the *cis*-form is the dominant species in aqueous solution at room temperature also concluded, from e.s.r. spectra in ethanol–water glasses at 77 K, that the *cis*- and *trans*-forms exist in comparable concentrations under these concentrations. From new e.s.r. evidence[220], however, it appears that the predominant species in the low temperature glasses are the disproportionation products $[Cu(dip)(H_2O)_4]^{2+}$ and $[Cu(dip)_3]^{2+}$. From an analysis of the various kinds of distortion which occur in square planar and 6-coordinate bisdipyridyl and bis-(1,10-phenanthroline) ML_2 and *trans*-ML_2X_2 complexes it has been argued that the most likely candidates for formation of such structures are the third row transition metals and others in low oxidation states[221]. New definitive information has been brought to bear on this problem through the determination of the crystal structure of $[Pd(dip)_2](NO_3)_2 \cdot H_2O$[222]. The PdN_4 skeleton is distorted in two ways from a regular square plane. Firstly, the 'bite' of the ligand is such as to impose a rectangular distortion reducing the chelate ring NPdN angle to *ca.* 80 deg and raising the non-chelate ring NPdN angle to *ca.* 102 °C. Secondly, a tetrahedral distortion (twist) occurs which relieves the interligand proton–proton interactions. A similar tetrahedral distortion occurs in $[Cu(dip)_2ClO_4]ClO_4$ which consists of infinite chains of cations linked by ClO_4 bridges[223]. In this complex the two pyridine rings of each dip molecule are twisted slightly towards each other. In $[Cu(dip)_3](ClO_4)_2$ the axial Cu—N bonds are of unequal length[224]. The crystal and molecular structures of *cis*-$[Co(dip)_2Cl_2]CoCl_4$[225] and *cis*-bis(2-methoxyphenyl)bis(dipyridyl)Cr^{III} iodide monohydrate[226] have been reported. The metal ion is 10-coordinate in $La(dip)_2(NO_3)_3$[227]. The preparation of *cis*-$[Rh(dip)_2Cl_2]^+$, like that of *trans*-$[Rh(py)_4Cl_2]^+$, from $RhCl_3$ in aqueous HCl in the presence of two-electron reducing agents is catalytic[228].

The presence of a substituent *ortho* to the nitrogen atoms in dip or phen may be expected to increase interligand steric interactions greatly and to cause reductions in normal coordination numbers. Five-coordination has been suggested, but not proved, for Rh^{III} and Ir^{III} in MLX_3 complexes where L = 2,9-dimethyl-1,10-phenanthroline[229]. The 2-methyl derivative, having only one *ortho* substituent, should have an intermediate steric effect. In accord with this expectation it is found that while tris-$[FeL_3]^{2+}$ complexes can be isolated they are unstable in aqueous solution. They also have high-spin ground states at room temperature[230]. Mössbauer effect experiments show, however, that the ClO_4 and BF_4 salts exist in a $^5T_2 \rightleftharpoons {}^1A_1$ spin equilibrium between 230 and 120 K[231]. The bis-complexes, FeL_2X_2, have magnetic properties and Mössbauer spectra which depend on the nature of X. Thus, when X = Cl, Br, NCS, N_3 or $\frac{1}{2}$malonate, the complexes are high-spin, while $FeL_2(CN)_2$ has a $^1A_1$ ground state at room-temperature (μ_{eff} = 1.02 B.M.)[232].

Reference has already been made to the π-acceptor potentiality of pyridine. Back donation in dip and phen complexes is more extensive. Recent i.r. studies (4000–100 cm^{-1}) by Nakamoto and co-workers[233] on tris-dipyridyl complexes (phen complexes also in some cases) of the metal ions or atoms, Cr^{III}, Cr^{II}, Cr^{I} and Cr^{0}; Fe^{III} and Fe^{II}; V^{II} and V^{0}; Ti^{0} and Ti^{-1}; Co^{III}, Co^{II}, Co^{I} and Co^{0}; Mn^{II}, Mn^{0} and Mn^{-1}, have shown that as the oxidation state of a given metal is varied there is little accompanying change in the frequency of the M—N stretching vibrations provided the series does not include a change from a d electron configuration where all the electrons are in t_{2g} orbitals to one where the e_g set becomes partly occupied. This result suggests that as the oxidation state of a metal is lowered an increasing fraction of the electron density resides on the ligand thereby maintaining a fairly constant electron density on the metal. In other words, the π-bond becomes stronger as the σ-bond weakens. The 41 000–7000 cm^{-1} electronic spectra of various mono-, bis- and tris-dipyridyl complexes of Fe, Ru and Os in the +2 and +3 oxidation states have been interpreted in terms of metal–oxidation and metal–reduction $\pi(dip) \rightarrow t_2(M)$ and $\pi(halide) \rightarrow t_2(M)$ charge-transfer transitions, depending on the oxidation state of M and the nature of the other ligands[234].

A number of interesting photochemical and electrochemical studies have been reported. The binuclear oxo-bridged complex $[Fe_2(phen)_4(H_2O)_2O]^{4+}$ is photochemically reduced to $[Fe(phen)_3]^{2+}$ in aqueous solution[235]. The process apparently continues after irradiation has stopped. It is suggested from kinetic evidence that a thermally labile peroxo-bridged binuclear complex may be produced initially which then undergoes thermal reduction and rearrangement to product. Photoreduction of $[Fe(phen)_3]^{3+}$ in acid solution has also been observed[236]. Irradiation of the mono-complex of 2,9-dimethylphen with Cu^{II} in methanol at 26 000–40 000 cm^{-1} gave the corresponding Cu^{I} complex[237]. The primary photoprocess is possibly a homolytic cleavage of a Cu^{II}–methanol coordination bond. Both Ni^{+1} and Ni^{-1} species appear to be formed in the electrochemical reduction of $[Ni(dip)_3]^{2+}$ [238].

A novel antiferromagnetic oxo-bridged Mn complex containing the metal in two oxidation states (III, IV) has been described[239]. The solid of composition $[Fe(dip)_2(NO_3)_2]ClO_4$ isolated from concentrated solutions of HNO_3 rapidly generates $[Fe(dip)_3]^{2+}$ in water in the presence of light but appears stable in solutions of oxidising agents[240]. Stability constants and calorimetric enthalpy changes for the addition of dip (and certain other bases) to some bis-β-diketonates of Cu^{II} and Zn^{II} [241] and tris-β-diketonates of several tripositive lanthanide metals[242] are reported. For the dipositive metals the data suggest that the lower the stability of the diketonate the greater the stability of the adduct. For the lanthanide complexes there is some evidence that the addition of a second molecule of dip may cause the dip molecule already present to become monodentate.

Reaction of $R_2Ni(dip)$ (R = alkyl) with various olefins at room temperature leads to cleavage of the Ni–alkyl bond and formation of π-olefin complexes of the type $(olefin)_nNi(dip)$ ($n = 1$ or 2)[243]. In the case where the olefin is acrylonitrile or acrolein the intermediates $R_2Ni(dip)(olefin)$ were isolated. Spectrophotometrically-determined stability constants indicate a linear relation between the stability of the olefin complex and its rate of formation.

Low metal–nitrogen stretching frequencies in complexes of 2,7-dimethyl-1,8-naphthyridine are attributed to strain in the 4-membered chelate ring. Assignments for the tris-ligand perchlorate salts of Mn^{II}, Co^{II}, Ni^{II} and Zn^{II} were made with the aid of the metal isotope technique[244]. Table 6.1 compares the values obtained with values for the corresponding tris-dipyridyl complexes.

Table 6.1 M—N stretching frequencies for ML_3^{2+} complexes of 2,7-dimethyl-1,8-naphthyridine and dipyridyl

	Mn	Co	Ni	Zn
dip	240, 182	266, 228	277.5, 260.0	235, 186
naphth	202, 152	214, 183	217.4, 197.7	193, 192

Up to six unsubstituted 1,8-naphthyridines can coordinate to certain M^{III} lanthanide perchlorates. I.R. and conductance measurements have shown that the ClO_4 groups are ionic while i.r. and ^1H n.m.r. spectra indicate that the ligands are bidentate[245]. Thus, $[ML_6](ClO_4)_3$ (La–Pr) and $[ML_5](ClO_4)_3$ (Nd–Eu) are 12- and 10-coordinate respectively. In the case of the substituted 2,7-dimethyl compounds, 10-coordinate $[ML_2(NO_3)_3]$ (La–Yb) are obtained[246]. The previously reported[247] Cu^{II} complex of this ligand was probably a Cu^{I} compound[248].

6.3.8.3 *Pyrazines*

Pyrazine having nitrogen atoms in the 1,4-positions of a planar 6-membered ring is well suited to a bridging function between adjacent metal atoms. Co^{II} and Ni^{II} complexes of the type ML_2X_2 (X = halide) are now known[249] to contain 6-coordinate metal ions linked by pyrazine bridges and not halide bridges as first postulated[250]. Similar bridged structures occur in some Cu^{II}, and Ru complexes. In the Cu^{II} compounds spin–spin coupling is transmitted via the bridging ligand ($J = -60$ cm^{-1}, $g = 2.22$)[251]. In the Ru complexes the pair of metal atoms may occur as (Ru^{II}, Ru^{II}), (Ru^{II}, Ru^{III}) or (Ru^{III}, Ru^{III})[252]. The Ag^{II} complex, $Ag(pz)_2S_2O_8$ also possesses magnetic properties (80–300 K) indicative of antiferromagnetic exchange interaction[253]. Since this compound also contains bridging pyrazine it seems probable that the exchange is transmitted by the organic ligand here also. From redox measurements on the systems

$$Ru(NH_3)_5L^{3+} + e \rightleftharpoons Ru(NH_3)_5L^{2+}$$

and

$$Ru(NH_3)_4L_2^{3+} + e \rightleftharpoons Ru(NH_3)_4L_2^{2+}$$

pyrazine has been shown to stabilise the lower oxidation state, presumably because of back coordination of d_π electrons. Reduction of the coordinated pyrazine has also been observed[254]. In these complexes pyrazine is, of course, acting as a monodentate ligand. Pyrazine (and quinoxaline) is also unidentate in MoL_2Cl_4 while 4,4′-dipyridyl appears to be, respectively, unidentate and bridging in MoL_2Cl_2 and $MoLCl_4$[255]. Some complexes of the isomeric pyridazine (also of phthalazine and 3,4-benzocinnoline) with several first row dipositive metal ions and Zn^{2+} and Cd^{2+} have been reported but few conclusions about structure could be drawn.

6.3.8.4 *Pyrazoles, imidazoles, thiazoles and tetrazoles*

Of the two isomeric 5-membered heterocycles containing two nitrogen atoms, pyrazole (31) has received less attention than imidazole (32)

(31) (32)

probably because of the biological importance of the latter. Earlier work on these two ligands and their derivatives is summarised in Series 1[1]. Some bis-, tetrakis-, but not hexakis-, Cu^{II} complexes of pyrazole and of its 3-methyl and 3,5-dimethyl derivatives have been prepared[257]. It has been suggested that due to the acidic nature of the imino-hydrogen in pyrazole the methyl group in 3(5)-methylpyrazole may be located at either the 3- or 5-positions. Consideration of accumulated data for various pyrazole complexes suggests that the 5-position is the favoured one in coordination compounds[258]. This is certainly true for $Mn(Me\text{-}pyrazole)_4Br_2$ which has approximate D_{4h} symmetry. Distortion parameters have been obtained from e.s.r. measurements for this complex and that of the unsubstituted ligand[259]. The e.s.r. spectra of several Fe^{III} complexes of some heterocyclic amines including pyrazole have been discussed[260]. The effect of coordination to Ni^{II} on the kinetics of iodination of pyrazole has been studied[261].

Imidazole may coordinate as the neutral ligand via the tertiary nitrogen atom or as the anion. Like pyrazole, and in contrast to pyridine, imidazole readily forms hexakis-ML_6^{n+} complex ions. Further detailed studies on Co^{II} complexes of the neutral and anionic ligand have been reported[262]. Much of the recent literature concerns complexes of substituted imidazoles. *N*-Substitution does not seriously affect the steric demands of the ligand and hexakis-(*N*-n-butylimidazole) complexes of Mn^{2+}, Fe^{2+}, Co^{2+}, Ni^{2+} and Cd^{2+} have been prepared[263, 264]. In the case of substitution at the 2-position, however, not more than four imidazole molecules are found to coordinate. The compounds obtained are mostly of the type ML_4X_2 and ML_2X_2. In the case of $M = Ni^{II}$ and $X = Cl$ or Br the products have a square pyramidal structure, $[NiL_4X]X$. When L = 1,2-dimethylimidazole, the complexes NiL_4X_2 (X = halide) are square planar while the complexes CoL_4I_2, $CoL_4(NO_3)_2$ and $ZnL_4(NO_3)_2H_2O$ are tetrahedral, as are also ML_2X_2 ($M = Co^{II}$, Ni^{II}, Zn^{II}; X = Cl, Br, I, NO_3)[265]. The complex CuL_3Cl_2 is a slightly distorted trigonal bipyramid in which the three nitrogen atoms occupy the two axial positions and one equatorial position; $Cu—N_{eq}$ (2.145 Å) is appreciably longer than $Cu—N_{ax}$ (2.005 Å)[266]. Only bis-complexes were obtained on reaction of 2-methylbenzimidazole[267] with Co^{II}, Ni^{II}, Cu^{II} or Zn^{II} salts, or on reaction of 1-benzyl-2-phenylbenzimidazole[268] with Ni^{II}; a variety of coordination geometries were found. In *trans*-$[Pt(imidazole)_2(NH_3)_2]Cl_2 \cdot 2H_2O$ the heterocyclic ligands are coplanar with the coordination square plane. Both the *trans*- and *cis*-isomers undergo a relatively easy solid-state photochemical reaction in which the NH_3 groups are replaced by Cl^- ions in the coordination sphere[269]. Two differing Cu–N distances occur in tetrakis(imidazole)bis(methoxy-acetato)Cu^{II} [270]. One pair of ligand donor atoms is at 1.913 Å and the other at 2.045 Å from the metal. It is suggested that in the case of the more closely bonded ligands the amino proton is displaced and associated equally with the nitrogen, the methoxy oxygen and the un-coordinated oxygen of the unidentate acetato-group. The centro-symmetric complex $Fe(dmg)_2L_2$ (dmg = dimethylglyoxime anion) is a distorted FeN_6 octahedron[271].

The histidine molecule is potentially tridentate having as possible bonding sites the imidazole nitrogen, the amino nitrogen and the carboxyl oxygen atoms. In the hope of establishing the sequence of binding sites in the Ni^{II}–histidine system the kinetics of reaction of Ni^{II} with histidine, histidine-methylester and imidazole were studied[272]. The results suggest that the imidazole nitrogen complexes first. There is no evidence that the exocyclic amino group is involved in coordination in Co^{II} and Ni^{II} complexes of 2-aminobenzimidazole[273]. A somewhat related situation occurs in complexes of the virus inhibitor 2-α-hydroxybenzylbenzimidazole (33) which sometimes coordinates as a unidentate ligand using only the tertiary nitrogen but more often as a chelate in which the hydroxyl oxygen is also bonded (to Co^{II} and Ni^{II})[274].

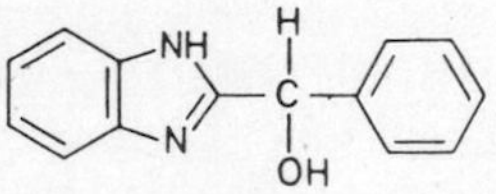

(33)

Thiazole (34) is known to coordinate through the nitrogen atom in its complexes with Co^{II} [275]. Isothiazole (35) apparently behaves similarly with Co^{II}, Ni^{II}, Cu^{II} and Pt^{II}[276]. The ligands 2-acetamidothiazole (36) and 2-acetamidobenzothiazole reveal a varied coordination behaviour[277]. They

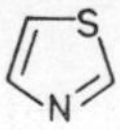

(34)  (35) 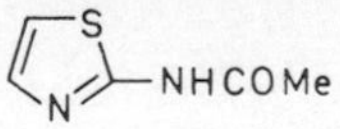 (36)

may form 6-membered chelate rings by use of the ring nitrogen and the amide oxygen or they may link adjacent metal atoms using the ring nitrogen (with or without coordination of the oxygen) and the amide nitrogen.

There are several potential donor atoms in tetrazole and its derivatives and there has been some speculation about the bonding site in different complexes. Clearly, these questions are best resolved by single-crystal structure determinations. This is now available for ZnL_2Cl_2 where L = 1-methyltetrazole (37)[278]. The coordination geometry is distorted tetra-

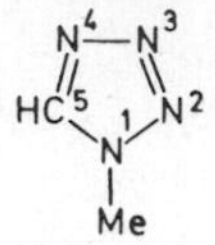

(37)

hedral, the metal atom being approximately co-planar with the tetrazole rings. Here, the bonding atom is N(4). The reason why the 4-position of the ring is chosen by the metal atom is not obvious and it may be that with other metal atoms in complexes of different stereochemistry other bonding sites may occur. An ^{1}H n.m.r. study[279] of the interaction of 1,5-dimethyltetrazole with Mg^{2+} and Ag^+ in MeCN and $MeNO_2$ gave some indication that the bonding site may be N(3).

6.3.9 Some chelating ligands derived from heterocyclic amines

Oximes derived from pyridine are effective in stabilising high oxidation states. Drago and Baucom[280] have characterised a Ni^{IV} complex of diacetylpyridinedioxime and Ni^{III} complexes of *syn*-benzolylpyridineoxime and *syn*-2-phenacylpyridineoxime and proposed a model to account for the stability of these systems. In these complexes the oxime ligands occur as the singly charged anions. Lees *et al.*[281] have described Ni^{II} complexes of pyridine 2-carbaldehyde oxime (and the corresponding quinoline and isoquinoline compounds) and pyridine 2-carbaldehyde *O*-methyloxime in which the bidentate ligands are uncharged. Dimeric *cis*-octahedral structures are ascribed to the NiL_2X_2 complexes of the unsubstituted oxime and a monomeric *trans*-octahedral structure to the NiL_2Br_2 complex of the *O*-methyl derivative. With the same ligand and with *syn*-phenyl-2-pyridylketoxime both neutral and anionic ligand species occur, sometimes in the same complex, as in [PdL(L—H)]Cl. The donor atoms are, as expected, the pyridine nitrogen and the oxime nitrogen. There is strong intramolecular H-bonding between the oxime proton and the π-cloud of the phenyl ring[282].

In the $[ML_2]X_2 \cdot nH_2O$ complexes of the amide (38) the amido-oxygen

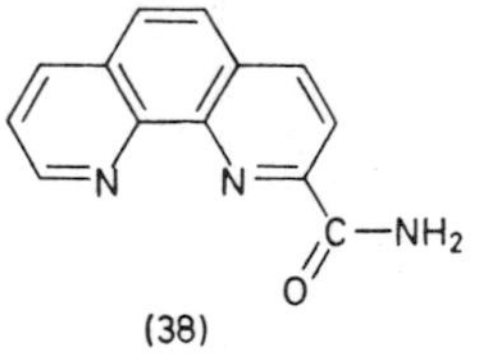

(38) 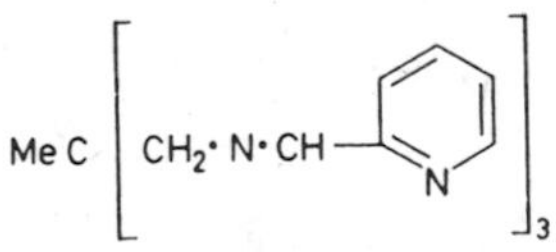 (39)

is coordinated making the ligand tridentate. The Fe^{II} complexes are high-spin reflecting the relatively weak nature of the ligand field[283]. Pyridine 2-carbaldehyde has been widely used in Schiff base condensations with primary amines to yield a variety of polydentate ligands. With 1,1,1-tris(aminomethyl)ethane in the presence of Fe^{II}, Co^{II}, Ni^{II} or Zn^{II}, 6-coordinate complexes of the sexadentate ligand (39) were obtained[284]. In the case of Mn^{II} and Cu^{II} only two molecules of aldehyde condense with the amine. With bis-(3,3'-aminopropyl)amine the pentadentate ligand (40) is formed, which with Ni^{II} gives complexes of the type $[NiLX]Y$ and $[NiLX]PF_6$ (Y = halide, NCS, or NO_3). Electronic spectra and magnetic and conductance measurements established an octahedral structure in these cases for both solid and solution species[285].

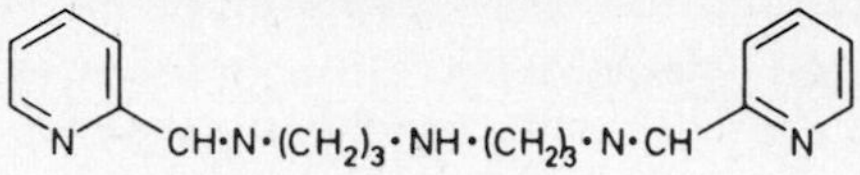

CH·N·(CH2)3·NH·(CH2)3·N·CH

(40)

The coordinated azomethine group is often observed to undergo addition reactions with H_2O or ROH where the metal stabilises an intermediate (41) in the hydrolysis (or alcoholysis) of the Schiff base. A recent example is reaction of H_2O or ROH with the Cu^{II} complex of the Schiff base prepared from two mols pyridine 2-carbaldehyde and one mol of en (or *C*-substituted en). The observation that only one of the two C=N linkages is reactive was rationalised in terms of relief of steric constraints[286]. A somewhat different case of addition to a coordinated C=N bond has been noted by Farrington and Jones[287] who found that decomposition of the tris-complex of (42) with Fe^{II} with EDTA provided a convenient route to the parent Schiff base. However, when the decomposition was carried out in water–acetone mixtures the acetone addition product (43) was isolated.

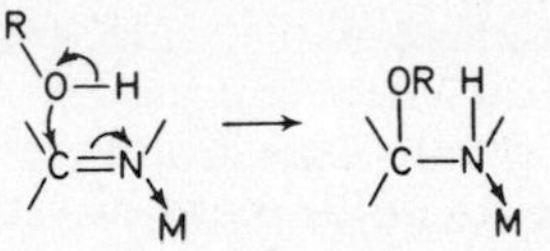

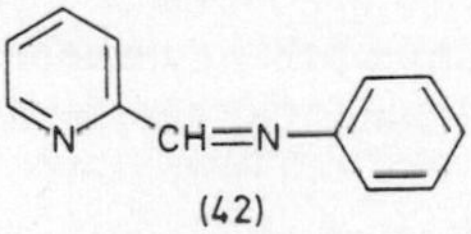

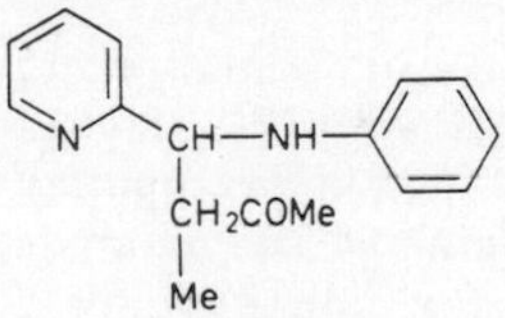

(41)

(42)

(43)

The hexadentate ligand (44) undergoes trigonal prismatic coordination with Mn^{II}, Co^{II} and Zn^{II}. When the acyclic C=N bonds are hydrogenated with BH_4^- the conformational restraints responsible for this stereochemistry are

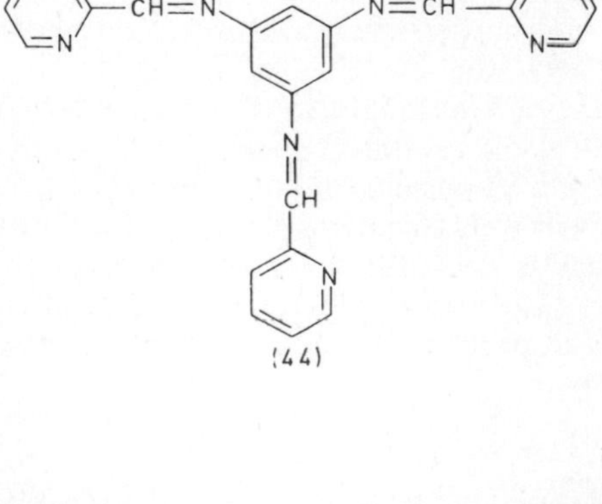

(44)

removed and the product complexes have octahedral coordination[288]. An interesting reaction of a coordinated imine is the free radical chain reaction of tris(biacetylbismethylimine)Fe^{II} in aqueous acid solution. The formation of free radicals was confirmed by polymerisation of acrylonitrile added to the reaction mixture. The suggested explanation for the reaction is that molecular oxygen attacks the methyl groups attached to the carbon atoms of the di-imine group[289]. The two forms of the tridentate amidine ((45), and (46)) exist in tautomeric equilibrium in $CHCl_3$ solution[290]. When complexed to Fe^{II} or Ni^{II} the tautomer (45) is stabilised probably because this is the structure that

(45) (46)

provides for α-di-imine conjugation within the one of the chelate rings leading to increased stabilisation of the complex. Electronic and Mössbauer spectra and other properties define the ML_2^{2+} complexes as 6-coordinate with approximate D_{2d} symmetry. The Fe^{II} salts exist in a ${}^5T_2 \rightleftharpoons {}^1A_1$ equilibrium between 90 and 400 K. When one or both of the pyridine rings are substituted by Me at the 6-positions the Fe^{II} complexes become fully high-spin. A variety of stereochemistries have been observed for the Co^{II} and Ni^{II} complexes of the ligands (47), (48) and (49).

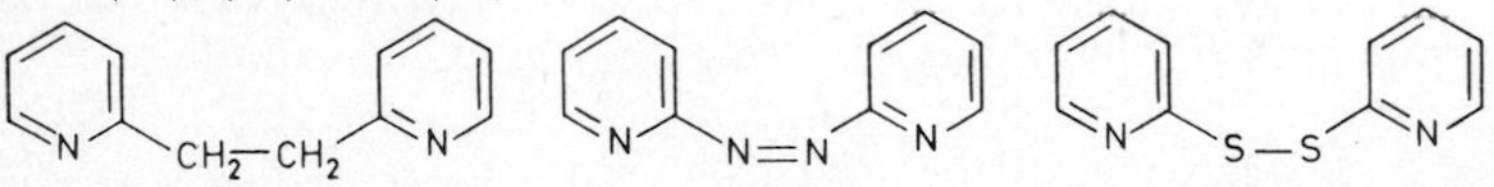

(47) (48) (49)

In all cases coordination is via the pyridine nitrogens only. However, in the complex $HgLCl_2$ of (49) only sulphur atoms appear to be bonded[291].

Table 6.2 summarises some of the recent work on chelate complexes of ligands containing pyridine, imidazole, thiazole and other residues in combination with other donor groups.

The steric barrier to square planar coordination in $M(dip)_2^{n+}$ complexes has already been discussed. The related ligand 2,2′-dipyridylamine in which the two pyridyl groups are linked by the saturated NH group might be expected to accommodate itself more easily to this geometry since coplanarity of the pyridine rings is no longer a requirement. In fact, $[Cu(dipyam)_2](ClO_4)$ was once thought to be square planar[303] though later spectroscopic work by the same author[304] suggested distortion towards a tetrahedral geometry. The crystal and molecular structure has now been determined by x-ray diffraction[305]. The complex cation is a distorted tetrahedron with a dihedral angle of 55.6° between the N—Cu—N planes of the two ligands. The analysis confirms the bidentate nature of the ligand, the Cu—NH distance being 3.19 Å. The single-crystal electronic and e.s.r. spectra have also been measured[306]. Tri-(2-pyridyl)amine is known to use all three pyridyl groups, or alternatively only two pyridyl groups, in coordination[1]. When one of the pyridine rings is substituted by Me at either the 4- or 6-position, isomeric complexes $[Cu(tripyam)_2(ClO_4)_2]$ and $[Cu(tripyam)_2](ClO_4)_2$ in which the ligand is respectively bi- and tri-dentate, have been obtained[307].

Table 6.2

Ligand	*Complexes prepared*	*Structure or other feature of interest*	*Ref.*
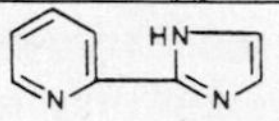	$[ML_3]X_2$ $M = Co^{II}, Ni^{II}, Zn^{II}$ X = Cl, Br, I, NCS	octahedral *Dq* (Ni) = 1140 cm^{-1}	292
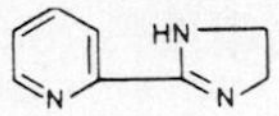	$[NiL_3]X_2 \cdot nH_2O$ $X = Cl, NO_3$ $NiL_2X_2 \cdot nH_2O$ X = Cl, Br, I	Octahedral salts X-bridged octahedral polymers	293
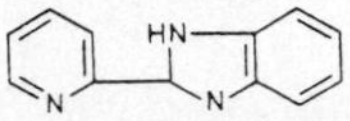	$[ML_3]X_2$ ML_2X_2	as above *Dq* for Ni in $NiL_3^{2+} = 1120\ cm^{-1}$	292
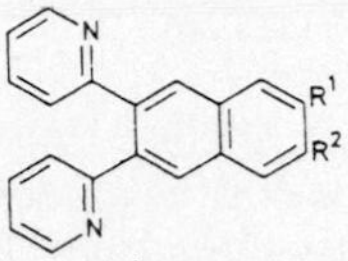 R^1 = Me, R^2 = H $R^1 = R^2$ = Me	$Co^{II}, Ni^{II}, Cu^{II}$	Bidentate non-bridging, bis-unidentate and bis-bidentate bridging modes in a variety of stereochemistries	294
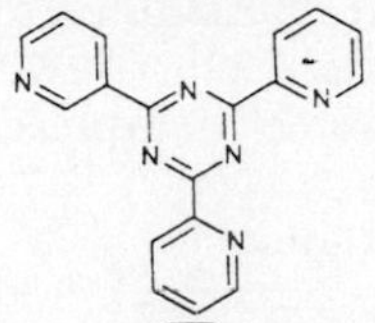	Co^{II}	Tridentate; formation constants of mono- and bis-complexes	295
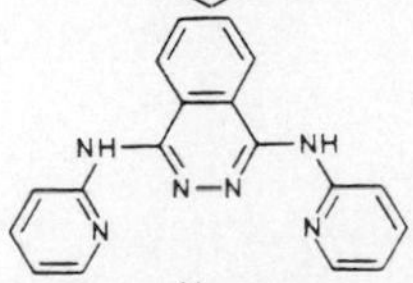	Co^{II}	Binuclear tetrahedral coordination. Weak antiferromagnetic exchange	296
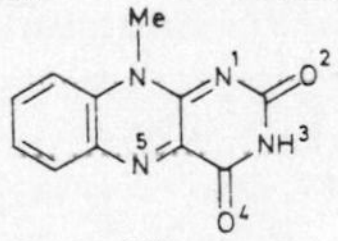	$CoL_2(ClO_4)_2 \cdot 4H_2O$	Bonding sites are O(4) and N(5)	297
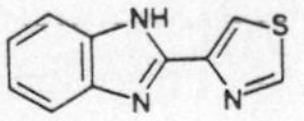	Mono-, bis- and tris-complexes of Mn^{II}-Zn^{II}, Fe^{II}	Mostly insoluble, no conclusions about bonding atoms	298
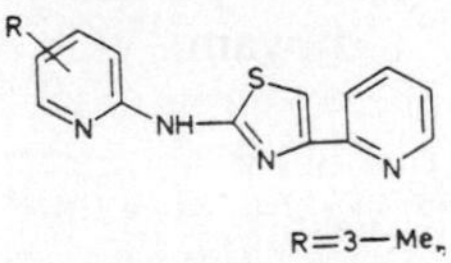 R=3—Me, 4—Me	Ni^{II}, Fe^{II}	$[FeL_2]X_2$ exists in $^5T_2 \rightleftharpoons {}^1A_1$ equilibrium. FeL_2X_2 complexes are high spin, with bidentate L	299
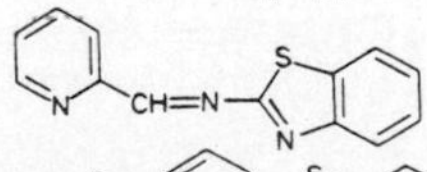	Co^{II}, Ni^{II}	Bidentate, pyridine N and imine *N*	300
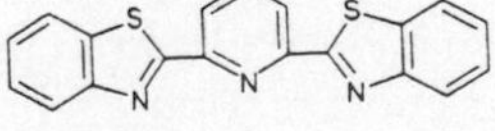	MLX_2 $M = Mn^{II}$–Ni^{II} $ML_2(ClO_4)_2$ $M = Fe^{II}, Co^{II}$	Tridentate, 5- and 6-coordination. $^5T_1 \rightleftharpoons {}^1A_1$ proposed for Fe^{II}	301
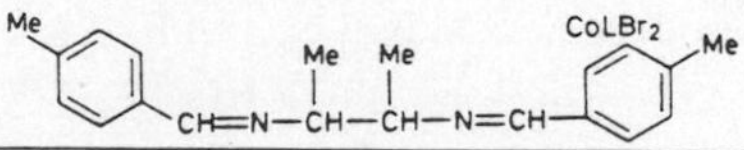	$CoLBr_2$	Tetrahedral	302

When the substituent is 5-NO_2, only bidentate coordination was observed. The crystal structures and magnetic interactions in some Cu^{II} complexes of 2-(alkylamino)pyridine containing, in some cases, Br^- or OH^- bridges, have been reported and discussed[308]. The sterically hindered ligand *NN*-bis-(6-methyl-2-pyridylmethyl)methylamine (50) is well suited to the formation of high-spin 5-coordinate (distorted trigonal bipyramidal)

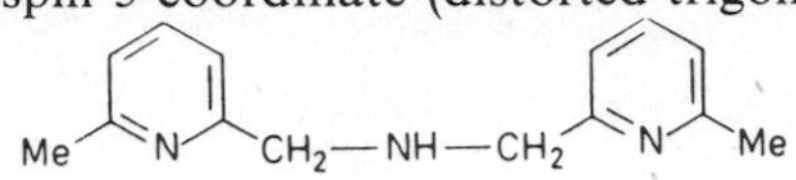

Me N CH_2—NH—CH_2 N Me

(50)

N CH_2— NH—$(CH_2)_n$—NH—CH_2 N

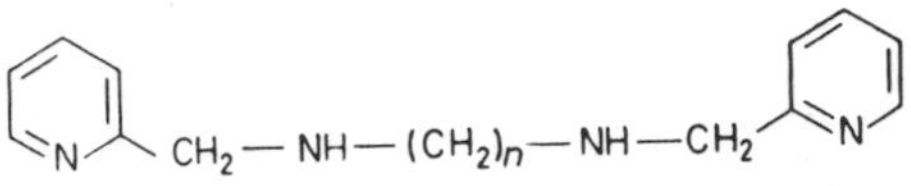

(51) = n = 2
(52) = n = 3

complexes of dipositive metal salts[309]. The two 6-methyl substituents are effective in blocking the approach of a fourth coordinating group in the *NNN* plane, rendering the formation of square planar or octahedral geometries difficult[310]. In order to examine how metal ion acceptors which have strong preferences for these stereochemistries would respond to this steric restriction, the x-ray structures of the $PdLCl_2$ and $TiLCl_3$ were determined[311]. In both cases the distorted geometry, planar and octahedral, respectively, is preferred to the less sterically strained trigonal-bipyramid.

The facultative character of aliphatic polyamines has been discussed briefly above. Gibson and McKenzie[312] have summarised the various kinds of strain in complexes of facultative ligands under three headings and discussed their application to the case of two facultative tetramines (51, 52) containing terminal pyridyl groups. The three types of strain defined are: (a) the strains along the chelate chain in a particular conformation; (b) strain arising from the bulkiness of a ligand (B-strain); and (c) cumulative ring strain (C-strain). From an examination of the i.r., electronic, n.m.r. spectra and x-ray powder patterns of complexes of (51) and (52) with Co^{III}, Ni^{II}, Cu^{II} and Pd^{II} it was concluded, in agreement with earlier predictions, that *cis*-MN_4X_2 species are favoured over the *trans* form for Co^{III} and Ni^{II} but not sufficiently to prevent the formation of square planar [PdN_4] (and possibly also [CuN_4] in distorted form). Where α-*cis* and β-*cis* configurations could be distinguished, as in the Co^{III} complexes, both were found for complexes of (51) but only the β-*cis* configuration for complexes of (52). Nine crystalline forms of $CuL(NCS)_2$ (L = (52)) have been identified and the crystal and molecular structures of two off them reported[313]. Both have 5-coordinate geometries, the two compounds differing mainly in the configuration at the aliphatic nitrogen atoms of the tetramine. When one amino hydrogen atom of ligand (51) is replaced by the 2-pyridylethyl group the resulting branched-chain pentadentate ligand is thermally unstable. However, it readily forms complexes, $ML(ClO_4)_2$ (M = Ni^{II}, Cu^{II} or Zn^{II}) whose properties suggest that all five nitrogen atoms are coordinated in a square-pyramidal arrangement[314]. The tetramine disulphide (53) gives an octahedral complex, [NiLCl]ClO_4 in which one sulphur atom of the disulphide group is coordinated. The con-

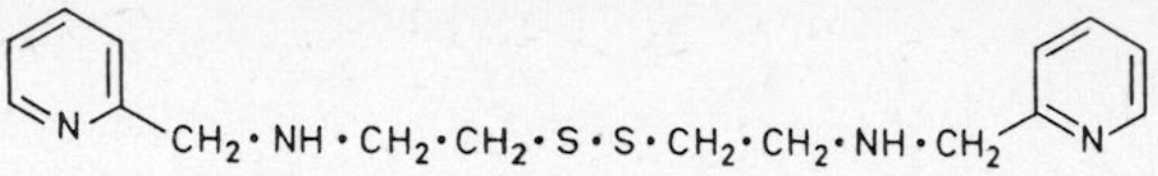

$N \quad CH_2 \cdot NH \cdot CH_2 \cdot CH_2 \cdot S \cdot S \cdot CH_2 \cdot CH_2 \cdot NH \cdot CH_2 \quad N$

(53)

formation of the ligand is such that the two pyridyl groups are *cis*, the two amino groups *trans* and the Cl and S atoms *cis*[315]. The *NSN* ligand (54) probably has a vicinal arrangement of the three donor atoms in its octahedral complexes[316].

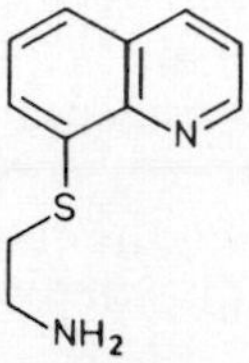

N S NH_2

(54)

6.3.10 Ligands containing the N—N and N═N groups

Hydrazine can readily function as a monodentate or as a bridging ligand but chelation has not been definitely established. However, chelate structures have been suggested for some 1:1 Fe^{II} and Co^{II} complexes of azines of the type $R_2C{:}N \cdot N{:}CR_2$ prepared by direct reaction of, e.g. Co^{II} halides with ketazines and aldazines or by condensation of bis(hydrazine) complexes with ketones and aldehydes. The electronic spectra and magnetic moments point to tetrahedral coordination for both metals. I.R. spectra indicate that both nitrogen atoms of the ligand are coordinated. However, the conclusion that these are bonded to the same metal atom rests only on molecular weight measurements in β-naphthol which alternatively may be interpreted in terms of a splitting of ligand bridges by solvent. Acyl- and aryl-hydrazines $R \cdot CO \cdot NHNH_2$ can employ the carbonyl oxygen and the terminal nitrogen as donor atoms. Such 5-membered chelate structures have been proposed for some cationic complexes of Co^{II} and Ni^{II} and confirmed in an x-ray structure determination of bis-(*N*-benzoylhydrazine)Cu^{II} perchlorate[318]. Other workers[319] have claimed, on the basis of i.r. spectra, monodentate coordination through the amide nitrogen. However, the i.r. evidence is far from conclusive and the stoichiometry and other properties of the complexes are more consistent with a bidentate coordination mode. Co^{II} and Ni^{II} complexes of several acyl- and aryl-thiocarbazides contain bidentate chelating ligands with the sulphur atom and the terminal hydrazine nitrogen as donor atoms. The product of reaction of hydrazine with *cis*-dichlorobis(methylisocyanide)Pd^{II} contains a novel ligand structure (55) derived from the addition of N_2H_4 to two coordinated isonitrile molecules[321]. The complex is planar and the C—N and N—N distances indicate some multiple bonding within the

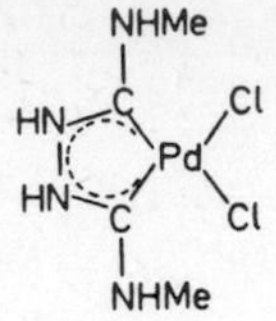

(55)

chelate ring and in the C—N bonds external to it. This study shows that the original binuclear formulation of Chugaev *et al.*[322] for the corresponding Pt^{II} complex must be incorrect.

A number of nitrogen- and sulphur-containing chelating ligands (56)–(59) derived from hydrazine thiocarboxylic acids have been investigated[323, 324]. From i.r. data (57) and (58) are considered to exist in the tautomeric form shown except in solution where they are probably in equilibrium with the thiol tautomer (60). In the latter form the thiol proton is readily lost and the anionic ligands produced coordinate bidentately (57) or tridentately (58). All

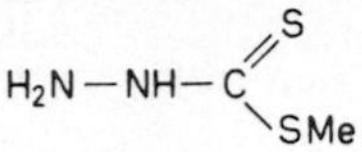

(56) 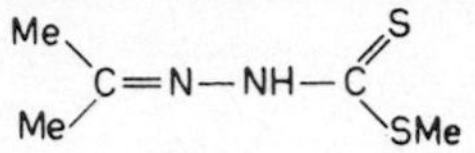 (57)

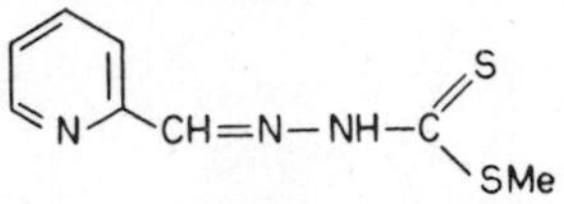

(58) 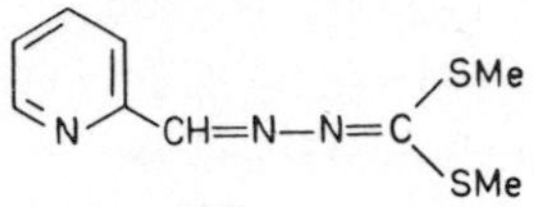 (59)

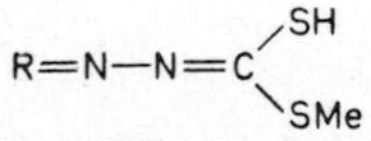

(60)

three ligands (56)–(58) also chelate in the un-ionised form. In the case of (56) and (57) bidentate chelation occurs via the imino nitrogen and a sulphur atom. Ligand (58) is tridentate in its complexes, as is ligand (59) in some cases, though bidentate (NN) in others. The complex CdL_2Cl_2 of ligand (61) has a *trans*-octahedral structure in which the coordination polyhedron is composed of two terminal hydrazinic nitrogen atoms, two sulphur atoms and two

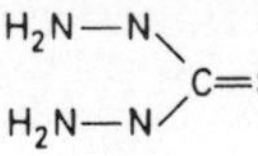

(61)

chlorine atoms, the second terminal hydrazine functions of each ligand being uncoordinated[325].

Osazones offer a number of coordination sites. The 1:1 complexes (62) with $PdCl_2$ are formulated as shown on the basis of spectroscopic and chemical properties[326]. Elimination of HCl occurs on treatment with SiO_2 or NEt_3

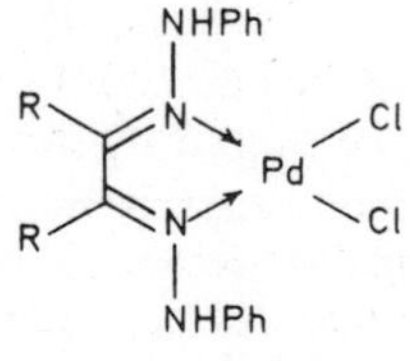

(62)

forming the Cl-bridged binucleur complex (63) containing Pd—N covalent bonds. The reaction is reversible.

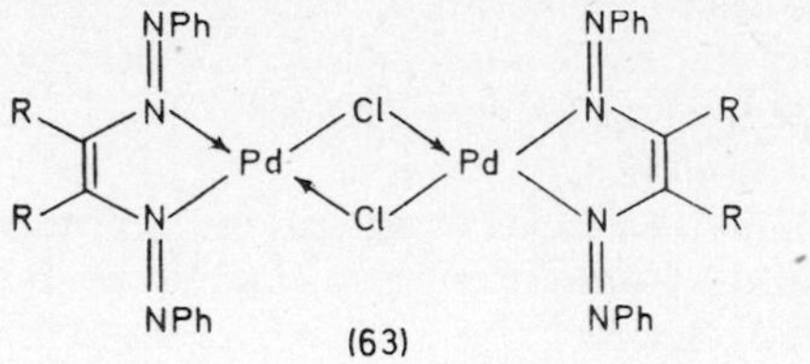

(63)

1,3,5-Triphenylformazan (64) is stoichiometrically oxidised to the corresponding tetrazolium salt (65) by Cu^{II} chloride or acetate. In contrast,

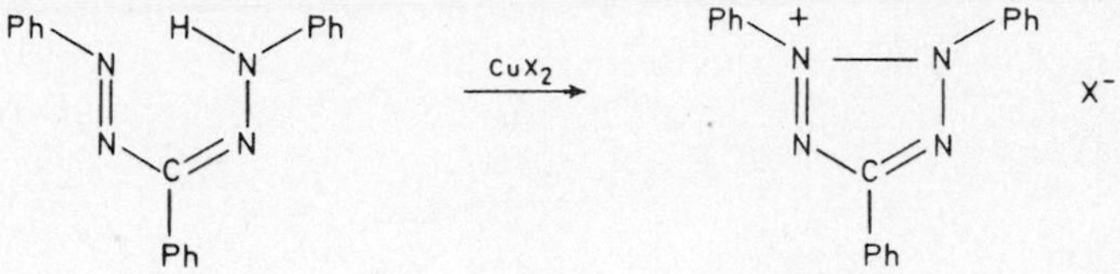

(64) (65)

the 1-(2-carboxyphenyl)-3,5-diphenyl- and 1-(2-hydroxyphenyl)-3,5-diphenylformazans are not oxidised; instead, they give 1:1 complexes. The different behaviour is attributed to an increase in denticity of the formazan from two to three[327].

Not unexpectedly the basicity of hydrazine is dramatically reduced by monoprotonation (from $pK_a = 7.93$ to $pK_a = -2.95$). Nonetheless, the 1,1,1-trimethylhydrazinium cation, $[(Me)_3NNH_2]^+$, is an effective ligand giving paramagnetic distorted octahedral complexes, $Ni(L^+)_2X_4$, with $NiCl_2$ and $NiBr_2$[328]. The yellow chloride undergoes a sharp change to a blue compound on heating to 418 K. The blue compound was identified as the hydrazinium salt of the tetrahedral $[NiCl_4]^{2-}$ anion. The yellow 6-coordinate complex re-forms on cooling.

There is strong evidence for all four modes of attachment (66)–(69) of the azo group to metal atoms or ions. In addition, reactions in which the N—N bond is cleaved may occur, while in azobenzene, for example, an *ortho*-

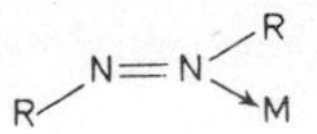

(66) 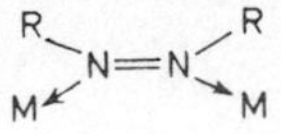 (67)

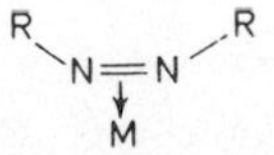

(68) 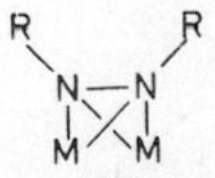 (69)

carbon as well as an azo-nitrogen may be involved in bonding. A new and convenient synthesis of azobenzene complexes of Ni^0 has been described in which $Ni(PR_3)_2Cl_2$ is treated with azobenzene and metallic Li in tetrahydrofuran below room temperature[329]. The product, $Ni(PR_3)$(azobenzene), was obtained in high yield. Various reactions have been reported. A similarly bonded azobenzene complex of Ni^0 has been reported by Otsuka *et al.*[330]. When an equimolar mixture of $Ni(Bu^tNC)_2$ and azobenzene in ether was stirred under N_2 at room temperature, orange red crystals of $Ni(Bu^tNC)_2$-(azobenzene) were obtained. A number of new complexes of $PdCl_2$ with the azo compounds of general formula R—N=N—R^2 have been prepared (R^1 = cyclohexyl or cyclohexenyl and R^2 = Me, Ph or *p*-tolyl)[331]. The complexes are of stoichiometry PdL_2Cl_2 and contain the azo ligands coordinated by electron pair bonds. Complexes of azo compounds with iron carbonyls have been described by Bennett[332] and by Herberhold and Golla[333]. An x-ray analysis of the 2,3-diazabicyclo[2.2.1]hept-2-ene complex has shown it to have the structure (70)[334]. The N—N bond length (1.404 Å) is close to that expected for a single bond.

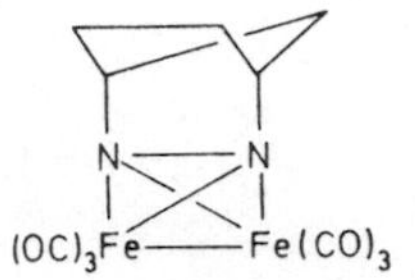

(70)

The 7-membered ring diazepines react with transition metal compounds in different ways. With $Fe_2(CO)_9$ and 3,5,7-triphenyl-4*H*-1,2-diazepine (71) cleavage of the *N*—*N* bond occurs with formation of a *N*-bridged metallo-

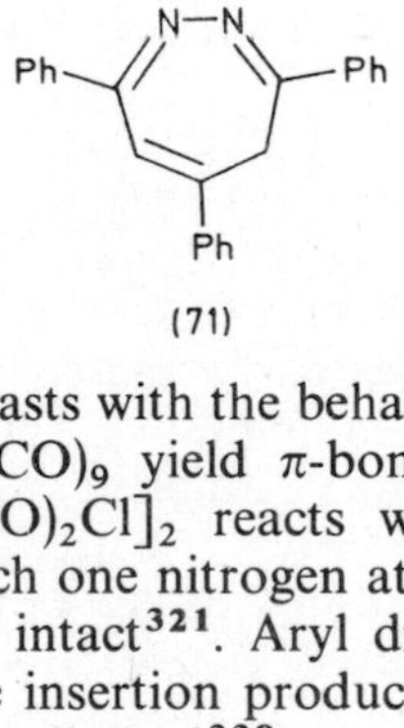

(71)

bicyclo system[335]. This contrasts with the behaviour of 1-substituted-*H*-1,2-diazepines which with $Fe_2(CO)_9$ yield π-bonded diene-type iron tricarbonyl complexes[336]. $[Rh(CO)_2Cl]_2$ reacts with (71) to give a product Rh(CO)Cl(diazepine) in which one nitrogen atom of the azo group is coordinated, the ring remaining intact[321]. Aryl diazonium cations react with *mer*-$IrH_3(PPh_3)_3$ to give the insertion product (72) in which one nitrogen atom of the azo group is coordinated[338].

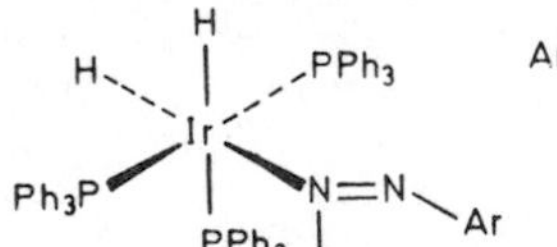

(72) 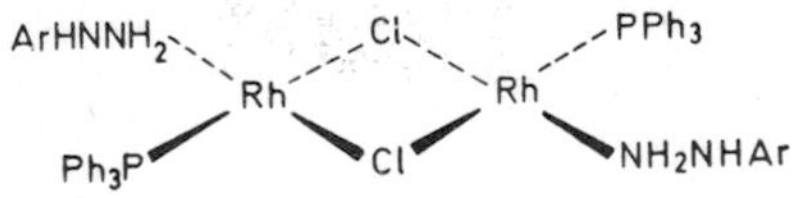 (73)

The presence of the proton bonded to nitrogen was confirmed by ^1H n.m.r. spectra. Reaction with $RhH_2Cl(PPh_3)_2$(solvent) gave, on the other hand, an aryl-substituted hydrazine complex (73). Presumably, in this case, the insertion process is followed by reduction. Reaction of some *p*-substituted phenyldiazonium BF_4 salts with Vaska's compound has yielded tetrazene complexes rather than the expected diarylazo derivatives[339]. From a single-crystal x-ray structure determination of the *p*-fluorophenyl derivative it has been shown that the metal atom has an approximately square pyramidal geometry having one phosphorus atom at the apex, the base (not perfectly planar) comprising two nitrogen atoms, one phosphorus atom and one carbon atom. Bond length considerations suggest a structure intermediate between the Valence Bond representations (74) and (75), but probably closer to (74).

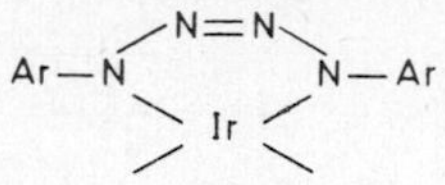

(74) 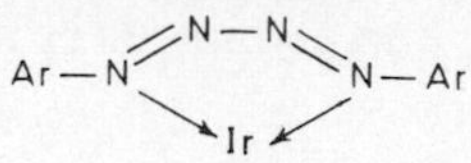 (75)

1-*H*-1,2,3-Benzotriazole (76) readily displaces benzonitrile from $Pd(PhCN)_2Cl_2$ in benzene to give the corresponding triazole complex[340]. A strong absorption at 3200 cm^{-1} confirmed that the NH proton is retained. The 1-benzyl- and 1-vinyl-compounds react similarly. They also react with π-$(C_3H_5)PdCl_2$, splitting the chloride bridges, to yield π-$(C_3H_5)PdLCl$. N.M.R. measurements strongly suggest that the donor atom is either N(2) or N(3)[340].

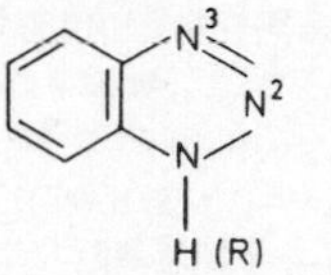

(76)

In principle, each of the four nitrogen atoms of the tetrazole ring is capable of acting as the donor atom in coordination compounds (see earlier).

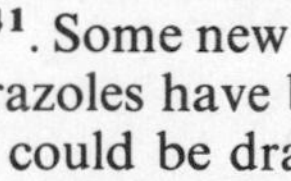

(77)

The few available structure determinations show that N(4), and N(4) and N(3), are the donor atoms when the tetrazole is acting as a unidentate and bridging ligand, respectively[341]. Some new complexes PdL_2Cl_2 and ZnL_2Cl_2 of 1-phenyl- and 1-*p*-tolyl-tetrazoles have been reported[342], but few conclusions about the bonding site could be drawn. The stereochemistries of the complexes of several 5-substituted tetrazoles with Co^{II}, Ni^{II}, Cu^{II}, Zn^{II} and Cr^{III} have been established, in most cases, from measurements of magnetic moments and from the i.r., electronic and e.s.r. spectra but once again nothing could be deduced about the atom(s) engaged in bonding[343]. The 5-membered heterocyclic tetrazaborolines (77) have both electron-accepting (the B atom) and electron-donating properties. The preparation, thermochemistry and some properties of the adducts of 1,4,5-substituted borolines with $TiCl_4$ and with a number of main group chlorides have been reported[344]. It was concluded that the tetrazoboroline ring behaves as a monodentate ligand towards BCl_3, $SbCl_5$ and $SnCl_4$ whilst bridged complexes are probably formed with $TiCl_4$.

6.3.11 Nitriles as ligands

Nitriles are ambidentate in that they may form coordinate σ-bonds via the nitrogen lone pair or they may coordinate through the π-system of the $C{\equiv}N$ triple bond. The former is the more commonly observed mode of bonding. Generally, but not always (see later) σ-bonded nitriles show an increase in the frequency of the $C{\equiv}N$ stretch over the free ligand of *ca.* 30–110 cm^{-1} [345]. π-Bonded nitriles on the other hand, show large decreases in ν(CN) of *ca.* 100–250 cm^{-1}. The latter type of bonding appears to occur mainly in complexes of chelating dinitriles[346]. Conflicting results have been reported for some dinitrile complexes of Mn^{I} and Re^{I} carbonyl halides of the type $M(CO)_3LX$. On the basis of, mainly, vibrational spectra Dunn and Edwards[347] have proposed binuclear halide-bridged structures involving monodentately *N*-coordinated dinitriles while Farona and Kraus[346] formulated the complexes as mononuclear containing chelating π-bonded nitriles. It now appears that both types of bonding may occur[348].

The bonding characteristics of nitrile ligands in some Ru^{II}, Ru^{III} and Rh^{III} pentammine complexes, $M(NH_3)_5L^{n+}$, are of interest[349]. When M = Rh^{III} the ν(CN) increases by 30–70 cm^{-1} while when M = the isoelectronic Ru^{II} ν(CN) is reduced by 13–93 cm^{-1}. The reductions in the CN frequency in the Ru^{II} compounds were attributed to $d_{\pi} \rightarrow p_{\pi*}$ back bonding sufficient to over-ride the normal Lewis acid coordination effect. Similar decreases in ν(CN) have been noted by Tolman[350] for some Ni^{0}–nitrile complexes, namely, $Ni(phos)_3L$ (phos = $P(o\text{-}C_6H_4Me)_3$). Some new Rh^{III}–nitrile complexes of formula $[RhL_2X_4]^-$ (X = Cl or Br) are considered to have *trans*-configurations on the basis of ^1H n.m.r. and i.r. spectra; the ν(CN) are increased (56–61 cm^{-1})[351]. Reaction of simple aliphatic nitriles with V^{V} oxide trinitrate, $VO(NO_3)_3$, which from i.r. and Raman spectra is believed to be 7-coordinate, causes one of the three bidentate NO_3 groups to become unidentate thereby maintaining the same coordination number for the product $VO(NO_3)_3L$ [352]. A single-crystal x-ray determination in the case of L = MeCN has shown that the nitrate oxygen atoms form an equatorial pentagon, the vanadyl oxygen and the nitrile nitrogen occupying apical positions[352]. The V—N bond is rather long (2.24 Å) and the increase in ν(CN) is 37 cm^{-1}. The e.s.r. and electronic spectra of some nitrile complexes of $TiCl_3$ have been discussed in terms of deviations from octahedral symmetry[353]. Treatment of $TiCl_3$ with an excess of mono-, di- and trichloro-acetonitrile or with 1-cyano-1,1,2-trichloroethane in sealed ampoules at 35–60 °C gives Ti^{IV} complexes of formula $TiLCl_4$ and TiL_2Cl_4 depending on the nature of L[354]. Products having identical i.r. spectra were also obtained by reaction of $TiCl_4$ with the appropriate ligand. Examination of the organic products of the reaction in the case of L = CCl_3CN suggested that the oxidation probably involves transfer of a chlorine atom from a coordinated CCl_3CN to the metal followed by expulsion of the CCl_2CN moiety. A similar metal oxidation occurs in the case of reaction of CCl_3CN with W^{V} chloride. One of the reaction products has the structure (77a)[355]. The short W—N distance

(1.71 Å) and the almost linear W—N—C skeleton led to the formulation of a triple W≡N bond with $p_\pi \rightarrow d_\pi$ bonding incorporating the nitrogen lone pair.

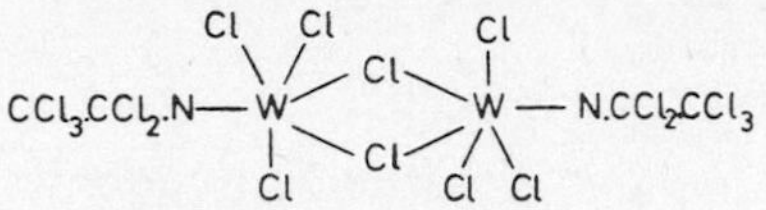

$CCl_3CCl_2.N—WCl_3(\mu-Cl)_2WCl_3—N.CCl_2CCl_3$

(77a)

Diaminomaleonitrile reacts with Pd^{II} and Pt^{II} chlorides in acidic solution to give complexes, ML_2Cl_2, of the neutral ligand[356]. A single-crystal x-ray study of the Pd^{II} complex has shown it to be *trans*-square planar with each ligand coordinated via one amino nitrogen atom. In basic or neutral solution reaction with Ni^{II}, Pd^{II} or Pt^{II} gave highly-coloured inner chelates $[M(C_4H_2N_4)_2]$ in which each ligand has lost two protons. In none of these complexes is the nitrile group coordinated. The relative labilities of the ligands in some complexes of the type $PtL^1L^2Cl_2$ (where L^1 = MeCN or PhCN and L^2 = CO or C_2H_4) have been estimated from the temperatures at which spin coupling between ligand protons and ^{195}Pt appeared in the ^1H n.m.r. spectra[357]. In the *trans*-complexes both nitrile and ethylene labilise each other while in the *cis*-configuration neither ligand exchanges with solvent. Acetonitrile is a better *trans*-labilising ligand than PhCN.

Coordination (via the nitrogen lone pair) of the nitrile group may enhance its susceptibility to nucleophilic attack at the nitrile carbon. *cis*-$[Co(en)_2(NH_2CH_2CN)Cl)^{2+}$ reacts in near neutral or basic solution to give a purple complex shown by x-ray analysis to contain the tridentate amidine (78). The mechanism proposed[358] is that an NH_2 group *trans* to Cl is deprotonated

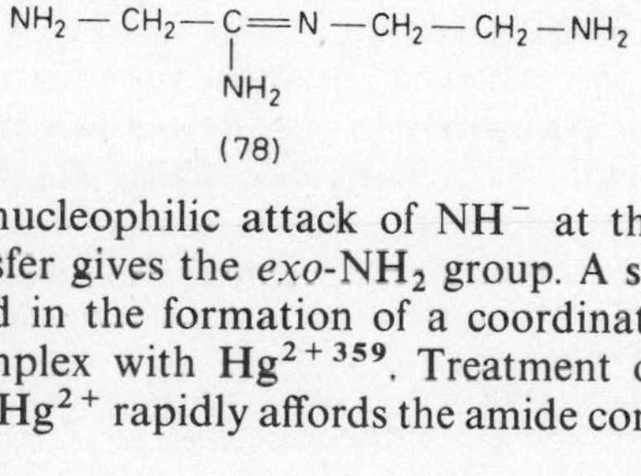

$$NH_2—CH_2—C(NH_2)=N—CH_2—CH_2—NH_2$$

(78)

this being followed by nucleophilic attack of NH^- at the nitrile carbon. Subsequent proton transfer gives the *exo*-NH_2 group. A somewhat similar reaction step is involved in the formation of a coordinated amide in the reaction of a nitrile complex with Hg^{2+}[359]. Treatment of the kinetically robust complex (79) with Hg^{2+} rapidly affords the amide complex (80) in 95% yield.

A kinetic, ^1H n.m.r. and chemical analysis of the overall reaction indicates three steps: a Hg^{2+} assisted abstraction of Br^-; rapid entry of solvent (H_2O) into the vacant coordination site; and Hg^{2+} assisted nucleophilic attack of coordinated OH^- at the nitrile carbon. The formation of imino–ether complexes via nucleophilic attack of alcohols on pentafluorobenzonitrile coordinated to Pt^{II} may involve π-bonded nitrile species as intermediates[360].

Cyanamide has two potential donor atoms. Balahura and Jordan[361] concluded that cyanamide (81) and N-cyanoguanidine (82) are coordinated via the nitrile nitrogen in complexes of the type $[Co(NH_3)_5L]^{3+}$ because of the increase in ν(CN), and, in the case of (82), because of the equivalence of the NH_2 protons in the ^1H n.m.r. spectra. Coordination to Co^{3+} alters the pK_a of cyanamide from 10.27 to 5.18. A greater increase in acidity might be expected to result from NH_2 coordination. Reaction of the complex with

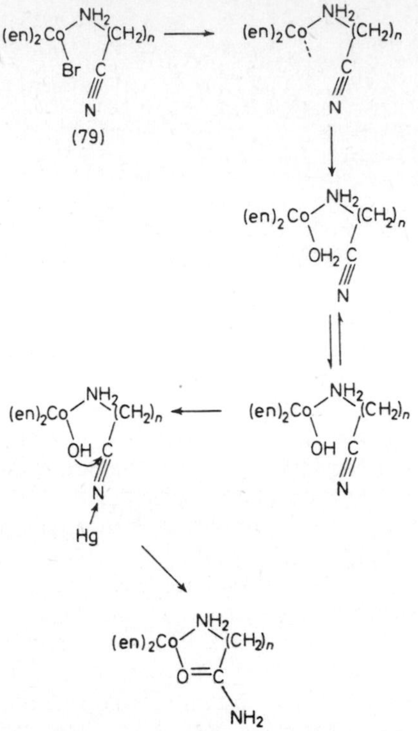

(79)

(80)

HNO_2 gave $[Co(NH_3)_5(NCO)]^{2+}$ and not $[Co(NH_3)_5(H_2O)]^{3+}$, the expected product if the coordinated atom was the amino nitrogen. However, nitrile-coordinated cyanamide has been suggested[361] for some Co^{II} and Ni^{II} complexes of formula $ML_4(SO_4)$ and $ML_6(NO_3)_2$. In these $\nu(CN)$ is unchanged from the value in the free ligand while $\nu(NH)$ is lowered by *ca.*

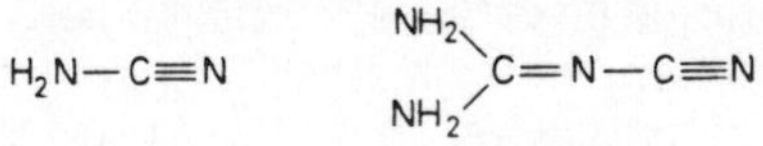

$H_2N—C≡N$ (81)

$(NH_2)_2C=N—C≡N$ (82)

30 cm^{-1}. Electronic spectra of these complexes suggest octahedral coordination, a result which is not consistent, in the case of the sulphate, with ionic SO_4^{2-} (single i.r. band at 1000–1200 cm^{-1}) and unidentate coordination by cyanamide.

References

1. Nelson, S. M. (1972). Transition Metals, Part 1. Chap. 5. *MTP Int. Rev. of Sci.* Inorg. Chem. Series 1. (Sharp, D. W. A. editor) (London: Butterworth)
2. See McKenzie, E. D. (1971). *Coord. Chem. Rev.*, **6,** 187 and Brubaker, G. R., Schaefer, D. P., Worrell, J. H. and Legg, J. I. (1971). *Coord. Chem. Rev.*, **7,** 161 for reviews on this subject
3. Black, D. St. C. and Hartshorn, A. J. (1972). *Coord. Chem. Rev.*, **9,** 219
4. Lindoy, L. F. (1971). *Quart. Rev.* (*Chem. Soc.*), **25,** 379
5. Calligaris, M., Nardin, G. and Randaccio, L. (1972). *Coord. Chem. Rev.*, **7,** 385
6. Holm, R. H. and O'Connor, M. J. (1971). *Progr. Inorg. Chem.*, **14,** 241

7. Boucher, L. J. (1972). *Coord. Chem. Rev.*, **7,** 331
8. Sacconi, L. (1972). *Coord. Chem. Rev.*, **8,** 351
9. Wood, J. S. (1972). *Progr. Inorg. Chem.*, **16,** 227
10. Hoskins, B. F. and Whillans, F. D. (1973). *Coord. Chem. Rev.*, **9,** 365
11. Barbucci, R., Fabrizzi, L. and Paoletti, P. (1972). *Coord. Chem. Rev.*, **8,** 31
12. Waltz, W. L. and Sutherland, R. G. (1972). *Chem. Soc. Rev.*, **1,**241
13. Simmons, E. L. and Wendlandt, W. W. (1971). *Coord. Chem. Rev.*, **7,** 11
14. Addison, C. C., Logan, N., Wallwork, S. C. and Garner, C. D. (1971). *Quart. Rev.* (*Chem. Soc.*), **25,** 289
15. Klonis, H. B. and King, E. L. (1972). *Inorg. Chem.*, **11,** 2933
16. Karraker, D. G. (1971). *J. Inorg. Nucl. Chem.*, **33,** 3713
17. Brannon, D. G., Morrison, R. H., Hall, J. L., Humphrey, G. L. and Zimmerman, D. S. (1971). *J. Inorg. Nucl. Chem.*, **33,** 981
18. Hughes, M. N., Waldron, B. and Rutt, K. J. (1972). *Inorg. Chim. Acta*, **6,** 619
19. Dotson, R. L. (1972). *J. Inorg. Nucl. Chem.*, **34,** 3131
20. Hughes, M. N., Underhill, M. and Rutt, K. J. (1972). *J. Chem. Soc. Dalton Trans.*, 1219
21. Fowles, G. W. A., Rice, D. A. and Wilkins, J. D. (1971). *J. Chem. Soc. A*, 1920
22. Coutts, R. S. P., Martin, R. L. and Wailes, P. C. (1971). *Aust. J. Chem.*, **24,** 2533
23. Borden, R. S., Loeffler, P. A. and Dyer, D. S. (1972). *Inorg. Chem.*, **11,** 2481
24. Karayannis, N. M., Mikulski, C. M., Speca, A. N., Cronin, J. T. and Pytlewski, L. L. (1972). *Inorg. Chem.*, **11,** 2330
25. Benner, L. S. and Root, C. A. (1972). *Inorg. Chem.*, **11,** 652
26. Westland, A. D. and Muriithi, N. (1972). *Inorg. Chem.*, **11,** 2971
27. Sellmann, D. (1972). *Angew. Chem. Int. Edn. Engl.*, **11,** 534
28. Madan, S. K. (1971). *J. Inorg. Nucl. Chem.*, **33,** 1025
29. Atlanasio, D., Collimati, I. and Ercolani, C. (1972). *J. Chem. Soc. Dalton Trans.*, 772, and earlier papers
30. Nakamura, Y., Isobe, K., Morita, H., Yamazaki, S. and Kawaguchi, S. (1972). *Inorg. Chem.*, **11,** 1573, and earlier papers
31. Bakker, J., Feller, M. C. and Robson, R. (1971). *J. Inorg. Nucl. Chem.*, **33,** 747
32. Green, R. W. and Rogerson, M. J. (1971). *Aust. J. Chem.*, **24,** 65
33. Baker, R. J., Nyburg, S. C. and Szymanski, J. T. (1971). *Inorg. Chem.*, **10,** 138
34. Brown, E. L. and Brown, D. B. (1971). *Chem. Commun.*, 67
35. Brown, J. N., Pierce, A. G. and Trefonas, L. M. (1972). *Inorg. Chem.*, **11,** 1830; Majeste, R. J. and Trefonas, L. M. (1972). *Inorg. Chem.*, **11,** 1834
36. Farona, M. F., Grasselli, J. G., Grossman, H. and Ritchey, W. H. (1969). *Inorg. Chim. Acta*, **3,** 495, 553
37. Reedijk, J. (1971). *Inorg. Chim. Acta*, **5,** 687
38. Gentile, P. S., Carfagno, P., Haddad, S. and Campisi, L. (1972). *Inorg. Chim. Acta*, **6,** 296
39. James, B. R., McMullan, R. S. and Ochai, E. (1972). *Inorg. Nucl. Chem. Lett.*, **8,** 239
40. Bour, J. J., Birker, P. J. and Steggerda, J. J. (1971). *Inorg. Chem.*, **10,** 1202
41. Nonayama, M. and Yamasaki, D. (1971). *Inorg. Chim. Acta*, **5,** 124
42. Burgess, C. M. and Toogood, G. E. (1971). *Inorg. Nucl. Chem. Lett.*, **7,** 761; Perrier, M. and Vicentini, G. (1971). *J. Inorg. Nucl. Chem.*, **33,** 1733, 2497; Airoldi, C. and Gushikem, Y. (1972). *J. Inorg. Nucl. Chem.*, **34,** 3925
43. Bagnell, K. W., du Preez, J. G. H. and Gibson, M. L. (1971). *J. Chem. Soc. A*, 2124
44. Kirksey, K. and Hamilton, J. B. (1972). *Inorg. Chem.*, **11,** 1945
45. Stone, M. E., Robertson, B. E. and Stanley, E. (1971). *J. Chem. Soc. A*, 3632
46. Singh, P. P. and Pande, I. M. (1972). *J. Inorg. Nucl. Chem.*, **34,** 591
47. Peyronel, G., Pellacani, G. C. and Pignedoli, A. (1971). *Inorg. Chim. Acta*, **5,** 627; Pellaconi, G. C. and Peyronel, G. (1972). *Inorg. Nucl. Chem. Lett.*, **8,** 299
48. Peyronel, G., Pellaconi, G. C., Pignedoli, A. and Beneti, G. (1971). *Inorg. Chim. Acta*, 5, 263; Fabretti, A. C., Pellacani, G. C. and Peyronel, G. (1971). *J. Inorg. Nucl. Chem.*, **33,** 4247
49. Furlani, C., Tarantelli, T. and Riccieri, P. (1971). *J. Inorg. Nucl. Chem.*, **33,** 1389
50. Tarantelli, T. and Furlani, C. (1972). *J. Inorg. Nucl. Chem.*, **34,** 999
51. De. Filippo, D., Devillanova, F., Preti, C. and Verani, G. (1971). *J. Chem. Soc. A*, 1465
52. Hughes, M. N. and Rutt, K. J. (1972). *J. Chem. Soc. Dalton Trans.*, 1311
53. Ackerman, L., du Preez, J. G. H. and Gibson, M. L. (1971). *Inorg. Chim. Acta*, **5,** 539
54. Morrison, A. and Nelson, S. M. (1973). *Inorg. Chem.*, **12,** 960
55. Bennett, L. E. (1970). *Inorg. Chem.*, **9,** 1941

56. Cunningham, D. W. and Workman, M. D. (1971). *J. Inorg. Nucl. Chem.*, **33,** 3861
57. Herlocker, D. W. (1972). *J. Inorg. Nucl. Chem.*, **34,** 389
58. Watt, G. W. and Strait, W. R. (1972). *J. Inorg. Nucl. Chem.*, **34,** 946
59. Karayannis, N. M., Cronin, J. T., Mikulski, C. M., Pytlewski, L. L. and Labes, M. M. (1971). *J. Inorg. Nucl. Chem.*, **33,** 3185, 4344
60. Prabhackaran, C. P. and Patel, C. C. (1972). *J. Inorg. Nucl. Chem.*, **34,** 2371
61. Al-Kharaghouli, A. R. and Wood, J. S. (1972). *Chem. Commun.*, 516
62. Barnes, J. A., Barnes, W. C. and Hatfield, W. E. (1971). *Inorg. Chim. Acta*, **5,** 276
63. Joltham, R. W., Kettle, S. F. A. and Marks, J. A. (1972). *J. Chem. Soc. Dalton Trans.*, 1133
64. Johnson, D. R. and Watson, W. H. (1971). *Inorg. Chem.*, **10,** 1281
65. Da Silva, J. J. R. F., Vilas Boas, L. F. and Wootton, R. (1971). *J. Inorg. Nucl. Chem.*, **33,** 2029
66. De Bolster, M. W. G., Kortram, I. E. and Groeneveld, W. L. (1972). *J. Inorg. Nucl. Chem.*, **33,** 575
67. Ciani, G., Manassero, M. and Sansoni, M. (1972). *J. Inorg. Nucl. Chem.*, **34,** 1760
68. Contreras, E., Riera, V. and Uson, R. (1972). *Inorg. Nucl. Chem. Lett.*, **8,** 287
69. Bertrand, J. A. and Kalganaraman, A. R. (1971). *Inorg. Chim. Acta*, **5,** 341
70. Goodgame, D. M. L. and Cotton, F. A. (1961). *J. Amer. Chem. Soc.*, **83,** 2298
71. Hart, F. A. and Van Nice, R. (1971). *Inorg. Chem.*, **10,** 115
72. Cotton, S. A. and Gibson, J. F. (1971). *J. Chem. Soc. A*, 859
73. De Bolster, M. W. G., Kortram, I. E. and Groeneveld, W. L. (1972). *Inorg. Nucl. Chem. Lett.*, **8,** 751
74. Scheide, E. P. and Guilbault, G. G. (1971). *J. Inorg. Nucl. Chem.*, **33,** 1689
75. Walmsley, J. A. and Tyree, S. Y. (1963). *Inorg. Chem.*, **2,** 312
76. Brisdon, B. J. (1972). *J. Chem. Soc. Dalton Trans.*, 2247
77. Sandhu, S. S. and Sandhu, R. S. (1972). *J. Inorg. Nucl. Chem.*, **34,** 2295
78. Matthew, M. and Palenik, G. J. (1971). *Inorg. Chim. Acta*, **5,** 573
79. Siddall, T. H. (1963). *J. Inorg. Nucl. Chem.*, **25,** 883
80. Stewart, W. E. and Siddall, T. H. (1971). *J. Inorg. Nucl. Chem.*, **33,** 2965
81. Joesten, M. D. and Chen, Y. T. (1972). *J. Inorg. Nucl. Chem.*, **34,** 257
82. De Bolster, M. W. G. and Groeneveld, W. L. (1971). *Rec. Trav. Chim.*, **90,** 477
83. Sylvanovitch, J. A. and Madan, S. K. (1972). *J. Inorg. Nucl. Chem.*, **34,** 1675; Vicenti, G. and Dunstan, P. O. (1972). **34,** 1303; (1971). **33,** 1749
84. Joesten, M. D., Koch, R. C., Martin, T. W. and Venable, J. H. (1971). *J. Amer. Chem. Soc.*, **93,** 1138
85. Miller, P. T., Lenhaert, P. G. and Joesten, M. D. (1972). *Inorg. Chem.*, **11,** 2221
86. Palmer, R. A. and Taylor, C. R. (1971). *Inorg. Chem.*, **10,** 2546
87. Joesten, M. D. and Chen, Y. T. (1972). *Inorg. Chem.*, **11,** 429
88. Speca, A. N., Karayannis, N. M. and Pytlewski, L. L. (1972). *Inorg. Chim. Acta*, **6,** 639; see also, Speca, A. N., Pytlewski, L. L. and Karayannis, N. M. (1972). *J. Inorg. Nucl. Chem.*, **34,** 3671
89. Berney, C. V. and Weber, J. H. (1971). *Inorg. Chim. Acta*, **5,** 375
90. Zinner, L. B. and Vicentini, G. (1971). *Inorg. Nucl. Chem. Lett.*, **7,** 967
91. Bertran, P. B. and Madan, S. K. (1972). *J. Inorg. Nucl. Chem.*, **34,** 3081
92. Savant, V. V. and Patel, C. C. (1972). *J. Inorg. Nucl. Chem.*, **34,** 1462
93. Potts, R. A. (1972). *J. Inorg. Nucl. Chem.*, **34,** 1749
94. Price, J. H., Williamson, A. N., Schramm, R. F. and Wayland, B. B. (1972). *Inorg. Chem.*, **11,** 1280
95. Donohue, J. and Tsai, C. C. (1972). Quoted in Ref. 94
96. Vigee, G. S. and Ng, P. (1971). *J. Inorg. Nucl. Chem.*, **33,** 2477
97. Frankel, L. S. (1971). *Inorg. Chem.*, **10,** 814
98. Jacob, S. W., Rosenbaum, E. E. and Wood, D. C. (1971). *Dimethylsulphoxide.* (New York: Marcel Dekker)
99. Laing, M., Baines, S. and Somerville, P. (1971). *Inorg. Chem.*, **10,** 1057
100. Stephens, F. S. (1972). *J. Chem. Soc. Dalton Trans.*, 1350
101. McPherson, G. I., Weil, J. A. and Kinnaird, J. K. (1971). *Inorg. Chem.*, **10,** 1574
102. Cleveland, J. M., Bryan, G. H. and Sironen, R. J. (1972). *Inorg. Chim. Acta*, **6,** 54
103. Boer, F. P., Carter, V. B. and Turley, J. W. (1971). *Inorg. Chem.*, **10,** 651
104. Jeter, D. Y., Hodgson, D. D. and Hatfield, W. E. (1972). *Inorg. Chem.*, **11,** 185
105. Drago, R. S. and Chiang, R. L. (1971). *Inorg. Chem.*, **10,** 453
106. Buckingham, A. D., Foxman, B. M. and Sargeson, A. M. (1970). *Inorg. Chem.*, **9,** 1790

107. Foxman, B. M. (1972). *Chem. Commun.*, 515
108. Rorabacher, D. B. and Melenclez-Cepeda, C. A. (1971). *J. Amer. Chem. Soc.*, **93,** 6071
109. Hansen, L. D. and Temer, D. J. (1971). *Inorg. Chem.*, **10,** 1439
110. Fine, D. A. (1971). *Inorg. Chem.*, **10,** 1825
111. Gans, P. and Marriage, J. (1972). *J. Chem. Soc. Dalton Trans.*, 1738
112. Essen, L. N. and Bukhtizarova, T. N. (1969). *Russ. J. Inorg. Chem.*, **14,** 242
113. Sawai, H. and Hirai, H. (1971). *Inorg. Chem.*, **10,** 2068
114. Bertrand, J. A., Howard, W. J. and Kalyanaraman, A. R. (1971). *Chem. Commun.*, 437
115. Guilbault, G. G. and Billedeau, S. M. (1972). *J. Inorg. Nucl. Chem.*, **34,** 1167
116. Goggin, P. L., Goodfellow, R. J. and Reed, F. J. S. (1972). *J. Chem. Soc. Dalton Trans.*, 1298
117. Greene, P. T., Russ, B. J. and Wood, J. S. (1971). *J. Chem. Soc. A*, 3636
118. McWhinnie, W. R., Miller, J. D., Watts, J. B. and Waddam, D. Y. (1971). *Chem. Commun.*, 629
119. Dudley, R. J., Hathaway, B. J. and Hodgson, P. G. (1971). *J. Chem. Soc. A*, 3355
120. Garner, C. D., Lambert, P., Mabbs, F. E. and Porter, J. K. (1972). *J. Chem. Soc. Dalton Trans.*, 320
121. Osterberg, R., Sjoberg, B. and Soderquist, R. (1970). *Chem. Commun.*, 1408
122. Velicko, F. K., Kuzmino, N. A. and Ermalova, L. D. (1965). *Zh. Priklad. Khim.*, **38,** 153
123. Nakahara, A., Hidaka, J. and Tsuchida, R. (1956). *Bull. Chem. Soc. Japan*, **29,** 925
124. Butcher, A. V., Phillips, D. J. and Redfern, J. P. (1971). *J. Chem. Soc. A*, 1640, 2104
125. Key, D. L., Larkworthy, L. F. and Salmon, J. E. (1971). *J. Chem. Soc. A*, 371, 2583
126. Baldwin, D. A. and Clark, R. J. H. (1971). *J. Chem. Soc. A*, 1725
127. Sariego, R. and Costamagna, J. A. (1971). *J. Inorg. Nucl. Chem.*, **33,** 1528
128. Birnbaum, E. R. (1971). *J. Inorg. Nucl. Chem.*, **33,** 3031
129. Wayland, B. B. and Wisniewski, M. D. (1971). *Chem. Commun.*, 1025
130. Angus, J. R., Woltermann, G. M., Vincent, W. R. and Wasson, J. R. (1971). *J. Inorg. Nucl. Chem.*, **33,** 3041
131. Allen, E. A., Johnson, N. P., Rosevear, D. T. and Wilkinson, W. (1971). *J. Chem. Soc. A*, 2141
132. Contreras, G. and Astigarrabia, E. (1971). *Inorg. Chim. Acta*, **5,** 54
133. Vallarino, L. M., Goedkin, V. L. and Quagliano, J. V. (1972). *Inorg. Chem.*, **11,** 1466
134. Murthy, A. S. N., Quagliano, J. V. and Vallarino, L. M. (1972). *Inorg. Chim. Acta*, **6,** 49
135. Pyke, S. C. and Linch, R. G. (1971). *Inorg. Chem.*, **10,** 2445
136. Johnson, D. A. and Delphin, W. H. (1971). *Inorg. Nucl. Chem. Lett.*, **7,** 717
137. Nakamoto, K. (1970). *Infrared Spectra of Inorganic and Coordination Compounds* 2nd edn. (New York: Wiley—Interscience)
138. Iwamoto, T. and Shriver, D. F. (1971). *Inorg. Chem.*, **11,** 2428
139. Dubicki, L. and Day, P. (1971). *Inorg. Chem.*, **10,** 2043
140. Lever, A. B. P. and Mantovani, E. (1971). *Inorg. Chim. Acta*, **5,** 429; *Inorg. Chem.*, **10,** 817
141. Lever, A. B. P., Mantovani, E. and Donini, J. C. (1971). *Inorg. Chem.*, **10,** 2424
142. Ponticelli, G. (1971). *Inorg. Chim. Acta*, **5,** 461; Ponticelli, G. and Preti, C. (1972). *J. Chem. Soc. Dalton Trans.*, 708
143. House, D. A., Ireland, P. R., Maxwell, I. E. and Robinson, W. T. (1972). *Inorg. Chim. Acta*, **5,** 397
144. Cannas, M., Carta, G. and Marongiu, G. (1971). *Chem. Commun.*, 623
145. Larkworthy, L. F., Patel, K. C. and Trigg, J. K. (1971). *J. Chem. Soc. A*, 2766
146. McLean, J. A. and Goorman, R. I. (1971). *Inorg. Nucl. Chem. Lett.*, **7,** 9
147. Barbucci, R., Fabrizzi, L. and Paoletti, P. (1972). *J. Chem. Soc., Dalton Trans.*, 740, 2593; Arenare, E., Paoletti, P., Dei, A. and Vacca, A. (1972). *J. Chem. Soc. Dalton Trans.*, 736
148. Turan, T. S. and Rorabacher, D. B. (1972). *Inorg. Chem.*, **11,** 288
149. Newman, M. S., Busch, D. H., Chesney, G. E. and Gustafson, C. R. (1972). *Inorg. Chem.*, **11,** 2890
150. Graybill, G. R., Wrathall, J. W. and Ihrig, J. L. (1972). *Inorg. Chem.*, **11,** 722
151. Patel, C. C. and Goldberg, D. E. (1972). *Inorg. Chem.*, 759
152. Fukuda, Y. and Sone, K. (1972). *Bull. Chem. Soc. Japan*, **45,** 465
153. Evilia, R. F., Young, D. C. and Reilly, C. N. (1971). *Inorg. Chem.*, **10,** 433
154. Belford, R., Fenton, D. E. and Truter, M. R. (1972). *J. Chem. Soc. Dalton Trans.*, 2345
155. Watt, G. W. and Cuddeback, I. E. (1971). *Inorg. Chem.*, **10,** 947
156. Goedkin, V. (1972). *Chem. Commun.*, 207
157. Schlapfer, C. W., Saito, Y. and Nakamoto, K. (1972). *Inorg. Chim. Acta*, **6,** 284
158. Ghilardi, C. A. and Orlandini, A. B. (1972). *J. Chem. Soc. Dalton Trans.*, 1698

159. Fereday, R. J. and Hathaway, B. J. (1972). *J. Chem. Soc. Dalton Trans.*, 197
160. Bew, M. J., Hathaway, B. J. and Fereday, R. J. (1972). *J. Chem. Soc. Dalton Trans.*, 1229
161. Dapports, P. and Di Vaira, M. (1971). *J. Chem. Soc. A*, 1891
162. Di Vaira, M. and Orlandini, A. B. (1972). *J. Chem. Soc. Dalton Trans.*, 1704; Sacconi, L. and Morassi, R. (1971). *J. Chem. Soc. A*, 1969, 2904
163. Searle, G. H. and Keene, F. R. (1972). *Inorg. Chem.*, **11**, 1006
164. Forsberg, J. H. and Wathen, C. A. (1971). *Inorg. Chem.*, **10**, 1379
165. Skelton, B. W., Waters, T. N. and Curtis, N. F. (1972). *J. Chem. Soc. Dalton Trans.*, 2133
166. Zido, R. F., Allen, M., Titus, D. D., Gray, H. B. and Dori, Z. (1972). *Inorg. Chem.*, **11**, 304
167. Brubaker, G. R. and Schaeffer, D. P. (1971). *Inorg. Chem.*, **10**, 968
168. Lewis, R. H. and Alexander, M. D. (1971). *Inorg. Chim. Acta*, **5**, 86
169. Payne, N. C. (1972). *Inorg. Chem.*, **11**, 1376
170. Alexander, M. D. (1972). *J. Inorg. Nucl. Chem.*, **34**, 387
171. Barbucci, R., Fabbrizzi, L. and Paoletti, P. (1972). *J. Chem. Soc. Dalton Trans.*, 745, 1099, 1529
172. Lincoln, S. F. and West, R. J. (1972). *Aust. J. Chem.*, **25**, 469
172a. Cara, E., Cristini, A., Diaz, A. and Ponticelli, G. (1972). *J. Chem. Soc. Dalton Trans.*, 527, 1361
173. Lincoln, S. F. and West, R. J. (1972). *Aust. J. Chem.*, **25**, 469
174. Turley, J. W. and Asberger, R. G. (1971). *Inorg. Chem.*, **10**, 558
175. Gibson, J. G. and McKenzie, E. D. (1971). *J. Chem. Soc. A*, 1029
176. Wilder, R. I., Kamp, D. A. and Garner, C. S. (1971). *Inorg. Chem.*, **10**, 1393
177. Snow, M. R. (1971). *J. Chem. Soc. Dalton Trans.*, 1627
178. Black, D. St. C. and McLean, I. A. (1971). *Aust. J. Chem.*, **24**, 1391
179. Carr, J. D. and Vasillades, J. (1972). *Inorg. Chem.*, **11**, 2104
180. Lui, C-H. and Grier, M. W. (1972). *Chem. Commun.*, 656
181. Forsberg, J. H., Kubik, T. M., Moeller, T. and Gucwa, K. (1971). *Inorg. Chem.*, **10**, 2656
182. Sime, R. S., Dodge, R. P., Zalkin, A. and Templeton, D. H. (1971). *Inorg. Chem.*, 537
183. Bertini, I., Ciampolini, M., Dapporto, P. and Gatteschi, D. (1972). *Inorg. Chem.*, **11**, 2254
184. Schafer, J. L. and Raymond, K. N. (1971). *Inorg. Chem.*, **10**, 1799
185. Maxwell, I. E. (1971). *Inorg. Chem.*, **10**, 1782
186. Collins, R. K. and Drew, M. G. B. (1972). *Inorg. Nucl. Chem. Lett.*, **8**, 975
187. Zanella, A. and Taube, H. (1971). *J. Amer. Chem. Soc.*, **93**, 7166
188. Zwickel, A. M. and Creutz, C. (1971). *Inorg. Chem.*, **10**, 2398
189. Tong, M. and Brewer, D. G. (1971). *Can. J. Chem.*, **49**, 102
190. Cameron, A. F., Taylor, D. W. and Nuttall, R. H. (1972). *J. Chem. Soc. Dalton Trans.*, **58**, 422, 1603
191. Bonamico, M., Dessy, G., Fares, V. and Scaramuzza, (1972). *J. Chem. Soc. Dalton Trans.*, 2477
192. Bushnell, G. W. (1971). *Can. J. Chem.*, **49**, 555
193. Davey, G. and Stephens, F. S. (1971). *J. Chem. Soc. A*, 1917, 2577
194. Shepperd, T. M. (1972). *J. Chem. Soc. Dalton Trans.*, 813
195. Caira, M. R., Haigh, J. M. and Nassimbeni, L. R. (1972). *J. Inorg. Nucl. Chem.*, **34**, 3171 (1972). *Inorg. Nucl. Chem. Lett.*, **8**, 109
196. Konig, E. and Thomas, G. (1972). *J. Inorg. Nucl. Chem.*, **34**, 1173
197. Lewin, A. H., Michl, R. J., Ganis, P., Lepore, U. and Avitabile, G. (1971). *Chem. Commun.*, 1400
198. Chen, K-L. H. and Iwamoto, R. T. (1971). *Inorg. Chim. Acta*, **5**, 92
199. Lewin, A. H., Michl, R. J., Ganis, P. and Lepore, U. (1972). *Chem. Commun.*, 661
200. Lang, M. and Carr, G. (1971). *J. Chem. Soc. A*, 1141
201. Raichart, D. W. and Taube, H. (1972). *Inorg. Chem.*, **11**, 999
202. Cheng, P-T., Loescher, B. R. and Nyburg, S. C. (1971). *Inorg. Chem.*, **10**, 1275
203. Merrithew, P. B., Rasmussen, P. G. and Vincent, D. H. (1971). *Inorg. Chem.*, **10**, 1401
204. Forster, D. and Dahm, D. J. (1972). *Inorg. Chem.*, **11**, 918
205. Long, G. J. and Baker, W. A. (1971). *J. Chem. Soc. A*, 2956
206. Long, G., Whitney, D. L. and Kennedy, J. E. (1971). *Inorg. Chem.*, **10**, 1406
207. Singh, P., Jeter, D. Y., Hatfield, W. E. and Hodgson, D. J. (1972). *Inorg. Chem.*, **11**, 1657
208. Duckworth, V. F. and Stephenson, N. C. (1969). *Acta Crystallogr. B*, **25**, 1795
209. Jeter, D. Y., Hodgson, D. J. and Hatfield, W. E. (1971). *Inorg. Chem. Acta*, **5**, 257. See also Hyde, K. E., Quinn, B. C. and Yang, I. P. (1971). *J. Inorg. Nucl. Chem.*, **33**, 2377

210. Jolly, K. W., Buckley, P. D. and Blackwell, L. F. (1972). *Aust. J. Chem.*, **25**, 1311
211. Farma, R. D. and Swinehart, J. H. (1972). *Inorg. Chem.*, **11**, 645
212. Goodgame, M. and Hayward, P. J. (1971). *J. Chem. Soc. A*, 3406
213. Watt, G. W., Thompson, L. K. and Pappas, A. J. (1972). *Inorg. Chem.*, **11**, 747
214. Vaska, L. and Yamaji, T. (1971). *J. Amer. Chem. Soc.*, **93**, 6673
215. Aug, H. G., Kow, W. E. and Mok, K. F. (1972). *Inorg. Nucl. Chem. Lett.*, **8**, 829
216. Sutton, G. J. (1971). *Aust. J. Chem.*, **24**, 919
217. Suzuki, S., Nakahara, M. and Watanabe, K. (1971). *Bull. Soc. Chem. Japan*, **44**, 1441
218. Sigel, H. (1972). *Inorg. Chim. Acta*, **6**, 195
219. Noack, M. and Gordon, G. (1968). *J. Chem. Phys.*, **48**, 2689
220. Walker, F. A. and Sigel, H. (1972). *Inorg. Chem.*, **11**, 1162
221. Berka, L. H., Edwards, W. T. and Christian, P. A. (1971). *Inorg. Nucl. Chem. Lett.*, **7**, 265
222. Carty, A. J. and Chieh, P. C. (1972). *Chem. Commun.*, 158; Chieh, P. C. (1972). *J. Chem. Soc. Dalton Trans.*, 1643; Hinamoto, J., Ooi, S. and Kuroya, H. (1972). *Chem. Commun.*, 356
223. Nakai, H. (1971). *Bull. Chem. Soc. Japan*, **44**, 2412
224. Anderson, O. P. (1972). *J. Chem. Soc. Dalton Trans.*, 2597
225. Hinamoto, M., Ooi, S. and Kuroya, H. (1971). *Bull. Chem. Soc. Japan*, **44**, 586
226. Daly, J. J. and Sanz, F. (1972). *J. Chem. Soc. Dalton Trans.*, 2584
227. Al-Karaghouli, A. R. and Wood, J. S. (1972). *Inorg. Chem.*, **11**, 2293
228. Gidney, P. M., Gillard, R. D. and Heaton, B. T. (1972). *J. Chem. Soc. Dalton Trans.*, 2621; Gillard, R. D., Heaton, B. T. and Vaughan, D. H. (1971). *J. Chem. Soc. A*, 1840
229. Kulasingham, G. C. (1971). *Inorg. Chim. Acta*, **5**, 180
230. Irving, H. and Mellor, M. J. (1962). *J. Chem. Soc.*, 5237
231. Konig, E., Ritter, G., Spiering, H., Kremer, S., Madeja, K. and Rosenkranz, A. (1972). *J. Chem. Phys.*, **56**, 3139
232. Konig, E., Ritter, G., Madeja, K. and Rosenkranz, A. (1972). *J. Inorg. Nucl. Chem.*, **34**, 2879
233. Saito, Y., Takemoto, J., Hutchinson, B. and Nakamoto, K. (1972). *Inorg. Chem.*, **11**, 2103; Takemoto, J., Hutchinson, G. and Nakamoto, K. (1971). *Chem. Commun.*, 1007
234. Bryant, G. M., Ferguson, J. E. and Powell, H. K. J. (1971). *Aust. J. Chem.*, **24**, 257, 275
235. David, P. G. (1972). *Chem. Commun.*, 1151; David, P. G., Richardson, J. G. and Wehry, E. L. (1971). *Inorg. Nucl. Chem. Lett.*, **7**, 251
236. Wehry, E. L. and Ward, R. A. (1971). *Inorg. Chem.*, **10**, 2660
237. Sundararajan, S. and Wehry, E. L. (1972). *J. Inorg. Nucl. Chem.*, **34**, 3699
238. Tanaka, N., Ogata, T. and Niizuma, S. (1972). *Inorg. Nucl. Chem. Lett.*,
239. Plaksin, P. M., Stoufer, R. C., Matthew, M. and Palenik, G. J. (1972). *J. Amer. Chem. Soc.*, **94**, 2121
240. Briscoe, G. B., Fernandopulle, M. E. and McWhinnie, W. R. (1972). *Inorg. Chim. Acta*, **6**, 598
241. Kassierer, E. F. and Kertes, A. S. (1972). *J. Inorg. Nucl. Chem.*, **34**, 3209
242. Kertes, A. S. and Kassierer, E. F. (1972). *Inorg. Chem.*, **11**, 2108
243. Yamamoto, T., Yamamoto, Y. and Ikeda, S. (1971). *J. Amer. Chem. Soc.*, **93**, 3360
244. Hutchinson, B. and Sutherland, A. (1972). *Inorg. Chem.*, **11**, 1948
245. Foster, R. J., Bodner, R. L. and Hendriker, D. G. (1972). *J. Inorg. Nucl. Chem.*, **34**, 3795
246. Hendricker, D. G. and Foster, R. J. (1972). *Inorg. Chem.*, **11**, 1949
247. Hendricker, D. G. and Bodner, R. L. (1970). *Inorg. Chem.*, **9**, 273
248. Emad, A. and Emerson, K. (1972). *Inorg. Chem.*, **11**, 2288
249. Goldstein, M., Taylor, F. B. and Unsworth, W. D. (1972). *J. Chem. Soc. Dalton Trans.*, 418; Carrick, P. W., Goldstein, M., McPartlin, E. M. and Unsworth, W. D. (1971). *Chem. Commun.*, 1634
250. Lever, A. B. P., Lewis, J. and Nyholm, R. S. (1962). *J. Chem. Soc.*, 1235
251. Villa, J. F. and Hatfield, W. E. (1971). *J. Amer. Chem. Soc.*, **93**, 4081; Inman, G. W. and Hatfield, W. E. (1972). *Inorg. Chem.*, **11**, 3085
252. Mayok, B. and Day, P. (1972). *J. Amer. Chem. Soc.*, **94**, 2885
253. Matthews, R. W. and Walton, R. A. (1971). *Inorg. Chem.*, **10**, 1433
254. Lim, H. S., Barclay, D. J. and Anson, F. C. (1972). *Inorg. Chem.*, **11**, 1460
255. Carmichael, W. M. and Edwards, D. A. (1972). *J. Inorg. Nucl. Chem.*, **34**, 1181
256. Allan, J. R., Barnes, G. A. and Brown, D. H. (1971). *J. Inorg. Nucl. Chem.*, **33**, 3765
257. Nicholls, D. and Warburton, B. A. (1971). *J. Inorg. Nucl. Chem.*, **73**, 1041
258. Reedijk, J., Stork-Blaisse, B. A. and Verschoor, G. C. (1971). *Inorg. Chem.*, **10**, 2594; Reedijk, J., Windhorst, J. C. A., van Ham, N. H. M. and Groeneveld, W. L. (1971). *Rec. Trav. Chim.*, **91**, 234

259. Dowsing, R. D., Nieuwenhuijse, B. and Reedijk, J. (1971). *Inorg. Chim. Acta*, **5**, 301
260. Cotton, S. A. and Gibson, J. F. (1971). *J. Chem. Soc. A*, 1696
261. Vaughan, J. D. and Smith, W. A. (1972). *J. Amer. Chem. Soc.*, **94**, 2460
262. Davis, W. J. and Smith, J. (1971). *J. Chem. Soc. A*, 317
263. Reedijk, J. (1971). *J. Inorg. Nucl. Chem.*, **33**, 179
264. Hitchman, M. A. (1972). *Inorg. Chem.*, **11**, 2389
265. Goodgame, D. M. L., Goodgame, M. and Rayner-Canham, G. W. (1971). *J. Chem. Soc. A*, 1923
266. Huq, F. and Skapski, A. C. (1971). *J. Chem. Soc. A*, 1927
267. Goodgame, D. M. L., Goodgame, M. and Rayner-Canham, G. W. (1972). *Inorg. Chim. Acta*, **6**, 245
268. Bose, K. S. and Patel, C. C. (1971). *J. Inorg. Nucl. Chem.*, **33**, 755
269. Carmichael, J. W., Chan, N., Cordess A. W., Fair, C. K. and Johnson, D. A. (1972). *Inorg. Chem.*, **11**, 1117
270. Prout, C. K., Allison, G. B. and Rosotti, F. J. C. (1971). *J. Chem. Soc. A*, 3331
271. Bowman, K., Gaughan, A. P. and Dori, Z. (1972). *J. Amer. Chem. Soc.*, **94**, 727
272. Letter, J. E. and Jordan, R. B. (1971). *Inorg. Chem.*, **10**, 2692
273. Campbell, M. J. M., Card, D. W. and Grzeskowiak, R. (1972). *J. Chem. Soc. Dalton Trans.*, 1687
274. Goodgame, M. and Piggott, B. (1971). *J. Chem. Soc. A*, 826
275. Eilbeck, W. J., Holmes, F. and Underhill, A. E. (1967). *J. Chem. Soc. A*, 757
276. Peach, M. E. and Ramaswamy, K. K. (1971). *Inorg. Chim. Acta*, **5**, 445; Rivest, R. and Weisz, A. *Can. J. Chem.*, **49**, 1750
277. Hughes, M. N. and Rutt, K. J. (1971). *Inorg. Nucl. Chem. Lett.*, **49**, 1750
278. Baenziger, N. C. and Schultz, R. J. (1971). *Inorg. Chem.*, **10**, 661
279. Bowers, D. M., Erlich, R. H., Policec, S. and Popov, A. I. (1971). *J. Inorg. Nucl. Chem.*, **33**, 81
280. Baucom, E. I. and Drago, R. S. (1971). *J. Amer. Chem. Soc.*, 6499; idem (1972). *Inorg. Chem.*, **11**, 2064
281. Lees, G., Holmes, F., Underhill, A. E. and Powell, D. B. (1971). *J. Chem. Soc. A*, 337
282. Sen, B. and Malone, D. (1972). *J. Inorg. Nucl. Chem.*, **34**, 3509; Holmes, F., Lees, G. and Underhill, A. E. (1971). *J. Chem. Soc. A*, 999
283. Goodwin, H. A. and Smith, F. E. (1972). *Aust. J. Chem.*, **25**, 37
284. Wandiga, S. O., Sarneski, J. E. and Urbach, F. L. (1972). *Inorg. Chem.*, **11**, 1349
285. Spencer, C. T. and Taylor, L. T. (1971). *Inorg. Chem.*, **10**, 2407
286. Cressey, M., McKenzie, E. D. and Yates, S. (1971). *J. Chem. Soc. A*, 2677. See also Mangia, A., Nardeli, M., Pelizzi, C. and Pelizzi, G. (1972). *J. Chem. Soc. Dalton Trans.*, 996
287. Farrington, D. J. and Jones, J. G. (1972). *Inorg. Chim. Acta*, **6**, 575
288. Wentworth, R. A. D. (1971). *Inorg. Chem.*, **10**, 2615
289. Chum, H-L, da Costa, A. M. and Krumholz, P. (1972). *Chem. Commun.*, 772
290. Nelson, S. M., Boylan, M. J. and Deeney, F. A. (1971). *J. Chem. Soc. A*, 976
291. Keeton, M. and Lever, A. B. P. (1971). *Inorg. Chem.*, **10**, 47
292. Goodgame, D. M. L. and Machado, A. A. (1972). *Inorg. Chim. Acta*, **6**, 317
293. Dosser, R. J. and Underhill, A. E. (1972). *J. Chem. Soc. Dalton Trans.*, 611
294. Cotton, D. F. and Geary, W. J. (1971). *J. Chem. Soc. A*, 2457; (1972). *J. Chem. Soc. Dalton Trans.*, 547
295. Prasad, J. and Peterson, N. C. (1971). *Inorg. Chem.*, **10**, 88; Frazer, F. H., Epstein, P. and Macero, D. J. (1972). *Inorg. Chem.*, **11**, 2031
296. Lever, A. B. P., Thompson, L. K. and Rieff, W. M. (1971). *Inorg. Chem.*, **10**, 104
297. Fritchie, C. J. (1972). *Chem. Commun.*, 1220
298. Kowala, C., Murray, K. S., Swan, J. M. and West, B. O. (1971). *Aust. J. Chem.*, **24**, 1369
299. Goodwin, H. A. and Mather, D. W. (1972). *Aust. J. Chem.*, **25**, 715
300. Montenero, A. and Pelizzi, C. (1972). *Inorg. Chim. Acta*, **6**, 88
301. Livingstone, S. E. and Nolan, J. D. (1972). *Chem. Commun.*, 218
302. Johnston, D. L., Rohrbaugh, W. L. and Horrocks, W. D. (1971). *Inorg. Chem.*, **10**, 547
303. McWhinnie, W. R. (1964). *J. Chem. Soc.*, 5165
304. Lancaster, J. C. and McWhinnie, W. R. (1971). *J. Chem. Soc. A*, 1742
305. Johnson, J. E., Beineke, T. A. and Jacobson, R. A. (1971). *J. Chem. Soc. A*, 1371
306. Dudley, R. J., Hathaway, B. J. and Hodgson, P. G. (1972). *J. Chem. Soc. Dalton Trans.*, 882
307. Lancaster, J. C. and McWhinnie, W. R. (1971). *Inorg. Chim. Acta*, **5**, 515

308. Copeland, V. C., Singh, P., Hatfield, W. E. and Hodgson, D. J. (1972). *Inorg. Chem.*, **11**, 1826; Jeter, D. Y., Lewis, D. L., Hempel, J. C., Hodgson, D. J. and Hatfield, W. E. *ibid.*, **11**, 1958; Lewis, D. L., Hatfield, W. E. and Hodgson, D. J. *ibid.*, 2216; Bailey, N. A. and Bowler, S. J. (1971). *J. Chem. Soc. A*, 1763
309. Da Mota, M. M., Rodgers, J. and Nelson, S. M. (1969). *J. Chem. Soc. A*, 2036
310. Rodgers, J. and Jacobson, R. A. (1970). *J. Chem. Soc. A*, 1826
311. Drew, M. G. B., Riedl, M. J. and Rodgers, J. (1972). *J. Chem. Soc. Dalton Trans.*, 234; Collins, R. K., Drew, M. G. B. and Rodgers, J. (1972). *J. Chem. Soc. Dalton Trans.*, 899
312. Gibson, J. G. and McKenzie, E. D. (1971). *J. Chem. Soc. A*, 1666
313. Bailey, N. A. and McKenzie, E. D. (1972). *J. Chem. Soc. Dalton Trans.*, 1566
314. Mazurek, W. and Casey, A. T. (1971). *Aust. J. Chem.*, **24**, 501
315. Riley, P. E. and Seff, K. (1972). *Inorg. Chem.*, **11**, 2993
316. Black, D. St. C. and McLean, I. A. (1971). *Aust. J. Chem.*, **24**, 1377
317. Stapfer, C. H. and D'Andrea, R. W. (1971). *Inorg. Chem.*, **10**, 1224; Stapfer, C. H., D'Andrea, R. W. and Herber, R. H. (1972). *Inorg. Chem.*, **11**, 204
318. Alcock, J. F., Baker, R. J. and Diamantis, A. A. (1972). *Aust. J. Chem.*, **25**, 289
319. Ahmed, A. D. and Chaudhuri, N. R. (1971). *J. Inorg. Nucl. Chem.*, **33**, 189
320. Dutt, N. K. and Chakder, N. C. (1971). *J. Inorg. Nucl. Chem.*, **33**, 393; (1972). *Inorg. Chim. Acta*, **5**, 188
321. Butler, W. M. and Enemark, J. H. (1971). *Inorg. Chem.*, **10**, 2416
322. Chugaev, L., Skanavy-Grigorieva, M. and Posniak, A. (1925). *Z. Anorg. Allgem. Chem.*, **143**, 37
323. Iskander, M. F. and El-Sayed, L. (1971). *J. Inorg. Nucl. Chem.*, **33**, 4253
324. Akbar, M., Livingstone, S. E. and Phillips, D. J. (1972). *Inorg. Chim. Acta*, **6**, 493, 552
325. Bigoli, F., Braibanti, A., Langfredi, A. M. M., Tiripicchio, A. and Camellini, M. T. (1971). *Inorg. Chim. Acta*, **5**, 392
326. Caglioti, L., Cattalini, L., Ghedini, M., Gasparrini, F. and Vigato, P. A. (1972). *J. Chem. Soc. Dalton Trans.*, 514
327. Price, R. (1971). *J. Chem. Soc. A*, 3379
328. Goedkin, V. L., Vallarino, L. M. and Quagliano, J. V. (1971). *Inorg. Chem.*, **10**, 2682
329. Klein, H-F. and Nixon, J. F. (1971). *Chem. Commun.*, 42
330. Otsuka, S., Yoshida, T. and Tatsuno, Y. (1971). *Chem. Commun.*, 67; see also Dickson, R. S. and Ibers, J. A. (1972). *J. Amer. Chem. Soc.*, **94**, 2988
331. Caglioti, L., Cattalini, L., Gasparrini, F., Marangoni, G. and Vigato, P. A. (1971). *J. Chem. Soc. A*, 324
332. Bennett, R. P. (1970). *Inorg. Chem.*, **9**, 2184
333. Herberhold, M. and Golla, W. (1971). *J. Organometal. Chem.*, **26**, C27
334. Little, R. G. and Doedens, R. J. (1972). *Inorg. Chem.*, **11**, 1392
335. Carty, A. J., Madden, D. P., Matthew, M., Palenik, G. and Birchall, T. (1970). *Chem. Commun.*, 1664
336. Streith, J. and Cassal, J. M. (1969). *Bull. Soc. Chim. France*, 2175
337. Smith, R. A., Madden, D. P. and Carty, A. J. (1971). *Chem. Commun.*, 427
338. Tonioto, L. and Eisenberg, R. (1971). *Chem. Commun.*, 455
339. Einstein, F. W. B., Gilchrist, A. B., Rayner-Canham, G. W. and Sutton, D. (1971). *J. Amer. Chem. Soc.*, **93**, 1826
340. Watanabe, Y., Mitsudo, T., Tanaka, M. and Yamamoto, K. (1972). *Bull. Chem. Soc. Japan*, **45**, 928
341. Bodner, R. L. and Popov, A. I. (1972). *Inorg. Chem.*, **11**, 1410; Baenziger, N. C. and Schultz, R. J. (1971). *Inorg. Chem.*, **10**, 661
342. Biefield, R. M. and Gilbert, G. L. (1971). *J. Inorg. Nucl. Chem.*, **33**, 3950
343. Labine, P. and Brubaker, C. H. (1971). *J. Inorg. Nucl. Chem.*, **33**, 3383
344. Hessett, B., Morris, J. H. and Perkins, P. G. (1971). *J. Chem. Soc. A*, 2466
345. Walton, R. A. (1965). *Quart. Rev.* (*Chem. Soc.*), 126; Reedijk, J., Zuur, A. P. and Groeneveld, W. L. (967). *Rec. Trav. Chim.*, **86**, 1103
346. Farona, M. F. and Kraus, K. F. (1970). *Inorg. Chem.*, **9**, 1700
347. Dunn, J. G. and Edwards, D. A. (1971). *Chem. Commun.*, 482
348. Farona, M. F. and Kraus, K. F. (1972). *Chem. Commun.*, 513
349. Clarke, R. E. and Ford, R. C. (1970). *Inorg. Chem.*, **9**, 227; Foust, R. D. and Ford, P. C. (1972). *Inorg. Chem.*, **11**, 899

350. Tolman, C. A. (1971). *Inorg. Chem.*, **10,** 1540
351. Catsikis, B. D. and Good, M. L. (1971). *Inorg. Chem.*, **10,** 1522
352. Einstein, F. W. B., Enwall, E., Morris, D. M. and Sutton, D. (1971). *Inorg. Chem.*, **10,** 678
353. Hoff, G. R. and Brubaker, C. H. (1971). *Inorg. Chem.*, **10,** 2063
354. Fowles, G. W. A., Moss, K. C., Rice, D. A. and Rolfe, N. (1972). *J. Chem. Soc. Dalton Trans.*, 915
355. Drew, M. G. B., Fowles, G. W. A., Rice, D. A. and Rolfe, N. (1971). *Chem. Commun.*, 231
356. Miles, M. G., Hursthouse, M. B. and Robinson, A. G. (1971). *J. Inorg. Nucl. Chem.*, **33,** 2015
357. Spaulding, L., Reinhardt, B. A. and Orchin, M. (1972). *Inorg. Chem.*, **11,** 2092
358. Buckingham, D. A., Foxman, B. M., Sargeson, A. M. and Zarella, A. (1972). *J. Amer. Chem. Soc.*, **94,** 1007
359. Buckingham, D. A., Sargeson, A. M. and Zanella, A. (1972). *J. Amer. Chem. Soc.*, **94,** 8246
360. Clark, H. C. and Manzer, L. E. (1971). *Chem. Commun.*, 387
361. Wolsey, W. C., Huestis, W. H. and Theyson, T. W. (1972). *J. Inorg. Nucl. Chem.*, **34,** 2359

7
Structural Chemistry of Transition Metal–Metalloid Compounds

WOLFGANG JEITSCHKO
E. I. du Pont de Nemours and Company, Wilmington, Delaware

7.1 INTRODUCTION

This review reports on compounds of transition metals with the elements B, C, Si, Ge, N, P, As, S, Se and Te. Both binary and ternary compounds containing these elements are treated. Most of the work included was published in the period from 1971 to early 1973. In some areas, where no recent reviews were available, literature from 1968 to 1970 is referred to also.

An attempt is made to give an account which can be read with profit by the non-specialist, e.g. a graduate student just beginning his thesis work. For that reason each subchapter begins with a few paragraphs in which general structural characteristics, properties, and applications are described briefly and previous reviews are cited.

The ultimate aim of the structural chemist is, of course, not only to determine crystal structures but also to 'understand' them: not only the position of the nuclei but also the state of the electrons needs to be described. For this latter task knowledge of near-neighbour coordinations and distances is frequently not sufficient since simple ionic or covalent bonding models are of little use for these compounds. Magnetic, electrical, optical and other physical data help in this difficult pursuit. Thus work dealing with physical properties is also referred to.

Nearly all binary systems of transition metals with metalloids have been investigated over the past 20 years and most simple crystal structures have been determined. Most of the recent work on binary compounds has been carried out on complicated structures especially on transition metal-rich phosphides, sulphides and selenides. Electron diffraction techniques have become valuable for the characterisation of ordered defect structures and polytypes which form at moderate temperatures in many phasefields previously thought to be single phase. Examples of such work are studies on transition metal-rich nitrides and oxides, on carbides with defect-NaCl type structures, and on sulphides with layer structures.

In ternary systems some of the most interesting work has been on new ternary borides with planar boron nets made up from 5-, 6- and 7-membered rings; on ternary compounds of silicon with rare-earth and late transition metals which combine structural characteristics of ionic and intermetallic compounds; on two-dimensional superconducting 'filled' CdI_2 type dichalcogenides where metal atoms or organic Lewis bases are intercalated between layers of chalcogen atoms; and on ternary superconductors with the filled Mo_3S_4 type structure. Many interesting new compounds belonging to now well established structural families—e.g. the σ-phase related structures or ternary carbides with octahedral carbon coordination—were characterised structurally. Work on physical properties dominates the literature on the late transition metal phosphides, arsenides, and especially chalcogenides where more localised bonding facilitates rationalisation of experimental results.

In this series transition metal–metalloid compounds were discussed by Ward[1]; Nowotny[2] reviewed borides and silicides; chalcogenides (sulphides, selenides, tellurides) were treated by Jellinek[3] and Flahaut[4]. Hulliger[5] reviewed the structural chemistry of chalcogenides and pnictides (phosphides, arsenides, antimonides). The literature up to 1965 is most readily found in Pearson's handbooks[6]. Excellent drawings of most structural types encountered for transition metal–metalloid compounds are given in monographs by Schubert[7] and Pearson[8]. More specialised and earlier review articles and books are referred to in the appropriate sections of this article.

7.2 BORIDES

Many properties of transition metal borides are similar to those of the carbides and silicides. They have high melting points (up to 3300 K), great hardness and brittleness, and metallic conductivity. They are chemically stable and inert even at high temperatures. They can, however, usually not compete with transition metal carbides and silicides, and technical uses are so far very limited in spite of much research especially during the past two decades. Preparation and properties of transition metal borides were summarised previously[9–12].

The structural chemistry of boron is most fascinating. Among the structures of the elements, the structures of the various boron modifications are the most complex ones. The variety of its compounds with hydrogen is rivalled only by the carbon hydrides. Among minerals, borates combine the diversity of carbonates and silicates. Finally among transition metal–metalloid compounds the number of borides is equalled only by silicides and phosphides. In these compounds boron is found in trigonal prisms and square antiprisms formed by the transition metal atoms. Octahedral coordination is rare. The trigonal prism allows for boron–boron linkages across the rectangular faces of face-sharing prisms. Since the pioneering work of Kiessling[13], transition metal borides are classified according to the linking of the boron atoms. In boron-rich compounds boron atoms form three-dimensional frameworks. With increasing metal content, boron atoms form nets, double chains, single chains, pairs, and finally in boron-poor compounds they are isolated from each other. The structural chemistry of transition metal borides was discussed extensively by Aronsson, Lundström, and Rundqvist[14,15], and more recently within this series by Nowotny[2] and Ward[1]. The crystal chemistry of binary transition metal borides is fairly well established and most current research is done on ternary systems.

7.2.1 Binary transition metal borides

The crystal structure of the boron-rich phase in the Mo—B system was determined through a careful study of the x-ray powder pattern[16]. Its composition which previously had been given as MoB_4 and MoB_{12} was formulated as $Mo_{1-x}B_3$ ($x = ca.0.20$) since one of the two Mo sites is only partially occupied. The boron atoms form graphite-like nets which previously were not observed at this composition but are found in diborides. It is most remarkable that the 'trigonal prism' of Mo atoms around the boron atom is not complete and two of the Mo atoms are missing. This results in a large hole. It has been proposed previously that in the corresponding tungsten compound the hole accommodates boron pairs[17] or boron octahedra[18]. Because of the small scattering power of boron a careful single crystal study would be needed to establish the structure beyond any doubt.

Preparation and (re-)crystallisation of ZrB_2, TaB_2, and MoB_2 with Cu, Sn, or Pb as solvents ('auxiliary metal bath', Lebeau process) was reported[19]. Another paper describes the preparation of Ti, V, Cr, Fe, Co, and Ni borides by reaction of a $BCl_3 + H_2$ mixture or by reaction of boron with the metal sulphides[20]. N.M.R. studies were carried out on Co_2B[21], several TB_2 compounds[22], and all phases in the V—B system[23].

7.2.2 Ternary borides in systems transition metal–transition metal–boron

The structure of the phase ~$TiCo_2B$ found by Stadelmaier *et al.*[24] was determined[25]. The ideal composition is $Ti_3Co_5B_2$. Atoms are arranged in layers perpendicular to the short *c* axis of the tetragonal cell (Figure 7.1).

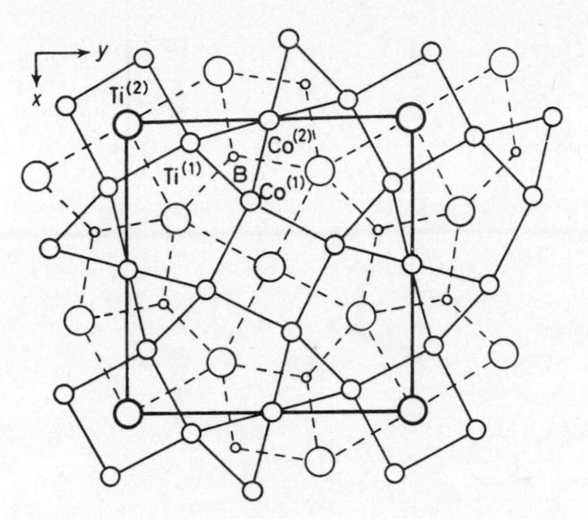

Figure 7.1 Projection of the $Ti_3Co_5B_2$ structure. Atoms connected by broken and continuous lines are separated by $z = \frac{1}{2}$. (From Kuz'ma and Yarmolyuk[25] by courtesy of *Zhurnal Structurnoi Khimii*)

Boron is situated in trigonal prisms formed by Co with three Ti atoms outside the rectangular faces of the prism. Two independent Co atoms are each 12-coordinate. The two independent Ti atoms have coordination numbers of 14 and 17 respectively, suggesting quaternary compounds where large metal atoms substitute for the 17-coordinate Ti atoms.

Two new structural types were described for the compositions $W_2Ir_3B_{6-x}$ ($x \approx 1$) and Mo_2IrB_2[26]. In accordance with the different metal : boron ratios of the two compounds four of the boron atoms of the formula unit $W_2Ir_3B_{\sim 5}$ form infinite chains and the fifth boron atom is isolated. In the structure of Mo_2IrB_2 (Figure 7.2) boron forms chain-like aggregations of four boron atoms.

Ternary phases (Cr,Ru)B and (Cr,Os)B were found[26] to crystallise with the FeB type structure. (Mo,Ru)B has a CrB type structure and $(Mo,Ru)_3B_4$ is isotypic with Ta_3B_4[26]. Cr_3NiB_6 and $Cr_2Ni_3B_6$ crystallise with V_2B_3 and V_5B_6 type structures[27]. Ti_2ReB_2 has a Mo_2FeB_2 (ordered U_3Si_2) type structure[28].

The crystal structure of the κ-phase Hf_9Mo_4B[29,30] shows that the boron atoms occupy trigonal prismatic holes formed by Hf and not the octahedral voids which are most likely filled in the corresponding carbides[31]. Zr_9Mo_4B and Zr_9W_4B are isostructural with Hf_9Mo_4B[29].

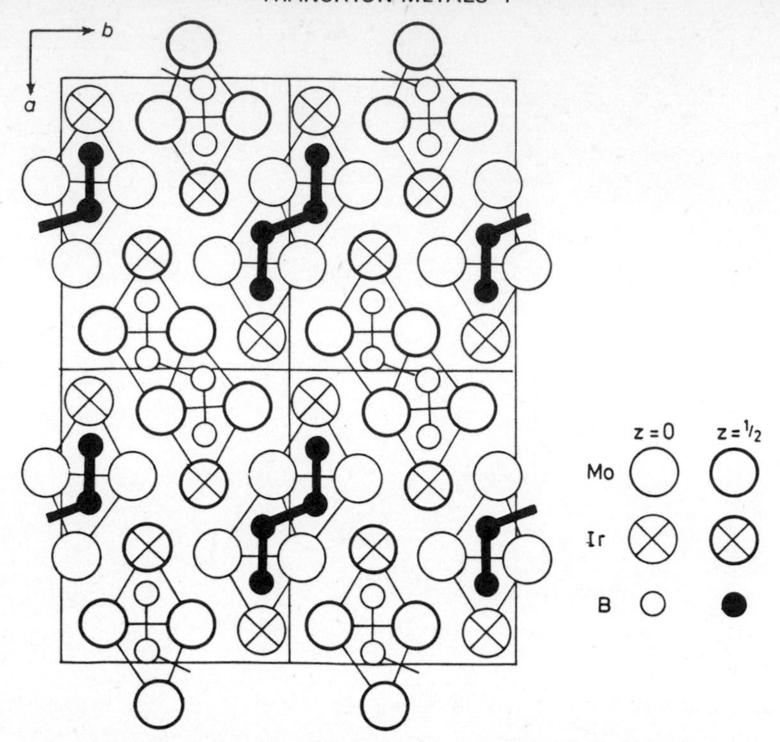

Figure 7.2 Structure of Mo_2IrB_2 projected along the orthorhombic *c* axis. (From Rogl, Benesovsky, and Nowotny[26] by courtesy of *Monatsh. Chem.*)

7.2.3 Ternary borides in systems lanthanide–transition metal–boron

Research in this area began only recently and virtually all published work originates from the very active Ukrainian group in L'vov. Many ternary phases with two transition metals and boron crystallise in structures found for binary phases. Because the rare earth and transition metals differ greatly in size and electropositivity, complex borides containing those metals have a greater tendency to form true ternary compounds.

Two series of compounds with a metal : boron ratio of 1 : 2 were found with closely related structural types. For the one with composition $LnCrB_4$ (Ln = Y,Gd,Tb,Dy,Ho,Er,Lu) the structure was determined for $YCrB_4$[32]. The other structure with composition Ln_2ReB_6 (Ln = Y,Gd,Tb,Dy,Ho,Er, Tm,Lu) was solved for Y_2ReB_6[33]. Both structures are orthorhombic with the same space group *Pbam*-D_{2h}^9 and have atoms arranged in two layers perpendicular to the short *c* axes. The metal atoms are confined to one layer forming trigonal prisms which contain the boron atoms. The boron atoms are bonded together across the rectangular faces of the prisms and in this way form a tessellation of 5- and 7-membered rings in $YCrB_4$ (Figure 7.3). In Y_2ReB_6 the boron network contains 6-membered rings as well as 5- and 7-membered rings. (Figure 7.4). Six-membered rings were found previously in the AlB_2 type structure which occurs for nearly all rare earth and transition

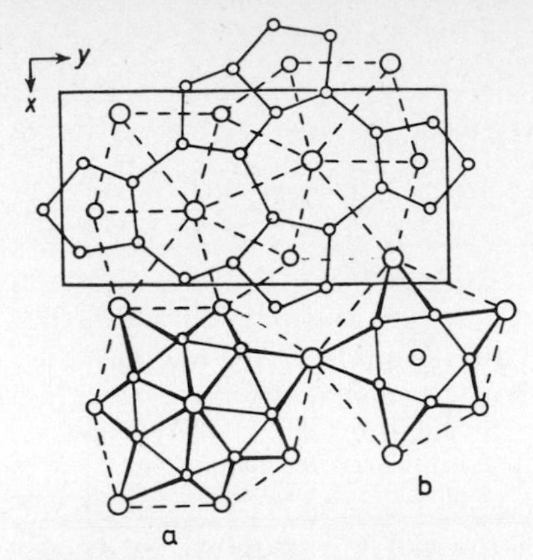

Figure 7.3 Structure of $YCrB_4$. The networks of Y and Cr atoms at $z=0$ (broken lines), networks of B atoms at $z=\frac{1}{2}$ (continuous lines), and coordination polyhedra of (a) Y and (b) Cr are shown. (From Kuz'ma[32] by courtesy of *Kristallografiya*)

metal diborides. Five- and seven-membered rings were previously found only in ScB_2C_2 where both boron and carbon form one network[34]. This structure is actually very closely related to $YCrB_4$; in ScB_2C_2 the positions above and below the five-membered rings remain unoccupied, while they are occupied by Cr atoms in $YCrB_4$.

The structure of $CeCo_4B_4$ is tetragonal ($P4_2/nmc$; $Z=2$; $a=5.06$, $c=7.06$ Å)[35]. The boron atoms are situated in deformed trigonal prisms and form pairs. This structure is also found for $LnCo_4B_4$ where Ln = Gd to

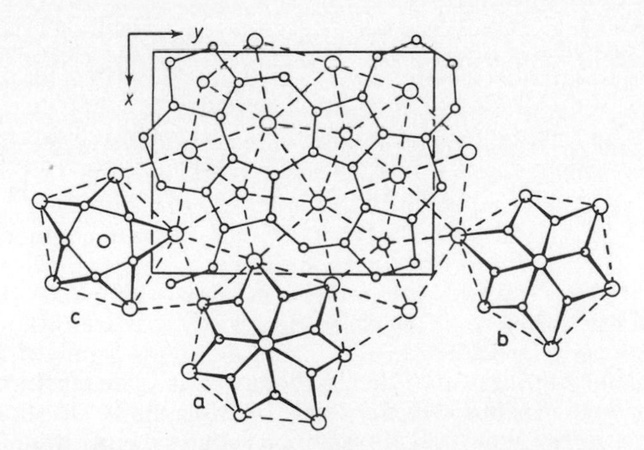

Figure 7.4 Structure of Y_2ReB_6. Y and Re atoms at $z=0$ are connected with dashed lines, B atoms at $z=\frac{1}{2}$ with solid lines. Coordination polyhedra of Y are shown in (a) and (b); (c) shows the Re coordination. (From Kuz'ma and Svarichevskaya[33] by courtesy of *Kristallografiya*)

Tm,Lu,Y. The unit cell volume of $CeCo_4B_4$ is only slightly larger than that of $GdCo_4B_4$, indicating the transition of Ce to the tetrapositive state.

The $ThCr_2Si_2$ type structure, an ordered version of the $BaAl_4$ structure, is frequently observed for silicides. YCo_2B_2 seems to be the first boride with that structural type[36].

A new structure type was found for $CeCr_2B_6$[37]. The boron atoms are on two sites. One is situated in a trigonal prism of 4Cr and 2Ce atoms. It has three additional boron atoms outside the rectangular faces of that prism at the single bond distance of *ca.* 1.75 Å. The coordination of the other boron atoms is unusual: it has five metal neighbours (2Ce + 3Cr), one boron atom at 1.74 Å, and two boron atoms at 2.02 Å. The boron network is three-dimensional. There are short Ce—Ce distances of 3.10 Å which, according to the authors, indicate that cerium is in the tetrapositive state.

The occurrence of the $Cr_{23}C_6$ structure in its ternary ordered version ($W_2Cr_{21}C_6$ type, τ-phase) was reviewed by Stadelmaier[38]. It is obtained most frequently with carbon as the metalloid component when the smaller transition metal is Cr or Mn. The borides are formed in ternary systems where Ni or Co is the small transition metal. $Sc_2Co_{21}B_6$ and $Sc_2Ni_{21}B_6$ were described earlier[39]. The recently found $Ce_2Ni_{21}B_6$ is the first $W_2Cr_{21}C_6$ type phase containing a lanthanide[40]. Another series of compounds with approximate composition $LnCo_{12}B_6$ and rhombohedral unit cell, but as yet unknown structure was announced recently[41]. Complex borides $LnCo_3B_2$ with ordered $CaCu_5$ type structure were also confirmed recently[42,43].

7.2.4 Ternary borides in systems transition metal–main group element–boron

The systems Re—Al—B and Re—Si—B were investigated. Only one ternary phase was found: Re_3Al_2B[44]. It crystallises with a Mn_3Ni_2Si type structure[45], a variant of the Ti_2Ni structure[46,47]. Ti_2Ni is the 'hostmetal' structure of the η-carbides, where carbon is found in octahedral 'interstitial' sites. It is remarkable that in Re_3Al_2B the boron atom occupies one of the Ni positions of the Ti_2Ni structure, thus reminding us of the dual character of boron in transition metal compounds. Like carbon it is frequently found on 'interstitial sites'; occasionally, however, it replaces metal atoms as is more frequently found for silicon. Boron takes the position of a silicon atom in $Ni_{4.6}Si_2B$ which has a W_5Si_3 type structure[48].

7.3 CARBIDES

The extreme hardness and high melting points of many transition metal carbides were recognised in the 1890s by Moissan[49]. These properties are the basis for their industrial applications as cutting tools, wire drawing dies, drilling tools in mining and other wear resistant materials. More recently, the high-temperature stability of transition metal carbides has become important for the reactor and aerospace technology. At low temperatures the superconducting properties are of interest. As a result of the many real and potential uses, research on physical and chemical properties has been extensive.

The preparation of transition metal carbides was treated by Windisch and Nowotny[50]. A review by Williams[51] on binary transition metal carbides emphasises physical properties. Phase relationships and electrical properties were reviewed recently in this series by Storms[52]. Monographs, with extensive coverage of transition metal carbides were written by Kieffer and Benesovsky[9,53], Goldschmidt[10], Storms[54], and Toth[55]. The book by Kosolapova[56] gives access to the Russian literature. Carbides which occur in steels were reviewed recently by Jack and Jack[57].

7.3.1 Coordination of the carbon atom

Some aspects of the crystal chemistry of transition metal carbides can be rationalised by geometrical arguments[58]. These are of course not the only structure determining parameters but they are the ones best understood. The carbon atom is much smaller than the transition metal atoms. The radius ratios are ideal for sixfold coordination of the carbon atom. In compounds with the larger early transition metals octahedral coordination occurs. The chromium group is transitional with both octahedral and trigonal prismatic coordination for carbon. In the carbides of the smaller iron and platinum group metals, trigonal prismatic coordination occurs, but not exclusively. When both metal–metal bonds and metal–carbon bonds are to be formed, the ideal radius ratios are $r_c/r_T = 0.41$ for octahedral and $r_c/r_T = 0.53$ for trigonal prismatic coordination. The empirical limits for compounds with the NaCl structure (octahedral carbon coordination) are 0.42 for CeC and 0.57 for VC*. They are even wider for compounds in which carbon is only a minor constituent.

Considering the many examples of boron-boron linkages in transition metal borides, the complete absence of carbon–carbon bonds in transition metal carbides is surprising. Carbon–carbon bonds could be formed in compounds where carbon is in octahedral metal-coordination whenever the octahedra share faces. Although face-sharing metal octahedra occur frequently in carbides, in no case has it been proven that carbon atoms occupy both of the face-sharing octahedra. On the other hand, there are numerous well established structures where only one of the face-sharing octahedra contains a carbon atom. The reason for this avoidance of carbon–carbon interaction in transition metal carbides is not known, although it has obviously to do with the fact that no transition metal carbides are known with a carbon: transition metal ratio >1. Carbon–carbon bonds occur in the acetylides of rare earth and actinide metals. In these compounds both carbon atoms are within the same metal-polyhedron. The compound ScB_2C_2 has a two-dimensional polyanion with C—C single bonds of 1.45 Å[34].

* These values depend of course on the choice of atomic radii.

7.3.2 Binary transition metal carbides

The binary systems of all transition metals (T) with carbon have been investigated fairly thoroughly during the last 20 years and phase diagrams have been established for most systems[52,54]. The structurally well characterised carbides are listed in Table 7.1. The tendency to form carbides decreases with increasing group number of the transition metal. HfC and TaC melt at *ca.* 4300 K and are among the highest melting compounds known, while the carbides of Fe, Co, and Ni are metastable, although well characterised. The Pt group metals do not form binary carbides, although fleeting carbide formation at surfaces, etc., is possible.

Table 7.1 Carbides of the transition metals

(a) carbides with close packed metal atoms and carbon in octahedral sites.

TiC*	V_2C	V_4C_3	VC*				Fe—C	Co_2C
ZrC*	Nb_2C	Nb_4C_3	NbC*	Mo_2C	Mo_3C_2	MoC_{1-x}	(martensite)	(metastable)
HfC	Ta_2C	Ta_4C_3	TaC	W_2C		WC_{1-x}		

* for ordered defect derivative structures see text

(b) carbides with carbon in trigonal prismatic and square antiprismatic sites.

$Cr_{23}C_6$	$Mn_{23}C_6$	Fe_3C	Co_3C	Ni_3C
C_7C_3	Mn_3C	Fe_5C_2	(metastable)	(metastable)
Cr_3C_2	Mn_7C_3	Fe_7C_3		
	Mn_5C_2	(all metastable)		
MoC				
WC				

7.3.2.1 Binary carbides with octahedral carbon coordination

In binary carbides with octahedral carbon coordination the metal atoms are close packed. In cubic close packed structures, octahedral voids do not share faces and all voids can be occupied resulting in the ideal composition TC (NaCl type). In hexagonal close packing, octahedral voids share faces and only every other octahedral site is occupied: composition T_2C. In more complex stacking sequences the optimal composition is in between these two extremes. For instance, Mo_3C_2, $(Ta_2V)C_2$, and V_4C_3 have stacking sequences, hcc, hcc, and hhcc respectively[59,60]. In carbides with hexagonal close packed metal atoms, vacant and occupied octahedral sites order at low temperatures giving rise to superstructures. The order–disorder temperatures have been determined in some cases by high temperature neutron diffraction[61]. Mo_2C transforms between 1450 and 1650 K, W_2C and Nb_2C at *ca.* 2000 and 1500 K respectively. The critical and comprehensive review on close packed transition metal carbides by Parthé and Yvon may be consulted for further structural details[62].

In addition to the systematic carbon defects related to hexagonal stacking sequences, carbon defects also occur by non-occupancy of 'allowed' octahedral sites. The compositions listed in Table 7.1 therefore correspond only to the upper limits of the homogeneity ranges. For instance, in cubic TiC the homogeneity range extends to *ca.* $TiC_{0.5}$. At high temperatures these defects are more or less randomly distributed. At lower temperatures the defects order. In most phases no detailed studies have been made; however, in VC_{1-x}, which is by far the best investigated system, ordering was found for compositions V_8C_7[63–66] and V_6C_5. For the latter composition, two closely related structures were described[67,68]. In both the arrangement of vacancies is such as to give each vanadium atom five carbon neighbours instead of the six as in an ideal NaCl type VC. As a consequence, the positions of the V atoms are also slightly altered: the empty V octahedra are bigger than the carbon filled ones[69]. This indicates either a strong electrostatic repulsion between the V atoms or a strong covalent V—C bond or (most likely) both. The phase diagram V—C was re-investigated[70] in the critical range of carbide formation to establish the temperatures for vacancy ordering of V_6C_5 (~1250 °C) and V_8C_7 (~1150 °C). These temperatures are considerably lower than the melting temperature of the disordered VC_{1-x} phase (~2650 °C). Both ordered superstructures have significant homogeneity ranges. Single crystals used for this work were grown by zone-melting rod-shaped, sintered, polycrystalline specimens[71]. They were investigated after varying heat treatments by electron diffraction. Samples with compositions close to $VC_{0.75}$ quenched from *ca.* 900 °C exhibit electron diffraction patterns which contain bands of diffuse intensity[72] characteristic for short-range order. The interpretation of these diffuse diffraction patterns shows that the short-range order is similar to the long-range order in the adjacent V_6C_5 phase[73]. The study of the NbC_{1-x} system revealed only one long-range ordered compound: Nb_6C_5 with the V_6C_5 type structure[72,74]. An ordered superstructure was also found for Ti_2C[75–77].

7.3.2.2 *Binary carbides with carbon in trigonal prismatic or square antiprismatic sites*

Only five structural types are well established for the carbides of Cr, Mn and the iron group metals; one for each of the five stoichiometries listed in Table 7.1b. In $Cr_{23}C_6$ and $Mn_{23}C_6$ carbon occupies square antiprismatic voids. This has been confirmed recently by a neutron diffraction study[78]. In the other types carbon is in trigonal prismatic coordination. No carbon–carbon interaction occurs since the occupied prisms do not share faces. MoC and WC are exceptional in that they are the only carbides where the occupied trigonal prisms share faces. They share the triangular faces in contrast to the borides where the boron–boron linkages occur across the rectangular faces. The distance between the carbon atoms in WC, however, is 2.84 Å and cannot be considered as bonding.

The structure of Cr_7C_3 and Mn_7C_3 was redetermined recently[79]. Although the structure was essentially confirmed, it was shown that the true symmetry is monoclinic and not trigonal as originally found by Westgren and Phragmén[80]. Extensive twinning mimics the higher symmetry. It seems possible that the trigonal cell is correct at high temperatures and the monoclinic cell is the result of a martensitic (diffusionless) phase transition.

The system Mn—C was re-investigated[81]. Besides the four compounds listed in Table 7.1 the compound $\sim Mn_{15}C_4$—with unknown structure and possibly stabilised by oxygen[57]—was again observed[81,82]. At high temperatures a new phase ε-Mn_3C_{1-x} was found. Its structure seems to be similar to ε-Fe_3N[81].

7.3.2.3 Bonding in carbides

Hägg's concept[58] of regarding transition metal carbides and nitrides as interstitial alloys has been refined over the years and two distinct bonding models have been developed[83]. The first model stresses the importance of the T—T bonds. An electron transfer from the interstitial atom to the T atom is assumed, thus filling T—T bonding orbitals. In the second model the high stability of many carbides and nitrides is attributed to the strength of the T—X bonds and electron transfer T→X is assumed. This model is in accord with chemical intuition, allowing for a charge transfer from the electropositive to the more electronegative element. It is supported by recent electron spectroscopic (ESCA) and x-ray emission and absorption studies which show that the electron transfer is from the metal to carbon or nitrogen[84–86]. Similar results were obtained in an x-ray diffraction study of the charge distribution in NbC[87] and in calculations of the electronic band structure for VC[88] and for NbN[89,90]. This model is also supported by systematic changes in bonding distances in complex carbides (see Section 7.3.3.1). The merits of the opposite model with electron transfer X→T were restated recently[91]. The remarkable continuous solid solutions Mo_2C–Re and W_2C–Re, now confirmed[92–94], may be cited in support of this model.

7.3.3 Ternary carbides

Although our knowledge of ternary systems of two metals with carbon has made rapid progress during the past decade, only about one half of the possible ternary combinations has been investigated to any extent. Nevertheless, the behaviour of the technically important transition metal carbides towards other metals and metalloids is fairly well known. Only scanty information is available about ternary carbides involving alkali, alkaline earth, and rare earth metals. Aside from the ternary carbides occurring in steels and the solid solutions of binary refractory carbides, so far no important applications have been found for ternary carbides.

7.3.3.1 Structural characteristics of ternary carbides

Considering the extreme stability of some binary transition metal carbides such as TiC, the formation of ternary compounds, like Ti_3AlC and Ti_2AlC, is surprising at first. From a structural point of view, however, this is plausible. In the ternary compound, the carbon atom is still surrounded by six transition metal atoms, thus forming the stable octahedral or trigonal prismatic T_6C groups already discussed for binary T carbides. It can be assumed that these T_6C groups exist in a more or less condensed form in the melt. Upon cooling they agglomerate, in most cases forming first binary TC and/or T_2C phases which in turn react (peritectically) with the remaining melt at lower temperatures, thus forming a ternary (complex) carbide. In the ternary phase the carbon atom is usually surrounded by those metal atoms forming the more stable (~higher melting) carbide. For instance in Mo_3Al_2C or W_6Fe_6C the carbon atom is situated in octahedral voids formed by Mo or W while Al and Fe form metal–metal bonds only.

The degree of condensation of the T_6C groups may be taken as a classifying principle. There are complex carbides with isolated T_6C octahedra (W_6Fe_6C), with T_6C octahedra linked by corners (Mn_3AlC: perovskite; Mo_3Al_2C: filled β-Mn), with T_6C octahedra linked by both corners and edges (V_3AsC: filled anti-$PuBr_3$), by edges (Cr_2AlC: *H* phase; Ti_3SiC_2), and finally T_6C octahedra linked by faces (Mo_5Si_3C: Nowotny phase; $T_6M_6C_3$: η-carbides). The latter are listed for the ideal composition, assuming all T octahedra filled. As was emphasised above for binary transition metal carbides, there is no structure for which full carbon occupancy of face sharing octahedra has been proven and it may equally well be assumed that only every other octahedron is occupied. A neutron diffraction study[47] of the Nowotny phase in the system Mo—Si—C resulted in a composition close to $Mo_{4.8}Si_3C_{0.6}$. Thus less than two-thirds of the face-sharing octahedra contain carbon atoms.

The importance of the transition metal–carbon interaction can be seen from the shortening of the T—C distances with decreasing carbon coordination of the T atom (Table 7.2). The decreasing carbon coordination also shortens the T—T bonds, since the T atoms retain more electrons to fill T—T bonding states. This interpretation of bonding distances thus supports bonding models for binary carbides which assume electron transfer from the transition metal to carbon.

In a recent review[96], the correspondence of many complex carbide structures to more or less ionic antitypes was emphasised. Examples are the Nowotny phases ~Mo_5Si_3C which correspond to the apatites[97,98] such as $Ca_5(PO_4)_3F$, or the perovskite carbides Mn_3AlC–$CaTiO_3$. From a mnemonic viewpoint it is of interest that most complex carbide structures have their analogues also among 'non-filled-up' types. For instance, the relative arrangements of the metal atoms in Mo_5Si_3C and Mn_3AlC correspond to Mn_5Si_3 and Cu_3Au.

Table 7.2 Correlation of Ti—C and Ti—Ti distances (Å) with the carbon coordination *n* of the Ti atom in complex carbides[95]

Compound	*n*	Ti—C	Ti—Ti	Ti—Al
TiC	6	2.16	3.06(12×)	
Ti_2AlC	3	2.12	3.06(6×), 2.94(3×)	2.85(3×)
Ti_3AlC	2	2.08	2.94(8×)	2.94(4×)

7.3.3.2 *Ternary carbides in systems transition metal–transition metal–carbon*

Historically these carbides—many of which occur in steels—were the first to be studied to any extent. Most of the phases found in systems: early transition metal–iron group metal–carbon have been characterised as 'η-carbides' through their face-centred cubic x-ray diffraction patterns. The exact composition is frequently not known. Complete structure determinations were done for W_3Fe_3C, W_6Fe_6C, and the oxide T_4Cu_2O. The positions of the metal atoms correspond in the three determinations; the ordering of the metal atoms and occupancy of different octahedral 'interstitial' sites are different of course. More recently, several of the ternary phase diagrams were determined and the following η-carbides were found or confirmed[99–101]: Mo_6Co_6C, Mo_6Ni_6C, W_6Fe_6C, W_6Co_6C, W_6Ni_6C, W_3Fe_3C, W_3Co_3C and W_4Co_2C.

7.3.3.3 *Ternary carbides in systems actinide or lanthanide–transition metal–carbon*

The structure of the phases U_2TC_2 (T = Ru,Rh,Os,Ir,Pt)[102–105] was determined for U_2IrC_2 by Bowman *et al.*[106]. It is described best as a defect-NaCl structure with ordered metal positions and one-third of the anion positions vacant. It is therefore closely related to UC (NaCl type). The carbon environment is octahedral with five U and one Ir atom. Ir is two-coordinated and U five-coordinated to carbon. The Ir—C bonds are remarkable since platinum metals do not form binary carbides. In $UMoC_2$[107] which also has octahedral coordination for carbon, the higher affinity of molybdenum to carbon is reflected in the larger number of Mo—C bonds: the two different C atoms have two and three Mo—C bonds respectively*.

Actinides and lanthanides form binary compounds AcT_3 and LnT_3 where T stands for platinum group metals. These are cubic close-packed structures and probably ordered (Cu_3Au type). It has been shown[108,109] that $LnPd_3$ phases exist but no $LnRh_3$ phases (except for $CeRh_3$; but Ce is often four valent). Recently, the formation of ternary carbides LnT_3C was investigated by Holleck[110]. No $LnRu_3C$ phases were found (except for Ln = Ce), while $LnRh_3C$ phases do form. Similar regularities were observed for AcT_3C phases[110,111]. The results were interpreted as to suggest that carbon raises the effective valency of the platinum metal. The position of the carbon atom in these phases is not known.

7.3.3.4 *Ternary carbides in systems transition metal–main group element–carbon*

This group of carbides comprises by far the largest number of representatives and structural types. It must be mentioned, however, that no sharp borderline can be drawn between main group elements (M) and transition metals (T). For instance, both η-carbides and perovskite carbides occur in systems

* It could be argued that this is merely a consequence of the different U : T ratio. This, however, does not explain why the compound is stable in the first place.

T—T—C and T—M—C. Recent reviews were written by Stadelmaier[38], Nowotny[112] and Ward[1].

The structure of ferromagnetic ($T_c = 284$ K) Mn_5SiC was determined[113]. The arrangement of the Mn atoms offers octahedral and trigonal prismatic sites for carbon. Although the accuracy of the data was somewhat limited, the evidence puts carbon in trigonal prismatic environments, as is also the case for binary manganese carbides. As usual, the metametal (Si) has only transition metal (Mn) near neighbours. The crystal structure of the phases $\sim Mn_8Si_2C$ and $\sim Fe_8Si_2C$ is not yet known[114].

The system Ti—Si—C which had been previously determined by powder metallurgical methods[115] was re-investigated[116] by a vapour deposition technique with the reaction system $TiCl_4(g) + SiCl_4(g) + CCl_4(g) + H_2$ (excess). A deposition diagram for all binary and the two ternary phases $Ti_5Si_3C_{1-x}$ and Ti_2SiC_2 was determined and the remarkable plasticity of plate-like Ti_3SiC_2[117] was studied[116].

The phases Cr_3GeC and V_3GeC with filled-up $PuBr_3$ (Re_3B) type structure were found by Boller[118] who refined this structure for the compositions Cr_3GeC and $Cr_3AsC_{0.75}N_{0.25}$. An interesting difference between the two compositions was found in the coordination arrangements of Ge and As. Although the covalent and atomic radii are very similar for both elements, bonding Ge—Ge distances are 2.90 Å in Cr_3GeC, while the corresponding distance in the As-isotype is 3.06 Å. This indicates higher polarity of the Cr—As (as compared to Cr—Ge) bonds and correspondingly weaker As—As interactions.

The ternary system V–P–C was investigated by Boller[119]. Besides V_3PC_{1-x} with a filled $PuBr_3(Re_3B)$ type structure, V_2PC with a Ti_2SC (*H* phase, Cr_2AlC) type structure, and the Nowotny phase V_5P_3C, two new ternary compounds were found. One has approximate composition V_6P_3C and an as yet unknown structure. The other, V_4P_2C, represents a new structural type (Figure 7.5) where V_6C octahedra are not in contact with each other. The P atoms are in a trigonal prismatic vanadium environment. There are three sites for the V atoms. One V atom is in a square pyramidal configuration

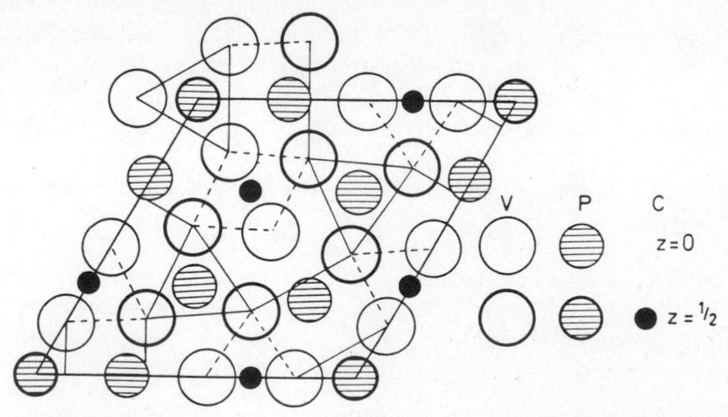

Figure 7.5 Crystal structure of hexagonal V_4P_2C projected along the short *c* axis. (From Boller[119], by courtesy of *Monatsh. Chem.*)

of 1C and 4P atoms, one has trigonal bipyramidal coordination from 3P and 2C, and the third V atom is in tetragonal bipyramidal coordination of 4P and 2C atoms. As usual for a metal-rich compound there are of course also a great number of short V—V interactions.

References for work on amorphous T–P–C alloys will be given in Section 7.6.3.

The investigation of the system Ta–S–C led to the discovery of the new complex carbide 3s—Ta_2S_2C[120]. Its structure consists of close-packed layers of Ta and S atoms with carbon in octahedral voids formed by Ta only. It combines building elements of TaC (NaCl type) and TaS_2 (Figure 7.6). As is well known for MoS_2, bonding between contacting sulphur layers is weak. Thus it was found[120] that grinding of 3s—Ta_2S_2C destroys long-range order and the powder pattern of the ground product reflects only the subcell: 1s—Ta_2S_2C. Intercalation compounds as suggested by the structural relation between Ta_2S_2C and TaS_2 have also been prepared and the magnetic behaviour of the first row transition metals (T = Ti to Ni) in the compounds $T_xTa_2S_2C$ ($0.25 < x < 0.33$) was determined[121]. The T metal occupies octahedral sites between the sulphur layers.

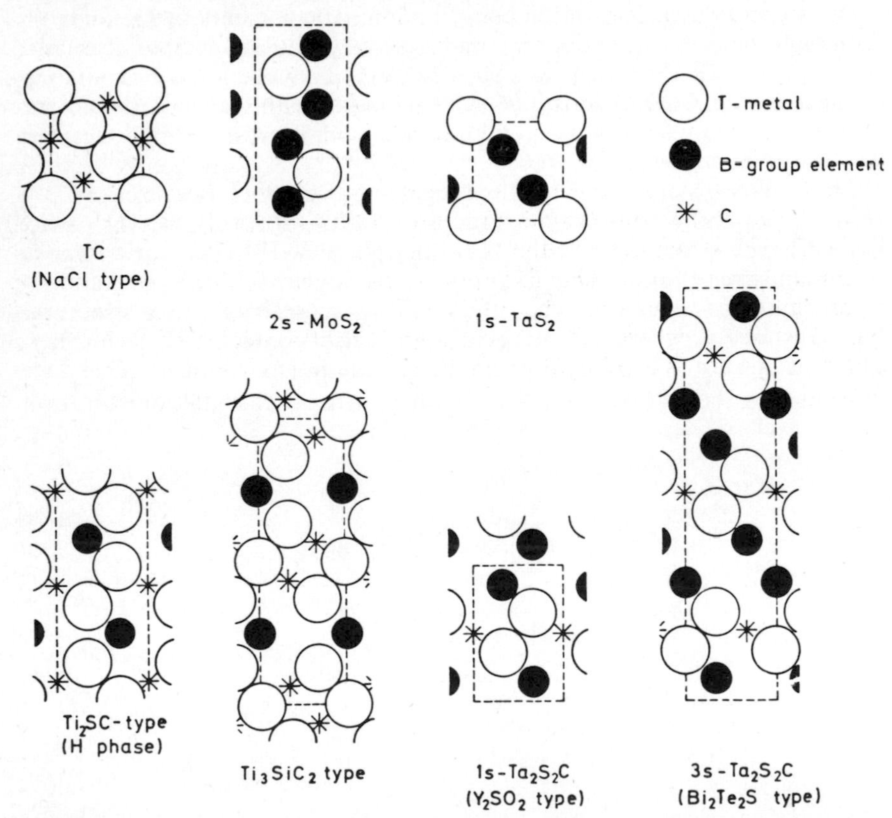

Figure 7.6 Stacking variations of close packed layers of transition metal and main (B) Group elements in complex carbides and related compounds. The structures are represented by cuts through the (110) plane of the hexagonal cell. (Modified from Beckmann, Boller, and Nowotny[120], by courtesy of *J. Solid State Chem.*)

7.4 SILICIDES AND GERMANIDES

Preparation, properties and phase diagrams of silicides were summarised by Kieffer and Benesovsky[9] and Goldschmidt[10]. More recently accounts of the preparation of silicides and germanides by chemical transport[122,123] or by the Lebeau technique with Hg, Cu, Ag, Sn and Pb as solvents[124,125] became available. The Bridgman method was used to grow single crystals of PtSi and PtGe[126]. Single crystals of V_3Si, which forms peritectically, were grown by repeated floating zone passes[127]. Directional solidification in the system Ti—Si was investigated by Nickl and Schweitzer[128].

The disilicides, especially $MoSi_2$, have found application as oxidation-resistant coatings (formation of SiO_2) for high melting transition metals. The high transition temperatures of superconducting Cr_3Si type silicides and germanides have also stimulated research on silicides and germanides. With the exception of some silicon-rich compounds the transition metal silicides and germanides are metallic conductors. $CrSi_2$, Mn_nSi_{2n-m}, and β-$FeSi_2$ have interesting semiconducting and thermoelectric properties[129–137].

The crystal structures of transition metal silicides and germanides are similar. Because of their larger atomic sizes Si and Ge cannot substitute to any large extent for carbon, and transition metal carbides are not isostructural with silicides. It is more difficult to draw a line between germanides and stannides, and isostructural series like Zr_5Si_3, Zr_5Ge_3, Zr_5Sn_3, or $FeGe_2$, $FeSn_2$ are known. Heats of formation and melting points decrease from carbides to plumbides; the latter frequently do not form at all.

In its tendency for self-bonding, silicon ranks between boron (which even forms 'inverse interstitial' compounds like TB_{12}) and carbon (no transition metal solid-state compound contains more than 50 atom % carbon). Near-neighbour environments of boron and silicon (and germanium) are similar. The trigonal prism—usually augmented by three additional ligands outside its rectangular faces—and the tetragonal antiprism occur most frequently. In compounds with high metal content, as in the σ-phase related structures, silicon also occurs in higher coordination (C.N. 12–14).

The structural chemistry of silicides was treated by Nowotny[138], Aronsson, Lundström, and Rundqvist[14], and more recently in this series by Nowotny[2] and Ward[1]. The crystal chemistry of both silicides and germanides was reviewed in the monograph by Gladyshevskii[139]. It is not available in English, and the reader is also referred to the critical compilation by Pearson[6], for the literature on germanides.

7.4.1 Binary transition metal-rich silicides and germanides

The Mn-rich portion of the Mn–Si system contains two phases with complex crystal structures[140,141]. The phase $\sim Mn_{86}Si_{14}$ has the so-called *R* phase structure[142]. The structure of the other, $Mn_{81.5}Si_{18.5}$, designated as ν or N phase[140,141], is now also known[143]. Both belong to the family of σ-phase related structures which are sometimes also called 'tetrahedrally close-packed' structures since they have only tetrahedral interstitial voids[144]. This is a consequence of the differing atomic sizes and the tendency to achieve high coordination for all atoms which is characteristic for the de-localised bonding in intermetallic phases. Like all σ-phase related structures, the ν-$Mn_{81.5}Si_{18.5}$ structure is quite complex with 29 different atomic sites. The slightly larger Mn atoms occupy all 16- and 15-coordinated and most of the 14-coordinate sites. Of the ten 12-coordinate sites some are only occupied by Mn or Si, while the others and also some of the 14-coordinate sites have mixed occupancies. Thus the choice of site is not determined by size only. It was found to be determined also by the kind of neighbour: the average percentage Mn in the first coordination shell increases with decreasing percentage Mn in the central site, and the two 12-coordinate sites that are occupied by Si only, have only Mn atoms in their first coordination shells[143]. Thus bonding between unlike neighbours is clearly preferred, suggesting some charge transfer.

The phase Mn_5Si_2 described earlier[145] was found to have the *D* phase structure[146] which is closely related to the δ-phase[147]. Both are σ-phase related structures.

The Cr_3Si type silicides continue to be of interest because of their low temperature martensitic (diffusionless) phase transitions which seem to be related to their superconducting properties[148–150].

A re-investigation of the Pd–Si system resulted in the identification of two new phases with tentative compositions Pd_5Si and Pd_4Si[151].

Long-range order in the solid solution of iron with 18 atomic % Ge which was observed earlier[152] was shown from Mössbauer data[153] to be similar to Fe_3Al.

7.4.2 Binary transition metal silicides and germanides with intermediate compositions

Although most binary T–Si,Ge systems were systematically investigated in the past, it became clear only recently that phases which previously were described as possessing wide homogeneity ranges actually form whole series of closely-related structures. This is also the case for the NiAs and the filled NiAs (Ni_2In) type phases which frequently occur in this compositional range.

A series of phases at approximate composition Ni_5Si_2 has structures closely related to Ni_2In. The structure of one, with ideal composition $Ni_{31}Si_{12}$, was determined by Frank and Schubert[154]. In order to describe the structure and its relation to Ni_2In, these authors assigned letters to the different atomic layers perpendicular to the hexagonal axis. The Ni_2In structure can then be described as A(Ni), B(NiIn), A(Ni), C(NiIn). Using the same symbolism, $Ni_{31}Si_{12}$ corresponds to a stacking sequence ABCBCACBCB. The structure refinement resulted in some irregular thermal parameters which possibly indicate mixed occupancies for some of the atomic sites.

Structures of NiAs, Ni_2In, and related types also exist in the Ni–Ge system which was re-investigated in the range Ni_3Ge to NiGe[155]. The following phases were identified and their structures determined: (a) Ni_5Ge_2 with a Pd_5Sb_2 type structure which is closely related to the $Ni_{31}Si_{12}$ type structure described above[156]; (b) Ni_5Ge_3 and $Ni_{19}Ge_{12}$ which both have partially-filled ordered NiAs type structures with low symmetry; (c) Ni_2Ge which has a Ni_2In-type structure at high temperatures. This phase can be obtained at low temperatures by fast quenching. If the quenching is not done fast enough it transforms martensiticly to anti-$PbCl_2$ type Ni_2Ge which is the stable phase at low temperatures. In the equilibrium diagram

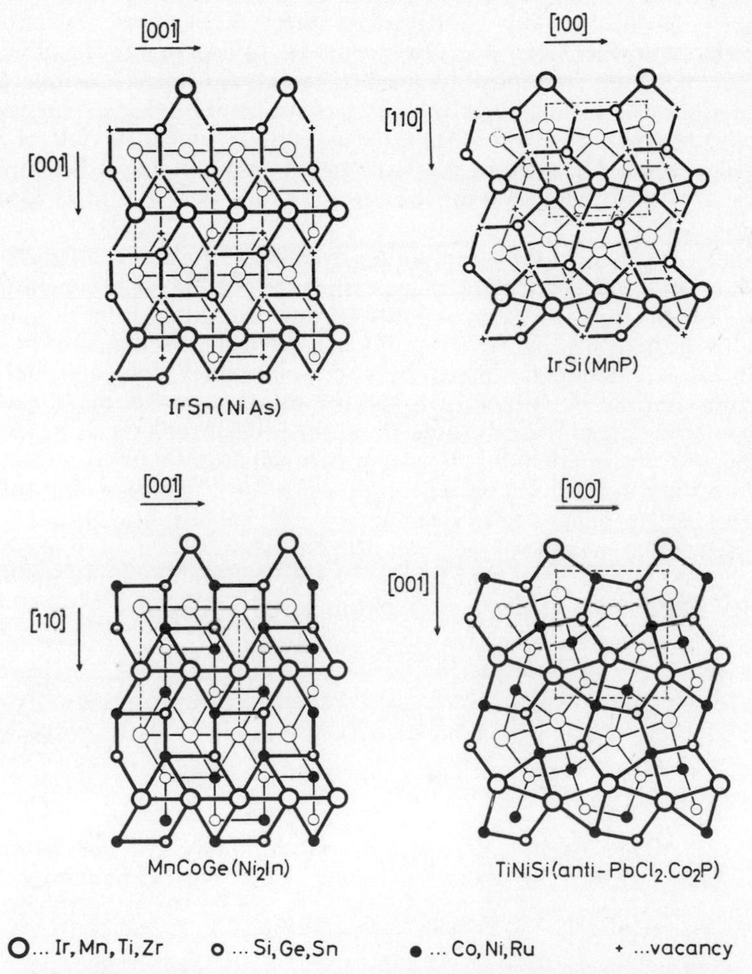

Figure 7.7 Structural relationships between the NiAs filled NiAs (Ni_2In), MnP, and filled MnP (anti-$PbCl_2$) type structures. Atoms connected by thin and thick lines are separated by half a translation period in the projection direction. Atoms and vacancies are connected to emphasise relationships between the structures. (From Johnson and Jeitschko[159], by courtesy of *NBS Publications*)

the two phases do not have exactly the same composition and the high temperature phase reacts peritectoidally with Ni_3Ge to give the low temperature phase. A diffusionless transformation from a high temperature Ni_2In to a low temperature anti-$PbCl_2$ type structure was also found for Ni_2Si[157]. The close relationship between the Ni_2In and the anti-$PbCl_2$ type structure[158] is shown for ordered ternary examples in Figure 7.7 which also shows the corresponding relationship between the NiAs and MnP type structures. A diffusionless phase transition between the latter two types is known for MnAs[160].

The Mn_5Si_3 structure which can also be derived from the NiAs type structure through distortion, filling, and vacancy formation[161] is common in this compositional range. Some of its better characterised variants and their relation to the Ni_2In structure are shown in Figure 7.8. It can be seen that the positions occupied by carbon atoms in the Nowotny phase $Mo_5Si_3C_{1-x}$ can also be filled by small post-transition metals. The recently refined $Nb_{10}Ge_7$ ($=Nb_5Ge_3Ge_{0.5}$) structure has only one half of these octahedral voids filled[162]. In the light of this new finding it might be of interest to refine occupancies for the corresponding positions in Ti_5Ga_4 and Hf_5Sn_3Cu.

Atomic ordering in the solid solutions Mn_3Si_3–Fe_5Si_3 was studied by magnetic and Mössbauer effect measurements[163–165]. The smaller, less electropositive atom (Fe) preferentially occupies the 4d position (Figure 7.8) as is also the case in $Ti_3Mo_2Si_3C_{1-x}$[166]. An extensive discussion of bonding in Mn_5Si_3 is given in the paper by Johnson *et al.*[165] which stresses the importance of the short bond between the metal atoms in the 4d position (the contact between the Fe atoms along the projection direction of Figure 7.8). Such strong interactions also occur between atoms on both sides of the sixfold vertices of the Frank–Kasper polyhedra[167,168] which are the building elements of the σ-phase related structures[144,169]. They are also quite common in intermetallic phases in general. Similar short bonding metal–metal distances are also present in V_6Si_5 and Ti_6Ge_5[170]. This structure was found first for $Nb_2Cr_4Si_5$ where the transition metal atoms are ordered[171]. It

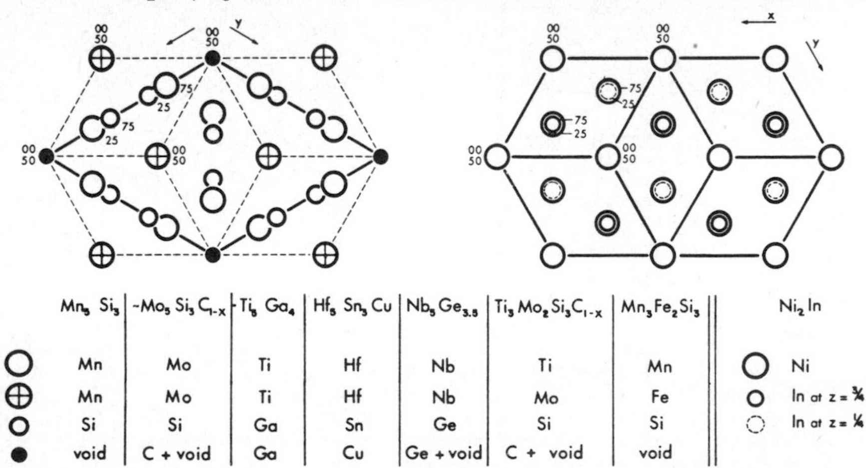

Figure 7.8 Relationships between Ni_2In, Mn_5Si_3, and various filled Mn_5Si_3 structures. (From Johnson and Jeitschko[159], modified, by courtesy of *NBS Publications*)

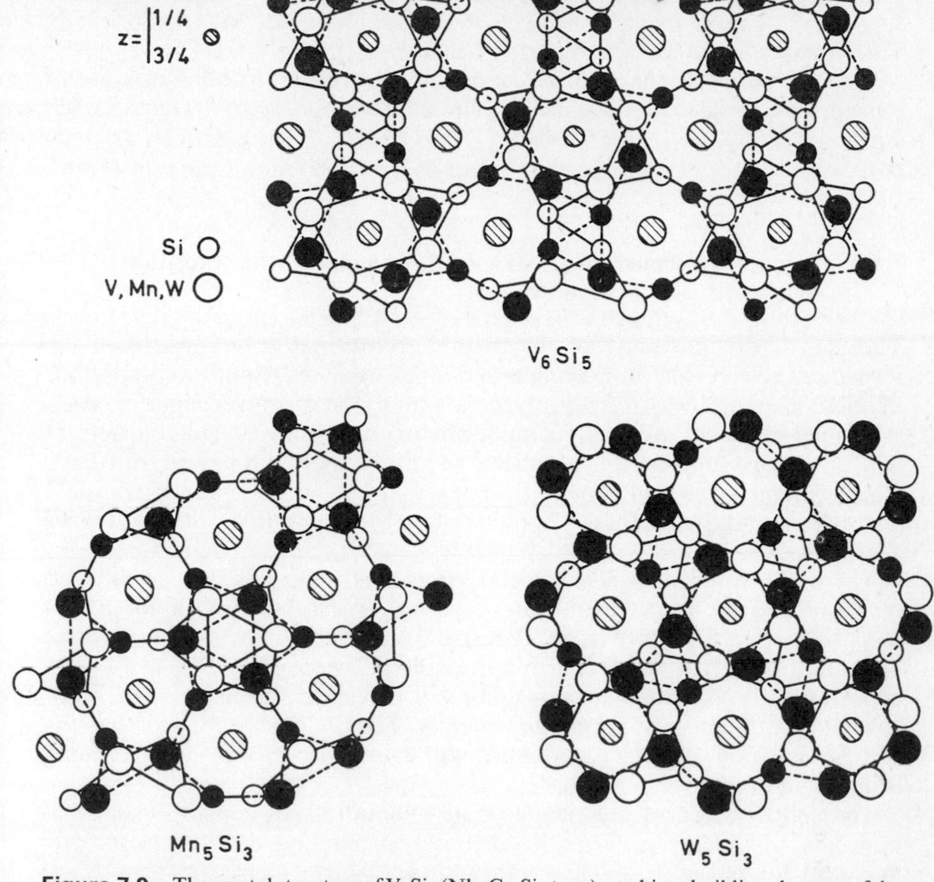

Figure 7.9 The crystal structure of V_6Si_5 ($Nb_2Cr_4Si_5$ type) combines building elements of the Mn_5Si_3 and W_5Si_3 type structures. Atoms connected by continuous and broken lines are separated by half a translation period in the projection direction. (From Spinat, Fruchart, and Herpin[170], by courtesy of *Bull. Soc. Fr. Minéral. Crystallogr.*)

combines building elements of the Mn_5Si_3 and W_5Si_3 type structures (Figure 7.9).

In a series of papers, magnetic, electrical and bonding characteristics of FeGe were described and discussed[172–177]. The structure of Fe_6Ge_5 was communicated briefly[178].

7.4.3 Binary transition metal silicides and germanides with high content of silicon or germanium

The structure of the low temperature phase β-$FeSi_2$ was described[179]. It has the distorted CaF_2 type structure[180]; $OsSi_2$ has the same structure[181]. $FeSi_2$ can also be prepared in an amorphous form. Its conductivity is smaller than metallic α-$FeSi_2$ but higher than semiconducting β-$FeSi_2$.

The structures of several phases in the Ir–Si system are still not known[183]. The structure proposed for $IrSi_3$ has unusually short Si—Si distances of 2.17 Å[184]. A refinement from neutron or x-ray counter data is desirable. The structure of $IrGe_4$ was reported briefly[185].

The atomic positions of $CrSi_2$ type $NbSi_2$, $NbGe_2$, and $TaGe_2$ were refined from single-crystal data[186]. The shortest Si—Si and Ge—Ge distances are 2.57 ($NbSi_2$), 2.66 ($NbGe_2$), 2.65 ($TaGe_2$) Å respectively as compared to 2.35 and 2.45 Å in elemental Si and Ge with diamond-type structures.

7.4.4 Ternary compounds of IA, IIA or IIIB elements with transition metals and silicon or germanium

The high electropositivity of the metals of the first three groups of the Periodic Table results in a predominance of ionic bonding in compounds of these elements with transition metals and silicon or germanium. As a consequence, many of these compounds crystallise in salt-like structures with medium coordination numbers, relatively high symmetry, and fixed stoichiometry. This was already recognised about 40 years ago by Zintl[187] and compounds with these characteristics are sometimes referred to as Zintl phases.

The alkali and earth alkali metal compounds are usually sensitive to moisture and for that reason they have only recently been explored systematically. On the other hand, they are interesting since bonding in these compounds is often more readily rationalised because the Group IA, IIA or IIIB metals contribute a known number of valence electrons. This is, of course, not the case for transition metals. Thus it can be hoped that the systematic study of these compounds will contribute to our understanding of bonding in intermetallic phases.

The coordination of the silicon or germanium atom consists usually of both kinds of metal atoms and sometimes, depending on the silicon (germanium) content, also of one or more Si(Ge) atoms. These compounds are frequently isostructural with simple ionic structures or can be derived from those.

The $MnCu_2Al$ (Heusler) type structure may be regarded as either an ordered superstructure of CsCl or a filled NaCl type structure. It occurs for $LiNi_2Si$, $LiNi_2Ge$[188], $LiCo_2Ge$, and $LiNi_2Sn$[189]. The structures of Li_2MnGe and Li_2MnSn are possibly tetragonal distorted versions of $MnCu_2Al$ with different atomic ordering[190]. The atomic ordering was not determined for the cubic compounds Li_2PdGe, $LiPd_2Ge$ and several homologous tin containing compounds, e.g. Li_2PtSn and Li_2IrSn[191]. They are, however, almost certainly related to or isotypic with $MnCu_2Al$. Their colour was reported as varying between silvery-white, yellow, reddish and brown. $MnCu_2Al$ type stannides where both electropositive components are transition metals, e.g. $ZrCo_2Sn$, $ZrNi_2Sn$, and $NbNi_2Sn$ have also been prepared[192].

The PbFCl structure can be derived from the NaCl structure by filling one-half of the tetrahedral holes formed by the large ions. It occurs for a large number of seemingly unrelated compositions like UNI[193], LiFeAs[194],

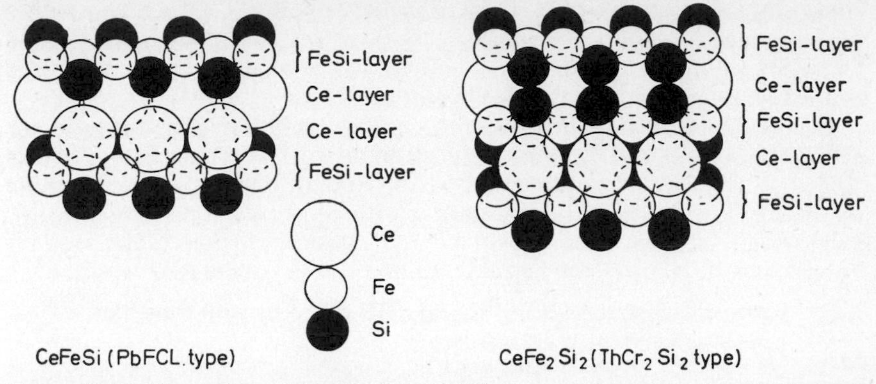

Figure 7.10 Comparison of the PbFCl and $ThCr_2Si_2$ type structures. The projections are normal to the fourfold axes. (Modified, from Eisenmann, May, Müller, Schäfer, Weiss, Winter, Ziegleder[205], by courtesy of *Naturforschung*)

MgCuAs[195] or ZrSiS[196] which were discussed by Bodak, Gladyshevskii and Kripyakevich[197]. These authors also reported the series LnFeSi (Ln = La,Ce,Pr,Nd,Sm,Gd,Tb,Dy,Ho,Yb) and LnCoSi (Ln = La,Ce,Pr,Nd,Sm) with this structure (Figure 7.10). MnAlGe and MnGaGe are ferromagnetic[198–200] and are of interest in the form of thin films[201–204]. Other representatives are discussed under ternary phosphides and arsenides.

The relation of the $ThCr_2Si_2$ (ordered $BaAl_4$ or $CeGa_2Al_2$) type structure to the PbFCl type structure is shown in Figure 7.10 for CeFeSi and $CeFe_2Si_2$. Schäfer, Weiss, and co-workers[205] compared the two structures for the examples BaMgSi and $BaMg_2Si_2$. They pointed out that in $BaMg_2Si_2$ the Si atoms are separated by only 2.48 Å and therefore may be described as forming pairs. In carrying this idea somewhat further, the Si atoms can be assigned oxidation state −3 and the formula can be written as polyanionic $Ba^{2+}Mg^{2+}Mg^{2+}[Si_2]^{6-}$ while BaMgSi can be rationalised as a normal valence compound $Ba^{2+}Mg^{2+}Si^{4-}$. This reasoning can also be applied for $CaNi_2Si_2$, $CaNi_2Ge_2$, and $SrNi_2Ge_2$ with $ThCr_2Si_2$ type structures[205,206]. This rationalisation does not always make sense for isostructural compounds where the differences in electronegativities are smaller, such as $SrAl_2Pb_2$[207] or the series $LnAu_2Si_2$ (Ln = La through Er) and $LnAg_2Si_2$ (Ln = La through Eu)[208]. This is also the case for the compounds LnT_2M_2 and AcT_2M_2 (Ln = rare earth; Ac = Th, U; T = Cr, Mn, Fe, Co, Ni, Cu; M = Si, Ge[206,209,210]. More recently YCr_2Si_2 was described with the $ThCr_2Si_2$ type structure[211]. The $ThCr_2Si_2$ type compounds ThT_2M_2 with T = Cr,Mn,Fe,Co,Ni,Cu, and M = Si or Ge were found to be alternately ferromagnetic and antiferromagnetic depending on the atomic number of the T metal[212]. The atomic and magnetic structure of $NdFe_2Si_2$ was studied by neutron diffraction[213].

The phase diagram of the ternary system Ce–Ni–Si shows a total of 13 ternary phases[214]. In most of the phases, Ni and Si—being of similar size—are exchangeable within the limits of 2–10 atomic %, while the Ce content remains practically constant. Phases with known crystal structures include $Ce(Ni,Si)_2$ with a AlB_2 type structure, $Ce_3Ni_6Si_2$ with a Ca_3Ag_8 type structure, $Ce_2Ni_{15}Si_2$ with a Th_2Ni_{17} type structure, $CeNi_{8.5}Si_{4.5}$ with a $NaZn_{13}$ type structure, $CeNi_{8.6}Si_{2.4}$ with a $BaCd_{11}$ type structure, and $CeNi_2Si_2$ with a $ThCr_2Si_2$ type structure. The crystal structures of several of the remaining phases have also been determined.

$CeNiSi_2$ belongs to a new type[215] combining building elements of the AlB_2, $ThCr_2Si_2$, and α-$ThSi_2$ type structures. It contains two kinds of silicon atoms. One is coordinated by four Ni and four Ce atoms, the other by six Ce, one Ni, plus two Si atoms with Si—Si distances close to the single-bond distance, thus forming a polyanionic Si-chain. The isotypic compounds $LnNiSi_2$ (Ln = La to Tm) and $CeNiGe_2$ were also reported[215].

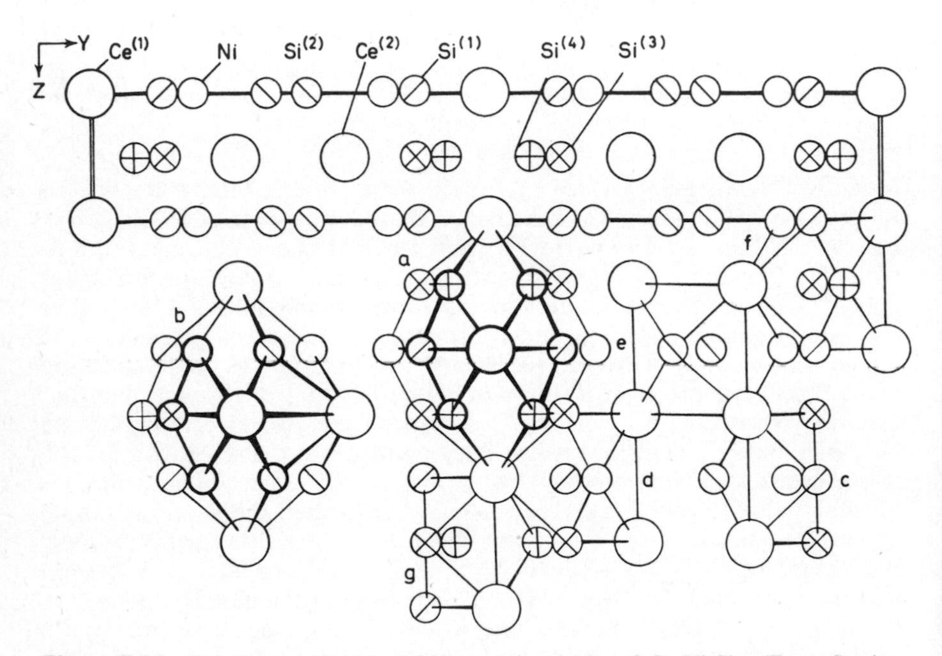

Figure 7.11 Structure and near neighbour environments of $Ce_3Ni_2Si_8$. (From Stepien, Lukaszewicz, Gladyshevskii and Bodak[216], by courtesy of *Bull. Acad. Polon. Science.*)

The structure of $Ce_3Ni_2Si_8$ is orthorhombic and may be considered as a combination of the $ThCr_2Si_2$ and AlB_2 type structures[216]. Four different Si atoms have coordination numbers of 7, 8 or 9 excluding Si—Si distances greater than 2.9 Å. Each Si atom has at least two close Si neighbours at distances varying between 2.38 and 2.45 Å which is only 0.2 Å more than the bonding distance in elemental Si. A projection is shown in Figure 7.11.

The phase with approximate composition Ce_2NiSi crystallises with a large hexagonal cell[217]. It is closely related to the $Rh_{20}Si_{13}$ structure[218]. In $Ce_2NiSi = Ce_{20}(Ni,Si)_{20}$ the Rh positions are taken by Ce, and the Si sites of $Rh_{20}Si_{13}$ are occupied by varying mixtures of Si and Ni. Trigonal prismatic sites which are void in $Rh_{20}Si_{13}$ are occupied in $Ce_{20}(Ni,Si)_{20}$. Compositions La_2NiSi and Pr_2NiSi have the same structure as Ce_2NiSi[217].

7.4.5 Ternary compounds of two transition metals with silicon or germanium

The crystal structure of a transition metal-rich phase in the system Mn–Co–Si was determined[219,220]. The phase, which has been called *X*, *Y* and *Q* in previous phase-diagram studies[221,222], occurs at composition $Mn_{39-45}Co_{40-45}Si_{15-19}$. The *X* phase—the name adopted in the structural studies—belongs to the family of σ-phase related structures. A projection of the orthorhombic structure is shown in Figure 7.12. The larger Mn atoms occupy mainly seven atomic sites: five with coordination number (C.N.) = 16 and one each with C.N. = 15 and C.N. = 14.

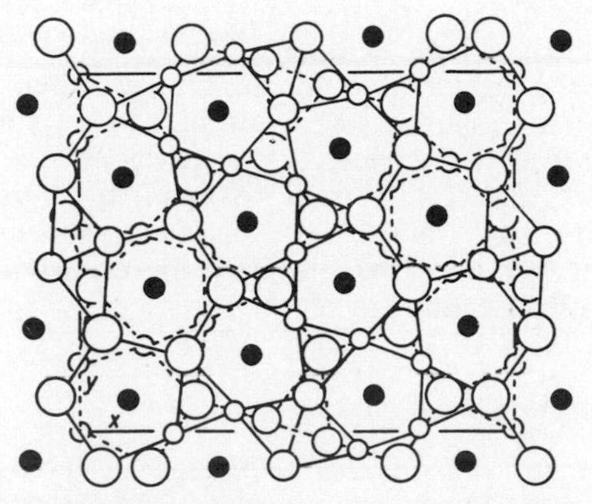

Figure 7.12 Projection of the *X* phase structure $Mn_{\sim 45}Co_{\sim 40}Si_{\sim 15}$. The net at $z = 0$ is indicated by solid lines, the net at $z = \frac{1}{2}$ by broken lines. Filled in circles are atoms at approximately $\frac{1}{4}$ and $\frac{3}{4}$. The sizes of the circles increase with increasing CN of atoms. (From Manor, Shoemaker, and Shoemaker[220], by courtesy of *Acta Crystallogr.*)

The smaller Co and Si atoms are situated on the nine C.N. = 12 sites. Three of them are nearly exclusively occupied by Si atoms, one has both Co and Si atoms and the remaining five C.N. = 12 sites contain predominantly Co atoms. As was found in the ν-$Mn_{81.5}Si_{18.5}$ structure[143], there are no Si atoms present in the first coordination shell around the 'Si only' positions, indicating an ionic charge on the Si atoms.

Several new ternary equiatomic transition metal silicides and germanides were reported. TiRhSi, TiPdSi, MnRhSi, ZrFeGe, ZrRhGe and NbRhGe were found[223] to crystallise with the TiNiSi (ordered $PbCl_2$) structure. ZrRuSi[223], MnRhGe[224] and MnPdGe[224] have the Fe_2P type structure. A TiFeSi type structure, which is a distorted variant of Fe_2P[225] was found for TiRuSi[223], NbFeGe, TaMnGe, TaMnSi[226] and ~TiMnGe[227]. In MnRhSi the larger Rh atoms occupy the position of the smaller Ni atom of the TiNiSi type structure, thus demonstrating the dominance of the electronic (valence) factor over the size factor for these compounds[159]. The phases $Zr_3Cu_4Si_2$, $Zr_3Cu_4Ge_2$, $Hf_3Cu_4Si_2$ and $Hf_3Cu_4Ge_2$ crystallise with an ordered variant of the Fe_2P type structure where the Zr atoms take the positions of the Fe atom with large C.N., three of the four Cu atoms occupy the position of the Fe atom with low C.N. and the remaining Cu as well as the Si atoms occupy the two P positions in an orderly manner[571].

The phase previously reported as '$TiCrSi_2$'[228] was found[229] to be isostructural with $Nb_2Cr_4Si_5$[171]. This structure also occurs in the system Ti–V–Si[229].

7.5 NITRIDES

The transition metal nitrides are in almost every respect closely related to the carbides. Those of the early transition metals are extremely hard with melting points almost as high as those of the carbides. They are, however, more brittle than the carbides and this has impeded their practical applications. They are also more cumbersome to prepare in the laboratory. For these reasons they are not as well explored as the carbides. The books by Kieffer and Benesovsky[9,53] and Goldschmidt[10] contain chapters on nitrides. The nitrides of the first row transition metals were thoroughly reviewed by Juza[230]. Recently, Jack and Jack discussed the occurrence and structural properties of the nitrides found in steels[57].

7.5.1 Binary transition metal nitrides

Although most binary transition metal–nitrogen systems have been investigated repeatedly, complete phase diagrams have been reported only for a few systems. Many phases have not been well characterised with respect to the equilibrium nitrogen pressure or the role of impurities like oxygen, carbon or hydrogen. The stability of the transition metal nitrides decreases with increasing Group number: in nitrogen at 1 atm TiN is stable up to 3100 K while Ni_3N decomposes rapidly at 700 K.

The structures of the nitrides—especially of the transition metal-rich nitrides—are frequently the same as those of the carbides and most of the discussion on the structural chemistry of carbides could be paraphrased here. For isostructural carbides and nitrides the nitride has the smaller lattice dimensions when the transition metal is of Groups IV, V or VI while they are larger for late transition metal nitrides.

Two phases have been known in the titanium–nitrogen system for some time: TiN_{1-x} with an NaCl type structure and the low temperature phase Ti_2N with a (anti-)rutile type structure. At high temperatures the cubic phase TiN_{1-x} includes the composition $TiN_{0.5}$. Recently it has been shown that at lower temperatures nitrogen atoms and vacancies in $TiN_{0.5}$ order[231] giving rise to a hexagonal superstructure.

Until recently the NaCl type phases ZrN_{1-x} and HfN_{1-x} were the only well characterised nitrides in their respective systems. Rudy[232] reported on Hf_3N_2 and Hf_4N_3. The metal atoms are close packed in both phases with stacking sequences like those found for $(Ta_2V)C_2$[59] and V_4C_3[60]. The positions of the nitrogen atoms could not be determined. There is, however, no doubt that they occupy octahedral interstices. According to the rule, discussed above for carbides, that both of the face-sharing octahedral voids cannot be occupied, the ideal compositions are Hf_3N_2 and Hf_4N_3. However, the experimentally-observed compositions $Hf_3N_{1.69}$ and $Hf_4N_{2.56}$ indicate even higher nitrogen deficiencies.

The vanadium–nitrogen system contains NaCl-type VN_{1-x} and a nitrogen deficient phase based on hexagonal close packed V atoms with ideal composition V_2N similar to V_2C. The ordering of the nitrogen atoms is not well established. Recently it was suggested to correspond to the Ni_3N type structure[233].

Substantial evidence was found for low temperature (*ca.* 600 K) vanadium-rich nitrides with approximate composition $V_{16}N$ where the arrangement of the V atoms is similar to elemental vanadium[234]. In that paper references to earlier work on similar phases in systems V–N, V–O, Nb–O, Ta–C and Ta–N are given. Metal-rich phases were also observed at or near the compositions Ta_2O[235], $Ta_{32}O_9$[236] and Zr_3O[237]. The most reliable structural work on these compounds was done on Ti_3O[238] and $V_{16}O_3$[239]. Due to the difficulty in obtaining untwinned single crystals of these low temperature phases, the structures are not refined and interatomic distances are not known. Bonding in these compounds is not understood. Because of the large differences in electronegativities and the truly interstitial character of the metalloid (the metal atoms have the same arrangement as in the element) knowledge of the exact atomic and electronic structures could greatly contribute to our general understanding of bonding in transition metal–metalloid compounds.

Niobium nitrides were reviewed by Terao[240]. According to that author five structures occur within the range from Nb_2N to NbN. Terao has recently reported [241] that the thin film nitrides Nb_4N_5 and Nb_5N_6 are isostructural with the corresponding tantalum compounds[242]. The proposed structure for Nb_4N_5 and Ta_4N_5 is derived from the NaCl type structure through ordering of vacancies on niobium sites. It was pointed out[243] that Ta_4N_5 is isotypic with Ti_4O_5[244]. Nb_5N_6 and Ta_5N_6 are hexagonal, but their structures are not fully established[241].

Three phases are known for composition TaN[245]. Besides ε-TaN with a CoSn type structure and a phase with NaCl type structure which is stable only at nitrogen pressures greater than 10 atm[246], the stoichiometry 1 : 1 was also reported for WC type TaN at high pressure[247]. At high temperatures and pressures NaCl type TaN forms continuous solid solutions with all IVA and VA transition metal carbides and nitrides of that structure[245]. A Ta-rich cubic low temperature phase was identified by electron diffraction and a structure corresponding to the ideal composition Ta_9N_2 was proposed[248].

Equilibrium pressures and homogeneity limits of the two known chromium nitrides CrN_{1-x} (NaCl type) and hexagonal Cr_2N_{1-x} were investigated as a function of temperature[249]. From these data a phase diagram was established and the reaction enthalpies and entropies for formation of the two nitrides were calculated[250]. The antiferromagnetic orthorhombic low temperature structure of CrN has been re-investigated[251].

The system molybdenum-nitrogen was studied by Ettmayer[252]. WC type MoN was confirmed. Paramagnetic tetragonal β-Mo_2N_{1-x} is a low temperature phase[252,253]. Cubic γ-Mo_2N_{1-x} is stable at high temperature. The importance of small amounts of oxygen for its stability is disputed[57,252].

Since the positions and possible ordering of the nitrogen atoms are not known for the subnitrides the true symmetries and unit cell sizes are uncertain.

Other work on binary transition metal nitrides includes electronic band structure calculations on NbN[89,90], a study of x-ray emission and absorption spectra of TiN and Ti_2N[254] and a magnetic study of the hexagonal first row transition metal nitrides T_2N[255]. Metallic conductivity of the IVA and VA transition metal nitrides with NaCl type structures was confirmed[256]. The pseudoternary system TiC–TiN–TiO forms a continuous series of solid solutions at high temperatures[257]. VN and VO show only partial solid solubility[258]. Vapor deposition of TiN[259] and other early transition metal nitrides[260] was described. Vanadium nitrides can be prepared by reaction of vanadium oxides with nitrogen-containing liquid lithium[261].

7.5.2 Ternary compounds containing transition metals and nitrogen

Li_2ZrN_2 was prepared from a mixture of the metal nitrides under ammonia[262]. Because of the small scattering factors of the light atoms, its structure could not be determined with certainty. It crystallises probably with the Mg_2MnAs_2 (ordered anti-La_2O_3) type structure[263]. Li_2ThN_2 and Li_2UN_2 have the same structure[262].

By heating alkaline earth nitrides with molybdenum or tungsten in nitrogen, compounds with the following tentative formulae were prepared: Ca_5MoN_5 and Ca_5WN_5, $Sr_{27}Mo_5N_{28}$ and $Sr_{27}W_5N_{28}$[264]. Ca_5MoN_5 is a paramagnetic insulator; $Sr_{27}W_5N_{28}$ is diamagnetic. The structures of the Ca and Sr compounds are different but unknown[264].

The crystal structures of the new compounds Th_2CrN_3, Th_2MnN_3, U_2CrN_3 and U_2MnN_3 have been determined[265]. They contain two independent nitrogen atoms in octahedral coordination to the metal atoms. The ratios of actinide metal to transition metal neighbours of the two nitrogens are 5 : 1 and 4 : 2 respectively. This predominance of the more electropositive metal in the environment of the metalloid is also found in the structures of U_2IrC_2[106] and $UMoC_2$[107].

A new phase with approximate composition $UCrN_2$ and unknown structure has been reported[266]. UVN_2 has a $UMoC_2$ type structure[267].

The investigation of the system Ti–Cr–N showed a complete series of solid solution between TiN and CrN. No ternary phase was found[268]. Hf_4Ni_2N has an η-carbide structure[269]. The crystal structure of NbCrN was determined[270,271]. In agreement with the rule that the metal with the higher affinity to the metalloid has more metalloid contacts the nitrogen atom is coordinated by five Nb atoms and one Cr atom (Figure 7.13).

Since Stadelmeier's review[38], the following nitrides with (anti-) perovskite type structures have been described: Cr_3RhN, Cr_3PdN, Cr_3IrN, Cr_3PtN, and Cr_3SnN[272]. The paramagnetic to antiferromagnetic phase transition of the new perovskite Mn_3NiN has been studied[273]. The perovskite type phases in the systems Ti–Zn,Cd,Hg–N seem to exist with some deviation from ideal stoichiometry[274].

The normal perovskites, like $BaTiO_3$, frequently show diffusionless phase transitions which lead to distorted versions of the cubic prototype. Two

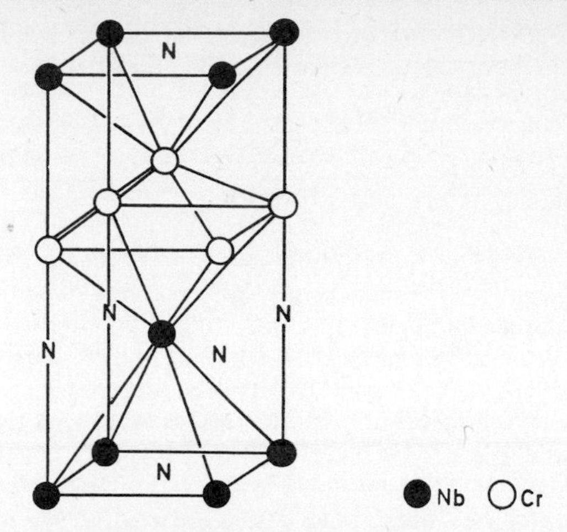

Figure 7.13 Crystal structure of NbCrN. (From Jack and Jack[271] by courtesy of *J. Iron Steel Inst.*)

distorted anti-perovskite structures are known. One was first determined for Cr_3AsN[275]. It was also found for Mn_3GeN, Fe_3GeN[275] and Mn_3AsN[276]. The other one was determined for Cr_3GeN[277]. They are shown together with the perovskite nitride Cr_3GaN in Figure 7.14. It has been shown recently that Cr_3GeN transforms on heating to the Cr_3AsN type structure[272]. Cr_3GeN and Cr_3AsN are tetragonal distortions of the perovskite structure. An orthorhombic distortion has been found for $Mn_3As_{0.5}Sb_{0.5}N$[276].

Several representatives of known structural types were found. V_3PN, V_3AsN, Cr_3PN[278], V_3GeN, and V_3GaN[118] crystallise with the V_3AsC type structure, Cr_2GaN has a Ti_2SC (H-phase) structure[274] and V_5P_3N is a Nowotny phase (filled Mn_5Si_3 structure)[279].

$MnSiN_2$ and $MnGeN_2$ are structurally not related to the previously-discussed complex nitrides. They crystallise with a β-$NaFeO_2$ or $BeSiN_2$

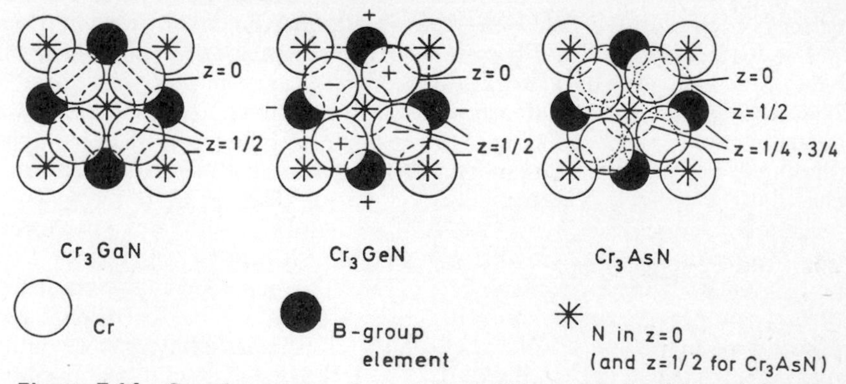

Figure 7.14 Crystal structures of Cr_3GaN ($CaTiO_3$ Type), Cr_3GeN, and Cr_3AsN (Filled U_3Si Type). (From Nowotny, Boller, and Beckmann[96], by courtesy of *J. Solid State Chem.*)

type structure[280,281] which is a tetrahedral structure derived from würtzite[282]. The antiferromagnetic structure of $MnGeN_2$ was determined from neutron-diffraction data[281].

7.6 PHOSPHIDES AND ARSENIDES

Several authoritative and comprehensive reviews on transition metal phosphides are available. Preparation, physical and chemical properties, and thermodynamic data of all solid-state phosphides were reviewed extensively by Wilson[283]. More than half of the article, with some 700 references, is devoted to transition metal phosphides. The structural chemistry of transition metal phosphides was treated well by Rundqvist[284], Lundström[285] and Ward[1]. Hulliger's article[5] includes arsenides, as well as the other pnictides.

Near-neighbour environments of phosphorus and arsenic in transition metal-rich compounds are similar to those of silicon and germanium with the exception, however, that the pnicogen elements (P,As) do not participate in the formation of σ-phase related structures. In compounds with a high and medium content of transition metal, the preferred coordination of P or As is the trigonal prism usually augmented by zero–three additional neighbours outside the rectangular faces of the prism. Corresponding to their places in the Periodic System, phosphorus and arsenic also form compounds with tetrahedral coordination of the pnicogen atom which is more frequently found for transition metal chalogenides. This coordination is found in pnicogen-rich compounds, especially of the late transition metals. The change in conductivity from metallic to semiconducting parallels the change in coordination from trigonal prismatic to tetrahedral, although there is not a one-to-one correspondence and tetrahedral coordination alone is of course not sufficient to predict semiconductivity.

Current research activity on binary transition metal pnictides may be divided along this line. Research on pnictides of the early transition metals is mainly concerned with the preparation of new compounds and their structural characterisation. Most of the new compounds found in these systems are metal-rich with the pnicogen atom in trigonal prismatic coordination. On the other hand, most of the research on late transition metal pnictides is done on known transition metal dipnictides where the pnicogen atom is tetrahedrally and the metal atom octahedrally coordinated. This more readily makes possible an explanation of bonding and permits an understanding of conductive and magnetic properties.

7.6.1 Binary compounds of early transition metals with phosphorus or arsenic

Although all binary systems of transition metals with phosphorus and arsenic have been investigated over the past four decades, the careful re-investigations of these systems by Rundqvist and his students in Uppsala show that new compounds can still be found.

Direct reaction of Hf and As powders in silica tubes yields HfAs and $HfAs_2$ which have been known for some time. More recently it was found that new phases are formed when the products obtained by the silica-tube technique are arc-melted[286]. Hf_2As has a Ta_2P type structure, Hf_3As_2 is isostructural with Hf_3P_2, and Hf_5As_3 was found with a structure later determined for Nb_5P_3[287]. Further intermediate phases were found in the Hf–As system but are not characterised fully[286].

$V_{12}P_7$ has an anti-Th_7S_{12} type structure[288]. The crystal structure of V_3P (Ti_3P type) was refined[289]. Like the other Ti_3P type structures it is distinguished by one very short metal–metal distance. VP with a NiAs type structure was confirmed and the structure of MnP type VAs was refined[290].

The structure of Nb_4As_3[291] was described as an array of interconnected Nb_6As triangular prisms (Figure 7.15). One additional arsenic atom (As_4) is surrounded by six niobium neighbours in a rather unusual arrangement.

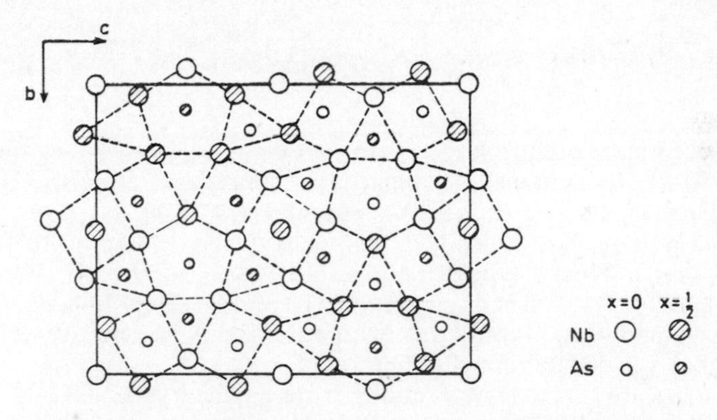

Figure 7.15 Projection of the Nb_4As_3 structure. (From Carlsson and Rundqvist[291], by courtesy of *Acta Chem. Scand.*). In this and in several of the following figures the trigonal prisms around the metalloid atoms are indicated by broken lines

Five independent Nb atoms are coordinated to between four and six As atoms each, with additional Nb—Nb interactions increasing the coordination of the Nb atoms to between seven and twelve. As could be expected from the composition no As—As bonds occur. This structure was also found for V_4P_3[119] and refined from single-crystal data for V_4As_3[292].

New structural types were also found for Nb_5P_3[287] and Nb_8P_5[293]. They are, like the Nb_4As_3-type structure, layered structures in the sense that atoms are confined to two planes. This results in one rather short axis which is ideal as a projection direction (Figures 7.16 and 7.17). Both structures have a large number of independent atoms. As usual the P atoms are all situated in trigonal prisms of the Nb atoms with additional Nb atoms outside the rectangular prism faces increasing coordination numbers to 7, 8, or 9. The coordinations of the Nb atoms are more diverse: the tetragonal pyramid of 5As atoms occurs most often; trigonal bipyramids and more or less distorted squares of As atoms are also found. As usual for a transition metal-rich compound, there are a large number of metal–metal interactions with

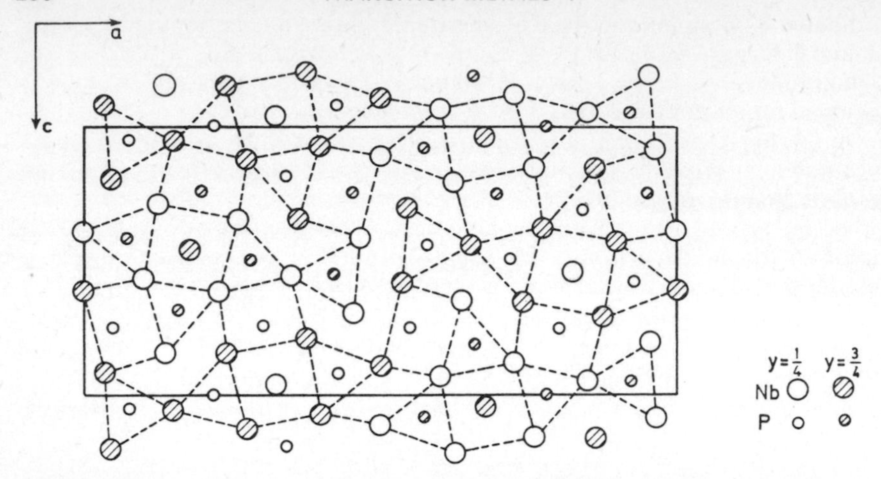

Figure 7.16 Crystal structure of Nb_5P_3. (From Hassler[287] by courtesy of *Acta Chem. Scand.*)

distances usually slightly larger than those found in the pure metal. The Nb_8P_5 structure contains one remarkably short Nb—Nb bond of 2.54 Å, while all other Nb—Nb distances are greater than 2.80 Å.

In the systems Nb–As and Ta–As several new phases were found[294]: Nb_7As_4 has a Nb_7P_4 type structure, Nb_5As_3 is of the Nb_5P_3 type and Ta_2As and Ta_5As_4 are isotypic with Ta_2P and Ti_5Te_4 respectively. Ta_3As which was previously thought to be of the Ti_3P type, was found to have a different as yet undetermined structure[294].

The structure of Cr_3P was refined from single-crystal data[295] and confirmed to have a Fe_3P type structure and not the very similar α-V_3S type structure which occurs for Mo_3P. The structure of $Cr_{12}P_7$ was shown in a single crystal investigation to be similar to $V_{12}P_7$ and Th_7S_{12}[296].

The phosphorus-rich section of the Cr–P system was investigated at high pressures and two new compounds with compositions CrP_2 and CrP_4 were found. CrP_4 is a metallic conductor and Pauli paramagnetic[297]. CrP_4 and isostructural high-pressure MoP_4[297] have the highest phosphorus content of known transition metal phosphides. The Cr atoms are octahedrally co-

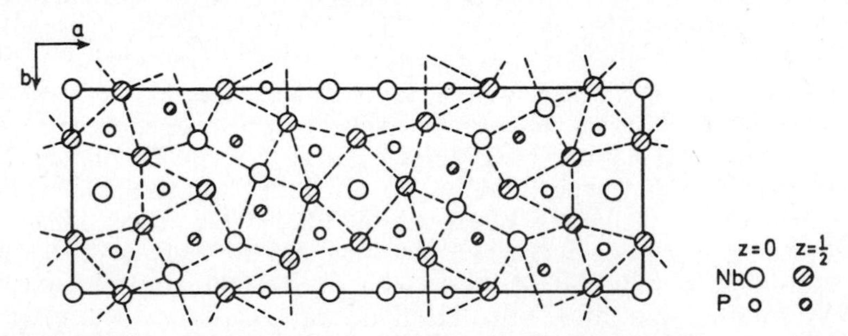

Figure 7.17 Crystal structure of Nb_8P_5. (From Annugul, Pontchour, and Rundqvist[293], by courtesy of *Acta Chem. Scand.*)

ordinated by P atoms and the P atoms are tetrahedrally coordinated by both Cr and P (Figure 7.18). All P—P interactions are close to 2.21 Å which is the bonding distance in the known P modifications. If two-electron bonds are assumed for all short P—P and Cr—P interactions, four valence electrons per formula unit remain unaccounted for. They were placed on the Cr atoms resulting in a formal oxidation state of +2 for Cr. Two were assigned to one nonbonding t_{2g} orbital, while the other two were accommodated in the two other t_{2g} orbitals, thus resulting in a Cr—Cr bonding interaction which accounts for the distortions of the edge-sharing [CrP_6] octahedra as well as for the non-magnetic behaviour of the compound[297]. The near-neighbour

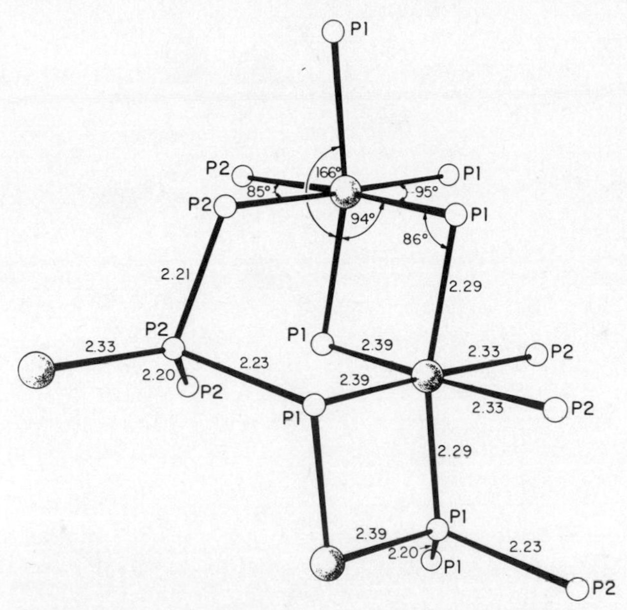

Figure 7.18 Near neighbour environment in CrP_4. (From Jeitschko and Donohue[297], by courtesy of *Acta Crystallog.*)

environments of CrP_2 (Figure 7.19) when compared with those of CrP_4 show higher coordination for all atoms[298]. If classical two-electron bonds are assumed in CrP_2 for all very short near-neighbour interactions, a deficit of one electron results. If second-nearest neighbour interactions—at least some of which are undoubtedly bonding—are also counted, the electron deficit becomes even greater. This lack of bonding electrons is of course characteristic of metals and intermetallic compounds. Thus the near-neighbour environment in CrP_2 was described as the result of a compromise between directional covalent and delocalised metallic bonding. Like high pressure CrP_2, high pressure $CrAs_2$ has a $OsGe_2$ type structure[298].

Cr_2As with a Cu_2Sb type structure was shown to be stoichiometric $Cr_{2.00\pm0.05}As$[299]. Its magnetic structure was re-studied[300]. At high temperatures Cr_2As has a Fe_2P type structure[301,302].

Cr_4As_3 crystallises with a new structural type which is related to MnP. One of the Cr atoms is coordinated by a square of As atoms plus 8Cr atoms

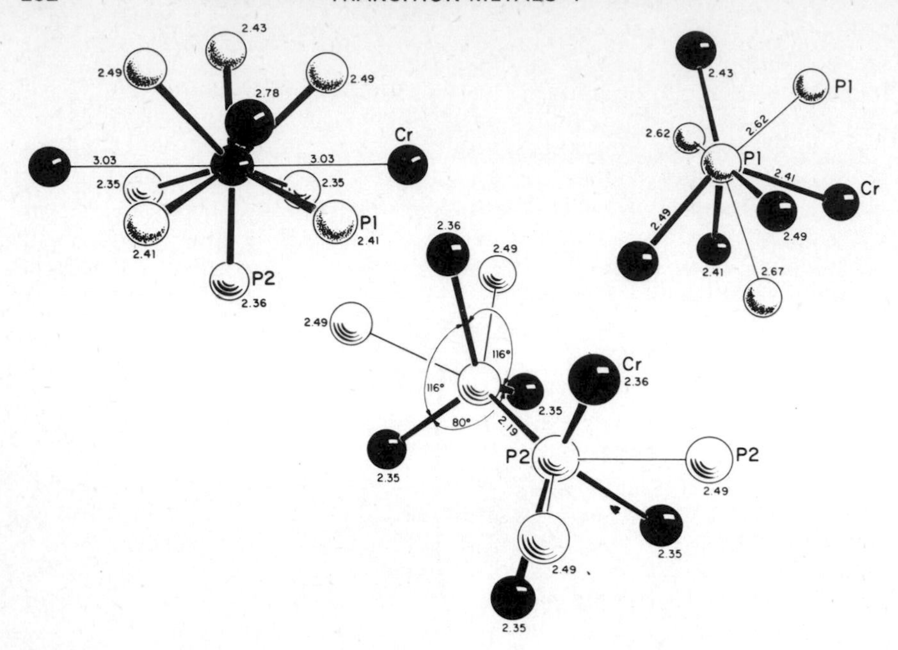

Figure 7.19 Near neighbour environments in CrP_2 with $OsGe_2$ type structure. (From Jeitschko and Donohue[298], by courtesy of *Acta Crystallog.*)

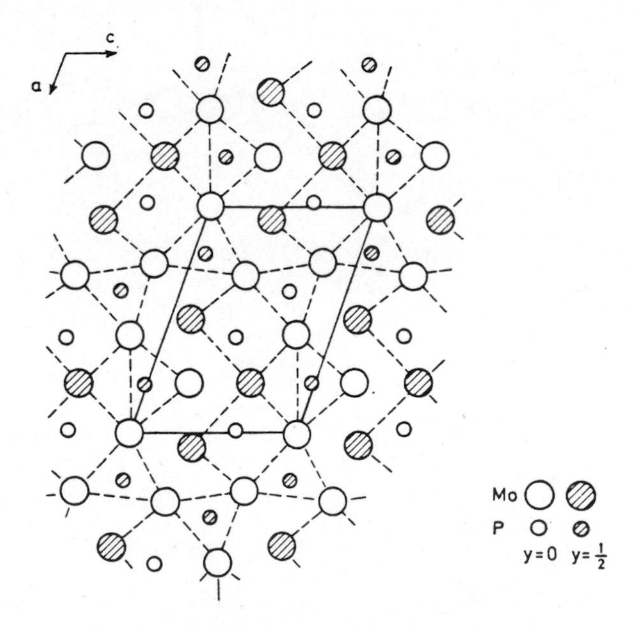

Figure 7.20 Crystal structure of Mo_8P_5. (From Johnsson[133], by courtesy of *Acta Chem. Scand.*)

forming a cube[303]. This building element occurs also in Ti_5Te_4, Nb_5P_3, Nb_8P_4, Nb_7P_4, Ta_2P, Nb_4As_3 and $Nb_{21}S_8$ and therefore represents a relatively stable configuration for IVA and VA metals in compounds with metalloids of that compositional range.

The magnetic structure and phase transition of CrAs was studied by various authors both in the binary compound[304–306] and in the solid solutions of CrAs with Sb[307], Se[308], Mn[309] and Fe[310]. The phase transition is between high temperature NiAs type and low-temperature MnP type which has a helimagnetic structure.

The structure of Mo_8P_5 was determined[311]. All P atoms have a trigonal prismatic metal environment with 1–3 additional metal atoms outside the quadrilateral faces of the prisms (Figure 7.20). The Mo coordinations are similar to those discussed above for Nb_5P_3 and Nb_8P_5.

Mn_2P with a Fe_2P type structure was shown to be an antiferromagnet below 103 K[312,313]. Its magnetic structure was determined[312]. The pressure dependence of the magnetic behaviour of MnP was studied[314]. Mn_2As, which is known with a Cu_2Sb type structure, has a Fe_2P type structure at high pressure[302]. It is antiferromagnetic below 50 K[302]. Thermodynamic properties and phase transitions of MnAs were studied in the range 5–700 K[315]. The crystal structure of Re_3As_7 (Ir_3Ge_7 type) was refined, and, in agreement with the generalised $(8 - N)$ rule, found to be metallic[316].

7.6.2 Binary compounds of late transition metals with phosphorus or arsenic

Structural characteristics of the early and late transition metal pnictides are similar for metal-rich compounds. Like all the other transition metal-metalloid compounds (except CrP_4) discussed so far, they show high coordination for all atoms and they do not possess enough valence electrons to form two-electron bonds for all near-neighbour interactions. This is of course characteristic for intermetallic (and ionic) structures. With the dipnictides of the late transition metals we encounter compounds where the number of valence electrons (defined as the number of electrons outside the inert gas configuration) is equal or greater than is needed to form (covalent) two-electron bonds for all near neighbour contacts. The diphosphides and diarsenides of the late transition metals have tetrahedral coordination for the pnicogen atoms and octahedral coordination for the metal atoms (except for NiP_2 and PdP_2 where the metal atoms are coordinated by four P atoms forming a square). These simple coordinations allow rationalisation of near-neighbour σ-bonding as resulting from interactions of d^2sp^3 orbitals of the metal with sp^3 orbitals of the non-metal. Additional valence electrons fill more or less localised d states of the metal. They may form metal–metal bonds in the marcasite-type structures where the TX_6 octahedra share edges. The deduced bonding schemes should of course reflect the magnetic and conductive behaviour of the compounds.

The late transition metal dipnictides are frequently semiconducting and have structures with pnicogen–pnicogen bonds. They are therefore usually reviewed together with other polyanionic semiconductors[5,8,317]. More recent papers deal with properties and bonding of compounds with the pyrite, marcasite, and arsenopyrite type structures[318–321] which are most frequently encountered for late transition metal dipnictides.

Compositions of the marcasite type compounds FeP_2, $FeAs_2$, β-$NiAs_2$, RuP_2, $RuAs_2$, OsP_2, and $OsAs_2$ and of the arsenopyrite compound $CoAs_2$, RhP_2, $RhAs_2$, IrP_2 and $IrAs_2$ were shown to be very close to MX_2[322,323]. The crystal structures of FeP_2, $FeAs_2$, β-$NiAs_2$, RuP_2, $RuAs_2$, OsP_2 and $OsAs_2$ with marcasite type structure were refined from x-ray powder data[324] and the magnetic properties reported[325]. Structures of pyrite type PtP_2 and marcasite type FeP_2 were refined from single-crystal data[326]. The compounds $CoAs_2$, RhP_2, $RhAs_2$, IrP_2 and $IrAs_2$ are diamagnetic; their structures were refined[323]. The preparation, electric and magnetic properties of $FeAs_2$ and $FeSb_2$ were reported[327,328].

The crystal structure of α-$NiAs_2$ was refined from powder data and found to be isotypic with brookite (TiO_2) and $AuSn_2$[329]. The structure is transitional between the marcasite and pyrite structures in that the $NiAs_6$ octahedra share one edge, compared to two in marcasite and none in pyrite. The structure was also found and refined from single-crystal data for the corresponding mineral pararammelsbergite which contains some Co and S as minor constituents[330].

New modifications of several transition metal dipnictides were found at high pressure: NiP_2 and $NiAs_2$ with pyrite structure[331], CoP_2 with arsenopyrite[332], and $MnCoP_4$ with marcasite structures[332]. Their magnetic and conducting properties were determined[331,332].

In the skutterudite ($CoAs_3$) structure, which occurs for about ten combinations of Co,Rh,Ir with pnicogen, the metal atom is also coordinated octahedrally by the anions. Corresponding to the large anion content, there are more anion–anion bonds: the arsenic atoms form a 4-membered ring. The skutterudite structure was refined once more[333].

The crystal structure of maucherite, $Ni_{11}As_8$, was determined[334]. It contains six non-equivalent Ni atoms of which five are in square-pyramidal coordination and one is in stretched octahedral coordination with As. Ni—Ni bonding certainly also occurs with nearest Ni—Ni distances ranging from 2.48 to 2.87 Å (Figure 7.21). There are no short As—As distances. Square-pyramidal coordination of metalloid atoms around the transition metal atom is frequently found in transition metal silicides and germanides with similar metal : metalloid ratios[335]. Examples are Ti in TiCoGe (ordered Fe_2P type) and TiNiSi (ordered anti-$PbCl_2$ type), and Nb in $Nb_5Cu_4Si_4$. It is also found in Nb_5P_3[287] and Nb_8P_5[293] and in chalcogenide structures like millerite (NiS) and α-Ni_7S_6[334,336].

At higher Ni : As ratios two phases have been characterised recently. Both are related to the Ni_2In structure in a way similar to that described above for $Ni_{31}Si_{12}$[154]. One, with composition Ni_5As_2, is isostructural with Pd_5Sb_2[156]; the other with a slightly higher metal-content was designated $Ni_{5+}As_{2-}$. It probably corresponds to a stacking variant of the Pd_5Sb_2 structure with a large unit cell[156].

At high temperatures Pd_2As has the Fe_2P-type structure. The low temperature structure of Pd_2As can be obtained only by prolonged annealing below the transition temperature (728 K). It is orthorhombic with eight formula units per cell (Figure 7.22) as was determined by Bälz and Schubert[337].

Three of the four independent Pd atoms are in square pyramidal configuration of As atoms which occurs often in this compositional range c f. $Ni_{11}As_8$. The fourth Pd atom, however, is coordinated by four As atoms forming a square, similar to Pd in the PdP_2 structure, where localised directional bonding certainly is predominant. One of the two independent As atoms in low-temperature Pd_2As is in the frequently-occurring ninefold coordination in which six metal atoms form a trigonal prism and three additional metal atoms lie outside the rectangular faces of the prism. The other As atom has an unusual eightfold coordination.

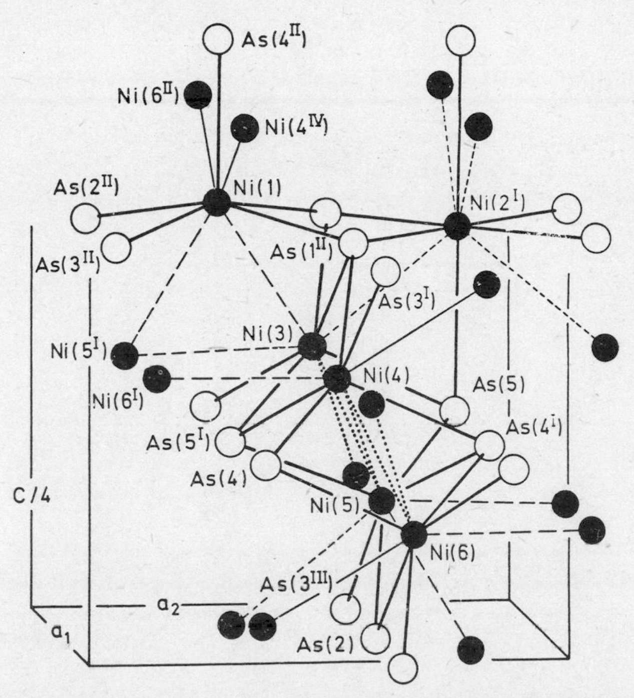

Figure 7.21 Near neighbour environments of the Ni atoms in $Ni_{11}As_8$. (From Fleet[334], by courtesy of *American Mineralogist*)

The crystal structures of MnP type FeP, FeAs, and CoAs were refined from single-crystal film data and their magnetic properties determined[338–340]. Some doubt was thrown on the correctness of space group Pnma which is normally accepted for the MnP type. Using Hamilton's significance test, the lower symmetry space group $Pna2_1$ was suggested which lacks the mirror plane perpendicular to the short axis of the MnP type cell. This test has sometimes led to unreliable results when systemmatic errors in the data correlated with the properties under examination. FeAs has a helimagnetic structure[341] which is similar to the one described for CrAs[304,305].

Several iron phosphides were studied by Mössbauer spectroscopy[342–344]. Preparation and properties of Fe_2P were reported[345]. Magnetic properties

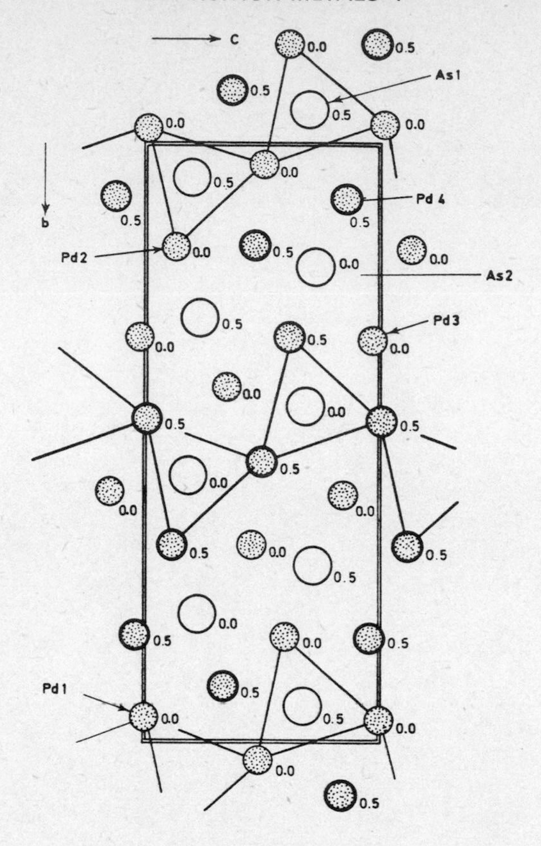

Figure 7.22 Crystal structure of the low-temperature modification of Pd_2As. (From Bälz and Schubert[337], by courtesy of *J. Less-Common Metals.*)

of the high- and low-temperature forms of Co_2As and Rh_2As were determined[299]. The magnetisation of amorphous and crystallised $Ni_{85}P_{15}$ was studied[346].

7.6.3 Ternary compounds containing transition metals and phosphorus or arsenic

Few ternary systems of transition metals with phosphorus or arsenic and a third element have been investigated. Considering the large number of ternary phases encountered in ternary systems with silicon and in view of the structural similarities between phosphides and silicides, many ternary phosphides should exist and remain to be discovered. Most of the work on ternary phosphides and arsenides has been an extension of similar work on silicides.

The only ternary phosphides which so far are not isostructural with silicides are $Hf_2Co_4P_3$ and $Zr_2Fe_{12}P_7$. Both structures were determined by Ganglberger[347,348]. In both compounds phosphorus has the usual trigonal prismatic coordination with three more metal atoms outside the rectangular faces of the prism. Corresponding to the higher electropositivity and the larger size of Zr and Hf the phosphorus atoms have more short interactions with Hf and Zr than would be expected from the Hf : Co and Zr : Fe ratios alone.

The PbFCl type structure is found with compounds comprising the range from salts to intermetallics. Several silicides and germanides with that structure were enumerated above (Section 7.4.4). Juza and co-workers characterised LiFeP, LiFeAs, and LiCoAs[349] and identified five phases in the system Li–Mn–As[350] with PbFCl-type and closely related structures. Magnetic properties of the five latter phases were also determined[351]. Although several of these compounds have detectable homogeneity ranges, they may still be considered as Zintl compounds since bonding can be rationalised with formal valencies corresponding to $Li^{1+}Fe^{2+}P^{3-}$. In the recently reported compounds NbSiAs, NbGeAs[352] and TaSiAs[353] with PbFCl-type structure, Si—Si and Ge—Ge bonding was assumed in accordance with earlier suggestions. Thus the formulae may be written $Nb^{5+}[Si]^{2-}As^{3-}$, where the numbers denote relative oxidation states and the brackets indicate bonding between the closely approaching silicon atoms. Slight deviations from the ideal 1 : 1 : 1 composition were reported for the PbFCl type compounds ZrAsS, ZrAsSe and ZrSiSe[354,355]. The system Zr–As–Te contains a phase with a relatively large homogeneity range and an orthorhombic distorted PbFCl-like structure[356].

Almost 200 ternary compounds with ordered Fe_2P and $PbCl_2$ type structures were reported at the equiatomic composition T_ET_LM where T_E is a rare earth, an actinide, or early transition metal, T_L is a late transition metal or Cu and M is a metalloid. Borides[357], aluminides and gallides[358], silicides and germanides[223,225,359,360] stannides[361], as well as phosphides[362] and arsenides[363] were found with these two structural types. The near-neighbour environments of the metal sites in the two structures are virtually identical: both have one site with pyramidal and one with tetrahedral metalloid coordination; additional metal neighbours increase the coordination numbers to 15 and 12 respectively. The site with the higher coordination is usually occupied by the larger more electropositive metal. In the very few cases where the more electropositive metal is smaller than the less electropositive metal, e.g. in MnRhSi, the ordering shows[159] that d-electron configuration or electropositivity is more important than atomic size. In CoNiP (Fe_2P), on the other hand, where atomic volume and electropositivity of the metals are similar, the smaller *effective* volume of Co[363,359] favours the tetrahedral site[364]. In FeCoP the Fe and Co atoms were found by Mössbauer spectroscopy to be randomly distributed at 1170 K[365]. Magnetic properties of the pseudobinary section Mn_2P–Fe_2P were investigated[366]; $PbCl_2$-type MnFeP orders antiferromagnetically at 330 K thus doubling its unit cell[367]. Several new ternary phosphides and arsenides were reported within the last few years; they include: Cr{Fe,Co,Ni}P, Mn{Fe,Co}P, FeCoP, and MnCoAs with ordered $PbCl_2$

type structures[368]; FeNiP, CoNiP, CrCoAs, CrNiAs, and FeCoAs with ordered Fe_2P type structures[368]; MnNiP and MnNiAs were found with both structural types[368]; VNiAs, VCoAs, and VFeP with $PbCl_2$ type and VFeAs, TiMnP, TiMnAs, and TiFeAs with Fe_2P type structures[369]. TiCrAs is also of the ordered Fe_2P type[227].

Ti_2PSb, Zr_2PSb, and Hf_2PSb crystallise with ordered TiP type (Ti_2SC, Cr_2AlC, H phase) structures[370]. The P atom occupies the octahedral and Sb the trigonal prismatic site. Ti_2PAs is completely miscible with both TiP and TiAs, while TiP and Ti_2PSb have a miscibility gap[370].

Phase diagrams of the ternary systems Fe–Ni–As[371] Fe–Co–As and Co–Ni–As[372] were investigated. The crystal structure of the mineral rhabdite, $\sim Fe_2NiP$, with a Ni_3P type structure was refined and the atomic distribution of the Fe and Ni atoms determined by anomalous dispersion of CoK_α diffraction data[373]. Solid solutions of hypothetical MoAs in TAs (T = Ti,V,Cr,Mn,Fe) were investigated[374].

Crystallisation of rapidly-quenched amorphous $Fe_{75}P_{15}C_{10}$ alloys occurs between 600 and 700 K[375]. The crystallisation process was studied by electron diffraction[376]. The activation energy for the growth of the Fe_2P crystallites is about one-half of that required for the diffusion of P in α-Fe[377]. Magnetic susceptibilities of amorphous metal-rich alloys in the systems Mn–P–C, Cr–Ni–Pt–P, and Fe–Pd–P were explored[378–380]. The radial distribution function of amorphous $Mn_{0.75}P_{0.15}C_{0.10}$ was investigated[381]. Electrical properties of metal-rich amorphous Pd–Ni–P alloys were studied[382,383].

7.7 SULPHIDES, SELENIDES, AND TELLURIDES

Transition metal chalcogenides were thoroughly reviewed in this Series by Flahaut[4] and Jellinek[3]. The present account is intended mainly to update these articles. For convenience the material is arranged according to the Periodic System and thus follows closely Jellinek's outline[3].

7.7.1 Some remarks on close-packed structures

Both ionic and metallic bonding have little directional character (at least in idealised model substances such a NaCl or Na) and packing becomes more important as a structure determining (and classifying) principle with increasing ionic or metallic bonding character. Hexagonal and cubic close* packing are well known. In Section 7.3.2.1 close-packed carbides were discussed. There the metalloid atom occupies the octahedral void formed by the close-packed metal atoms. In the sulphides the atomic size ratios are

* The words 'dense' or 'close' packing are somewhat misleading since they draw attention to space filling only. It is frequently overlooked that hexagonal or cubic close packing also provides for the largest separations of the 'close' packed atoms for a given volume. This latter consideration becomes of course important if the 'dense' packed atoms carry repelling charges.

inversed and it is the metal atom which occupies the octahedral sites formed by the close-packed sulphur atoms. Besides hexagonal and cubic close packing, a great variety of more complex stacking sequences occur both in carbides and sulphides.

In general, it can be expected that simple stacking sequences are the most stable ones, especially at high temperatures. At lower temperatures the vibrational energy* becomes less important and more complicated stacking sequences may become stable. Structures of this type are likely to have stacking faults. If one occurs about every 1000 Å, long-range order may not be detectable by x-rays (superstructure reflections become diffuse). Since the more complicated stacking sequences are expected to occur only at low temperatures, very long annealing times are required to establish order which can be detected by x-ray techniques. If stacking faults can be stabilised by impurities, x-ray detectable long-range order may not occur at all. Stacking faults may also be stabilised by slight deviations from ideal compositions. At even lower temperatures these stacking faults may order and give rise to structures with even greater repeat units. In many sulphides the possibilities for structural variations are increased even more by ordering of vacant and occupied 'interstitial' sites of the metal atoms. Partially occupied metal sites may order at lower temperatures and again stacking faults may prevent detection by conventional x-ray techniques. It is therefore not surprising that seemingly conflicting results have been reported for compounds with close-packed sulphur atoms.

For the carbide structures, it was noted that carbon atoms avoid occupying both of the two face-sharing octahedral sites. For the 'anti-type' sulphides, such a generalisation cannot be made; in fact in many sulphides metal-metal bonding across face-sharing octahedra occurs.

7.7.2 Chalcogenides of Group IV transition metals

7.7.2.1 Binary chalcogenides of Group IV transition metals

Many phases are known[3,4] in the range $Ti_{0.5}S$–TiS, all based on close packed sulphur atoms with Ti in octahedral holes. Recently some stacking variants with repeat units of up to almost 2000 Å were reported[384,385] for compositions in the range $Ti_{0.55}S$–$Ti_{0.75}S$ and annealing temperatures of 1050–1200 K. Tilley[386] has used transmission electron microscopy for phase analysis in the $Ti_{0.5}S$-TiS system. He observed a number of previously unreported phases as well as structural variations in phases which are well characterised by x-ray techniques. Because of the limitations of x-ray techniques discussed in the previous paragraphs, electron diffraction and imaging techniques[387] are superior for the identification of phases in such systems. More detailed structural information (stacking of sulphur atoms, atomic coordinations, bonding distances) is of course obtained through single-crystal x-ray studies whenever they can be accomplished.

* Small unit cells and high symmetry give more freedom for thermal vibrationals. Thus at high temperatures, small cells and high symmetry are favoured.

Ti_8Se_9 with an ordered defect NiAs type structure was characterised[388]. Stoichiometric TiSe with a NiAs type structure does not seem to exist[388].

The wide homogeneity range of $Zr_{1+x}Te_2$ with a partially-filled CdI_2-type structure was studied by x-ray powder methods and density determinations[389]. It was found to extend between the limits Te/Zr = 1.45 and Te/Zr = 1.735. For the corresponding phase $Hf_{1+x}Te_2$, the Te/Hf ratios were found between 1.40 and 1.60 (Ref. 390). This conflicts with the value of Te/Hf = 1.94 found in another study of this phase[391]. Single-crystal growth by vapour-transport was described[392] for all Group IV transition metal chalcogenides in the CdI_2-type structure. Brattas and Kjekshus[393] investigated the Zr-rich region of the Zr–Te system. They found Zr_5Te_4 with a Ti_5Te_4 type structure and confirmed a compound with a WC type structure and composition close to ZrTe. These authors also reported[390] unit cell dimensions for two new phases in the Hf–Te system: $HfTe_3$ with a $ZrSe_3$ type and $HfTe_5$ with an as yet unreported structure. They also studied[394] properties of the $ZrSe_3$ type compounds TiS_3, ZrS_3, HfS_3, $ZrSe_3$, $HfSe_3$, $ZrTe_3$ and $HfTe_3$.

The crystal structure of Zr_9S_2 was determined by Chen and Franzen[395]. As can be expected for a compound with high metal content, all atoms are highly coordinated. Five independent Zr atoms have coordinations identical or closely related to Kasper–Frank polyhedra. Four of them have two sulphur neighbours each, while the fifth has only Zr—Zr bonds. The co-ordination of the S atom is square-antiprismatic. The shortest Zr—Zr bonds are 2.89 Å which is considerably shorter than the sum of the metallic radii (3.20 Å). The authors pointed out that such short bonds are found also in many other metal-rich phases and—as was discussed also by Shoemaker and Shoemaker[144,169]—they occur between the atoms which form the sixfold vertices of two interpenetrating coordination polyhedra ('major ligand'). The *average* Zr—Zr distances, however, are greater than the sum of the 12 coordinate radii indicating weaker metal–metal interactions in the compound than in elemental Zr. This lengthening of the average metal–metal distances is also found in other metal-rich transition metal–metalloid compounds[225]. It suggests that electron transfer from the metal to the metalloid atoms not only draws off electrons from metal–metal bonding states but also creates repulsive positive partial charges on the metal atoms.

7.7.2.2 Ternary chalcogenides of Group IV transition metals

Most of the recent work in this category is on 'filled' CdI_2-type dichalcogenides where the third component atoms are intercalated between the close-packed layers of chalcogen atoms. The resulting structures are therefore ternary variants of the binary structures in the NiAs–CdI_2 family as discussed in Sections 7.7.1 and 7.7.2.1. The distinguishing name *intercalation* compound is justified, however, since only mild reaction conditions are needed to insert the third component between the slabs of the TS_2 host. Reaction of TiS_2 and ZrS_2 with alkali metals dissolved in liquid ammonia gives the formation of at least three phases in each of the systems Na_xTiS_2, K_xTiS_2, Rb_xTiS_2, Cs_xTiS_2[396–399], Na_xZrS_2 and K_xZrS_2[400,401] with x between 0.05 and 1.0. Li_xTiS_2[399] and the stoichiometric phase $(NH_3)ZrS_2$[401] were also prepared. Because of the poor degree of crystallisation, structural work is difficult. Partially occupied sites show no long-range order within the layers. Large translation periods perpendicular to the layers are due to variations in stacking sequences of the chalcogen atoms thus creating trigonal prismatic or octahedral sites for the intercalated components. Intercalation compounds of Group IV transition metal dichalcogenides are also discussed in Section 7.7.3.

The order–disorder transition and magnetic properties of $FeTi_2S_4$ were re-investigated[402,403]. The structures proposed[403] for $FeTi_4S_8$ and $FeTi_3S_6$ need to be proven. New phases, also based on partially filled CdI_2 type structures, were prepared[404] by solid-state reaction of Ni and TiS_2. The structure of $Ni_{0.25}TiS_2$ is not known. $Ni_{0.33}TiS_2$ ($NiTi_3S_6$) and $Ni_{0.40}TiS_2$ ($Ni_{1+x}Ti_3S_6$) crystallise with rhombohedral and trigonal structures similar to the two modifications of Cr_2S_3 and Cr_2Se_3. It is of interest that the trigonal structure always occurs when the metal : non-metal ratio exceeds 2 : 3. The structure of the earlier reported phase $NiTi_2S_4$ is in doubt. No structure was determined for $Ni_{0.75}TiS_2$[404]. Solid solutions T_xZrS_2 were reported[405] where x is from 0.00 to 0.37, 0.45 and 0.66 for T = Fe, Co, Ni respectively.

The ambient-pressure compound $PbZrS_3$ reported previously[406] was prepared at high pressure and was shown[407] to be isostructural with Sn_2S_3[408]. CdI_2-type ZrS_2 and SnS_2 form a complete series of solid solution[409]. Simultaneous vapour transport of ZrS_2 and SnS_2 was described[410].

7.7.3 Chalcogenides of Group V transition metals

Many phases are known in the range VS–$V_{0.6}S$ with superstructures of the type NiAs–CdI_2 as discussed above for the Ti–S system. The number of possible ordering schemes seems to be especially large for compositions close to V_7S_8: the lattice dimensions reported recently[411] for V_7S_8 (and V_7Se_8) are again different from those found earlier[412,413]. Magnetic susceptibility and n.m.r. spectra were studied[414] for the compounds in the compositional range VS–V_5S_8. The crystal structure of patronite, $V(S_2)_2$, was refined[415]. The sulphur–sulphur distances in the S_2 pairs are 2.03 and 2.04 Å.

The crystal structure of $Nb_{14}S_5$ was determined[416]. Five independent S atoms are in trigonal prismatic Nb coordination with one or two additional Nb atoms off the rectangular faces of the prism (Figure 7.23). The structure exhibits a large number of octahedral voids formed by Nb atoms, which possibly could accommodate small interstitial atoms. The 14 independent Nb atoms are coordinated to between one and five S atoms. The number of octahedral voids around the Nb atoms is inversely proportional to the number of S near-neighbours. Octahedral voids are necessarily present in close-packed structures with a majority of equally sized atoms. Several Nb atoms have capped cubic (coordination number 14) near-neighbour environments reminiscent of the body centred cubic near-neighbour coordination in elemental Nb[416].

Depending on composition, semiconducting 2s-$Nb_{1+x}S_2$ changes from p-type to n-type at $x = 0.5$[417]. The extended homogeneity range of

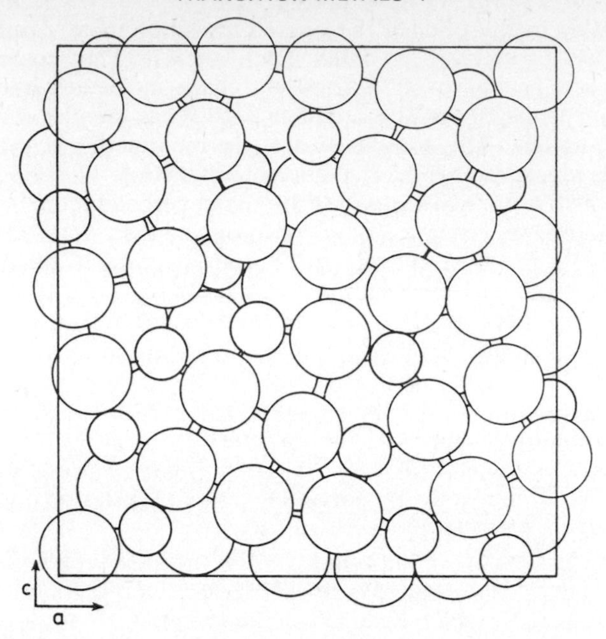

Figure 7.23 Projection of the orthorhombic $Nb_{14}S_5$ structure along the short axis (small circles: S). (From Chen, Tuenge, and Franzen[416], by courtesy of *Inorganic Chemistry*)

2s-$Nb_{1+x}S_2$ at high temperatures was also found during a study of the ternary system Nb–S–O[418].

Superconducting 2s-$NbSe_2$ has a low-temperature displacive phase transition[419–421]. The pressure dependence of the superconductive transition temperature was determined[422]. The electrical conductivity of 2s-$NbSe_2$ at low temperatures was re-investigated[423]. X-Ray K-absorption studies were made on NbX_2 and Nb_3X_4 (X = Se, Te)[424]. Electrical, magnetic, and optical properties of 1s- and 2s-TaS_2 were investigated on single crystals prepared by vapour transport[425]. A semiconductor-metal transition was reported for 1s-TaS_2[426]. X-Ray patterns of compositions in the $NbTe_2$–Te partial system failed to show any new phases although decomposition vapour pressure measurements seemed to indicate new phases[427].

The conductive and magnetic behaviour of Cr_3Se_4 type V_2FeTe_4 was investigated[428]. New ternary phases in the systems Cu,Tl–V,–S were prepared[429]. The new 'intercalation' compounds Al_xNbSe_2 ($x = 0$ to $\frac{2}{3}$)[430] and Tl_xNbS_2 ($x \sim 1.0$)[431] were reported. The study of the systems T_xNbS_2 (T = Ti, V, Cr, Mn, Fe, Co, Ni) showed two types of compounds[432]: one with approximate composition $T_{\frac{1}{6}}NbS_2$ has a superstructure derived from 3s-NbS_2 with T atoms in octahedral voids formed by the NbS_2 slabs[433]; the other type is derived from 2s-NbS_2; it occurs for compositions $T_{\frac{1}{4}}NbS_2$ and $T_{\frac{1}{2}}NbS_2$ and the ordering of the T atoms is as yet unknown[432]. Varied stacking sequences of TSe_2 layers were observed for several compositions in the pseudobinary system $NbSe_2$–WSe_2[434].

Intercalation compounds of the layered transition metal disulphides with organic Lewis bases have attracted much interest. The separation of the TX_2 slabs can be up to 60 Å for long chain molecules like octadecylamine[435] but is usually of the order of 5–10 Å for smaller molecules like pyridine or aniline. The complexes are superconducting if the TX_2 compound from which they are formed is superconducting. The critical temperature is insensitive to the separation of the superconducting layers which indicates that the superconductivity is two-dimensional and restricted to the TX_2 layers[435]. Electron diffraction[436] and imaging techniques[437] were used for the structural characterisation of these compounds. A chemical bond is assumed between the TX_2 layer and the intercalate involving lone-pair electrons of the intercalated Lewis base and a partially filled band of the TX_2 host[435,436]. This explains why only Group IV and Group V transition metal dichalcogenides form these complexes with weak Lewis bases: in semiconducting MoS_2 and WS_2 this band is filled. A large number of these compounds was prepared[435,438,439]. In no case, not even with alkali metals[440] and ammonia[441] as intercalates, was ordering of the intercalate within the layers reported.

7.7.4 Chalcogenides of Group VI transition metals

Electrical conductivity of Cr_2S_3 and its variation with composition were investigated[442,443]. Magnetic properties of Cr_2S_3[442,444], Cr_2Te_3[445] and Cr_3Te_4[446] were studied on single crystals. The pressure dependence of the electrical and magnetic properties of Cr_3Te_4 and Cr_7Te_8 was determined[447].

Single-crystal growth of MoS_2, $MoSe_2$, WSe_2, and $TaSe_2$ by direct vapour transport, and of $MoTe_2$ and WTe_2 by bromine transport was described[448]. The occurrence of stacking faults in deformed $MoSe_2$ was studied by x-ray line broadening[449]. Long-wavelength lattice vibrations of $MoSe_2$[450] and infrared optical properties of $MoSe_2$[451] were investigated.

A large amount of theoretical and experimental work is devoted to the elucidation of the band structure of the molybdenum and tungsten dichalcogenides. Electron paramagnetic resonance spectra of Nb-[452] and As-[453] doped MoS_2 were discussed. Optical absorption spectra of several molybdenum and tungsten dichalcogenides were studied[454,455]. Photoemission spectra were reported for dichalcogenides of Groups IV V, and VI transition metals[456,457]. The band structure of MoS_2 was calculated and compared with experimental results[458-461].

The crystal structure of Mo_3Se_4 is rhombohedral[462]. Mo is five-coordinated to Se atoms forming a square pyramid. Two kinds of Se atoms can be visualised as tetrahedrally coordinated: one to four Mo atoms, the other to three Mo atoms and a lone pair of electrons. There is no primary Se—Se bonding as all Se—Se distances are greater than 3.38 Å. However, Mo—Mo bonds occur, with bonding Mo—Mo distances varying between 2.68 and 2.84 Å, forming octahedral clusters of six Mo atoms (Figure 7.24). The Mo_6 octahedra are big enough to accommodate interstitial atoms like C, N or O. This is not meant to imply that they are filled in Mo_3Se_4; they could, however, be filled in derived ternary structures. A large number of

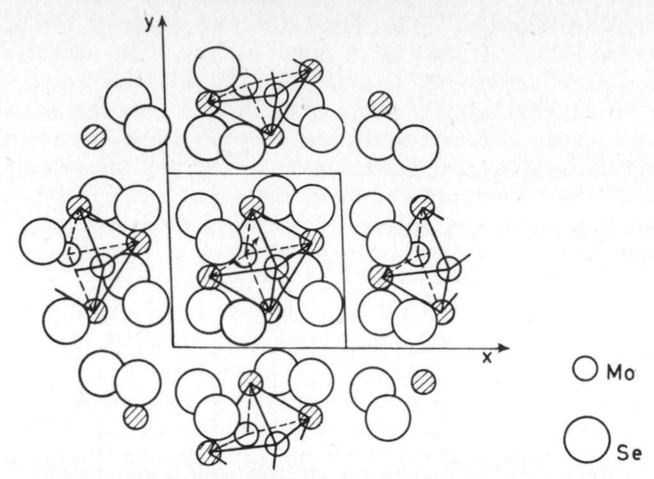

Figure 7.24 Crystal structure of rhombohedral Mo_3Se_4 projected along a translation period. The octahedral Mo clusters are outlined. (From Bars, Guillevic, and Grandjean[462], by courtesy of *J. Solid State Chem.*)

ternary phases, has been found, however, for compositions $M_xMo_3S_4$[463] and $M_xMo_3Se_4$[464,465] where M stands for a large number of Main Group or transition metal elements and x is between 0 and 1. All are derived from the Mo_3Se_4 structure and structural work on $Ni_{0.33}Mo_3Se_4$[466] and $NiMo_3S_4$[467] shows Ni atoms filling voids formed by the Se atoms of the Mo_3Se_4 type structure. The Ni atoms in $Ni_{0.33}Mo_3Se_4$ ($NiMo_9Se_{12}$) fill one site of the trigonally-distorted Mo_3Se_4 structure. In $NiMo_3S_4$ they occupy one-sixth of two sites each. No long-range order was reported for this structure. In the structure of $\sim PbMo_6S_8$ ($=Pb_{0.5}Mo_3S_4$)[468] the Pb atoms occupy one site (the corner of the rhombohedral cell) which corresponds to the centre of the six partially-occupied positions of one Ni site in $NiMo_3S_4$. The Pb atoms are surrounded by a distorted sulphur cube. The tetrahedral S atom of Mo_3S_4 is now bonded to one Pb and four Mo atoms, while the other S atom of Mo_3Se_4 is now tetrahedrally coordinated to one Pb and three Mo atoms. Several compounds of this series were found to be superconductors[469,470], with critical temperatures as high as 15 K for $PbMo_6S_8$.

The crystal structures of the rhombohedral compounds $ACrX_2$ (A = Na, K, Cu, Ag; X = S, Se) were refined[471,472] and the magnetic structures at 4 K of $NaCrS_2$, $NaCrSe_2$, $AgCrSe_2$[471] and $KCrS_2$[472] determined. Magnetic properties of the solid solution series $CuCr_2S_{4-x}Se_x$ were reported[473].

Magnetic, optical and electrical properties of ferro- and ferrimagnetic semiconducting thiospinels[474] continue to be of interest because of potential device applications. Crystal growth[475] is difficult because they melt incongruently. Chemical vapour transport of $FeCr_2S_4$, $CoCr_2S_4$[476] and $CdCr_2S_4$[477] was described. $CoCr_2S_4$[478] and $CdCr_2S_4$[479] were also prepared by hot pressing. Magnetoreflectance[480] and Kerr rotation[481] of $CoCr_2S_4$ were investigated. Magnetoresistance was determined for $FeCr_2S_4$ and

$CoCr_2S_4$ single crystals[482,483]. Structural and magnetic parameters of solid solutions $MCr_2S_{4-x}Se_x$ (M = Mn, Fe, Co) were reported[484]. Non-stoichiometry and possible presence of Cr^{2+} in $CdCr_2S_4$ was investigated[485–487]. Magnetic ordering and its influence on infrared phonons in $CdCr_2S_4$ was studied[488]. Magnetic field dependence of optical absorption in single-crystal $HgCr_2Se_4$ was reported[489]. Conductivity and thermoelectric power of single crystals of the series $ZnCr_2(S_{1-x}Se_x)_4$ were measured[490]. The pressure dependence of the Neél temperature of $ZnCr_2Se_4$ was determined[491].

Phase equilibria in the system Cr–Ni–S were studied[492,493]. Magnetic and electrical properties were investigated for a series of phases $LnTX_3$ (Ln = rare earth; T = Cr, Mn, Fe, Co; X = S, Se) with as yet unreported structures[494–497]. Semiconducting ferrimagnetic $SnCr_2S_4$[498] belongs to a series of phases[499] with unknown hexagonal structures. Magnetic and electrical properties of high-pressure pyrite type $Cr_xCo_{1-x}S_2$ were determined[500]. On intercalating semiconducting MoS_2 with alkali metals, superconducting ternary phases are obtained[501]. Thiotungstates Cu_2WS_4 and $CuNH_4WS_4$ were prepared[502].

7.7.5 Chalcogenides of Group VII transition metals

Paramagnetic sphalerite type MnSe was confirmed[503]. At high pressure MnSe has an NiAs type structure[504]. Magnetic and thermal properties of MnTe were determined[505]. Its single-crystal growth was described[506].

$Cs_2Mn_3S_4$ has an orthorhombic structure with sheets of edge-sharing MnS_4 tetrahedra interspaced with eight coordinate Cs atoms[507]. It has some remarkably short Mn—Mn distances of 2.68 Å. $Rb_2Mn_3S_4$ has the same structure[507]. The structure of Ba_2MnS_3 and Ba_2MnSe_3 is of K_2AgI_3 type[508] with parallel linear chains of corner-sharing $MnSe_4$ tetrahedra separated by seven-coordinate Ba atoms. The Mn atoms order antiferromagnetically within the chains[508]. The crystal structure proposed for the 'filled' spinel Cu_2MnTe_2[509] needs to be confirmed. At high pressure, semiconducting, paramagnetic Al_2MnS_4 has a spinel structure[510]. High pressure phases in the system Ga–Mn–S need characterising[511]. Magnetic properties of spinel type compounds Yb_2MnS_4 and Yb_2MnSe_4 were reported[512]. Antiferromagnetic PdMnTe has a MgAgAs type structure[513].

7.7.6 Chalcogenides of the iron-group metals

7.7.6.1 Binary and pseudobinary chalcogenides of the iron-group metals

At least six phases with long-range order plus several regions with short-range order were identified in a recent study[514] of the partial system FeS–Fe_7S_8. They all derive from the NiAs type structure through vacancy formation and are known as pyrrhotites[3,4,515]. The structure of the monoclinic 4C-type Fe_7S_8, originally described by Bertaut, was refined[516]. The sulphur atoms around the iron vacancies show a slight contraction of the vacant space by comparison with the space around iron atoms. This may be taken as an indication for the small ionicity of the Fe—S bonding (in defect NiAs-like $Re_{1.16}O_3$, the unoccupied sites are considerably bigger than the occupied ones[517]). On the other hand some charge transfer still takes place: the distances of the Fe atoms to the vacant site are significantly smaller than the Fe—Fe distances. This observation was also made for the 3C-type pyrrhotite structure[518]. According to a recent study[514], the NiAs (1C) cell is thermodynamically not stable in the system $Fe_{1-x}S$ at temperatures less than 350 K. The displacements of sulphur atoms from the ideal positions of the NiAs type structure found in a structure refinement of an $Fe_{1-x}S$ crystal[519] probably correspond to the displacements of the sulphur atoms in the ferroelectric 2C structure determined by Bertaut. Enthalpies of formation of $Fe_{1-x}S$ ($x = 0.00$–0.05) were determined[520]. Smithite, a Ni-containing modification of Fe_3S_4, has a hexagonal or monoclinic superstructure[521]. The pseudobinary section FeS_2–$FeSe_2$ was investigated[522,523].

The phase diagrams of the systems Co–Se[524] and Co–Te[525] and thermodynamic properties of Co–Te alloys[526] were determined. The defect-NiAs type regions CoSe–Co_3Se_4 and $Co_{1-x}Te$, which at low temperatures probably contain a large number of superstructures, were not further investigated. Several compounds in the systems of Fe, Co, Ni and Ru with Te were studied with the ^{125}Te Mössbauer effect[527,528].

At high temperatures, stoichiometric NiS has an NiAs-type structure; below 650 K it has the more complicated millerite structure. The NiAs type structure can be obtained at room temperature by quenching and is stable for long periods of time*. On further cooling it undergoes a transition from a Pauli paramagnetic metallic (or semimetallic) to an antiferromagnetic non-metallic state. The transition temperature is 265 K for stoichiometric NiS and falls with increasing Ni-defects. At $Ni_{0.96}S$ it disappears completely[530]. High pressure also lowers the transition temperature[531]. Conductivity, magnetic susceptibilities and the specific heat of the phase transition were studied in several laboratories[532–536]. The transition was also monitored by Mössbauer spectra of Fe-doped samples[537]. The electronic band structure of NiS was calculated[538]. It agrees with earlier rationalisations[539].

A structure determination of a high-temperature phase, which was previously variously described as Ni_9S_8, Ni_7S_6 and Ni_6S_5, resulted in the ideal composition Ni_6S_5[336]. It has five non-equivalent Ni positions. Three of them are rather close to each other; they were found with occupancies of about one-half (Figure 7.25). Evidence for ordering of Ni atoms in the half-occupied sites was found. The coordination of the Ni to the S atoms are square-pyramidal and tetrahedral and thus the structure is intermediate between Ni_3S_2 (tetrahedral Ni) and millerite NiS (square pyramidal Ni). In all those structures additional Ni—Ni bonds occur[336].

Electrical properties of pyrite type $Ni_{1-x}S_2$ depend strongly on composition[540,541]. This is also the case for the solid solution series $NiSe_xS_{2-x}$

* Mild cold working (crushing) of NiAs type NiS, however, introduces dislocations which act as nuclei for the growth of the millerite structure, to which the material transforms almost quantitatively over a period of several months[529].

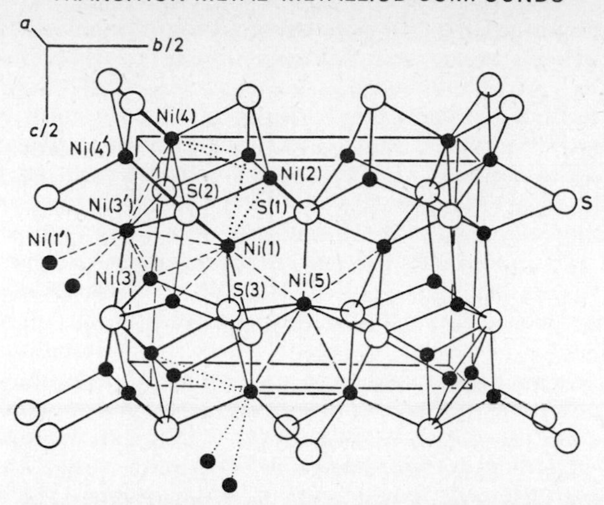

Figure 7.25 Part of the unit cell of Ni_6S_5. Selected Ni—Ni bonds are indicated by broken lines; bonds between sites with nearly full occupancy: large dashes; bonds between sites with half occupancy: small dashes; bonds greater than 2.6 Å: stipples. (From Fleet[336] by courtesy of *Acta Crystallogr.*)

which has a semiconductor–metal transition for $x \sim 0.5$[542]. Band models for solid solutions of that type were discussed[543].

Phase diagrams for the systems Ni–Se[544], Ni–Te[545], Fe–Ni–S[546], Co–Ni–Se[544] and Fe–Ni–Te[547] were determined. Ni_6Se_5 has a structure similar to Ni_6S_5[548]. No long-range order was found yet for the NiAs–CdI_2 type phase NiTe–$NiTe_2$[545,549]. Thermodynamic properties of the phases $Ni_{3\pm x}Te_2$ were investigated[550].

$Ni_{1.3}Te$ is hexagonal at high temperature. Cooling through 990 K leads to an orthorhombic distortion[551]. The high-temperature form can be stabilised at room temperature by substitution of Te by Se. It has close packed chalcogen atoms with the stacking sequence ABAC and Ni atoms in octahedral and tetrahedral sites[551]. The structure was refined for the composition $Ni_{1.25}Te_{0.82}Se_{0.18}$. The tetrahedral Ni position has an occupancy of less than 0.5[552]. The rhombohedral structure of $Ni_{1.5}Fe_{1.5}Te_2$ has also close packed Te atoms with the sequence ABCABC[553]. The metal atoms occupy tetrahedral and octahedral holes between every other Te layer thus forming infinite two-dimensional clusters. The layer character of the structure is also apparent from its graphite-like cleavage[553].

7.7.6.2 *Ternary chalcogenides of the iron-group metals*

The structures of several new phases in the system Ba–Fe–S were determined[554]. Ba_2FeS_3 has the same structure as Ba_2ZnS_3 (K_2CuCl_3 type). It consists of chains of corner-sharing FeS_4 tetrahedra separated by seven-coordinate Ba atoms. This structure was also found for Ba_2CoS_3 and

Ba_2FeSe_3[554]. Double chains of edge-sharing FeS_4 tetrahedra are formed in $BaFe_2S_3$ ($CsCu_2Cl_3$ type) and $BaFe_2Se_3$ ($CsAg_2I_3$ type)[554]. These two structures are very similar with corresponding orthorhombic unit cells. The cell of $BaFe_2S_3$ is base centred; $BaFe_2Se_3$ has a primitive cell. Thus $BaFe_2Se_3$ corresponds to a low-temperature superstructure of $BaFe_2S_3$. The structures of $Ba_6Fe_8S_{15}$ (tetragonal) and $Ba_3Fe_3Se_7$ (hexagonal) are new types with Fe in unique positions and average formal oxidation states of +2.25 and 2.66 respectively[554]. In both structures Fe is again tetrahedrally coordinated to sulphur. The FeS_4 tetrahedra share edges and corners to form double-chain-like aggregates in $Ba_6Fe_8S_{15}$. This structure has one sulphur atom which is octahedrally coordinated to Ba only. In $Ba_3Fe_3Se_7$, the $FeSe_4$ tetrahedra are connected by corners to form three-membered rings. Ba_2FeS_3, $Ba_6Fe_8S_{15}$, and $Ba_7Fe_6S_{14}$[555] are antiferromagnets[556].

In $Bi_2Ni_3S_2$ (parkerite) the Ni atoms occupy two sites with octahedral coordination to four Bi and two S atoms[557]. Although one of the two Ni sites is only half occupied no evidence for long-range order was observed. $Rb_2Co_3S_4$ and $Cs_2Co_3S_4$ have $Cs_2Mn_3S_4$ type structures[507]. $TlFeS_2$ and $TlFeSe_2$ have isotypic monoclinic undetermined structures[558]. The section EuS–FeS is eutectic with no ternary phase[559].

$CuFeS_2$, chalcopyrite, has a superstructure of sphalerite (cubic ZnS) caused by ordering of Cu and Fe atoms on the Zn sites. The structure was refined recently[560]. Talnakhite, $Cu_9Fe_8S_{16}$ ($M_{17}S_{16}$), has a 'filled' tetrahedral[282] structure with a large cubic cell corresponding to eight sphalerite cells ($M_{16}S_{16}$) and additional metal atoms on interstitial tetrahedral sites formed by the sulphur atoms[561]. As a consequence one of the sulphur atoms has trigonal-bipyramidal metal coordination. The superstructure is due to the 'filled' metal atoms. The ordering of the Cu and Fe atoms is not known. The structure is essentially the same as that of the earlier described 'β-phase' in Cu–Fe–S[562]. The structure of mooihoekite, $Cu_9Fe_9S_{16}$, corresponds to the 'γ-phase'[562]; it is not fully known[563]. In the structure of cubanite, $CuFe_2S_3$, recently refined[564], Fe^{2+} and Fe^{3+}—on identical crystallographic sites—occupy pairs of edge-sharing tetrahedral positions. It is a semiconductor and the iron atoms become equivalent by forming two-membered 'clusters'[565].

With decreasing temperature, paramagnetic Fe_2GeS_4 (olivine type) orders to antiferromagnetic and ferrimagnetic structures[566]. Cu_2FeSnS_4 is an antiferromagnet at low temperatures[567].

7.7.7 Chalcogenides of the platinum metals

K_2PtS_2 forms hygroscopic red needle-shaped, orthorhombic crystals with square-planar sulphur coordination of the Pt^{2+} atoms[568]. The PtS_4 squares share edges, thus forming PtS_2 chains separated by K atoms in sixfold coordination to sulphur atoms (Figure 7.26). $RbPtS_2$ has the same structure[568]. $Pb_2Pt_3S_4$ and $Cs_2Pt_3S_4$ form also dark-red hygroscopic crystals with square-planar sulphur coordination[569]. The structure is closely-related to the structure of $Cs_2Pd_3S_4$[570].

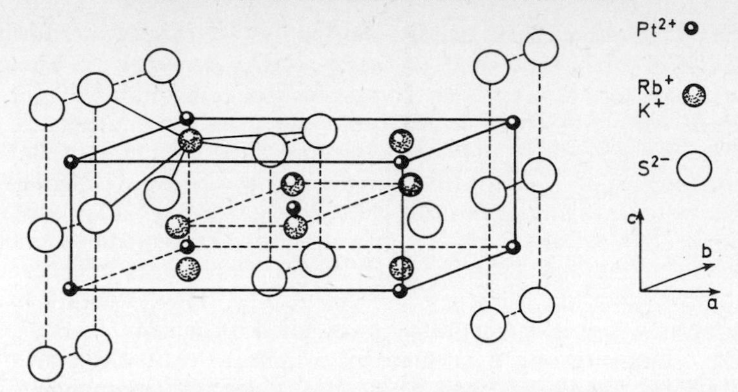

Figure 7.26 Crystal structure of K_2PtS_2. (From Bronger and Günther[568], by courtesy of *J. Less-Common Metals*)

References

1. Ward, R. (1972). *MTP International Review of Science. Inorganic Chemistry*, Series 1, Vol. **5**, p. 93 (London: Butterworths)
2. Nowotny, H. (1972). *MTP International Review of Science. Inorganic Chemistry*, Series 1, Vol. **10**, p. 151 (London: Butterworths)
3. Jellinek, F. (1972). *MTP International Review of Science. Inorganic Chemistry*, Series 1, Vol. **5**, p. 339 (London: Butterworths)
4. Flahaut, J. (1972). *MTP International Review of Science. Inorganic Chemistry*, Series 1, Vol. **10**, p. 189 (London: Butterworths)
5. Hulliger, F. (1968). *Structure and Bonding*, Vol. **4**, p. 83 (New York: Springer)
6. Pearson, W. B. (1958 and 1967). *A Handbook of Lattice Spacings and Structures of Metals and Alloys, Vols.* **1** *and* **2**. (Oxford: Pergamon Press)
7. Schubert, K. (1964). *Kristallstrukturen zweikomponentiger Phasen*. (Heidelberg: Springer)
8. Pearson, W. B. (1972). *The Crystal Chemistry and Physics of Metals and Alloys*. (New York: Wiley)
9. Kieffer, R. and Benesovsky, F. (1963). *Hartstoffe*. (Vienna: Springer)
10. Goldschmidt, H. J. (1967). *Interstitial Alloys*. (New York: Plenum Press)
11. Powell, C. F. (1967). *High-Temperature Materials and Technology*. p. 349. (I. E. Campbell and E. M. Sherwood, editors) (New York: Wiley)
12. Thompson, R. (1970). *Progress in Boron Chemistry*. Vol. **2** (R. J. Brotherton, and M. Steinberg, editors) (New York: Pergamon Press)
13. Kiessling, R. (1950). *Acta Chem. Scand.*, **4**, 209
14. Aronsson, B., Lundström, T. and Rundqvist, S. (1965). *Borides, Silicides and Phosphides* (London: Methuen)
15. Lundström, T. (1969). *Arkiv Kemi*, **31**(19), 227
16. Lundström, T. and Rosenberg, I. (1973). *J. Solid State Chem.*, **6**, 299
17. Romans, P. A. and Krug, M. P. (1966). *Acta Crystallogr.*, **20**, 313
18. Nowotny, H., Haschke, H. and Benesovsky, F. (1967). *Monatsh. Chem.*, **98**, 547
19. Jangg, G. and Kieffer, R. (1973). *Monatsh. Chem.*, **104**, 266
20. Cueilleron, J., Lahet, G., Thevenot, F. and Pâris, R. A. (1971). *J. Less-Common Metals*, **24**, 317
21. Kasaya, M., Hihara, T. and Koi, Y. (1973). *J. Phys. Soc. Japan*, **34**, 63
22. Carter, G. C. and Swartz, J. C. (1971). *J. Phys. Chem. Solids*, **32**, 2415
23. Barnes, R. G., Creel, R. B. and Torgeson, D. R. (1970). *Solid State Commun.*, **8**, 1411
24. Stadelmaier, H. H., Schöbel, J. D. and Jones, R. A. (1967). *Metallurgica*, **21**, 17
25. Kuz'ma, Yu. B. and Yarmolyuk, Ya. P. (1971). *Zhurnal Strukturnoi Khimii*, **12**, 458
26. Rogl, P., Benesovsky, F. and Nowotny, H. (1972). *Monatsh. Chem.*, **103**, 965

27. Chepiga, M. V., Krivutskii, V. P. and Kuz'ma, Yu. B. (1972). *Izv. Akad. Nauk SSSR, Neorgan. Mat.* **8,** 1059
28. Kuz'ma, Yu. B. (1971). *Izv. Akad. Nauk SSSR, Neorgan. Mat.*, **7,** 514
29. Rogl, P., Nowotny, H. and Benesovsky, F. (1973). *Monatsh. Chem.*, **104,** 182
30. Rogl, P., Nowotny, H. and Benesovsky, F. (1971). *Monatsh. Chem.*, **102,** 971
31. Reiffenstein, E., Nowotny, H. and Benesovsky, F. (1965). *Monatsh. Chem.*, **97,** 499
32. Kuz'ma, Yu. B. (1970). *Sov. Phys-Crystallogr.*, **15,** 312
33. Kuz'ma, Yu. B. and Svarichevskaya, S. I. (1972). *Sov. Phys.-Crystallogr.*, **17,** 569
34. Smith, G. S., Johnson, Q. and Nordine, P. C. (1965). *Acta Crystallogr.*, **19,** 668
35. Kuz'ma, Yu. B. and Bilonizhko, N. S. (1972). *Sov. Phys.-Crystallogr.*, **16,** 897
36. Niihara, K., Shishido, T. and Yajima, S. (1971). *Bull. Chem. Soc. Japan*, **44,** 3214
37. Kuz'ma, Yu. B. and Svarichevskaya, S. I. (1973). *Sov. Phys.-Crystallogr.*, **17,** 830
38. Stadelmaier, H. H. (1969). *Developments in the Structural Chemistry of Alloy Phases* (New York: Plenum Press) (AIME Publication, B. C. Giessen, editor)
39. Ganglberger, E., Nowotny, H. and Benesovsky, F. (1966). *Monatsh. Chem.*, **97,** 101
40. Kuz'ma, Yu. B. and Bilonizhko, N. S. (1971). *Izv. Akad. Nauk SSSR. Neorgan. Mater.*, **7,** 620
41. Niihara, K. and Yajima, S. (1972). *Chem. Letters* (*Japan*), 875
42. Kuz'ma, Yu. B., Kripyakevič, P. I. and Bilonizhko, N. S. (1969). *Doklady Akad. Nauk Ukr. SR. Ser. A*, **10,** 939
43. Niihara, K. and Yajima, S. (1973). *Bull. Chem. Soc. Japan.*, **46,** 770
44. Chaban, N. F. and Kuz'ma, Yu. B. (1972). *Izv. Akad. Nauk. SSSR, Neorgan. Mater*, **8,** 1065
45. Gladyshevskii E. I., Kuz'ma, Yu. B. and Kripjakevič, P. I. (1963). *Ž. Strukt. Khim.*, **4,** 372
46. Mueller, M. H. and Knott, H. W. (1963). *Trans. AIME*, **227,** 674
47. Parthé, E., Jeitschko, W. and Sadagopan, V. (1965). *Acta Crystallogr.*, **19,** 1031
48. Uraz, A. A. and Rundqvist, S. (1970). *Acta Chem. Scand.*, **24,** 1843
49. See Frad, W. A. (1968). *Advances in Inorganic Chemistry and Radiochemistry*, **11,** 153. (H. J. Emeléus and A. G. Sharpe, editors) (New York: Academic Press)
50. Windisch, S. and Nowotny, H. (1972). *Preparative Methods in Solid State Chemistry.* (P. Hagenmuller, editor) (New York: Akademic Press)
51. Williams, W. S. (1971). *Progress in Solid State Chemistry*, **6,** 57 (H. Reiss and J. O. McCaldin, editors) (New York: Pergamon Press)
52. Storms, E. K. (1971). *MTP International Review of Science. Inorganic Chemistry*, Series 1, Vol. **10,** p. 37 (London: Butterworths)
53. Kieffer, R. and Benesovsky, F. (1965). *Hartmetalle.* (Vienna: Springer)
54. Storms, E. K. (1967). *The Refractory Carbides* (New York: Academic Press)
55. Toth, L. E. (1971). *Transition Metal Carbides and Nitrides.* (New York: Academic Press)
56. Kosolapova, T. Ya. (1968). Translated by N. B. Vaughan (1971) *Carbides.* (New York: Plenum Press)
57. Jack, D. H. and Jack, K. H. (1973). *Materials Science and Engineering*, **11,** 1
58. Hägg, G. (1931). *Z. Phys. Chem.*, **B12,** 33
59. Rudy, E. (1970). *J. Less-Common Metals*, **20,** 49
60. Yvon, K. and Parthé, E. (1970). *Acta Crystallogr.*, **B26,** 149
61. Bowman, A. L. and Arnold G. P. (1971). *Advances High Temp. Chem.*, **4,** 243
62. Parthé, E. and Yvon, K. (1970). *Acta Crystallogr.*, **B26,** 153
63. De Novion, C. H., Lorenzelli, R. and Costa, P. (1966). *C. R. Acad. Sci. Paris.* **263,** 775
64. Froidevaux, C. and Rassier, D. (1967). *J. Phys. Chem. Solids*, **28,** 1197
65. Kordes, D. (1968). *Phys. Stat. Sol.*, **26,** K103
66. Henfrey, A. W. and Fender, B. E. F. (1970). *Acta Crystallogr.*, **B26,** 1882
67. Venables, J. D., Kahn, D. and Lye, R. G. (1968). *Phil. Mag.*, **18,** 177
68. Billingham, J., Bell, P. S. and Lewis, M. H. (1972). *Phil. Mag.*, **25,** 661
69. Arbuzov, M. P., Fak, V. G. and Khaenko, B. V. (1970). *Soviet Physics-Crystallogr.*, **15,** 164
70. Lewis, M. H., Billingham, J. and Bell, P. S. (1971). *Proceedings, Fifth International Materials Symposium*, Berkley, California, p. 1084
71. Billingham, J., Bell, P. S. and Lewis, M. H. (1972). *J. Cryst. Growth.*, **13/14,** 693
72. Billingham, J., Bell, P. S. Lewis, M. H. (1972). *Acta Crystallogr.*, **A28,** 602
73. Sauvage, M. and Parthé, E. (1972). *Acta Crystallogr.*, **A28,** 607

74. Venables, J. D. and Meyerhoff, M. H. (1972). NBS Special Publication 364. *Proceedings of 5th Mater. Res. Symposium*, p. 583
75. Goretzki, H. (1967). *Phys. Stat. Sol.*, **20,** K141
76. Chermant, J.-L., Delavignette, P. and Deschanvres, A. (1970). *J. Less-Common Met.*, **21,** 89
77. Bell, P. S. and Lewis, M. H. (1971). *Phil. Mag.*, **24,** 1247
78. Bowman, A. L., Arnold, G. P., Storms, E. K. and Nereson, N. G. (1972). *Acta Crystallogr.*, **B28,** 3102
79. Rouault, M. A., Herpin, P. and Fruchart, M. R. (1970). *Ann. Chim.*, **5,** 461
80. Westgren, A. and Phragmén, G. (1935). *Jernkont Ann.*, **118,** 231
81. Benz, R., Elliott, J. F. and Chipman, J. (1973). *Met. Trans.*, **4,** 1449
82. Bouchaud, J.-P. and Fruchart, R. (1964). *Bull. Soc. Chim. Fr.*, 1579
83. Nowotny, H. and Neckel, A. (1969). *J. Inst. Metals*, **97,** 161
84. Ramqvist, L., Hamrin, K., Johansson, G., Gelius, U. and Nordling, C. (1970). *J. Phys. Chem. Solids*, **31,** 2669
85. Ramqvist, L., Ekstig, B., Källne, E., Noreland, E. and Manne, R. (1971). *J. Phys. Chem. Solids*, **32,** 149
86. Ramqvist, L. (1971). *J. Appl. Phys.*, **42,** 2113
87. Merisalo, M., Inkinen, O., Järvinen, M., and Kurki-Suonio, K. (1969). *J. Phys. C*, Ser. 2, Vol. **2,** 1984
88. Neckel, A., Rastl, P., Weinberger, P. and Mechtler, R. (1972). *Theoret. Chim. Acta (Berlin)*, **24,** 170
89. Schwarz, K. (1971). *Monatsh. Chem.*, **102,** 1400
90. Mattheiss, L. F. (1972). *Phys. Rev.*, **B5,** 315
91. Lye, R. G. (1972). NBS Special Publication 364. *Proceedings of 5th Mat. Res. Symposium*, p. 567
92. Kuz'ma, Yu. B., Lakh, V. I., Markiv, V. Ya., Stadnyk, B. I. and Gladyshevskii, E. I. (1963). *Soviet Powder Met. Metal Ceram.*, **4,** 286
93. Fackelmann, J. M., Getz, R. W. and Mock, D. P. (1965) *AIME Symp.* French Lick, Indiana, Oct. 1965
94. Raub, Ch. J., Mons, W. and Lawson, A. C. (1972). *J. Less-Common Metals*, **26,** 319
95. Structural data for computation of interatomic distances in a nearly stoichiometric Ti_2AlC sample ($a = 3.059$, $c = 13.66$ Å, $z_{Ti} = 0.086$) are unpublished and were taken from a thesis: Jeitschko, W. (1964) University of Vienna
96. Nowotny, H., Boller, H. and Beckmann, O. (1970). *J. Solid State Chem.*, **2,** 462
97. Wondratschek, H.,Merker, L., and Schubert, K. (1964). *Z. Kristallogr.* **120,** 393
98. Parthé, E. and Rieger, W. (1968). *J. Dental Res.* **47,** 829
99. Fraker, A. C. and Stadelmaier, H. H. (1969). *Trans. TMS-AIME*, **245,** 847
100. Ettmayer, P. and Suchentrunk, R. (1970). *Monatsh. Chem.*, **101,** 1098
101. Pollok, C. B. and Stadelmaier, H. H. (1970). *Metall. Trans.*, **1,** 767
102. Krikorian, N. H., Wallace, T. C., Krupka, M. C. and Radosevich, C. L. (1967). *J. Nucl. Mater.*, **21,** 236
103. Holleck, H. (1968) *J. Nucl. Mater.*, **28,** 339
104. Haines, H. R. and Potter, P. E. (1969). *Nature (London)*, **221,** 1238
105. Holleck, H. (1971). *Monatsh. Chem.*, **102,** 1699
106. Bowman, A. L., Arnold, G. P. and Krikorian, N. H. (1971). *Acta Crystallogr.*, **B27,** 1067
107. Cromer, D. T., Larson, A. C. and Roof, R. B. (1964). *Acta Crystallogr.*, **17,** 272
108. Harris, I. R. and Raynor, G. V. (1965). *J. Less-Common Metals*, **9,** 263
109. Harris, I. R. and Norman, M. (1967). *J. Less-Common Metals*, **13,** 629
110. Holleck, H. (1972). *J. Nucl. Mater.*, **42,** 278
111. Holleck, H. and Kleykamp, H. (1970) *J. Nucl. Mater.*, **35,** 158
Nowotny, H. (1972). *Angew. Chem. Internat. Edn*, **11,** 906
113. Spinat, P., Fruchart, R., Kabbani, M. and Herpin, P. (1970). *Bull. Soc. Fr. Minéral. Cristallogr.*, **93,** 171
114. Spinat, P., Sénateur, J.-P., Fruchart R. and Herpin, P. (1972)., *Compt. Rend. Acad. Sci. Paris*, **274C,** 1159
115. Brukl, C. E. (1965). Techn. Rep. No. AFML-TR-65-2, Part II, Vol. VII
116. Nickl, J. J., Schweitzer, K. K. and Luxenburg, P. (1972). *J. Less-Common Metals*, **26,** 335
117. Jeitschko, W. and Nowotny, H. (1967). *Monatsh. Chem.*, **98,** 329

118. Boller, H. (1971). *Monatsh. Chem.*, **102,** 431
119. Boller, H. (1973). *Monatash. Chem.*, **104,** 48
120. Beckmann, O., Boller, H. and Nowotny, H. (1970). *Monatsh. Chem.*, **101,** 945
121. Boller, H. and Sobczak (1971). *Monatsh. Chem.*, **102,** 1226
122. Horyń, R. and Dryś, M. (1971). *Kristall and Technik.*, **6,** K85
123. Nickl, J. J. and Koukoussas, J. D. (1971). *J. Less-Common Metals*, **23,** 73
124. Mayer, I., Shidlovsky, I. and Yanir, E. (1967). *J. Less-Common Metals*, **12,** 46
125. Jangg, G., Kieffer, R., Blaha, A. and Sultan, T. (1972). *Z. Metallkde.*, **63,** 670
126. Baughman, R. J. and Quinn, R. K. (1972). *Mater. Res. Bull.*, **7,** 1035
127. Seeber, B. and Nickl, J. (1973). *Phys. Stat. Sol.*, **A15,** 73
128. Nickl, J. J. and Schweitzer, K. K. (1970). *Z. Metallkde.*, **61,** 541
129. Gol'dberg, A. I., Lipatova, V. A. and Gel'd, P. V. (1964). *Refractory Transition Metal Compounds* p. 201 (G. V. Samsonov, editor) (New York: Academic Press)
130. Birkholz, U. and Schelm, J. (1968). *Phys. Stat. Sol.*, **27,** 413
131. Birkholz, U. and Schelm, J. (1969). *Phys. Stat. Sol.*, **34,** K177
132. Abrikosov, N. Kh., Ivanova, L. D. and Murav'ev, V. G. (1972). *Izv. Akad. Nauk SSSR, Neorgan. Mater.*, **8,** 1194
133. Nikitin, E. N., Tarasov, V. I. and Tamarin, P. V. (1969). *Sov. Phys. Solid State*, **11,** 187
134. Nishida, I. (1972). *J. Mater. Sci.* **7,** 435
135. Nishida, I. (1972). *J. Mater. Sci.*, **7,** 1119
136. Levinson, L. M. (1973). *J. Solid State Chem.*, **6,** 126
137. Waldecker, G., Meinhold, H. and Birkholz, U. (1973). *Phys. Stat. Sol.*, **A15,** 143
138. Nowotny, H. (1963). *Electronic Structure and Alloy Chemistry of the Transition Elements*, p. 210, (P. A. Beck, editor) (New York: Interscience)
139. Gladyshevskii, E. I. (1971). *Crystal Chemistry of Silicides and Germanides* (Russ.) (Moscow: Izd. Metallurgia)
140. Kuz'ma, Yu. B. and Gladyshevskii, E. I. (1964). *Russ. J. Inorg. Chem.*, **9,** 373
141. Bardos, D. I. and Beck, P. A. (1965). *Trans. Met. Soc. AIME*, **233,** 1446
142. Komura, Y., Sly, W. G. and Shoemaker, D. P. (1960). *Acta Crystallogr.*, **13,** 575
143. Shoemaker, C. B. and Shoemaker, D. P. (1971). *Acta Crystallogr.*, **B27,** 227
144. Shoemaker, C. B. and Shoemaker, D. P. (1969). In *Developments in the Structural Chemistry of Alloy Phases* p. 107. B. C. Giessen, editor. (New York: Plenum Press)
145. Sénateur, J.-P. and Fruchart, R. (1964). *Compt. Rend.* Acad. Sci. Paris, **258,** 1524
146. Shoemaker, C. B. and Shoemaker, D. P. (1971). *Met. Trans.*, **2,** 2296
147. Shoemaker, C. B. and Shoemaker, D. B. (1963). *Acta Crystallogr.*, **16,** 997
148. Testardi, L. R. (1972). *Phys. Rev.*, **B5,** 4342
149. Maita, J. P. and Bucher, E. (1972). *Phys. Rev. Lett.*, **29,** 931
150. Larsen. R. E. and Ruoff, A. L. (1973). *J. Appl. Phys.*, **44,** 1021
151. Röschel, E. and Raub, C. J. (1971). *Z. Metallk*, **62,** 840
152. Brossard, L., Fatseas, G. A., Dormann, J. L. and Lecocq, P. (1971). *J. Appl. Phys.*, **42,** 1306
153. Rimlinger, L. and Lecorre, Ch. (1973). *Compt. Rend. Acad. Sci Paris* **276C,** 735
154. Frank, K. and Schubert, K. (1971). *Acta Crystallogr.*, **B27,** 916
155. Ellner, M. Gödecke, T. and Schubert, K. (1971). *J. Less-Common Metals*, **24,** 23
156. El-Boragy, M., Bhan, S. and Schubert, K. (1970). *J. Less-Common Metals*, **22,** 445
157. Frolov, A. A., Putintsev, Yu. V., Sidorenko, F. A., Gel'd, P. V. and Krentsis, R. P. (1972). *Izv. Akad. Nauk SSSR, Neorgan. Mater*, **8,** 468
158. Nowotny, H. (1947). *Z. Anorg. Allg. Chem.*, **254,** 31
159. Johnson, V. and Jeitschko, W. (1972). NBS Special Publication 364, Solid State Chem. Proceedings. p. 613
160. Wilson, R. H. and Kasper, J. S. (1964). *Acta Crystallogr.*, **17,** 95
161. Jellinek, F. (1959). *Österr. Chem. Z.*, **60,** 311
162. Horyń, R. and Kubiak, R. (1971). *Bull. Acad. Polon. Scienc. Ser. Scienc. Chim.*, **19,** 185
163. Narasimhan, K. S. V. L., Reiff, W. M., Steinfink, H. and Collins, R. L. (1970). *J. Phys. Chem. Solids.* **31,** 1511
164. Reiff, W. M., Narasimhan, K. S. V. L. and Steinfink H. (1972). *J. Solid. State Chem.*, **4,** 38
165. Johnson, V., Weiher, J. F., Frederick, C. G. and Rogers, D. B. (1972). *J. Solid State Chem.*, **4,** 311
166. Schachner, H., Cerwenka, E. and Nowotny, H. (1954). *Monatsh. Chem.*, **85,** 245
167. Frank, F. C. and Kasper, J. C. (1958). *Acta Crystallogr.*, **11,** 184

168. Frank, F. C. and Kasper, J. C. (1959). *Acta Crystallogr.*, **12,** 483
169. Shoemaker, C. B. and Shoemaker, D. P. (1971). *Monatsh. Chem.*, **102,** 1643
170. Spinat, P., Fruchart, R. and Herpin, P. (1970). *Bull. Soc. Fr. Minéral. Cristallogr.*, **93,** 23
171. Kripyakevich, P. I., Yarmolyuk, Ya. P. and Gladyshevskii, E. I. (1969). *Sov. Phys. Crystallogr.*, **13,** 677
172. Beckmann, O., Carrander, K., Lundgren, L. and Richardson, M. (1972). *Physica Scripta*, **6,** 151
173. Sundström, L. J. (1972). *Physica Scripta*, **6,** 158
174. Stenström, B. and Sundström, L. J. (1972). *Physica Scripta*, **6,** 164
175. Stenström, B., Sundström, L. J. and Sagredo, V. (1972). *Physica Scripta*, **6,** 209
176. Beckman, O., Sundström, L. J., Carrander, K. and Lundgren, L. (1973). *Solid State Comm.*, **12,** 1061
177. Malaman, B. Courtois, A., Protas, J. and Roques, B. (1973). *Compt. Rend. Acad. Sci. Paris*, **276C,** 323
178. Malaman, B., Courtois, A., Protas, J. and Roques, B. (1973). *Compt. Rend. Acad. Sci. Paris*, **276C,** 665
179. Dusausoy, Y., Protas, J., Wandji, R. and Roques, B. (1971). *Acta Crystallogr.*, **B27,** 1209
180. Wandji, R., Dusausoy, Y., Protas, J. and Roques, B. (1969). *Compt. Rend. Acad. Sci. Paris*, **269C,** 907
181. Engström, I. (1970). *Acta Chem. Scand.*, **24,** 2117
182. Geserich, H. P., Sharma, S. K. and Theiner, W. A. (1973). *Phil. Mag.*, **27,** 1001
183. Engström, I. and Zackrisson, F. (1970). *Acta Chem. Scand.*, **24,** 2109
184. White, J. G. and Hockings, E. F. (1971). *Inorg. Chem.*, **10,** 1934
185. Panday, P. K. and Schubert, K. (1969). *J. Less-Common Metals*, **18,** 175
186. Kubiak, R., Horyń, R., Broda, H. and Lukaszewicz, K. (1972). *Bull. Acad. Polon. Scienc. Ser. scienc. chim.*, **20,** 429
187. Zintl, E. (1939). *Angew. Chem.*, **52,** 1
188. Schuster, H.-U. and Mewis, A. (1969). *Z. Naturforsch.*, **24b,** 1190
189. Mewis, A. and Schuster, H.-U. (1971). *Z. Naturforsch.*, **26b,** 62
190. Schuster, H.-U. and Kistrup, C.-J. (1972). *Z. Naturforsch.*, **27b,** 80
191. Kistrup, C.-J. and Schuster, H.-U. (1972). *Z. Naturforsch.*, **27b,** 324
192. Jeitschko, W. (1970). *Met. Trans.*, **1,** 3159
193. Juza, R. and Meyer, W. (1969). *Z. Anorg. Allgem. Chem.*, **366,** 43
194. Juza, R. and Langer, K. (1968). *Z. Anorg. Allgem. Chem.*, **361,** 58
195. Nowotny, H. and Sibert, W. (1941). *Z. Metallkde.*, **33,** 391
196. Onken, H., Vierheilig, K. and Hahn, H. (1964). *Z. Anorg. Allgem. Chem.*, **333,** 267
197. Bodak, O. I., Gladyshevskii and Kripyakevich, P. I. (1970). *Zh. Strukt. Khim.*, **11,** 305
198. Murthy, N. S. S., Becum, R. J., Somanathan, C. S. and Murthy, M. R. L. N. (1969). *J. Appl. Phys.*, **40,** 1870
199. Shibata, K., Shinohara, T. and Watanabe, H. (1972). *J. Phys. Soc. Japan*, **32,** 1431
200. Shibata, K., Shinohara, T. and Watanabe, H. (1972). *J. Phys. Soc. Japan*, **33,** 1328
201. Lee, K., Sawatzky, E. and Suits, J. C. (1972). *J. Appl. Phys.*, **44,** 1756
202. Wieder, H. and Burn, R. A. (1972). *J. Appl. Phys.*, **44,** 1774
203. Sawatzky, E. and Street, G. B. (1972) *J. Appl. Phys.*, **44,** 1789
204. Wieder, H. and Burn, R. A. (1973). *Appl. Phys. Lett.*, **22,** 188
205. Eisenmann, B., May, N., Müller, W., Schäfer, H., Weiss, A., Winter, J. and Ziegleder, G. (1970). *Z. Naturforsch.*, **25b,** 1350
206. Rieger, W. and Parthé, E. (1969). *Monatsh. Chem.*, **100,** 444
207. May, N. and Schäfer, H. (1972). *Z. Naturforsch.*, **27b,** 864
208. Mayer, I., Cohen, J. and Felner, I. (1973). *J. Less-Common Metals*, **30,** 181
209. Bodak, O. I., Gladyshevskii, E. I. and Kripyakevich, P. I. (1966). *Izv. Akad. Nauk SSSR. Neorgan. Mater.* **2,** 2151
210. Ban, Z. and Sikirica, M. (1967). *Z. Anorg. Allgem. Chem.*, **356,** 96
211. Sobolev, A. S., Bodak, O. I. and Gladyshevskii, E. I. (1971). *Izv. Akad. Nauk. SSSR, Neorgan. Mater.* **7,** 41
212. Omejec, L. and Ban, Z. (1971). *Z. Anorg. Allgem. Chem.*, **380,** 111
213. Pinto, H. and Shaked, H. (1973). *Phys. Rev.*, **B7,** 3261
214. Bodak, O. I. and Gladyshevskii, E. I. (1969). *Izv. Akad. Nauk SSSR, Neorgan. Mat.*, **5,** 2060
215. Bodak, O. P. and Gladyshevskii, E. I. (1970). *Sov. Phys. Crystallogr.*, **14,** 859

216. Stepien, J. A., Lukaszewicz, K., Gladyshevskii, E. I. and Bodak, O. (1972). *Bull. Acad. Polon. Scienc. Ser. scienc. chim.*, **20**, 1029
217. Bodak, O. I., Gladyshevskii, E. I. and Mis'kiv, M. G. (1972). *Sov. Phys. Crystallogr.*, **17**, 439
218. Engström, I. (1965). *Acta Chem. Scand.*, **19**, 1924
219. Yarmolyuk, Y. P., Kripyakevich, P. I. and Gladyshevskii, E. I. (1970). *Sov. Phys. Crystallogr.*, **15**, 226
220. Manor, P. C., Shoemaker, C. Brink and Shoemaker, D. P. (1972). *Acta Crystallogr.*, **B28**, 1211
221. Kuz'ma, Yu. B. and Gladyshevskii, E. I. (1964). *Russ. J. Inorg. Chem.*, **9**, 373
222. Bardos, D. I., Malik, R. K., Spiegel, F. X. and Beck, P. A. (1966). *Trans. A. I. M. E.* **236**, 40
223. Johnson, V. and Jeitschko, W. (1972). *J. Solid State Chem.*, **4**, 123
224. Masumoto, H., Watanabe, K. and Mitera, M. (1973). *J. Phys. Soc. Japan*, **34**, 1414
225. Jeitschko, W. (1970) *Acta Crystallogr.*, **B26**, 815
226. Deyris, B., Roy-Montreuil, J., Rouault, A., Fruchart, R. and Michel, A. (1971). *Compt. Rend. Acad. Sci. Paris*, **273C**, 47
227. Johnson, V. (1973). *Mater. Res. Bull.* **8**, 1067
228. Markiv, V. Y. Lysenko, L. A., and Gladyshevskii, E. I. (1966). *Izv. Akad. Nauk SSSR, Neorgan, Mat.*, **2**, 1980
229. Lysenko, L. A., Markiv, V. Y., Tsybukh, O. V. and Gladyshevskii, E. I. (1971). *Izv. Akad. Nauk SSSR, Neorgan. Mat.*, **7**, 157
230. Juza, R. (1966). In *Advances in Inorganic and Radiochemistry*. **9**, 81. H. J. Emeléus and A. G. Sharpe, editors (New York: Academic Press)
231. Lobier, G. and Marcon, J.-P. (1969). *Compt. Rend. Acad. Sci. Paris*, **268C**, 1132
232. Rudy, E. (1970). *Met. Trans.*, **1**, 1249
233. Potter, D. I. (1973). *J. Less-Common Metals*, **31**, 299
234. Potter, D. and Altstetter, C. (1971). *Acta Met.*, **19**, 881
235. Steeb, S. and Renner, J. (1965). *J. Less-Common Metals*, **9**, 181
236. Steeb, S. and Renner, J. (1966). *J. Less-Common Metals*, **10**, 246
237. Dubertret, A. and Lehr, P. (1969). *Compt. Rend. Acad. Sci. Paris*, **268C**, 501
238. Jostsons, A. and Malin, A. S. (1968). *Acta Crystallogr.*, **B24**, 211
239. Hiraga, K. and Hirabayashi, M. (1973). *J. Phys. Soc. Japan*, **34**, 965
240. Terao, N. (1965). *Japan. J. Appl. Phys.*, **4**, 353
241. Terao, N. (1971). *J. Less-Common Metals*, **23**, 159
242. Gilles, J.-C. (1968). *Compt. Rend. Acad. Sci. Paris*, **266C**, 546
243. Oya, G. and Onodera, Y. (1971). Japan. *J. Appl. Phys.*, **10**, 1485
244. Watanabe, D., Terasaki, O., Jostsons, A. and Castles, J. R. (1968). *J. Phys. Soc. Japan*, **25**, 292
245. Brauer, G. and Ettmayer, P. (1973). Fourth International Conference on Solid Compounds of Transition Elements, Geneva, Abstracts, page 17
246. Kieffer, R., Ettmayer, P., Freudhofmaier, M. and Gatterer, J. (1971). *Monatsh. Chem.*, **102**, 483
247. Brauer, G., Mohr, E., Neuhaus, A. and Skokan, A. (1972). *Monatsh. Chem.*, **103**, 794
248. Geils, R. H. and Potter, D. I. (1973). *Met. Trans.*, **4**, 1469
249. Mills, T. (1970). *J. Less-Common Metals*, **22**, 373
250. Mills, T. (1972). *J. Less-Common Metals*, **26**, 223
251. Eddine, M. N., Sayetat, F. and Bertaut, E. E. (1969). *Compt. Rend. Acad. Sci. Paris*, **269B**, 574
252. Ettmayer, P. (1970). *Monatsh. Chem.*, **101**, 127
253. Karam, R. and Ward, R. (1970). *Inorg. Chem.* **9**, 1385
254. Zhurakovskii, E. A., Nikitin, L. V. and Lyutaya, M. D. (1972). *Izv. Akad. Nauk SSSR, Neorgan. Mat.*, **8**, 708
255. Mekata, M., Yoshimura, H. and Takaki, H. (1972). *J. Phys. Soc. Japan*, **33**, 62
256. Shvedova, L. K. (1971). *Izv. Akad. Nauk SSSR, Neorgan. Mat.*, **7**, 517
257. Neumann, G., Kieffer, R. and Ettmayer, P. (1972). *Monatsh. Chem.*, **103**, 1130
258. Brauer, G. and Reuther, H. (1973). *Z. Anorg. Allg. Chem.*, **395**, 151
259. Synielnikowa, W., Niemyski, T. Panczyk, J. and Kierzek-Pecold, E. (1971). *J. Less-Common Metals*, **23**, 1

260. Kieffer, R., Fister, D., and Heidler, E. (1972). *Metallurgica*, **25**, 128
261. Addison, C. C., Barker, M. G. and Bentham, J. (1972). *J. Chem. Soc. Dalton Trans.*, 1035
262. Palisaar, A.-P. and Juza, R. (1971). *Z. Anorg. Allg. Chem.*, **384**, 1
263. Juza, R. and Kroebel, R. (1964). *Z. Anorg. Allgem. Chem.*, **331**, 187
264. Karam, R. and Ward, R. (1970). *Inorg. Chem.*, **9**, 1849
265. Benz, R. and Zachariasen, W. H. (1970). *J. Nucl. Mat.*, **37**, 109
266. Spear, K. E. and Leitnaker, J. M. (1971). *High Temp. Sci.*, **3**, 26
267. Spear, K. E. and Leitnaker, J. M. (1971). *High Temp. Sci.*, **3**, 29
268. Kieffer, R., Ettmayer, P. and Petter, F. (1971). *Monatsh. Chem.*, **102**, 1182
269. Kotyk, M. and Stadalmaier, H. H. (1970). *Metallurg. Trans.*, **1**, 899
270. Ettmayer, P. (1971). *Monatsh. Chem.*, **102**, 858
271. Jack, D. H. and Jack, K. H. (1972). *J. Iron Steel Inst.* p. 790
272. Nardin, M., Lorthioir, G., Barberon, M., Madar, R., Fruchart, E. and Fruchart, R. (1972). *Compt. Rend. Acad. Sci. Paris*, **274C**, 2168
273. Fruchart, D., Bertaut, E. F., Madar, R., Lorthioir, G. and Fruchart, R. (1971). *Solid State Comm.*, **9**, 1793
274. Beckmann, O., Boller, H., Nowotny, H. and Benesovsky, F. (1969). *Monatsh. Chem.*, **100**, 1465
275. Boller, H. (1968). *Monatsh. Chem.*, **99**, 2444
276. Barberon, M., Fruchart, E., Fruchart, R., Lorthioir, G., Madar, R. and Nardin, M. (1972). *Mat. Res. Bull.*, **7**, 109
277. Boller, H. (1969). *Monatsh. Chem.*, **100**, 1471
278. Boller, H. and Nowotny, H. (1968). *Monatsh. Chem.*, **99**, 721
279. Boller, H. and Nowotny, H. (1968). *Monatsh. Chem.*, **99**, 672
280. Maunaye, M., Marchand, R., Guyader, J., Laurent, Y. and Lang, J. (1971). *Bull. Soc. Fr. Minéral. Crystallogr.*, **94**, 561
281. Wintenberger, M., Guyader, J. and Maunaye, M. (1972). *Solid State Comm.*, **11**, 1485
282. Parthé, E. (1972). *Cristallochemie des structures tetraedriques.* New York: Gordon and Breach
283. Wilson, A., (1971). In *Supplement to Mellor's Comprehensive Treatise on Inorganic and Theoretical Chemistry*. Vol. 8, Supplement 3. pp. 289–363. (New York: Wiley)
284. Rundqvist, S. (1962). *Ark. Kemi*, **20**, 67
285. Lundström, T. (1969). *Ark. Kemi*, **31**, 227
286. Rundqvist, S. and Carlsson, B. (1968). *Acta Chem. Scand.* **22**, 2395
287. Hassler, E. (1971). *Acta Chem. Scand.*, **25**, 129
288. Olofsson, O. and Ganglberger, E. (1970). *Acta Chem. Scand.*, **24**, 2389
289. Jawad, H., Lundström, T. and Rundqvist, S. (1971). *Physica Scripta*, **3**, 43
290. Selte, K., Kjekshus, A. and Andresen, A. F. (1972). *Acta Chem. Scand.*, **26**, 4057
291. Carlson, B. and Rundqvist, S. (1971). *Acta Chem. Scand.*, **25**, 1742
292. Yvon, K. and Boller, H. (1972). *Monatsh. Chem.* **103**, 1643
293. Anugul, S., Pontchour, C.-O. and Rundqvist, S. (1973). *Acta Chem. Scand.*, **27**, 26
294. Rundqvist, S., Carlsson, B. and Pontchour, C.-O. (1969). *Acta Chem. Scand.*, **23**, 2188
295. Owusu, M., Jawad, H., Lundström, T. and Rundqvist, S. (1972). *Physica Scripta*, **6**, 67
296. Baurecht, H. E., Boller, H., and Nowotny, H. (1971). *Monatsh. Chem.*, **102**, 373
297. Jeitschko, W. and Donohue, P. C. (1972). *Acta Crystallogr.*, **B28**, 1893
298. Jeitschko, W. and Donohue, P. C. (1973). *Acta Crystallogr.*, **B29**, 783
299. Kjekshus, A. and Skaug, K. E. (1972). *Acta Chem. Scand.*, **26**, 2554
300. Yamaguchi, Y., Watanabe, H., Yamauchi, H. and Tomiyoshi, S. (1972). *J. Phys. Soc. Japan*, **32**, 958
301. Wolfsgruber, H., Boller, H., and Nowotny, H. (1968). *Monatsh. Chem.*, **99**, 1230
302. Jeitschko, W. and Johnson, V. (1972). *Acta Crystallogr.* **B28**, 1971
303. Baurecht, H.-E., Boller, H. and Nowotny, H. (1970). *Monatsh. Chem.*, **101**, 1696
304. Boller, H. and Kallel, A. (1971). *Solid State Comm.*, **9**, 1699
305. Selte, K., Kjekshus, A., Jamison, W. A., Andresen, A. F. and Engebretsen, J. E. (1971). *Acta Chem. Scand.*, **25**, 1703
306. Kazama, N. and Watanabe, H. (1971). *J. Phys. Soc. Japan*, **31**, 943
307. Kallel, A., Nasr-Eddine, M. and Boller, H. (1973). *Solid State Comm.*, **12**, 665
308. Kjekshus, A. and Jamison, W. E. (1971). *Acta Chem. Scand.*, **25**, 1715
309. Kazama, N. and Watanabe, H. (1971). *J. Phys. Soc. Japan*, **30**, 1319

310. Kazama, N. and Watanabe, H. (1971). *J. Phys. Soc. Japan*, **30,** 578
311. Johnsson, T. (1972). *Acta Chem. Scand.*, **26,** 365
312. Yessik, M. (1968). *Phil. Mag.*, **17,** 623
313. Malik, S. K. and Vijayaraghavan, R. (1969). *Physics Lett.*, **28A,** 648
314. Banus, M. D. (1972). *J. Solid State Chem.*, **4,** 391
315. Gronvold, F., Snildal, S., and Westrum, E. F. (1970). *Acta Chem. Scand.*, **24,** 285
316. Jensen, P., Kjekshus, A. and Skansen, T. (1969). *J. Less-Common Metals*, **17,** 455
317. Hulliger, F. and Mooser, E. (1965). *Progress in Solid State Chem.*, **2,** 330
318. Brostigen, G. and Kjekshus, A. (1970). *Acta Chem. Scand.*, **24,** 2983
319. Brostigen, G. and Kjekshus, A. (1970). *Acta Chem. Scand.*, **24,** 2993
320. Kjekshus, A. and Nicholson, D. G. (1971). *Acta Chem. Scand.*, **25,** 866
321. Goodenough, J. B. (1972). *J. Solid. State Chem.*, **5,** 144
322. Holseth, H. and Kjekshus, A. (1968). *Acta Chem. Scand.*, **22,** 3273
323. Kjekshus, A. (1971). *Acta Chem. Scand.*, **25,** 411
324. Holseth, H. and Kjekshus, A. (1968). *Acta Chem. Scand.*, **22,** 3284
325. Holseth, H. and Kjekshus, A. (1968). *J. Less-Common Metals*, **16,** 472
326. Dahl, E. (1969). *Acta Chem. Scand.*, **23,** 2677
327. Rosenthal, G., Kershaw, R. and Wold, A. (1972). *Mat. Res. Bull.*, **7,** 479
328. Fan, A. K. L., Rosenthal, G. H., McKinzie, H. L. and Wold, A. (1972). *J. Solid State Chem.*, **5,** 136
329. Stassen, W. N. and Heyding, R. D. (1968). *Canadian J. Chem.*, **46,** 2159
330. Fleet, M. E. (1972). *American Mineralogist*, **57,** 1
331. Donohue, P. C., Bither, T. A. and Young, H. S. (1968). *Inorg. Chem.*, **7,** 998
332. Donohue, P. C. (1972). *Mat. Res. Bull.*, **7,** 943
333. Mandel, N. and Donohue, J. (1971). *Acta Crystallogr.*, **B27,** 2288
334. Fleet, M. E. (1973). *American Mineralologist*, **58,** 203
335. Jeitschko, W. (1970). *Acta Crystallogr.*, **B26,** 815
336. Fleet, M. E. (1972). *Acta Crystallogr.*, **B28,** 1237
337. Bälz, U. and Schubert, K. (1969). *J. Less-Common Metals*, **19,** 300
338. Selte, K. and Kjekshus, A. (1972). *Acta Chem. Scand.*, **26,** 1276
339. Selte, K. and Kjekshus, A. (1969). *Acta Chem. Scand.*, **23,** 2047
340. Selte, K. and Kjekshus, A. (1971). *Acta Chem. Scand.*, **25,** 3277
341. Selte, K., Kjekshus, A. and Andresen, A. (1972). *Acta Chem. Scand.*, **26,** 3101
342. Sénateur, J.-P., Roger, A., Fruchart, R. and Chappert, J. (1969). *Compt. Rendus Acad. Sci. Paris*, **269,** Serie C, 1385
343. Wäppling, R., Häggström, L., Rundqvist, S. and Karlsson, E. (1971). *J. Solid State Chem.*, **3,** 276
344. Maeda, Y. and Takashima, Y. (1973). *J. Inorg. Nucl. Chem.*, **35,** 1219
345. Bellavance, D., Mikkelsen, J. and Wold, A. (1970). *J. Solid State Chem.*, **2,** 285
346. Simpson, A. W. and Brambley, D. R. (1972). *Phys. Stat. Sol.* (b), **49,** 685
347. Ganglberger, E. (1968). *Monatsh. Chem.*, **99,** 566
348. Ganglberger, E. (1968). *Monatsh. Chem.*, **99,** 557
349. Juza, R. and Langer, K. (1968). *Z. Anorg. Allg. Chem.*, **361,** 58
350. Juza, R., Dethlefsen, W., Seidel, H. and Benda, K. (1968). *Z. Anorg. Allg. Chem.*, **356,** 253
351. Juza, R., Trapp, H.-D. and Seidel, H. *Z. Anorg. Allg. Chem.*, **356,** 273
352. Johnson, V. and Jeitschko, W. (1973). *J. Solid State Chem.*, **6,** 306
353. Hulliger, F. (1973). *J. Less-Common Metals*, **30,** 397
354. Barthelat, J. C. and Jeannin, Y. (1972). *J. Less-Common Metals*, **26,** 273
355. Jeannin, Y. and Mosset, A. (1972). *J. Less-Common Metals*, **27,** 237
356. Mosset, A. and Jeannin, Y. (1972). *Compt. Rend. Acad. Sci. Paris, Ser. C*, **275,** 877
357. Jeitschko, W. (1968). *Acta Crystallogr.*, **B24,** 930
358. Dwight, A. E. Mueller, M. H., Conner, R. E., Downey, J. W. and Knott, H. (1968). *Trans. Amer. Inst. Min. (Metall.) Engrs.*, **242,** 2075
359. Jeitschko, W., Jordan, A. G. and Beck, P. A. (1969). *Trans. Amer. Inst. Min. (Metall.) Engrs.*, **245,** 335
360. Jeitschko, W. (1970). *Met. Trans.*, **1,** 2963
361. Dwight, A. E., Harper, W. C., and Kimball, C. W. (1973). *J. Less-Common Metals*, **30,** 1
362. Rundqvist, S. and Nawapong, P. C. (1966). *Acta Chem. Scand.*, **20,** 2250

363. Rundqvist, S. and Tansuriwongs, P. (1967). *Acta Chem. Scand.*, **21,** 813
364. Sénateur, J. P., Rouault, A., L'Heritier, P., Krumbügel-Nylund, A. and Fruchart, R. (1973). *Mater. Res. Bull.*, **8,** 229
365. Maeda, Y. and Takashima, Y. (1973). *J. Inorg. Nucl. Chem.*, **35,** 1963
366. Nagase, S., Watanabe, H. and Shinohara, T. (1973). *J. Phys. Soc. Japan*, **34,** 908
367. Suzuki, T., Yamaguchi, Y., Yamamoto, H. and Watanabe, H. (1973). *J. Phys. Soc. Japan*, **34,** 911
368. Nylund, A., Roger, A. Sénateur, J. P. and Fruchart, R. (1971). *Monatsh. Chem.*, **102,** 1631 and *J. Solid State Chem.*, (1972), **4,** 115
369. Roy-Montreuil, Deyris, B., Michel, A., Rouault, A., L'Héritier, P., Nylund, A., Sénateur, J. P. and Fruchart, R. (1972). *Mater. Res. Bull.*, **7,** 813
370. Boller, H. (1973). *Monatsh. Chem.*, **104,** 166
371. Maes, R. and Strycker, R. de (1967). *Trans. Met. Soc. AIME*, **239,** 1887
372. Naud, J. and Breckpot, R. (1972). *Bull Soc. Chim. Belges*, **81,** 247
373. Doenitz, F.-D. (1970). *Z. Kristallogr.*, **131,** 222
374. Guérin, R., Sergent, M., and Prigent, J. (1972). *Comptes Rend. Acad. Sci. Paris*, **274C,** 1278
375. Rastogi, P. K. and Duwez, P. (1970). *J. Non-Crystalline Solids*, **5,** 1
376. Rastogi, P. K. (1973). *J. Mater. Science*, **8,** 140
377. Rastogi, P. K. (1970). *Scripta Metallurgica*, **4,** 939
378. Hasegawa, R. (1971). *Phys. Rev.*, **B3,** 1631
379. Sinha, A. K. (1971). *J. Appl. Phys.*, **42,** 5184
380. Sharon, T. E. and Tsuei, C. C. (1972). *Phys. Rev.* **B5,** 1047
381. Sinha, A. K. and Duwez, P. (1972). *J. Appl. Phys.*, **43,** 432
382. Maitrepierre, P. L. (1970). *J. Appl. Phys.*, **41,** 498
383. Boucher, B. Y. (1972). *J. Non-Crystalline Solids*, **7,** 277
384. Tronc, E. and Huber, M. (1971). *Compt. Rend. Acad. Sci. Paris*, **272C,** 1018
385. Tronc, E. and Huber, M. (1973). Fourth Intern. Conf. on Solid Comp. of Trans. Elem., Geneva. Abstracts, p. 181
386. Tilley, R. J. D. (1973). *J. Solid State Chem.*, **7,** 213
387. Allpress, J. G. and Sanders, J. V. (1973). *J. Appl. Crystallogr.*, **6,** 165
388. Brunie, S. and Chevreton, M. (1972). *Compt. Rend. Acad. Sci. Paris*, **274B,** 278
389. Gleizes, A. and Jeannin, Y. (1972). *J. Solid State Chem.*, **5,** 42
390. Brattas, L. and Kjekshus, A. (1971). *Acta Chem. Scand.*, **25,** 2783
391. Smeggil, J. G. and Bartram, S. (1972). *J. Solid State Chem.*, **5,** 391
392. Rimmington, H. P. B., Balachin, A. A. and Tanner, B. K. (1972). *J. Cryst. Growth*, **15,** 51
393. Brattas, L. and Kjekshus, A. (1971). *Acta Chem. Scand.*, **25,** 2350
394. Brattas, L. and Kjekshus, A. (1972). *Acta Chem. Scand.*, **26,** 3441
395. Chen. H.-Y. and Franzen, H. F. (1972). *Acta Crystallogr.*, **B28,** 1399
396. Rüdorff, W. (1965). *Chimia*, **19,** 489
397. Danot, M., LeBlanc, A. and Rouxel, J. (1969). *Bull. Soc. Chim. Fr.*, 2670
398. Rouxel, J., Danot, M. and Bichon, J. (1971). *Bull. Soc. Chim. Fr.*, 3930
399. Bichon, J., Danot, M. and Rouxel, J. (1973). *Compt. Rend. Acad. Sci. Paris*, **270C,** 1976
400. Rouxel, J., Cousseau, J. and Trichet, L. (1971). *Compt. Rend. Acad. Sci. Paris*, **273C,** 243
401. Cousseau, J., Trichet, L. and Rouxel, J. (1973). *Bull. Soc. Chim. Fr.* 872
402. Muranaka, S. (1973). *Mater. Res. Bull.*, **8,** 679
403. Takahashi, T. and Yamada, O. (1973). *J. Solid State Chem.*, **7,** 25
404. Danot, M., Bichon, J. and Rouxel, J. (1972). *Bull. Soc. Chim. Fr.*, **3063**
405. Trichet, L., Cousseau, J. and Rouxel, J. (1972). *Compt. Rend. Acad. Sci. Paris*, **274C,** 394
406. Sterzel, W. and Horn, J. (1970). *Z. Anorg. Allgem.Chem.*, **376,** 254
407. Yamaoka, S. (1972). *J. Amer. Ceram. Soc.*, **55,** 111
408. Mootz, D. and Puhl, H. (1967). *Acta Crystallogr.*, **23,** 471
409. Rimmington, H. P. B. and Balchin, A. A. (1971). *Phys. Status Solidi A*, **6,** K 47
410. Al-Alamy, F. A. S. and Balchin, A. A. (1973). *Mater. Res. Bull.*, **8,** 245
411. Brunie, S., Chevreton, M. and Kauffmann, J.-M. (1972). *Mater. Res. Bull.*, **7,** 253
412. Gronvold, F., Haraldsen, H., Pedersen, B. and Tufte, T. (1969). *Rev. Chim. Minér.*, **6,** 215
413. De Vries, A. B. (1972). *Dissertation*, Univ. of Groningen
414. De Vries, A. B. and Haas, C. (1973). *J. Phys. Chem. Solids*, **34,** 651
415. Kutoglu, A. and Allmann, R. (1972). *Neues Jahrb. Miner. Monatsh.*, 339
416. Chen. H.-Y., Tuenge, R. T. and Franzen, H. F. (1973). *Inorg. Chem.*, **12,** 552

417. Delmaire, J.-P. and Le Brusque, H. (1972). *Compt. Rend. Acad. Sci. Paris*, **275C**, 889
418. Hodouin, D. (1973). *J. Less-Common Metals*, **30**, 127
419. Ehrenfreund, E., Gossard, A. C., Gamble, F. R. and Geballe, T. H. (1971). *J. Appl. Phys.*, **42**, 1491
420. Marezio, M., Dernier, P. D., Menth, A. and Hull, Jr., G. W. (1972). *J. Solid State Chem.*, **4**, 425
421. Yamaya, K. and Sambongi, T. (1972). *Solid State Commun.*, **11**, 903
422. Yamaya, K. and Sambongi, T. (1972). *J. Phys. Soc. Japan*, **32**, 1150
423. Edwards, J. and Frindt, R. F. (1971). *J. Phys. Chem. Solids*, **32**, 2217
424. Bhide, V. G. and Bahl, M. K. (1971). *J. Phys. Chem. Solids*, **32**, 1001
425. Conroy, L. E. and Pisharody, K., R. (1972). *J. Solid State Chem.*, **4**, 345
426. Thompson, A. H., Gamble, F. R. and Revelli, J. F. (1971). *Solid State Commun.*, **9**, 981
427. Smith, D. L., Mochel, A. R., Banewicz, J. J. and Maguire, J. A. (1972). *J. Less-Common Metals*, **26**, 139
428. Plovnick, R. H. (1972). *J. Solid State Chem.*, **5**, 153
429. Kom, J. K. and Fournès, L. (1973). *Compt. Rend. Acad. Sci. Paris*, **276C**, 1521
430. Voorhoeve-van den Berg, J. M. (1972). *J. Less-Common Metals*, **26**, 399
431. Schmidt, V. and Rüdorff, W. (1973). *Z. Naturforsch.*, **28b**, 25
432. Rouxel, J., Le Blanc, A. and Royer, A. (1971). *Bull. Soc. Chim. Fr.*, 2019
433. Royer, A., Le Blanc-Soreau, A. and Rouxel, J. (1973). *Compt. Rend. Acad. Sci. Paris*, **276C**, 1021
434. Kalikhman, V. L., Gladchenko, E. P. and Pravoverova, L. L. (1972). *Izv. Akad. Nauk SSSR, Neorgan. Mater.*, **8**, 1163
435. Gamble, F. R., Osiecki, J. H., Cais, M. Pisharody, R., Di Salvo, F. J., Geballe, T. H. (1971). *Science*, **174**, 493
436. Beal, A. R. and Liang, W. Y. (1973). *Phil. Mag.*, **27**, 1397
437. Fernández-Morán, H., Ohstuki, M. Hibino, A. and Hough, C. (1971). *Science*, **174**, 498
438. Schöllhorn, R. and Weiss, A. (1972). *Z. Naturforsch.*, **27b**, 1277, 1278, and 1428
439. Ruthardt, R., Schöllhorn, R. and Weiss, A. (1972). *Z. Naturforsch.*, **27b**, 1275
440. Bayer, E. and Rüdorff, W. (1972). *Z. Naturforsch.*, **27b**, 1336
441. Schöllhorn, R. and Weiss, A. (1972). *Z. Naturforsch.*, **27b**, 1273
442. Mikami, M., Igaki, K. and Nobumitu, O. (1972). *J. Phys. Soc. Japan*, **32**, 1217
443. Le Brusq. H., Delmaire, J.-P. and Marion, F. (1971). *Compt. Rend. Acad. Sci. Paris*, **272C**, 1034
444. Sugiura, T., Iwahashi, K. and Masuda, Y. (1972). *J. Phys. Soc. Japan*, **33**, 1172
445. Hashimoto, T., Hoya, K., Yamaguchi, M. and Ichitsubo, I. (1971). *J. Phys. Soc. Japan*, **31**, 679
446. Yamaguchi, M. and Hashimoto, T. (1972). *J. Phys. Soc. Japan*, **32**, 635
447. Ozawa, K., Yoshimi, T., Irie, M. and Yanagisawa, S. (1972). *Phys. Status Solidi*, **11A**, 581
448. Al-Hilli, A. A., and Evans, B. L. (1972). *J. Cryst. Growth*, **15**, 93
449. Pratap, R. and Bhattacharya, D. L. (1972). *Phys. Status Solidi*, **12A**, 61
450. Wieting, T. J. (1973). *Solid State Commun.*, **12**, 931
451. Garg, A. K., Sehgal, H. K. and Agnihotri, O. P. (1973). *Solid State Commun.*, **12**, 1261
452. Title, R. S. and Shafer, M. W. (1972). *Phys. Rev. Lett.*, **28**, 808
453. Title, R. S. and Shafer, M. W. (1973). *Phys. Rev.*, **8B**, 615
454. Davery, B. and Evans, B. L. (1972). *Phys. Status Solidi*, **13A**, 483
455. Beal, A. R., Knights, J. C. and Liang, W. Y. (1972). *J. Phys.* C, **5**, 3531 and 3540
456. Williams, P. M. and Shepherd, F. R. (1973). *J. Phys.* C, **6**, L 36
457. Williams, R. H. (1973). *J. Phys.* C, **6**, L 32
458. Mattheiss, L. F. (1973). *Phys. Rev. Lett.*, **30**, 784
459. Kasowski, R. V. (1973). *Phys. Rev. Lett.*, **30**, 1175
460. Edmondson, D. R. (1972). *Solid State Commun.*, **10**, 1085
461. White, R. M. and Lucovsky, G. (1972). *Solid State Commun.*, **11**, 1369
462. Bars, O., Guillevic, J. and Grandjean, D. (1973). *J. Solid State Chem.*, **6**, 48
463. Chevrel, R., Sergent, M., Prigent, J. (1971). *J. Solid State Chem.*, **3**, 515
464. Sergent, M. and Chevrel, R. (1972). *Compt. Rend. Acad. Sci. Paris*, **274C**, 1965
465. Sergent, M. and Chevrel, R. (1973). *J. Solid State Chem.*, **6**, 433
466. Bars, O., Guillevic, J. and Grandjean, D. (1973). *J. Solid State Chem.*, **6**, 335

467. Guillevic, J., Bars, O. and Grandjean, D. (1973). *J. Solid State Chem.*, **7**, 158
468. Marezio, M., Dernier, P. D., Remeika, J. P., Corenzwit, E. and Matthias, B. T. (1973). *Mater. Res. Bull.*, **8**, 657
469. Matthias, B. T., Marezio, M., Corenzwit, E., Cooper, A. S. and Barz, H. E. (1972). *Science*, **175**, 1465
470. Lawson, A. C. (1972). *Mater. Res. Bull.*, **7**, 773
471. Engelsman, F. M. R., Wiegers, G. A., Jellinek, F. and van Laar, B. (1973). *J. Solid State Chem.*, **6**, 574
472. van Laar, B. and Engelsman, F. M. R. (1973). *J. Solid State Chem.*, **6**, 384
473. Belov, K. P., Tret'yakov, Yu. D., Gordeev, I. V., Koroleva, L. I., Ped'ko, A. V., Smirnovskaya, E. I., Alferov, V. A. and Saksonov, Yu. G. (1973). *Soviet Physics-Solid State*, **14**, 1862
474. Wojtowicz, P. J. (1969). *IEEE Trans. Mag.*, **MAG-5**, 840
475. Philipsborn, H. v. (1971). *J. Cryst. Growth*, **9**, 296
476. Watanabe, T. (1972). *J. Phys. Soc. Japan*, **32**, 1443
477. Barraclough, K. G. and Meyer, A. (1972). *J. Cryst. Growth*, **16**, 265
478. Carnall, Jr. E., Pearlman, D., Coburn, T. J., Moser, F. and Martin, T. W. (1972). *Mater. Res. Bull.*, **7**, 1361
479. Pearlman, D., Carnall, Jr. E. and Martin, T. W. (1973). *J. Solid State Chem.*, **7**, 138
480. Ahrenkiel, R. K., Lee, T. H., Lyu, S. L. and Moser, F. (1973). *Solid State Commun.*, **12**, 1113
481. Ahrenkiel, R. K. and Coburn, T. J. (1973). *Appl. Phys. Lett.*, **22**, 340
482. Watanabe, T. (1973). *Solid State Commun.*, **12**, 355
483. Watanabe, T. (1973). *J. Phys. Soc. Japan*, **34**, 1695
484. Gibart, P., Robbins, M. and Lambrecht, Jr., V. G. (1973). *J. Phys. Chem. Solids.*, **34**, 1363
485. Lotgering, F. K. and van der Steen, G. H. A. M. (1971). *J. Solid State Chem.*, **3**, 574
486. Hoekstra, B. and van Stapele, R. P. (1973). *Phys. Status Solidi*, **55b**, 607
487. Masumoto, K., Kiyosawa, T. and Nakatani, I. (1973). *J. Phys. Chem. Solids*, **34**, 569
488. Wagner, V. (1973). *Phys. Status Solidi*, **55b**, K 29
489. Arai, T., Wakaki, M., Onari, S., Kudo, K., Satoh, T. and Tsushima, T. (1973). *J. Phys. Soc. Japan*, **34**, 68
490. Pickardt, J. and Riedel, E. (1971). *J. Solid State Chem.*, **3**, 67
491. Fujii, H., Kamigaichi, T. and Okamoto, T. (1973). *J. Phys. Soc. Japan*, **34**, 1689
492. Kirkaldy, J. S., Bolze, G.-M., Mc Cutcheon, D. and Young, D. J. (1973). *Met. Trans.*, **4**, 1519
493. Lutz, H. D. and Bertram, K.-H. (1972). *Z. Anorg. Allg. Chem.*, **393**, 59
494. Gorochov, O., Barthélemy, E. and Dung, N. H. (1971). *Compt. Rend. Acad. Sci. Paris*, **273C**, 368
495. Mc Kinzie, H., Gorochov, O., Nguyen, H. D., Dagron, C. (1971). *Compt. Rend. Acad. Sci. Paris*, **273B**, 1040
496. Takahashi, T., Oka, T., Yamada, O. and Ametani, K. (1971). *Mater. Res. Bull.*, **6**, 173
497. Takahashi, T., Osaka, S. and Yamada, O. (1973). *J. Phys. Chem. Solids*, **34**, 1131
498. Sleight, A. W. and Frederick, C. G. (1973). *Mater. Res. Bull.*, **8**, 105
499. Omloo, W. P. F. A. M., Bommerson, J. C., Heikens, H. H., Risselada, H., Vellinga, M. B., van Bruggen, C. F., Haas, C. and Jellinek, F. (1971). *Phys. Status Solidi*, **5a**, 349
500. Donohue, P. C., Bither, T. A., Cloud, W. H. and Frederick, C. G. (1971). *Mater. Res. Bull.*, **6**, 231
501. Somoano, R. B., Hadek, V. and Rembaum, A. (1973). *J. Chem. Phys.*, **58**, 697
502. Müller, A. and Menge, R. (1972). *Z. Anorg. Allgem. Chem.*, **393**, 259
503. Murray, R. M., Forbes, B. C. and Heyding, R. D. *Canadian J. Chem.*, **50**, 4059
504. Cemič, L. and Neuhaus, A. (1972). *High Temp.-High Pressures*, **4**, 97
505. Gronvold, F., Kveseth, N. J., Marques, F. D. S. and Tichey, J. (1972). *J. Chem. Thermodynam.*, **4**, 795
506. Mateika, D. (1972). *J. Cryst. Growth*, **13/14**, 698
507. Bronger, W. and Böttcher, P. (1972). *Z. Anorg. Allgem. Chem.*, **390**, 1
508. Grey, I. E. and Steinfink, H. (1971). *Inorg. Chem.*, **10**, 691
509. Lotgering, F. K. and van der Steen, G. H. A. M. (1972). *J. Phys. Chem. Solids*, **33**, 2071
510. Donohue, P. C. (1970). *J. Solid State Chem.*, **2**, 6
511. Yokota, M., Syono, Y., and Minomura, S. (1971). *J. Solid State Chem.*, **3**, 520

512. Suchow, L. and Ando, A. A. (1970). *J. Solid State Chem.*, **2,** 156
513. Masumoto, H., Watanabe, K., and Ohnuma, S. (1972). *J. Phys. Soc. Japan*, **32,** 570
514. Nakazawa, H. and Morimoto, N. (1971). *Mater. Res. Bull.*, **6,** 345
515. Ward, J. C. (1970). *Rev. Pure Appl. Chem.*, **20,** 175
516. Tokonami, M. Nishiguchi, K., and Morimoto, N. (1972). *Amer. Mineral.*, **57,** 1066
517. Jeitschko, W. and Sleight, A. (1972). *J. Solid State Chem.*, **4,** 324
518. Fleet, M. E. (1971). *Acta Crystallogr.*, **B27,** 1864
519. Fasiska, E. J. (1972). *Phys. Status Solidi*, **10a,** 169
520. Bugli, G., Abello, L. and Pannetier, G. (1972). *Bull. Soc. Chim. France*, 497
521. Taylor, L. A. and Williams, K. L. (1972). *Amer. Mineral.*, **57,** 1571
522. Franz, E.-D. (1972). *N. Jb. Miner. Mh.*, 276
523. Franz, E.-D. (1971). *N. Jb. Miner. Mh.*, 436
524. Komarek, K. L. and Wessely, K. (1972). *Monatsh. Chem.*, **103,** 896
525. Klepp, K. O. and Komarek, K. L. (1973). *Monatsh. Chem.*, **104,** 105
526. Geffken, R. M., Komarek, K. L. and Miller, E. M. (1972). *J. Solid State Chem.*, **4,** 153
527. Kjekshus, A. and Nicholson, D. G. (1972). *Acta Chem. Scand.*, **26,** 3241
528. Ortalli, I. Fano, V. and Gibart, P. (1973). *Phys. Status Solidi*, **15a,** K 45
529. Jeitschko, W. (1973). Unpublished results.
530. Koehler, Jr., R. F., Feigelson, R. S., Swarts, H. W. and White, R. L. (1972). *J. Appl. Phys.*, **43,** 3127
531. Mc Whan, D. B., Marezio, M., Remeika, J. P. and Dernier, P. D. (1972). *Phys. Rev.* **B., 5,** 2552
532. Horwood, J. L. and Townsend, M. G. (1971). *Solid State Commun.*, **9,** 41
533. Townsend, M. G., Tremblay, R., Horwood, J. L. and Ripley, L. J. (1971). *J. Phys.*, **C4,** 598
534. Pauwels, L. J. and Maervoet, G. (1972). *Bull. Soc. Chim. Belges*, **81,** 385
535. Trahan, J. and Goodrich, R. G. (1972). *Phys. Rev.*, **B6,** 199
536. Koehler, Jr. R. F. and White, R. L. (1973). *J. Appl. Phys.*, **44,** 1682
537. Grosselin, J. R., Townsend, M. G., Tremblay, R. J., Ripley, L. G. and Carson, D. W. (1973). *J. Phys.*, **C6,** 1661
538. Kasowski, R. V. (1973). *Phys. Rev.* **38,** 1378
539. White, R. M. and Mott, N. F. (1971). *Phil. Mag.*, **24,** 845
540. Gautier, F., Krill, G., Lapierre, M. F. and Robert, C. (1972). *Solid State Commun.*, **11,** 1201
541. Kautz, R. L., Dresselhaus, M. S., Adler, D. and Linz, A. (1972). *Phys. Rev.*, **B6,** 2078
542. Bouchard, R. J., Gillson, J. L. and Jarrett, H. S. (1973). *Mater. Res. Bull.*, **8,** 489
543. Goodenough, J. B. (1971). *J. Solid State Chem.*, **3,** 26
544. Komarek, K. L. and Wessely, K. (1972). *Monatsh. Chem.*, **103,** 923
545. Klepp, K. O. and Komarek, K. L. (1972). *Monatsh. Chem.*, **103,** 934
546. Misra, K. C. and Fleet, M. E. (1973). *Mater. Res. Bull.*, **8,** 669
547. Rost, E. and Åkesson, G. (1972). *Acta Chem. Scand.*, **26,** 3662
548. Rost, E. and Haugsten, K. (1971). *Acta Chem. Scand.*, **25,** 3194
549. Carbonara, R. S. and Hoch, M. (1972). *Monatsh. Chem.*, **103,** 695
550. Gronvold, Fr., Kveseth, N. J. and Sveen, A. (1972). *J. Chem. Thermodynamics*, **4,** 337
551. Stevels, A. L. N. and Jellinek, F. (1971). *Monatsh. Chem.*, **102,** 1679
552. Haugsten, K. and Rost, E. (1972). *Acta Chem. Scand.*, **26,** 410
553. Åkesson, G. and Rost, E. (1973). *Acta Chem. Scand.*, **27,** 79
554. Hong, H. Y. and Steinfink, H. (1972). *J. Solid State Chem.*, **5,** 93
555. Grey, I. E., Hong, H. and Steinfink, H. (1971). *Inorg. Chem.*, **10,** 340
556. Steinfink, H., Hong, H. and Grey, I. (1972). *Nation. Bur. Stand. Special Publication*, **364,** 681
557. Fleet, M. E. (1973). *Amer. Mineral.*, **58,** 435
558. Wandji, R. and Kom, J. K. (1972). *Compt. Rend. Acad. Sci. Paris*, **275C,** 813
559. Meyer, A. and Pink, H. (1973). *J. Less-Common Metals*, **30,** 314
560. Hall, S. R. and Stewart, J. M. (1973). *Acta Crystallogr.*, **B29,** 579
561. Hall, S. R. and Gabe, E. J. (1972). *Amer. Mineral.*, **57,** 368
562. Hiller, J. E. and Probsthain, K. (1956). *Z. Kristallogr.*, **108,** 108
563. Cabri, L. J. and Hall, S. R. (1972). *Amer. Mineral.*, **57,** 689
564. Fleet, M. E. (1970). *Z. Kristallogr.*, **132,** 276
565. Sleight, A. W. (1973). *J. Solid State Chem.*, **8,** 29

566. Vincent, H. and Bertaut, E. F. (1973). *J. Phys. Chem. Solids*, **34,** 151
567. Ganiel, U., Hermon, E. and Shtrikman, S. (1972). *J. Phys. Chem. Solids*, **33,** 1873
568. Bronger, W. and Günther, O. (1972). *J. Less-Common Metals*, **27,** 73
569. Günther, O. and Bronger, W. (1973). *J. Less-Common Metals*, **31,** 255
570. Bronger, W. and Huster, J. (1971). *J. Less-Common Metals*, **23,** 67
571. Sprenger, H. and Nickl, J. J. (1972). *J. Less-Common Metals*, **27,** 163

Index